Global Issues in Water Policy

Volume 15

More information about this series at http://www.springer.com/series/8877

Quentin Grafton • Katherine A. Daniell
Céline Nauges • Jean-Daniel Rinaudo
Noel Wai Wah Chan
Editors

Understanding and Managing Urban Water in Transition

Springer

Editors
Quentin Grafton
Crawford School of Public Policy
The Australian National University
Canberra, Aust Capital Terr, Australia

Katherine A. Daniell
ANU Centre for European Studies
The Australian National University
Canberra, Australia

Céline Nauges
School of Economics
University of Queensland
Brisbane St Lucia, QLD, Australia

Jean-Daniel Rinaudo
Water Department
BRGM, French Geological Survey
Montpellier, France

Noel Wai Wah Chan
Crawford School of Public Policy
The Australian National University
Canberra, Aust Capital Terr, Australia

ISSN 2211-0631 ISSN 2211-0658 (electronic)
Global Issues in Water Policy
ISBN 978-94-017-9800-6 ISBN 978-94-017-9801-3 (eBook)
DOI 10.1007/978-94-017-9801-3

Library of Congress Control Number: 2015939582

Springer Dordrecht Heidelberg New York London

Printed on acid-free paper

Springer Science+Business Media B.V. Dordrecht is part of Springer Science+Business Media (www.springer.com)

Preface

This book has its origins in a meeting of Australian and French researchers in Montpellier in June 2011, a meeting that was preceded by an approach by Fritz Schmul at Springer to Quentin Grafton to write a book on urban water. On the Australian side, the French–Australian collaboration was initiated by Katherine Daniell and Quentin Grafton from the Australian National University, and on the French side, by Olivier Barreteau and Nils Ferrand from IRSTEA and the Embassy of France in Australia. The vision of the principals was to link across disciplines, distance, and language to develop meaningful collaborations and insights that would otherwise not be possible.

Several research initiatives grew out of the 2011 workshop and have led to various outcomes and outputs. One of the outcomes is this volume on urban water in transition. The book initially began as a series of ideas in a breakout session chaired by Quentin Grafton at the Montpellier workshop, and then, after the event, was developed further by all the editors. As editors, our goal has been broad: to develop a single framework, applicable to both rich countries and developing and emerging economies, for understanding and acting on urban water issues, despite the manifold shifts and transitions underway. We wanted to understand how urban water is valued, supplied, managed, delivered, consumed, and treated.

This volume is the outcome of a 3-year gestation and much hard work following the 2011 workshop. All the editors realized that the original group in Montpellier did not have sufficient diversity of knowledge and experience to deliver on what was intended to be a book on global urban water. Consequently, many additional experts, practitioners, and researchers were invited to contribute, and almost all accepted the invitation.

The end result of the collaboration of 51 contributors is a volume that provides a range of insights, case studies, summaries, and analyses of urban water from a global perspective. Collectively, the diverse contributions provide a major step forward in demonstrating how the challenges of delivering water to over half of humanity can be achieved safely, efficiently, and equitably.

Canberra, Australia — Quentin Grafton
Canberra, Australia — Katherine A. Daniell
Brisbane St. Lucia, Australia — Céline Nauges
Montpellier, France — Jean-Daniel Rinaudo
Canberra, Australia — Noel Wai Wah Chan

Acknowledgments

This volume has been made possible by the support of many, and not just the contributors and editors. Special thanks go to Vivienne Seedsman for husbanding the multiple chapters through to final publication. In terms of financial support, without the efforts of the Australian National University–UNESCO Chair in Water Economics and Transboundary Water Governance, many of the costs of putting the book together, including paying for the copyediting and indexing, could not have been met. Our gratitude also goes to the Crawford School of Public Policy at the Australian National University for hosting Vivienne and others as the book progressed. Final thanks go to Fritz Schmuhl at Springer for supporting the vision of the editors and for giving us the time to develop a comprehensive guide to urban water.

Contents

Part III Water Governance and Integrated Management

Chapter 1
Understanding and Managing Urban Water in Transition

Katherine A. Daniell, Jean-Daniel Rinaudo, Noel Wai Wah Chan, Céline Nauges, and Quentin Grafton

1 Introduction

Understanding and managing water in the urban context is of vital global importance. Over half the world's population now lives in urban environments (United Nations 2013) and the percentage is set to increase over coming decades. Quality urban living, like life anywhere, requires adequate quantities and qualities of water to support a range of social well-being, economic development, and environmental health. Managing water in cities, along with their linked energy, food, materials, environmental systems, and socio-economic systems is, therefore, an integral component of global sustainability challenges (Sheehan 2007; see also Kenway and Lant 2015, Chap. 28, this volume).

K.A. Daniell (✉)
ANU Centre for European Studies, The Australian National University, Building 67C, 1 Liversidge St, Canberra 0200, Australia
e-mail: katherine.daniell@anu.edu.au

J.-D. Rinaudo
Water Department, BRGM, French Geological Survey, Montpellier, France
e-mail: jd.rinaudo@brgm.fr

N.W.W. Chan • Q. Grafton
Crawford School of Public Policy, The Australian National University, Canberra, Aust Capital Terr, Australia
e-mail: noel.chan@anu.edu.au; quentin.grafton@anu.edu.au

C. Nauges
School of Economics, University of Queensland, Brisbane St Lucia, QLD, Australia
e-mail: c.nauges@uq.edu.au

Q. Grafton et al. (eds.), *Understanding and Managing Urban Water in Transition*, Global Issues in Water Policy 15,
DOI 10.1007/978-94-017-9801-3_1

These challenges are significant and extremely important; in particular, but not only for, the almost 130 million people globally who do not have access to safe drinking water and 714 million who do not have access to adequate sanitation in urban areas (UNICEF/WHO 2012). The current global water system is also vulnerable to increasing global water demands, changing urbanisation patterns, and increasing climate risk and uncertainty (e.g. Frederick and Major 1997; Niemczynowicz 1999; Vörösmarty et al. 2000; Jenerette and Larsen 2006). There are different but still substantial challenges in developed countries, where populations are benefiting from high quality water and sanitation services. What is often at stake here is the sustainability of these services, including the social, economic, environmental and governance dimensions of urban water management activities. In line with these challenges, many authors and global institutions consider that there is an urgent need to support the transitions of urban water systems towards more sustainable configurations that can cope with these and other global changes (Larsen and Gujer 1997; Hellström et al. 2000; Daniell et al. 2005; Mitchell 2006; UN-WATER 2012).

The definition of urban water systems and their management that we consider in this book is, thus, intentionally multi-faceted and includes all components of the urban system where water is a primary concern for managers, residents, community groups and businesses. This includes issues of water supply, demand, use, valuation, sanitation systems, storm water management, flood management and management of water dependent ecosystems affected by urban water use, as well as their associated governance structures and processes.

It is the premise of this book that there are a range of changes and transitions occurring in the way water is being managed in many urban settlements around the world. These are taking place as development patterns, climate, social preferences and values are changing, including the relative importance of those related to growing environmental concerns, the financial value of water, social well-being, and the role that access to water and sanitation plays in alleviating poverty.

Here we take the concept of transition to convey the idea of a progressive adaptation or transformation of a system in response to particular stimuli (drivers). Specifically, a transition describes the situation of a system where it lies between points of relative equilibrium. This implies that when a system is "in transition" we can be aware that it is or has rapidly changed from a previous state, but may not be able to predict or know how it will stabilise in the future. Such definitions and understandings of transitions (e.g. Rotmans et al. 2001; Geels 2002, 2004, 2010; Geels and Schot 2007; van der Brugge and Rotmans 2007) are based in complex adaptive systems theory (Prigogine and Stengers 1984; Holland 1995; Kauffman 1995; Gunderson and Holling 2002; Holling 2004) and a range of other literature including innovation and technological transitions, governance and evolutionary economics (e.g. Rogers 1983; Arthur 1988; Smith et al. 2005, 2010; Sabatier 1988; Nelson and Winter (1982).

In urban water systems, over time there have been a number of key water system management objectives linked to social values that have driven transitions to different configurations of water systems that are designed to provide specific types of

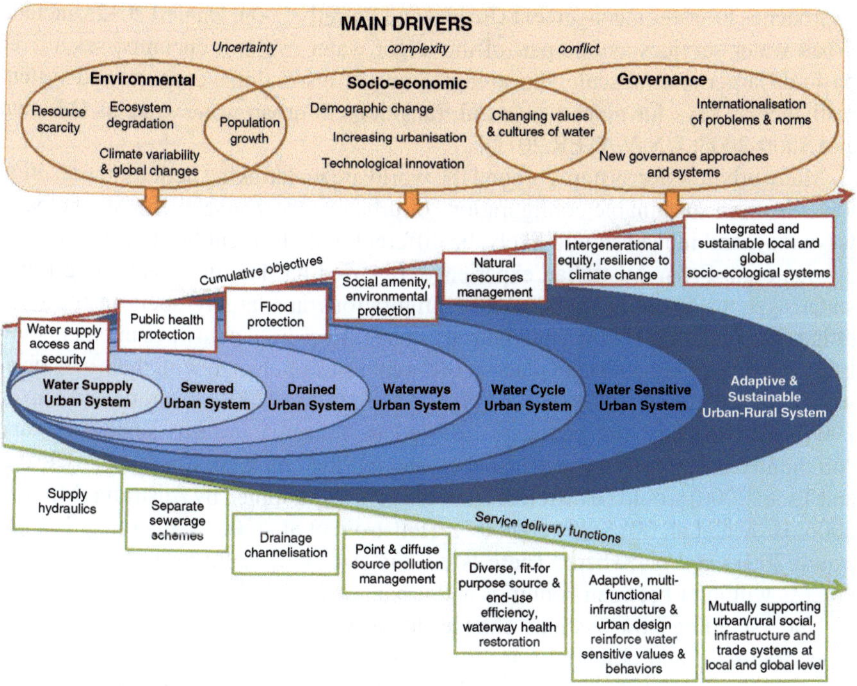

Fig. 1.1 Drivers and resulting evolutions in urban water systems (Adapted and extended from Brown et al. (2009))

service delivery functions. These water system management objectives and resultant changes in idealised urban water system types have come about as a result of a number of socio-economic, environmental and governance-related drivers as represented in Fig. 1.1.

From Fig. 1.1, we contend that the process of transitioning urban water systems towards more integrated, adaptive and sustainable configurations is a multi-dimensional process. Different objectives need to be increasingly integrated into urban water system management in order to be able to provide a variety of service delivery functions. The scope and dimensions of urban water systems that need to be considered also expand along with these objectives: from local to regional and regimes; from an intra-generational to inter-generational scale; and from sectoral to trans-disciplinary expertise requirements and approaches. As a result, the 'traditional, technical, linear management' approach that has been successfully applied in many urban water systems around the world and that focuses on urban water services to just be concerned with their local population and environment is now typically found to be insufficient and ineffective to respond to the increasing complexity, uncertainty and conflict that arises in water systems. Shifting paradigms and approaches towards more 'integrated, adaptive, coordinated and participatory' approaches to water management (Brown and Farrelly 2009; Daniell 2012), include urban water services as one part of the larger water cycle. It encompasses a river basin or larger special scale, the strengthening of institutional capacity and system resilience as keys for more successful transitions in urban water systems (Milman and Short 2008; UN-WATER 2012).

Many urban water systems around the world are undertaking actions to transition towards more sustainable configurations of urban water management, but the pathways may, and indeed should likely, be different with different barriers and hurdles encountered. From Australian experience, reported barriers in transitioning urban water systems to more sustainable configurations are largely socio-institutional rather than technical (Brown and Farrelly 2009). Transition pathways in other countries or regions may also take place at different rates of change with their own challenges (see examples in Tejada-Guibert and Maksimović 2001; Jenerette and Larsen 2006). For instance, the pathways adopted in Tanzania towards sustainable and community-based urban water management and the challenges they face (Cleaver and Toner 2006) would be different from the pathway adopted by Australia (Mitchell 2006; Brown et al. 2011), Sweden (e.g. Hellström et al. 2000), China (e.g. Bai and Imura 2001) and elsewhere. There is no one-size-fit-all solution to the transition process within or between countries and urban areas.

Understanding how systems change, identifying drivers of change and directions of current and potential transitions supports more strategic decision-making for the adoption of more sustainable and flexible technical infrastructure and adaptive institutions (Pahl-Wostl et al. 2007; Pearson et al. 2010). Using different countries' experiences, the purpose of this book is to illuminate and identify the challenges and opportunities that arise through urban water system transition journeys, as well as how these systems can be piloted towards a more resilient and sustainable futures.

To support the challenges of understanding and managing urban water in transition, we have developed a multi-authored and multi-disciplinary book that brings together a range of perspectives from various parts of the world. In total, the book comprises 28 chapters written by 51 contributors. These chapters have been grouped into three general sections on:

1. Water supply and sanitation;
2. Water demand and water economics; and
3. Water governance and integrated management.

There is thematic overlap across the sections due to the inherent inter-connections of these facets of urban water systems. Each section also includes both theoretical, historical and case-study based investigations of transitions in different areas of the world.

Collectively the book is much more than the sum of its parts. It offers innovative insights, practical guidance and best practice insights for managing water in the twenty-first century.

2 Drivers of Transition in Urban Water

In this section, we describe factors (internal and external) linked to those outlined in Fig. 1.1 that have, or will have, an important impact on the functioning of urban water systems in different parts of the world. For each of the drivers, we provide a brief description of issues involved and the resultant challenges they present for the effective functioning of current technical, economic or institutional models for urban water management. The general drivers of complexity, uncertainty and conflict are integrated into the descriptions of each of the drivers where most relevant, but interested readers are referred to the discussions on managing wicked or unstructured problems for more specific discussion on these issues (e.g. Rosenhead and Mingers (2001) for a general overview or Daniell (2012) specific to water management). Section 3 examines how transitions in urban water systems are linked to these drivers.

2.1 Population Growth, Demographic Change and Increasing Urbanisation

Population growth in many countries and migration of rural population to urban centres has resulted in rapid urban demographic growth. In developing (and to a lesser extent in developed) countries, this rapid expansion of cities provides many challenges, including for urban water management, such as how utilities in charge of water, sanitation and storm water management services are managed.

The first challenge confronting water managers is the need to increase water supply in order to meet fast growing water demand. Satisfying urban water demand is particularly challenging in emerging economies where the effect of population growth on water use is accentuated by the increase in living standards. Local resources, in particular groundwater, are often overexploited that can generate a series of negative impacts including seawater intrusion, the decline of groundwater dependent ecosystems or land subsidence (e.g. Mexico City, Bangkok, Beijing, Jakarta). As local resources become insufficient to sustain urban development, cities typically seek to tap increasingly distant resources, through the development of dams, canals or piped transfers. This often leads to conflicts with stakeholders of rural territories, who oppose the predatory behaviour of cities that they consider to be grabbing "their" water resources (see Rinaudo and Barraqué (2015), Chap. 8 this volume, on inter-basin transfers). Non-conventional resources such as desalination or wastewater recycling can also be used to meet growing demand in the developed world, but their cost remains prohibitive for most cities of emerging economies. Water supply thus remains a major challenge in many parts of the world.

The second challenge is how to ensure that development of sanitation and wastewater treatment infrastructure keeps pace with the galloping demographic growth and urban sprawl. The issue is particularly sensitive in developing countries where largely unplanned, informal or slum settlements host a very large share of the population (see for instance Makaudze and Gelles (2015), Chap. 5 this volume, for a South African illustration). What is at stake here is not only comfort and the quality of life but also health, as is illustrated by the Zimbabwean case study of Nhapi (2015), Chap. 4 this volume and also by White and Falkland (2015), Chap. 23 this volume, in their Pacific Islands case study. It is also an issue of ecological sustainability as water borne pollution from cities can contaminate large parts of river systems located downstream, groundwater and/or estuary and marine ecosystems, threatening the resource base of rural communities in those areas.

Urbanisation alters the hydrological cycle. The increase in impervious areas (e.g. rooves, roads, paved areas) reduces infiltration and groundwater recharge, accentuating groundwater depletion. It also increases volumes of run-off, flash-flood risk and accelerates the transport of pollutants from urban areas to wastewater systems and downstream rivers and environments. As metropolitan areas grow, the size of the centralised infrastructures required to manage storm water become increasingly costly and space consuming, leading urban water managers to consider the use of small scale decentralised technologies for on-site retention, recharge and natural treatment of polluted stormwater in ponds, constructed wetland or root-zone treatment facilities. The development of these decentralised technologies is increasingly integrated into urban development planning and not only considered as a water management issue (see Chap. 27 by Hussey and Kay (2015), this volume). Such approaches may promote ecological sustainability, though the production of ecological services (biodiversity, filtration, artificial groundwater recharge), turning stormwater from nuisance into a resource. Urbanisation also alters the hydrological cycle through the increasing importation of "embedded water" in food and materials bought, used and consumed by urban populations. For example, 2002 estimations in Sydney showed that almost half of water consumed by households was embedded in their food compared to just over 10 % that was from the city water supply used for drinking, gardening and other household activities (ABS 2002).

Financing the development, maintenance and renewal of water supply, sanitation and storm water management of growing cities is another major challenge. As health and environmental standards in many areas of the world become more stringent, the cost of water supply production and waste water treatment increases. In parallel, managers of urban water utilities are increasingly expected (by national regulators or by external donors) to recover investment and recurring costs thought higher user fees and charges. This cost-recovery objective is particularly difficult to achieve in poor countries where users' capacity to pay for water services remains low (e.g. Makaudze and Gelles (2015), Chap. 5 this volume). In emerging and developed countries alike, there is also a risk that a non-negligible percentage of the population stops using services, as they cannot afford to pay for access to them. Higher water fees may also provide incentives for the better-off users to bypass public water systems which they find to expensive, turning towards private supply solutions, such as private bore-well, rainwater harvesting and grey-water recycling systems (Montginoul and Rinaudo 2011).

This cost recovery imperative may be in contradiction with the social dimension of sustainability of urban water systems. Population growth is often accompanied by demographic changes such as increased inequality and poverty. A key challenge faced by utilities consists in developing innovative poverty alleviation schemes that ensure fair access to water supply and sanitation without jeopardizing the economic sustainability of the system.

2.2 *Increasing Resource Scarcity, Including Water*

The second main driver of urban water transitions, after demographic growth, is the increasing scarcity of resources needed to support human activities and lifestyles. Global analyses show that we are rapidly heading towards the limits of easily exploitable or renewable resources, and that this growing exploitation is having serious impacts on many human and ecological communities (e.g. Rockström 2009). For example, maintaining access to freshwater, clean air, healthy soil, phosphorus, energy and many other chemicals and minerals required for life, including food production (e.g. Lymbery and Oakeshott 2014), is an increasing challenge at many local, and sometimes, global scales; presenting significant challenges for urban (and rural) populations.

Looking specifically at water resources, scarcity is currently not an issue when looking at global statistics (e.g. Palaniappan and Gleick 2009; Rockström 2009), but is locally often significant due to a mismatch between freshwater and population distribution. This leads to conflict at a local and regional level and to the need for cooperation over scarce water resources between different users (Delli Priscoli and Wolf 2009), including urban and rural populations or different states (e.g. California – Arizona dispute over Colorado water). This scarcity will be reinforced in the future by climate change and its hydrological consequences, including the potential for reduced reliability of reservoirs and reduced groundwater recharge in some locations. Scarcity, at least in the sense perceived by human populations, is also due to strengthening environmental regulations (e.g. the Water Framework Directive in Europe or Endangered Species Act in the USA) which have led to increased environmental water allocations in order to maintain ecosystem health and services, often at the expense of existing diversion infrastructures, which have their capacities reduced.

2.3 *Technological Innovation*

Technological innovation is an important driver of change in urban water systems. In the 1980s and 1990s, transformations induced by technological innovation in the urban water sector mainly concerned water utilities, through the development of new treatment technologies. The development of membrane filtration techniques freed urban water managers not only from the water scarcity constraints (e.g. desalination, reuse) but also from environmental constraints (e.g. urban effluent treatment). Recent innovations in that domain include the integration of new treatment technologies into river basin management (spatial and temporal up-scaling) through groundwater replenishment programs (e.g. Orange county in California, Adelaide in South Australia, see also Nelson et al. (2015), Chap. 22 this volume) or indirect reuse (e.g. Singapore, see Kog (2015), Chap. 26 this volume). Simultaneously, there is a shift from chemical engineering to ecological engineering with the development of innovative biological treatment technologies which provide a wider range of ecological services including water purification, biodiversity, landscapes and storm water retention (Barraqué (2015), Chap. 9 this volume). This innovation seeks greater integration of water and land use planning (see also Sect. 3.5, this chapter). The uptake of such innovation is limited by their costs, which make them unaffordable for many developing countries, by consumer acceptance problems (reuse in particular); and health-related regulatory constraints.

In parallel with large scale technological innovations, the water industry is developing innovative small scale technologies, targeting individual users (industrial, commercial and domestic water users). From reverse osmosis to grey water recycling, technologies are, and will increasingly become affordable to individual users, who could choose to invest and by-pass collective public water services. Examples abound in Australia, where decentralised storm water management, drinking water treatment and wastewater treatment and recycling systems were promoted by public policies during the Millennium drought. The decentralisation process challenges the current technical and financial organisation of water systems, based on the principal of universal public services (see Barraqué et al. (2015), Chap. 2 this volume).

Change in urban water systems is driven by organisational management innovations in the field of computing, information and communication technologies (ICT) that can be used for monitoring (e.g. sensors, smart meters, smart devices) and automation of increasingly interconnected systems (e.g. water grids). This is expected to lead to increased flexibility of demand and supply in time and space. The development of market instrument innovations (e.g. tradable water rights, option markets, spot pricing), by increasing water reallocation possibilities, will reinforce this evolution. It is, thus, the combination of technological and institutional innovation that will allow increasing flexibility and theoretically resilience of urban water systems. New ICT tools are also likely to drive changes in the relationship between users of urban water systems and those who operate them. The development of smart phones, home automation technologies and smart meters allow two-way communication between the managers and customers. This opens up new opportunities in terms of short term demand forecasting or use of spot water pricing for instance, such as what is already widely practiced in the energy sector. In the longer term such innovations may allow improve tracking of water provenance/quality testing, and monitoring the risk and signs of potentially disruptive events (e.g. contamination attacks, ruptured or leaking water systems) over the long term.

Ongoing technological innovations described above offer opportunities for improving the performance of urban water services. Their uptake will, however, depend on their cost and perceived added-value they provide compared to existing systems, as consumers and local politicians of most developed countries are reluctant to implement changes that significantly increase water bills or result in citizen discontent over the short term. Innovation uptake, thus, depends on organisational transformation in the water sector, as discussed in the next section.

2.4 New Water Governance Approaches and Systems

Technological changes are typically accompanied by changes in terms of governance of urban water systems. The progressive development of decentralised technologies and infrastructures challenges traditional top-down and technocratic decision-making, requiring more collaboration and harmonisation across fragmented systems. The automation of water systems, from large supply networks to smart metering, can shift influence to experts capable of deciphering increasing mountains of data, whether at the large or local scale. In long-term planning, water utility managers now need to take into account the fact that hundreds of individuals or small groups can take independent investment decisions that are possibly inconsistent with their own objectives. They thus need to provide incentives (moral, economic, infrastructural) to influence those decisions in their desired direction. This evolution drastically changes the relationship between urban water managers and users, and necessitates changes in communication strategies between these groups. Managers have to meet a growing demand for transparency and to effectively involve citizen, communities and other private actors in the regulation of water utilities. In practice, consumer participation ranges from common sense customer care (one-way process) to cooperative ownership and management. Intermediate forms of participation include structured consultation procedures (such as public hearing, often mandated by law) and the participation to advisory committees or regulatory boards (Muller et al. 2008). This demand for transparency and participation by a range of stakeholders is particularly strong with public-private partnerships (PPPs), as private sector participation can be contested by citizens, consumer and environmental associations and even by some elected politicians. Nevertheless, there remains a significant gap between the recognition of participation as a citizenship right, as outlined in legal frameworks like the Aarhus Convention (UNECE 1998) and the actual practices to allow citizens expressing their preferences and grievances. Bridging this gap is one of the main challenges urban water utilities will have to take-up in the coming years.

Another major governance change is the progressive integration of urban water systems with urban land use planning (vertical integration) and river basin management planning (horizontal integration). Vertical integration takes place through technical, legal and economic cross-linkages between land use, water supply, sanitation and storm water planning documents. Horizontal integration requires urban water planners to consider broader environmental objectives at river basin level. In other words, river basin level environmental constraints "enter the city" and simultaneously, the city is increasingly accountable for its impact outside its direct control area. A consequence is that urban water managers need to construct new coalitions with stakeholders with whom they did not previously have to interact.

2.5 *Changing Water Values and Cultures*

Water is valued in many different ways by people. As a vital and non-substitutionable element of life for both humans and the environments they depend upon, people often have both spiritual and emotional attachments to water, as well as relationships to it for specific needs (e.g. drinking, washing, growing food, recreation). As economic wealth and living standards of urban populations increase, water service beneficiaries' mindsets often shift, as can the relative importance of some of these different values and cultural relationships to water. This can drive changes in values and cultures, for example from that of users to customers who, paying non-negligible fees for water access and use, have growing expectations in terms of quality of the service and their rights, and those of the urban administration to use water for purposes they consider important (e.g. maintaining green lawns, washing cars or filling swimming pools). Customers of urban water systems typically expect greater reliability of supply, intermittent supply being perceived as an unbearable infringement on domestic comfort for which they pay. Consumers also expect water quality (bacteriological and chemical) to improve as their sensitivity to health risks progressively rises. Meeting this new demand is a major challenge for utilities, as they may not have the technical capacities to improve the performance of the technical systems they operate. Public utilities may also lack the financial means to carry out the required investments, partly because only a fraction of the customer base has the required capacity to pay for improved services.

The increasing use of water pricing as a tool to recover cost or to promote efficient water use has also resulted in a growing opposition to the commodification of water, with a growing demand from some social groups to treat water supply and sanitation as a basic right which should be freely accessible to all. Reconciling economic efficiency, cost recovery and equity is a major challenge which some utilities attempt to address social water rates while other prefer to treat equity outside the bill through traditional social programs.

Last but not least, many urban citizens are increasingly aware of the environmental impacts of urban water services and are sensitive to the aesthetic and recreational values of water in the city. Particularly in developed countries, this results in a growing demand for low environmental impact technologies (storm and wastewater treatment) and in some cases for scenic urban river ecosystems, which contrast with the strict and rigid urban environment. Water planners thus need to reconsider the design of infrastructure which no longer needs to be made invisible, but which can become a constructive part of the urban landscape (e.g. replacing underground stormwater storage with ponds and wetlands, and making river banks accessible for pedestrians) and valued for both its aesthetics and utility by residents.

2.6 Climate Variability and Global Changes

Climate variability and other global changes, including climate change, are significant drivers of urban water management decisions. For example, urban water sector managers frequently need to make decisions concerning long-lived investments and to consider expected long term trends (and uncertainties) in terms of climate variability (e.g. lengths of likely dry/wet periods, and level and frequency of extreme hydrological events), demography, urban development, water demand, energy prices, technology and climate.

Climate change is a key driver of change, expected to have far reaching consequences for urban water systems. Changes in temperature and rainfall patterns will affect available water resources (rivers and groundwater), in particular, during low flow periods; the quality of surface water resources (loads of contaminants and suspended solids); flood risks and inundation patterns; and the state of water dependent ecosystems in general (Whitehead et al. 2009). Most urban water infrastructure will need to be adapted (capacity expansion), or operated differently, to cope with the new weather and hydrological conditions, including the reservoirs; inter-basin transfer schemes; wastewater treatment systems; storm water management schemes and other water infrastructure (e.g. desalination and water recycling plants; rainwater tanks). In coastal areas, many urban systems will be affected by sea level rise which will threaten systems in different ways, including: local water resources (e.g. sea water intrusion in coastal aquifers), low-lying supply and sanitation infrastructure; and flood protection infrastructure. The challenges of climate change are not only technical, but economic, as the required adaptation will influence the cost of urban water services.

A major concern of urban planners and water managers is that the uncertainty attached to future climate predictions is significant. This uncertainty is not likely to disappear as knowledge and climate models improve, implying that decision makers need to learn to live better with this uncertainty. A consequence is that new infrastructure should be designed to cope with a larger range of climatic conditions than in previous centuries and that this range will remain highly uncertain (Hallegatte 2009). This calls for new approaches that seek to either identify the most robust solutions (Graveline et al. 2014), defined as the most insensitive to climate conditions, or very flexible approaches that can be adjusted with changing conditions (Gordon 2013; Daniell 2013). Adopting such robust and/or flexible decision making methods represents a shift in paradigm for water practitioners whose actions have long been driven by the search for optimal solutions, but their use is already emerging in the water sector (Groves et al. 2008; Daniell 2013). The difficulties associated with the design of such flexible strategies are illustrated with the case of Melbourne, Australia (Grant et al. 2013).

Climate is not the sole source of uncertainty that confronts urban water planners and managers. Population and economic growth, technology, water demand, energy prices and prevailing political conditions are also unpredictable variables. In this new context, the use of contrasted scenarios is increasingly used to identify robust solutions (Graveline et al. 2014). Optimisation is progressively replaced with different practical decision making strategies. Hallegatte (2009) distinguished five such strategies. The first strategy consists in choosing "no regret" options which positive outcome over a wide range of future developments. The second is the "reversible strategy" which aims to keep the cost of making the wrong decision as low as possible, for instance by developing water infrastructures which can be upgraded at relatively low cost, or removed and relocated as sea-levels rise (Gordon 2013). The third is the "safety margin strategy", which consists of calibrating new infrastructure assuming upper-bound estimates for some of the assumptions made, so that the solution implemented is able to cope with almost every possible future development (within current imagination); this solution is only feasible when the marginal cost of oversizing infrastructure is small compared to its total cost. The fourth strategy consists of developing institutional frameworks that promote flexibility and adaptation of individual agents to changing conditions, the "soft strategy" or "soft path" (Wolff and Gleick 2002). This includes long term planning strategies such as those included in water resource management plans in the UK or California, which impose thinking several decades ahead (see also Rinaudo (2015), Chap. 11 this volume, on long run water demand forecasting). The use of economic instruments such as water markets, option contracts and insurance are also illustrative of this strategy. They imply much less inertia, reduced risk of sunk costs in case of wrong decisions, than with hard adaptation strategies relying on infrastructure. The fifth and last strategy consists of reducing decision making time horizons, acknowledging that uncertainty increases rapidly with time. This implies opting for technical solutions that have a shorter lifespan.

2.7 Ecosystem Degradation

Ecosystem degradation is another significant driver of the need for transformation in urban water systems. Decades of urbanisation and economic development have greatly increased the pressures on water resources and dependent ecosystems, including long term contamination, water resources depletion and loss of habitats and biodiversity in urban areas and their connected regions. The loss of ecosystem functions (e.g. water filtration/purification; seasonal flow regulation; erosion and sediment control; and habitat preservation) has led to new costs to urban areas, as damaged natural assets have had to be replaced with artificial infrastructure such as wastewater treatment plants or storm water retention infrastructure. Despite growing social concerns and movements over environmental matters and halting degradation since at least the 1970s, there is now a growing awareness that more effort should be dedicated to protecting river ecosystems within urban areas (rather than just those in "pristine" wilderness areas), to rebuilding and conserving remaining urban biodiversity (especially in the peri-urban fringe and along water corridors) and to managing pollution and increasing toxicity of the urban and linked rural environments. Designing institutional mechanisms which encourage higher levels of protection of watershed hydrological services of protection of watershed hydrological services (e.g. Postel and Thompson 2005) is thus one of the main challenges urban water managers will have to address in the coming decades, both in developed and developing countries.

2.8 *Political Ideology and Development of International Norms*

Political ideologies, linked to different values related to water discussed in Sect. 2.5, drive the development of different kinds of urban water systems, water policies and their governance structures. Ideological motivations have a great influence on policy choices, in particular concerning what services should be provided to the "public", and how these are financed, implemented and regulated. In urban water there are a few different political ideologies vying for attention and translation into the development of specific forms of urban water infrastructure, policy instruments and governance structures. In much of the world, there is a strong political ideology that water, including in urban areas, is and should be treated as a public good, leading to policies and international statements enshrining the "Human right to water" into law (Republic of South Africa 1998; United Nations 2010).

Since the early 1990s, the recognition of the economic value of water (e.g. ICWE 1992) also laid the foundation for a neoliberal political ideology and model of water policy based on market centred governance of water resources and services. In the water and sanitation sector, private sector participation (PSP) was pushed by many national and international agencies and water associations, including the World Bank, IMF and World Water Council, based on the assumption that private institutions are intrinsically superior to public institutions for the delivery of goods and services. The development of this ideology, which still underlies current policies in many parts of the world, resulted in the emergence of multinational or transnational private water companies who manage water systems and services in both developed and developing countries (e.g. Veolia, Suez). Despite this ideology dominating thinking in many urban centres, there is a growing recognition that the expected outputs of PSP in terms of efficiency gains and extension of coverage towards the poorest social groups have not materialized. Some authors even suggest that "*policies based on a commodification of water and sanitation services are intimately related to the increasing inequality that has been recorded in developing countries since the 1990s*" (Castro 2007). PSP has also been increasingly contested by citizens and politicians, as such evidence has highlighted the weakness of States to exercise regulatory control over private operators, particularly in weak democracies (Castro 2007; Ohemeng and Grant 2011). The ideological movement that seeks to reverse private sector involvement in water services delivery to citizens, born in Latin America, is now gaining momentum in Europe and increasingly becoming an important local political issue (Barraqué 2012).

A third significant political ideology that attempts to drive urban water reforms but has not yet gained sufficient power to force widespread policy changes is the green or ecologically rooted ideology that recognises that there are limits to growth and use of resources on the planet (see also Sect. 2.7 this chapter). Much of the environmental engineering paradigm (see Barraqué (2015), Chap. 9 this volume) and concepts like "water sensitive urban design" (see Hussey and Kay (2015), Chap. 27 this volume) stem from this ideology. Such an ideology also underlines the need for a reworking of the neo-liberal economic system and power structures to decouple economic growth from resource use, leading to it to support businesses based on renewable resources or environmental protection. The more moderate subscribing to such an ideology might also support economic instruments such as payments for ecological services, for example to protect drinking water catchments and reducing the need for chemical treatment and the energy costs associated with it. The final ideology that is gaining some ground and driving change, but still relatively marginal in urban water, is that of deliberative democracy and the rights of all people to participate in decision-making processes that affect them. Such an ideology is evident in documents such as the Dublin Statement (ICWE 1992) or the Aarhus Convention (UNECE 1998) and has been translated into other policy documents like the European Union's Water Framework Directive (EU 2000; see also EU 2002) or South Africa's national water policy (Republic of South Africa 1998).

3 Adaptations and Transformations in Urban Water Systems

Following on from the drivers of change in urban water systems and the challenges that these lead to, this section describes the resultant changes, adaptations and transformations that are taking place in urban water systems. Here we describe these changes or directions of potential transition in terms of the desired movement in the urban water system. Many of these relate to the objectives in Fig. 1.1. We also outline how the contributions of this book relate to these transitions or need for system movement in that direction.

3.1 Improvements in Public Health and Equality of Service

Health impacts resulting from a lack of easy access to potable water and sanitation of an acceptable quality are still significant in many places around the world. There are large inequities both between and within different countries on who is able to access water services of an acceptable quality, which is why one of the most significant transitions in urban (and rural) water systems sought on a global level is for improvements in public health and equality of service. Such objectives have previously resulted in transitions to centralised water supply and sewerage systems, piped into individual homes, as discussed for example by Troy (2008) and Bichai and Smeets (2015) (Chap. 6 this volume). As Bichai and Smeets (2015) outline, this is specifically as it has been easier to manage water quality systems in centralised, rather than decentralised systems, due to the ease of implementation of monitoring and treatment regimes. However, in response to other drivers such as increasing resource scarcity and alternative water governance systems and approaches, strains can appear in centralised systems in both developed and developing countries for different reasons. As Nhapi (2015) (Chap. 4 this volume) outlines, well performing water and sanitation system can rapidly cease to perform their key functions if the conditions for their effective management is not maintained. This was the case following a major economic and political crisis, where a water and sanitation system in Zimbabwe fell quickly into disrepair leading to a major cholera outbreak. In some developed countries, water security issues due to climate variability, change and potential reductions in water availability from traditional sources (e.g. dams, inter-basin transfers) in some regions, including Australia and Singapore have prompted suggestions for and implementation of recycled greywater and sewerage to be reinjected into potable water supply systems (see Kog (2015), Chap. 26 this volume on the Singapore case). This has led to concerns from some researchers and the public over the heightening of health risks (e.g. due to potential contaminants such as endocrine disrupting chemicals that are difficult for treatment systems to remove) in these centralised systems and led to the investigation of alternative decentralised and centralised water management options including "fit-for-purpose" use of different water sources, which may carry their own health risks, as outlined by Bichai and Smeets (2015) (Chap. 6 this volume), Rinaudo et al. (2015) (Chap. 7 this volume) when investigating use of bore water, and Reynaud and Garcia-Valiñas (2015) (Chap. 16 this volume) who demonstrate how households react to information on water quality by changing their consumption habits. Despite the potential issues that can develop around alternative systems and behaviours, Nelson et al. (2015) (Chap. 22 this volume) discuss how systems of governance and regulation can be put in place to ensure adequate water quality across both urban surface and groundwater services, providing an example in California, USA.

Adoption of alternative water systems in the community can also lead to inequities in the quality and quantity of water accessible for use occurring in these affected areas. Inequity in access to water and sanitation is obviously also a major challenge in much of the developing world with many millions lacking these basic rights, as outlined in our introduction. Makaudze and Gelles (2015) (Chap. 5 this volume) speaking about working towards rectifying this situation in South Africa and Nhapi (2015) (Chap. 4 this volume) on Zimbabwe, provide some insight into the challenges of service provision to all in the community, particularly some of the poorest urban residents living in slums. White and Falkland (2015) (Chap. 23 this volume) also provide some insights into the challenges in the Pacific Islands where traditional governance systems are challenged by increasing urbanisation. For areas that do have functioning centralised systems, there are ways of designing water tariffs in a way to ensure access to the poorest is possible, as outlined by Nauges et al. (2015) (Chap. 17 this volume) related to Egypt and Chan (2015, Chap. 15 this volume), related to many systems around the world, including the specifics of what occurs in Australian tariffs in order to integrate social and environmental goals, and responsibility for them, into the urban water sector.

3.2 Protection of Life, Livelihoods and Ensuring Well-Being (of Humans and the Environment)

A number of the drivers and the changes and challenges they induce in water systems have an impact on people's lives, their livelihoods, their health and wellbeing and that of the environment. These have led to transitions in the urban water systems to protect and reduce the risk of loss of life, for example due to a lack of access to adequate water and sanitation, as outlined in the last section, or through special provisions for flood and drought management. For example, van Vliet and Aerts (2015, Chap. 25 this volume) highlight how initial modifications in urban environments at risk of riverine, estuarine or coastal flooding were made to protect populations from recurrent events through hydro-technic infrastructure like dams, dykes and levees. They show how we are seeing a transition past this paradigm of "build and protect" to one where it is important to acknowledge the importance of non-structural, adaptive measures such as "dry-proofing" (in the case of Rotterdam in the Netherlands) to reduce flood risk. Other authors (e.g. Hallegatte 2009; Wenger et al. 2013) acknowledge the need to leave room for floodwater and to reduce the risks associated with the failure of flood-defence infrastructure in more extreme climate events through the implementation of a range of both structural and non-structural measures (Daniell 2013). For urban areas built on deltas, these areas are also often rich agricultural lands that can benefit from the water and sediments transported by floods if adequate compensation and insurance systems for crop and other material losses can be developed for land-owners who make their land accessible for "purposeful" flooding. This kind of transition to developing more non-structural measures to support urban water cycle management is a result of social and environmental drivers including amenity, access to, and quality of riverine environments for well-being. In some cases, as demonstrated by the case of the St Charles River in Quebec, Canada (Brun 2015, Chap. 24 this volume), this has actually led to the removal of water infrastructure and "renaturation" of rivers in urban environments.

3.3 Encouraging Resource Efficiency or 'Doing More With Less'

Environmental and social imperatives, as well as economic ones have also led to a transition in urban water systems in terms of seeking high levels of resource efficiency and "doing more with less" rather than just staying in the paradigm of increasing supply to match demand. This evolution is sometimes supported by regulations imposing water use efficiency thresholds. The evolution towards a more resource efficient society challenges the prevailing culture of water experts and consumers alike, who tend to consider water resources as unlimited and demand as uncompressible. The objectives in these new systems are to attempt to develop means of reducing water consumption, demand and waste (Guy et al. 2001), as well as ultimately to decouple economic growth from resource use (water, energy and other materials, minerals and chemicals) (e.g. Hargroves and Smith 2004; see also Kenway et al. (2015), Chap. 28 this volume, and Hussey et al. 2013).

In order to effectively transition to such systems, Troy (2015, Chap. 13 this volume) highlights how more significant changes than the water education and efficiency programs currently implemented (e.g. low-flow shower-heads, dual-flush toilets and water-efficient dishwashers and washing machine) will be needed for larger gains. These alternatives could include using potable supplies for only kitchen and bathroom uses then using treated wastewater or sewage for laundry and outside (e.g. garden) uses and installing dry-composting toilets. The role of water metering and pricing mechanisms is also acknowledged by Troy (2015, Chap. 13 this volume) and Garcia-Valiñas et al. (2015, Chap. 12 this volume) in inciting behavioural change that is required support a successful transition in this direction. Other issues of waste reduction are also important, adding considerably to urban water use and costs. For example, Dimova et al. (2015, Chap. 3 this volume) looks at how better management of extraneous water can lead to better economic and environmental outcomes. There is widespread and growing interest in promoting conservation in the industrial and commercial sectors. As Renzetti (2015, Chap. 14 this volume) outlines, governments in different countries are developing water conservation plans and manuals for firms seeking to reduce water use (e.g. target setting and benchmarking, and providing subsidies for water efficiency measure in of small and medium-sized firms). Other governance mechanisms can also be implemented to seek efficiencies in water management systems such as separation of water service functions (e.g. water and sewerage) or installing independent economic regulators with water services oversight (Reinhardt and Guérin-Schneider (2015), Chap. 20 this volume).

3.4 *Commodification and Economic Valuation of Water*

Linked to the important transition to conserving and using water more efficiently, and to improvements in health and equality of service is the increasingly widespread transition to the commodification and the economic valuation of water (see also Sect. 2.8 this chapter on the political driver behind this transition). With the acknowledgement of social and environmental values for water, and the limits to water access for many in urban systems from their own locally available sources (e.g. rooftop, well/bore, local river or lake) comes the possibility and often need to monetarise the purchase of water. Additional factors that lead to a transition in urban water systems to different economic valuations of water are: the need to recover costs of water supply and sanitation infrastructure development and maintenance; to manage demand; to ensure adequate returns for public or private stakeholders; or to create water transfer and trading systems. Although a few urban areas around the world still provide water for no cost to residents (see Chan (2015),

Chap. 15 this volume, for details), many urban residents have to pay for water, either from local water authorities, or legal or illegal vendors, as occurs often in slums or some other disadvantaged urban districts. Often these water charges, typically paid by the poorest, are much higher per litre than what richer residents would pay for a much better service (see Swyngedouw (2004), for a discussion of these issues in Bolivia). However, in many areas where there is a culture of free access to water (which is typical in many rural areas around the world), and water services are provided to urban residents through government or donor-financed programs, water authorities can struggle to enforce payment of water bills and hence face difficult challenges for cost recovery of the services (see for example Makaudze and Gelles (2015), Chap. 5 this volume, on the challenge of implementing this transition, desired by some but not by others, in the South African context).

Economic valuation and tariff setting of water typically varies on a number of factors, as Chan (2015), this volume, outlines that include water pricing: principles and objectives; processes and tariff structures; and desired outcomes, linked to other policies. Prices may not be relatively static like different types of block tariffs, but be more flexible and dynamic, like for example spot pricing, seasonal water rates or scarcity pricing (Grafton et al. (2015), Chap. 19 this volume; Grafton and Ward 2010) or those that reflect the cost of current consumption and opportunity costs of future supply (Sibly and Tooth (2015), Chap. 18 this volume). Grafton et al. (2015), Chap. 19 this volume, show that the economic benefits of efficient volumetric pricing in net present value terms would have been worth about $A1,900 per Sydney household if it had been instituted before the decision to construct a desalination plant in 2007.

The ability to use price as a demand and supply augmentation control variable is facilitated by the spread of technological innovation. For instance, smart meters allow consumers to better understand and monitor their own consumption and adapt their practices to respond to changes in tariffs. More flexible pricing systems, like those that represent the prevailing market value in a region, as in some trading systems, also present opportunities and challenges under this new paradigm. For example, Nelson et al. (2015), Chap. 22 this volume, show how innovative legal frameworks can help to overcome difficulties in rural-urban water trading and encourage more efficient use of scarce water resources. Although some consider that there are potentially ecological and social externalities created by a re-engineering of the water cycle through trading and the physical infrastructure that allows it between typically non-hydrologically communicating systems, this transition pathway is starting to accelerate in many places, including those where it may not have been expected, like in China (see Squires et al. (2015) for greater discussion on these challenges and the resulting transitions).

A side effect of this transition to commodification is that it increases the cost of water services for large users or residents and who may decide to develop their own water supply or sanitation systems. While the development of independent water supply systems remains a limited phenomenon, it could threaten the long term technical and financial sustainability of public water services in the future. The main question is how to cover fixed costs when the customer base erodes?

3.5 *Low Impact Development or "Water Sensitive Urban Design"*

In order to enhance environmental and aesthetic outcomes in urban settings, and to overcome many of the challenges highlighted in Sect. 2, there is a strong push by a number of water academics and stakeholder groups to transition urban water systems to forms of lower impact development or "water sensitive urban design" through use of more ecosystem-based approaches and technologies such as wetlands and renaturation of rivers, coupled with decentralised collection, treatment and fit-for purpose reuse strategies (e.g. NWC 2011; Hussey and Kay 2015, Chap. 27 this volume; Brun 2015, Chap. 24 this volume). The "Water Sensitive Cities" Cooperative Research Centre in Australia,[1] are seeking to champion these concepts, as are other groups within governments in a less direct manner, like the Office of Living Victoria in Australia with its new Melbourne water strategy based on Integrated Water Cycle Management analyses and the resultant mix of centralised and decentralised systems that are intended to increase urban system sustainability and water use effectiveness and efficiency (Coombes 2012; OLV 2013; see also Reinhardt and Guérin-Schneider 2015, Chap. 20 this volume, on the context of these reforms). Although there are increasing numbers of relatively small-scale projects, showing the potential for a transition to these forms of water cycle management, in most cases, generalised uptake of these kinds of systems are only in pre-development or just starting to "take-off" (see van der Brugge and Rotmans 2007). Further implementation faces a range of challenges and political struggles as outlined by Keath and Brown 2009; Brown et al. 2011; Daniell et al. 2014; Farrelly and Brown 2011; and Hussey and Kay 2015, Chap. 27 this volume.

3.6 *Integrated or Whole of Systems Approaches*

An extension of the "water sensitive urban design" concept is a transition to integrated or whole of systems approaches that implies that all resources and issues within a specific boundary will be treated managed holistically towards having the most sustainable and self-sufficient system possible. The kinds of urban resources, stocks and flows considered in these approaches cross multiple boundaries and include water/energy/materials/food/waste/land/social/economic nexus issues which are all vital to the sustainability of urban systems (Sheehan 2007). One concept that represents this kind of urban systems approach is the "urban metabolism" that looks at inputs and outputs to cities and the quantities of "embodied resources" used in the development of other resources: for example, the quantity of energy used in Water production (Kenway and Lant 2015, Chap. 28 this volume) or water used in energy production (Hussey et al. 2013). There are also a range of other

[1] http://watersensitivecities.org.au

"integrated" approaches that seek to address issues previously siloed in urban water management collectively, such as surface water and groundwater (Nelson et al. 2015, Chap. 22 this volume) or seeking to make multi- rather than single- objective evaluations, for example of economic, social and environmental costs and benefits to urban water system infrastructure renewals and development (see Marlow et al. 2015, Chap. 10 this volume).

Transitions to integrated approaches are becoming more widespread in the urban water industry (see Barraqué 2015, Chap. 9 this volume on the growing paradigm shift to "environmental engineering" from previous "civil engineering" and "sanitary engineering" approaches. Nevertheless, there is still typically more work required to achieve more seamless integration of urban resource management, planning and maintenance systems that are required for the effective understanding and implementation of whole of systems approaches.

3.7 Resilient and Adaptive Systems

Moving from the structure and content of the urban system and its management, to its ability to respond to a range of drivers in Sect. 2, the next transition of urban water management systems we consider is one to more resilient and adaptive systems. Previous urban water paradigms have focussed on the robustness and reliability of urban water systems to respond to population demands and climate extremes. There is an increasing recognition that the robustness of systems may actually lead to more catastrophic failures if the system design parameters are exceeded by extreme events.

A transition to developing more resilient systems that able to bounce back from extreme events and more effectively adapt to changes in urban and their linked rural and global systems is sought in many urban water management systems around the world. This can, hopefully, occur by ensuring a requisite variety of water systems and capacity to respond effectively under such pressures as population and political changes, floods, droughts, storms, sea level rise, pollution, contamination, or disease outbreaks, or other significant events such as climate step changes, dam breaks, earthquakes, tsunamis, electricity failures, or fires (e.g. Daniell 2013; van Vliet and Aerts 2015, Chap. 25 this volume).

There are many tools that are available to support the acceleration of this transition. These include: improving long-term water demand forecasting modelling (see Rinaudo 2015, Chap. 11 this volume) and understanding the determinants of not only residential but industrial, commercial and institutional water demands (Renzetti 2015, Chap. 14 this volume); understanding how price and other non-economic factors influence water users' behaviours, including their choice of uptake or installation of their own "water security" measures (e.g. bores, rainwater tanks) to ensure their own resilience to extreme events and changes in the urban water landscape (Troy 2015, Chap. 13 this volume; Garcia-Valiñas et al. 2015, Chap. 12 this volume, Rinaudo et al. 2015, Chap. 7 this volume); the adoption of pricing mechanisms for water that are themselves more adaptive to prevailing conditions (e.g. Sibly and Tooth 2015, Chap. 18 this volume; Grafton et al. 2015, Chap. 19 this volume; Grafton and Ward 2010); and by encouraging increased hybridisation and potentially redundancy in urban water systems that leave them potentially less vulnerable and more resilient and adaptive to a range of foreseen and unforeseen events (see also Sect. 3.9 and Hashimoto 1982).

3.8 Participatory Democracy

An important transition that is 'taking off' in urban water systems is one to a more stakeholder-inclusive and participatory democracy approach to decision-making related to urban water and planning for its future (see also Sect. 2.8 on the political ideology underlying this transition). This transition is manifested through the integration of a wide range of stakeholders in decision-making and engagement processes around water management, typically for reasons of equity, empowerment, developing shared understandings of values, cultures and problems in order to have a platform for the development of more broadly accepted and legitimated urban water decisions, which will suffer less opposition to implementation and (e.g. Creighton 2005; Dryzek 1990, 2010; Daniell 2012).

Participation can take place in a variety levels, from lower levels of interaction in simple information provision and consultation (seeking feedback on proposals/policies/plans), to more interactive and engaged joint analysis and decision-making, or even citizen-controlled decision-making, as in the case of some community collectives or individuals managing decentralised urban water systems. Participation initiatives and stakeholder inclusive approaches to urban water management can involve just individual or a variety of urban water issues in the same process. Single issue examples include initiatives such as citizen's juries or consensus conferences like the Australian Weather Channel's televised citizen jury in 2006 (which also enabled viewers to vote via mobile phone text message) on whether Sydney should build a desalination plant or not, the French consensus conference in 2003 on what should be done with wastewater treatment sludge, and the public debate in Paris, France, in 2009 on the future of their double pipe water network system (Bedu 2010).

Examples of processes that have a wider remit, such as most participatory planning or more creative empowerment and education initiatives, include "re-imagining" urban or degraded rivers through field visits and creative writing in South Yorkshire, UK (Selman et al. 2010), understanding the role of water and why it is valued in urban residents' lives through participatory photography in Nantes, France (Bedu 2010) and multi-level stakeholder processes for developing strategies to better cope with flood and drought risks in Sofia in Bulgaria, or estuarine and lagoon planning in the Lower Hawkesbury (northern Sydney, Australia) and the Thau Basin (next to Sète in France) (Daniell 2012; Plant et al. 2014). Even if such processes are not yet really widespread, the increased sensitivity and awareness of urban populations in many parts of the world to environmental and social issues related to water (water pollution, scarcity issues, cost of living pressures) in the population puts additional pressure on water utility managers and governments to disclose information on water quality, as well as on industrial users to protect water sources, to reduce pollution discharge into the water bodies, and to increase water recycling/recirculation. The issue of increasing household awareness and sensitivity to such issues is discussed in Reynaud and Garcia-Valiñas (2015, Chap. 16 this volume) and Reinhardt and Guérin-Schneider (2015, Chap. 20 this volume) and social pressure on industrial firms is discussed in Renzetti (2015, Chap. 14 this volume).

In some cases, social opposition to particular urban water decisions, such as the planned building of new dams, or of inter-basin transfers (see Rinaudo and Barraqué (2015), Chap. 8 this volume) can ultimately lead to a reversal of these decisions unless views of the large majority can be changed. For smaller disputes such as between customers of water authorities and these authorities, sometimes formal and informal mechanisms can be put in place to manage and/or resolve these in an orderly fashion such as customer service obligations (CSO), consumer advocacy groups or Ombudsman agencies, like the Independent Energy and Water Ombudsman in Australia. In some places around the world public participation in all water management, including in urban areas is mandated, as for example through the European Union's Water Framework Directive and the Aarhus Convention, yet exactly what this implies and its implementation in reality is far from widespread in many urban water decision-making processes.

3.9 Decentralisation, Diversification and Hybridisation of Water Systems

One of the most important transitions in urban water management, visible in recent decades, is the diversification and hybridisation of water management systems. Combinations of decentralised and centralised systems are increasingly being adopted, as well as a range of public/private and even community-sector financed, managed, and operated systems. These aspects are discussed by a number of our authors (Reynaud 2015, Chap. 21 this volume; Reinhardt and Guérin-Schneider 2015, Chap. 20 this volume; Hussey and Kay 2015, Chap. 27 this volume; Barraqué 2015, Chap. 9 this volume). This transition to a diversified or hybrid system (from a largely centralised, or largely decentralised, system) has developed in relationship to a range of competing objectives, including enhancing water security through the development of system robustness or resilience. This diversity reflects the innate tensions in the different societal values that underline constructions of sustainability (e.g. efficiency, equity, ecosystem and population health, participation, prosperity) and their relationships to urban water management systems.

Issues of water supply security due to climate variability and change, can also lead to a diversification of centralised systems from for example rain-fed dam water to both centralised non-rain-dependent systems like desalination and waste-water reuse and decentralised options where households and districts develop additional water systems to self-provide additional security (e.g. rain-water tanks, groundwater bores, local greywater recycling and sewer water mining and treatment). Likewise, diversification is seen in waste and storm-water systems to deal with changing environmental, social and economic conditions. Water-sensitive urban design pilot programs now sit nestled alongside centralised systems, and there are increasing efforts to "renaturalise" water ways for aesthetic and flood attenuation reasons (e.g. Hussey and Kay 2015, Chap. 27 this volume, Brun 2015, Chap. 24 this volume), as to septic tanks and ecological treatment systems in some urban areas.

Changing governance systems, from purely publicly or locally operated water systems to the increasing participation of the private sector in water services and infrastructure provision, maintenance and governance has also led to the hybridization and diversification of water systems, through for example public-private partnerships (PPPs) (see Reynaud 2015, Chap. 21 this volume). Urban households may also have the choice to either self-manage water, have it supplied buy either public utilities or private vendors, as is becoming increasingly possible in both developed and developing countries (Gleick 2005; Swyngedouw 2004). Such hybrid models are not always easy to manage or provide equitable or sustainable outcomes for users of the system. For example on PPPs, there are many different models with alternative allocations of risks and responsibilities under different arrangements. Although many in the water sector had hoped that PPPs would lead to efficiency gains, there is only mixed empirical evidence that PPPs have so far managed to significantly improve technical efficiency, cost efficiency or reduce water prices. Nevertheless, as these systems now play an important role in many of the world's urban water systems, concentrating on improving governance, accountability, transparency in the contractual arrangements will be important for achieving more long term sustainable outcomes (Reynaud 2015, Chap. 21 this volume chapter, Reinhardt and Guérin-Schneider 2015, Chap. 20 this volume). Increasing citizen interest in water matters is also leading to the decentralisation (or perhaps more correctly localisation) of issues and the development of "fit-for-community" urban water systems and programs, as discussed in Sect. 3.8.

Other hybrid systems are developing in response to managing competing political ideologies (Sect. 2.8). For example, private sector participation in urban water systems can be accompanied by strong regulatory bodies to regulate prices and environmental impacts, to guarantee yardstick competition and to allow public pressure through citizen advisory groups. The British solution embodied by OFWAT (The Water Services Regulation Authority) is maybe a good illustration of this "middle strategy". Alternatives to State or market models are also likely to emerge, involving greater cooperation driven by voluntary association and more decentralised organisation. Hybrid models are also likely to emerge, with private, public and community actors being simultaneously involved in various forms of water supply services and possibly competing for clients. The delivery of water services is not a simple choice between private or public, distributed governance and hybrid forms of organisations will need to be invented. Developing such alternative models will require radical changes in the way national and international policy makers approach this matter. This will only happen if the level of effort and political commitment are substantial enough to counter balance the inertial forces set in motion by the neo-liberal model and/or government operated monopolies that currently share most of the power in urban water around the world.

4 Perspectives: Challenges and Future Uncertainties for Urban Water Systems in Transition

The two previous sections have shown that the transformation processes in which most urban water systems are engaged are characterised by great uncertainty concerning the final outcome. Today, the main uncertainties that confront water managers and impact their decisions in terms of infrastructure dimensioning or phasing of investments include those around future water demands, water resource availability (for example due to climate and competition) and technology costs.

While we are witnessing social, economic and environmental imperatives influencing some management choices, tomorrow's water managers might have to make decisions which could set in motion radical shifts in term of business models, conceptions of social justice, organisational set-up and trade-offs between environmental, economic and social objectives. Managers will need to anticipate transition in urban water systems and navigate conflicts in systems of political and social values in order to avoid options that too strongly lead to system "lock-in" and failure if the conditions that led to that option being successful change. It is possible that "no regret strategies", if they can be identified, may make good first choices, but some kinds of lock-in and system inertia are likely regardless of the chosen paths. Challenges will remain significant and different between developing countries and rich nations. Technological innovations and management strategies will be required to alleviate some of the most significant health and social issues associated with a lack of access to adequate water and sanitation.

Our key intentions for this book is to show that the drivers of change and resulting transitions that are occurring in urban water management are not only related to the choice of water technologies and infrastructure systems, but also to organisational configurations, legislation, aid programs, social habits and values, policies, political ideologies and the management of societies. We believe that this volume provides many insights into this crucial global challenge and inspiration for the creation of more sustainable and resilient societies in the future.

References

ABS. (2002). *The influence of lifestyles on environmental pressure, 1301.0 – Year Book Australia, 2002*. Canberra: Australian Bureau of Statistics.

Arthur, B. W. (1988). Competing technologies. In G. Dosi (Ed.), *Technological change and economic development*. London/New York: Printer.

Bai, X., & Imura, H. (2001). Towards sustainable urban water resource management: a case study in Tianjin, China. *Sustainable Development, 9*(1), 24–35.

Barraqué, B. (2012). Return of drinking water supply in Paris to public control. *Water Policy, 14*, 903–914.

Barraqué, B. (2015). Three engineering paradigms in the historical development of water services: More, better and cheaper water to European cities. In R. Q. Grafton, K. A. Daniell, C. Nauges, J.-D. Rinaudo, & N. W. W. Chan (Eds.), *Understanding and managing urban water in transition*. Dordrecht: Springer. (pp. 201–216).

Barraqué, B., Isnard, L., & Souriau, J. (2015). How water services manage territories and technologies: History and current trends in developed countries. In R. Q. Grafton, K. A. Daniell, C. Nauges, J.-D. Rinaudo, & N. W. W. Chan (Eds.), *Understanding and managing urban water in transition*. Dordrecht: Springer. (pp. 33–60).

Bedu, C. (2010). *Quand une citadelle technique se (sou)met à l'épreuve de l'« impératif délibératif ». Récit et analyse pragmatique d'une procédure de type « mini public » dans le domaine de l'eau potable*. PhD thesis, Université de Strasbourg, France.

Bichai, F., & Smeets, P. (2015). Integrating water quality into urban water management and planning while addressing the challenge of water security. In R. Q. Grafton, K. A. Daniell, C. Nauges, J.-D. Rinaudo, & N. W. W. Chan (Eds.), *Understanding and managing urban water in transition*. Dordrecht: Springer. (pp. 135–154).

Brown, R. R., & Farrelly, M. A. (2009). Delivering sustainable urban water management: a review of the hurdles we face. *Water Science & Technology, 59*(5), 839–846.

Brown, R. R., Keath, N., & Wong, T. H. F. (2009). Urban water management in cities: historical, current and future regime. *Water Science & Technology, 59*(5), 847–855.

Brown, R. R., Ashley, R., & Farrelly, M. A. (2011). Political and professional agency entrapment: an agenda for urban water research. *Water Resources Management, 25*(15), 4037–4050.

Brun, A. (2015). The "renaturation" of urban rivers: The case of the St Charles River in Quebec. In R. Q. Grafton, K. A. Daniell, C. Nauges, J.-D. Rinaudo, & N. W. W. Chan (Eds.), *Understanding and managing urban water in transition*. Dordrecht: Springer. (pp. 527–548).

Castro, J. E. (2007). Poverty and citizenship: sociological perspectives on water services and public–private participation. *Geoforum, 38*, 756–771.

Chan, N. W. W. (2015). Integrating social aspects into urban water pricing: Australian and international perspectives. In R. Q. Grafton, K. A. Daniell, C. Nauges, J.-D. Rinaudo, & N. W. W. Chan (Eds.), *Understanding and managing urban water in transition*. Dordrecht: Springer. (pp. 311–336).

Cleaver, F., & Toner, A. (2006). The evolution of community water governance in Uchira, Tanzania: the implications for equality of access, sustainability and effectiveness. *Natural Resources Forum, 30*, 207–218.

Coombes P. J. (2012). *Integrated systems analysis to create evidence based policies for water cycle reform in Greater Melbourne*. 34th hydrology and water resources symposium, engineers Australia, Sydney.

Creighton, J. L. (2005). *The public participation handbook: Making better decisions through citizen involvement*. San Francisco: Jossey-Bass.

Daniell, K. A. (2012). *Co-engineering and participatory water management: Organisational challenges for water governance* (UNESCO international hydrology series). Cambridge, UK: Cambridge University Press.

Daniell, K. A. (2013). Practical responses to water and climate policy implementation challenges. *Australian Journal of Water Resources, 17*(2), 111–125. doi:10.7158/W13-WC01.2013.17.2.

Daniell, T. M., Foley, B. A., & Daniell, K. A. (2005). Sustainability in Water Resources Management. *Journal of Hydrologic Environment, 1*(1), 3–18.

Daniell, K. A., Coombes, P. J., White, I. (2014). Politics of innovation in multi-level water governance systems. *Journal of Hydrology 519, Part C*, 2415–2435.

Delli Priscoli, J., & Wolf, A. T. (2009). *Managing and transforming water conflicts*. Cambridge, UK: Cambridge University Press.

Dimova, G., Ribarova, I., & de Carné, F. (2015). Coping with extraneous water in sewerage systems. In R. Q. Grafton, K. A. Daniell, C. Nauges, J.-D. Rinaudo, & N. W. W. Chan (Eds.), *Understanding and managing urban water in transition*. Dordrecht: Springer. (pp. 61–90).

Dryzek, J. S. (1990). *Discursive democracy: Politics, policy and political science*. New York: Cambridge University Press.

Dryzek, J. S. (2010). *Foundations and frontiers of deliberative governance*. Oxford, UK: Oxford University Press.

EU. (2000). Directive 2000/60/EC of the European Parliament and of the Council, of 23 October 2000: Establishing a framework for community action in the field of water policy, L 327, 22.12.2000, EN. *Official Journal of the European Communities*, 1–72.

EU. (2002). *Guidance on public participation in relation to the water framework directive: Active involvement, consultation, and public access to information.* Final version after the Water Directors' meeting, Dec 2002. European Union. Copenhagen.

Farrelly, M., & Brown, R. (2011). Rethinking urban water management: experimentation as a way forward? *Global Environmental Change, 21*, 721–732.

Frederick, K. D., & Major, D. C. (1997). Climate change and water resources. *Climate Change, 37*, 7–23.

Garcia-Valiñas, M. A., Martínez-Espiñeira, R., & To, H. (2015). The use of non-pricing instruments to manage residential water demand: What have we learned? In R. Q. Grafton, K. A. Daniell, C. Nauges, J.-D. Rinaudo, & N. W. W. Chan (Eds.), *Understanding and managing urban water in transition*. Dordrecht: Springer. (pp. 269–282).

Geels, F. W. (2002). Technological transitions as evolutionary reconfiguration processes: a multi-level perspective and a case-study. *Research Policy, 31*, 1257–1274.

Geels, F. W. (2004). From sectoral systems of innovation to sociotechnical systems. Insights about dynamics and change from sociology and institutional theory. *Research Policy, 33*, 897–920.

Geels, F. W. (2010). Ontologies, socio-technical transitions (to sustainability), and the multi-level perspective. *Research Policy, 39*, 495–510.

Geels, F. W., & Schot, J. (2007). Typology of sociotechnical transition pathways. *Research Policy, 36*, 399–417.

Gleick, P. H. (2005). The myth and reality of bottled water. In P. H. Gleick (Ed.), *The world's water: The biennial report on freshwater resources 2004–2005*. Washington, DC: Island Press.

Gordon, A. D. (2013). Disposable infrastructure including relocatable buildings: Adapting to climate change. *Australian Journal of Water Resources, 17*(2), 152–160.

Grafton, R. Q., & Ward, M. B. (2010). *Dynamically efficient urban water policy.* Centre for Water Economics, Environment and Policy (CWEEP) research paper 2010–13. The Australian National University, Canberra. http://cweep.anu.edu.au/pdf/publications/research_papers/10-13_dynamically_efficient.pdf. Accessed 2 June 2014.

Grafton, R. Q., Chu, L., Kompas, T., & Ward, M. (2015). Volumetric water pricing, social surplus and supply augmentation. In R. Q. Grafton, K. A. Daniell, C. Nauges, J.-D. Rinaudo, & N. W. W. Chan (Eds.), *Understanding and managing urban water in transition*. Dordrecht: Springer. (pp. 401–420).

Grant, S. B., Fletcher, T. D., Feldman, D., Saphores, J.-D., Cook, P. L. M., Stewardson, M., Low, K., Burry, K., & Hamilton, A. J. (2013). Adapting urban water systems to a changing climate: lessons from the millennium drought in Southeast Australia. *Environmental Science & Technology, 47*, 10727–10734.

Graveline, N., Aunay, B., Fusillier, J. L., & Rinaudo, J. D. (2014). Coping with urban & agriculture water demand uncertainty in water management plan design: the interest of participatory scenario analysis. *Water Resources Management*. doi:10.1007/s11269-014-0656-5.

Groves, D. G., Lempert, R. J., Knopman, D., & Berry, S. H. (2008). *Preparing for an uncertain future climate in the Inland Empire. Identifying robust water-management strategies*. Santa Monica: RAND Corporation.

Gunderson, L. H., & Holling, C. S. (2002). *Panarchy: understanding transitions in human and natural systems*. Washington, DC: Island Press.

Guy, S., Marvin, S., & Moss, T. (2001). *Urban Infrastructure in transition: Networks buildings and plans*. London: Earthscan.

Hallegatte, S. (2009). Strategies to adapt to an uncertain climate change. *Global Environmental Change, 19*(2), 240–247.

Hargroves, C., & Smith, M. (Eds.). (2004). *The natural advantage of nations: Business opportunities, innovation and governance in the 21st century*. London: Earthscan.

Hashimoto, T., Stedinger, J., & Loucks, D. P. (1982). Reliability, resiliency and vulnerability criteria for water resource system performance evaluation. *Water Resources Research, 18*, 14–20. doi:10.1029/WR018i001p00014.

Hellström, D., Jeppsson, U., & Kärrman, E. (2000). A framework for systems analysis of sustainable urban water management. *Environmental Impact Assessment Review, 20*(3), 311–321.

Holland, J. H. (1995). *Hidden order: How adaptation builds complexity*. Cambridge, MA: Helix books/Perseus Books.

Holling, C. S. (2004). From complex regions to complex worlds. *Ecology and Society, 9*(1), 11. http://www.ecologyandsociety.org/vol9/iss1/art11.

Hussey, K., & Kay, E. (2015). The opportunities and challenges of implementing 'water sensitive urban design': Lessons from stormwater management in Victoria, Australia. In R. Q. Grafton, K. A. Daniell, C. Nauges, J.-D. Rinaudo, & N. W. W. Chan (Eds.), *Understanding and Managing Urban Water in Transition*. Dordrecht: Springer. (pp. 593–614).

Hussey, K., Carter, N., & Reinhardt, W. (2013). Energy sector transformation: Implications for water governance. *Australian Journal of Water Resources, 17*(2), 170–179. doi:10.7158/W13-025.2013.17.2.

ICWE. (1992). *The Dublin statement on water and sustainable development*. Dublin: International Conference on Water and the Environment.

Jenerette, G. D., & Larsen, L. (2006). A global perspective on changing sustainable urban water supplies. *Global and Planetary Change, 50*, 202–211.

Kauffman, S. (1995). *At home in the universe: The search for laws of complexity*. Oxford: Oxford University Press.

Keath, N., & Brown, R. (2009). Extreme events: being prepared for the pitfalls with progressing sustainable urban water management. *Water Science and Technology, 59*, 1271–1280.

Kenway, S. J., & Lant, P. A. (2015). How does energy efficiency affect urban water systems? In R. Q. Grafton, K. A. Daniell, C. Nauges, J.-D. Rinaudo, & N. W. W. Chan (Eds.), *Understanding and managing urban water in transition*. Dordrecht: Springer. (pp. 615–632).

Kog, Y. C. (2015). Transboundary urban water: The case of Singapore and Malaysia. In R. Q. Grafton, K. A. Daniell, C. Nauges, J.-D. Rinaudo, & N. W. W. Chan (Eds.), *Understanding and managing urban water in transition*. Dordrecht: Springer. (pp. 575–592).

Larsen, T. A., & Gujer, W. (1997). The concept of sustainable urban water management. *Water Science and Technology, 35*(9), 3–10.

Lymbery, P., & Oakeshott, I. (2014). *Farmageddon: The true cost of cheap meat*. London: Bloomsbury.

Makaudze, E. M., & Gelles, G. M. (2015). The challenges of providing water and sanitation to urban slum settlements in South Africa. In R. Q. Grafton, K. A. Daniell, C. Nauges, J.-D. Rinaudo, & N. W. W. Chan (Eds.), *Understanding and managing urban water in transition*. Dordrecht: Springer. (pp. 121–134).

Marlow, D. R., Beale, D. J., Cook, S., & Sharma, A. (2015). Investing in the water infrastructure of tomorrow. In R. Q. Grafton, K. A. Daniell, C. Nauges, J.-D. Rinaudo, & N. W. W. Chan (Eds.), *Understanding and managing urban water in transition*. Dordrecht: Springer. (pp. 217–236).

Milman, A., & Short, A. (2008). Incorporating resilience into sustainable indicators: An example for the urban water sector. *Global Environmental Change, 18*(4), 758–767.

Mitchell, G. V. (2006). Applying integrated urban water management concepts: A review of Australian experience. *Environmental Management, 37*(5), 589–605.

Montginoul, M., & Rinaudo, J.-D. (2011). Controlling households' drilling fever in France: An economic modeling approach. *Ecological Economics, 71*, 140–150.

Muller, M., Simpson, R., & van Ginneken, M. (2008). Ways to improve water services by making utilities more accountable to their users: A review. In World Bank (Ed.), *Water working notes*. Washington, DC: World Bank.

Nauges, C., Whittington, D., & El-Alfy, M. (2015). A simulation model for understanding the consequences of alternative water and wastewater tariff structures: A case study of Fayoum, Egypt. In R. Q. Grafton, K. A. Daniell, C. Nauges, J.-D. Rinaudo, & N. W. W. Chan (Eds.), *Understanding and managing urban water in transition*. Dordrecht: Springer. (pp. 359–382).

Nelson, R. R., & Winter, S. G. (1982). *An evolutionary theory of economic change*. Cambridge, MA: Belknap Press (Harvard University Press).

Nelson, R., Bischel, H. N., Luthy, R. G., & Thompson, B. H., Jr. (2015). Issues of governance, policy, and law in managing urban–rural and groundwater–surface water connections. In R. Q. Grafton, K. A. Daniell, C. Nauges, J.-D. Rinaudo, & N. W. W. Chan (Eds.), *Understanding and managing urban water in transition*. Dordrecht: Springer. (pp. 463–488).

Nhapi, I. (2015). Challenges for water supply and sanitation in developing countries: case studies from Zimbabwe. In R. Q. Grafton, K. A. Daniell, C. Nauges, J.-D. Rinaudo, & N. W. W. Chan (Eds.), *Understanding and managing urban water in transition*. Dordrecht: Springer. (pp. 91–120).

Niemczynowicz, J. (1999). Urban hydrology and water management – present and future challenges. *Urban Water, 1*(1), 1–14.

NWC. (2011). *The National Water Initiative – securing Australia's water future: 2011 assessment*. Canberra: National Water Commission (NWC).

Ohemeng, F. L. K., & Grant, J. K. (2011). Has the bubble finally burst? A comparative examination of the failure of privatization of water services delivery in Atlanta (USA) and Hamilton (Canada). *Journal of Comparative Policy Analysis: Research and Practice, 13*, 287–306.

OLV. (2013). *Melbourne's water future*. Consultation paper. Office of Living Victoria (OLV), Melbourne.

Pahl-Wostl, C., Sendzimir, J., Jeffrey, P., Aerts, J., Berkamp, G., & Gross, K. (2007). Managing change toward adaptive water management through social learning. *Ecology and Society, 12*(2), 30.

Palaniappan, M., & Gleick, P. H. (2009). Peak water. In P. H. Gleick (Ed.), *The world's water: The biennial report on freshwater resources 2008–2009*. Washington, DC: Island Press.

Pearson, L., Coggan, A., Proctor, W., & Smith, T. F. (2010). A sustainable decision support framework for urban water management. *Water Resources Management, 24*, 363–376.

Plant, R., Maurel, P., Barreteau, O., & Bertacchini, Y. (2014). The role of territorial intelligence: The case of the Thau territory, Southern France. In V. R. Squires, H. M. Milner, & K. A. Daniell (Eds.), *River basin management in the twenty-first century: Understanding people and place* (pp. 446–466). Boca Raton: CRC Press.

Postel, S. L., & Thompson, B. H. (2005). Watershed protection: capturing the benefits of nature's water supply services. *Natural Resources Forum, 29*, 98–108.

Prigogine, I., & Stengers, I. (1984). *Order out of chaos: Man's new dialogue with nature*. Boulder: New Science Library.

Reinhardt, W., & Guérin-Schneider, L. (2015). Governance and regulation of the urban water sector: Quoi de neuf? In R. Q. Grafton, K. A. Daniell, C. Nauges, J.-D. Rinaudo, & N. W. W. Chan (Eds.), *Understanding and managing urban water in transition*. Dordrecht: Springer. (pp. 423–446).

Renzetti, S. (2015). Non-household water demand: the industrial and commercial sectors. In R. Q. Grafton, K. A. Daniell, C. Nauges, J.-D. Rinaudo, & N. W. W. Chan (Eds.), *Understanding and managing urban water in transition*. Dordrecht: Springer. (pp. 297–310).

Republic of South Africa. (1998). *National Water Act*, Act no 36 of 1998, Government Gazette, 26 Aug 1998, Government Printers, Pretoria.

Reynaud, A. (2015). Public–private partnerships and their ownership in the urban water sector. In R. Q. Grafton, K. A. Daniell, C. Nauges, J.-D. Rinaudo, & N. W. W. Chan (Eds.), *Understanding and managing urban water in transition*. Dordrecht: Springer. (pp. 447–462).

Reynaud, A., & Garcia-Valiñas, M. A. (2015). Does residential water use depend on water quality? Some answers from a French case study. In R. Q. Grafton, K. A. Daniell, C. Nauges, J.-D. Rinaudo, & N. W. W. Chan (Eds.), *Understanding and managing urban water in transition*. Dordrecht: Springer. (pp. 337–358).

Rinaudo, J.-D. (2015). Long-term water demand forecasting. In R. Q. Grafton, K. A. Daniell, C. Nauges, J.-D. Rinaudo, & N. W. W. Chan (Eds.), *Understanding and managing urban water in transition*. Dordrecht: Springer. (pp. 239–268).

Rinaudo, J.-D., & Barraqué, B. (2015). Inter-basin transfers as a supply option: The end of an era? In R. Q. Grafton, K. A. Daniell, C. Nauges, J.-D. Rinaudo, & N. W. W. Chan (Eds.), *Understanding and managing urban water in transition*. Dordrecht: Springer. (pp. 175–200).

Rinaudo, J.-D., Montginoul, M., & Desprats, J.-F. (2015). The development of private bore-wells as independent water supplies: Challenges for water utilities in France and Australia. In R. Q. Grafton, K. A. Daniell, C. Nauges, J.-D. Rinaudo, & N. W. W. Chan (Eds.), *Understanding and managing urban water in transition*. Dordrecht: Springer. (pp. 155–174).

Rockström, J., Steffen, W., Noone, K., Persson, Å., Chapin, F. S., III, Lambin, E., et al. (2009). Planetary boundaries: Exploring the safe operating space for humanity. *Ecology and Society, 14*(2), 32. http://www.ecologyandsociety.org/vol14/iss2/art32/.

Rogers, E. M. (1983). *Diffusion of innovations*. New York: Free Press.

Rosenhead, J., & Mingers, J. (Eds.). (2001). *Rational analysis for a problematic world revisited*. Chichester: Wiley.

Rotmans, J., Kemp, R., & van Asselt, M. (2001). More evolution than revolution: transition management in public policy. *Foresight, 3*, 15–31.

Sabatier, P. A. (1988). An advocacy coalition framework of policy change and the role of policy-oriented learning therein. *Policy Sciences, 21*, 129–168.

Selman, P., Carter, C., Lawrence, A., & Morgan, C. (2010). Re-connecting with a neglected river through imaginative engagement. *Ecology and Society, 15*(3), 18. http://www.ecologyandsociety.org/vol15/iss3/art18/.

Sheehan, M. (2007). *2007 state of the world: Our urban future*. Washington, DC: Worldwatch Institute.

Sibly, H., & Tooth, R. (2015). Managing water variability issues. In R. Q. Grafton, K. A. Daniell, C. Nauges, J.-D. Rinaudo, & N. W. W. Chan (Eds.), *Understanding and managing urban water in transition*. Dordrecht: Springer. (pp. 383–400).

Smith, A., Stirling, A., & Berkhout, F. (2005). The governance of sustainable socio-technical transitions. *Research Policy, 34*(10), 1491–1510.

Smith, A., Voß, J.-P., & Grin, J. (2010). Innovation studies and sustainability transitions: The allure of the multi-level perspective and its challenges. *Research Policy, 39*, 435–448.

Squires, V. R., Milner, H. M., & Daniell, K. A. (Eds.). (2015). *River basin management in the twenty-first century: Understanding people and place*. Boca Raton: CRC Press.

Swyngedouw, E. (2004). *Social power and the urbanization of water: Flows of power*. Oxford: Oxford University Press.

Tejada-Guibert, José Alberto & Čedo Maksimović. (2001). *Frontiers in urban water management: Deadlock or hope?* In Symposium Proceedings, 18–20 June 2001, Marseille, France. IHP-V|Technical Documents in Hydrology, No. 45, UNESCO, Paris.

Troy, P. (Ed.). (2008). *Troubled waters: Confronting the water crisis in Australia's cities*. Canberra: ANU E-Press.

Troy, P. (2015). The Social determinants of water consumption in Australian cities. In R. Q. Grafton, K. A. Daniell, C. Nauges, J.-D. Rinaudo, & N. W. W. Chan (Eds.), *Understanding and managing urban water in transition*. Dordrecht: Springer. (pp. 283–296).

UNECE. (1998). *Aarhus convention on access to information, public participation in decision-making, and access to justice in environmental matters*. United Nations Economic Commission for Europe. http://www.unece.org/env/pp/documents/cep43e.pdf.

United Nations. (2010). *A/RES/64/292 The human right to water and sanitation: United Nations General Assembly*. http://www.un.org/waterforlifedecade/human_right_to_water.shtml.

United Nations. (2013). A new global partnership: Eradicate poverty and transform economies through sustainable development, United Nations, New York. http://www.un.org/sg/management/pdf/HLP_P2015_Report.pdf

UNICEF/WHO. (2012). Progress on drinking water and sanitation: 2012 update, UNICEF and World Health Organization, New York.

UN-WATER. (2012). *Managing water under uncertainty and risk*. The United Nations World Water Development Report 4. UNESCO, Paris.

van der Brugge, R., & Rotmans, J. (2007). Towards transition management of European water resources. *Water Resources Management, 21*, 249–267.

van Vliet, M., & Aerts, J. C. J. H. (2015). Adapting to climate change in urban water management: Flood management in the Rotterdam–Rijnmond area. In *Understanding and managing urban water in transition*. Dordrecht: Springer. (pp. 549–574).

Vörösmarty, C. J., Green, P., Salisbury, J., & Lammers, R. B. (2000). Global water resources: Vulnerability from climate change and population growth. *Science, 289*(5477), 284–288.

Wenger, C., Hussey, K., & Pittock, J. (2013). *Living with floods: Key lessons from Australia and abroad*. Gold Coast: National Climate Change Adaptation Research Facility.

White, I., & Falkland, T. (2015). Integrated management of urban water supply and water quality in developing Pacific Island countries. In R. Q. Grafton, K. A. Daniell, C. Nauges, J.-D. Rinaudo, & N. W. W. Chan (Eds.), *Understanding and managing urban water in transition*. Dordrecht: Springer. (pp. 489–526).

Whitehead, P. G., Wilby, R. L., Battarbee, R. W., Kernan, M., & Wade, A. J. (2009). A review of the potential impacts of climate change on surface water quality. *Hydrological Sciences Journal, 54*, 101–123.

Wolff, G. H., & Gleick, P. H. (2002). The soft path for water. In P. H. Gleick, W. C. G. Burns, E. L. Chalecki, et al. (Eds.), *The world's water: The biennial report on freshwater resources 2002–2003*. Washington, DC: Island Press.

Part I
Water Supply and Sanitation

Chapter 2
How Water Services Manage Territories and Technologies: History and Current Trends in Developed Countries

Bernard Barraqué, Laure Isnard, and Julien Souriau

1 Introduction

In recent decades urban water services have been confronted by new challenges questioning their sustainability. As detailed in previous chapters of this book, there have been environmental issues and climate change issues which require more efficient use of energy and water resources; and now the rise of 'water poverty' provokes renewed interest into social sustainability. At the same time, full cost recovery is broadly advocated, public subsidies are coming to an end, precisely when water services are facing calls for huge investment to renew their assets. Technical, economic, and social solutions have been provided by urban water industries and regulatory agencies, demonstrating that some of these challenges can be addressed. But the institutional and functional organisation of water services is also affected by these changes, which underlines the important governance issues that arise when crises occur and change is needed in urban water management.

This chapter aims at setting the historical framework of the evolution of urban water services governance. It looks at various countries and sets out a description of their past technical and territorial evolutions within a context of institutional path dependencies (Putnam 1993). We develop here a socio-historical comparison of

B. Barraqué (✉)
Centre International de Recherches sur l'Environnement et le Développement, Paris, France
e-mail: bernard.barraque@agroparistech.fr

L. Isnard
National Center for Scientific Research (CNRS), CIRED,
CIRED, AgroPrisTech, 19 avenue du Maine, Paris 75 015, France
e-mail: laure.isnard@agroparistech.fr

J. Souriau
AgroParisTech, CIRED, CIRED, AgroPrisTech, 19 avenue du Maine, Paris 75 015, France
e-mail: Julien.souriau@agroparistech.fr

Q. Grafton et al. (eds.), *Understanding and Managing Urban Water in Transition*, Global Issues in Water Policy 15,
DOI 10.1007/978-94-017-9801-3_2

urban water services that was initiated almost 30 years ago by a research lab which focused on innovation in urban networks (Dupuy 1984). Each country develops a specific organisation of its water services, depending upon the role of the state and its relations with local authorities, or upon geography and jurisdictional aspects (Guerin-Schneider et al. 2002). However, a historical analysis of the evolution of water services in developed countries, where the water industry has reached a certain maturity, can provide an opportunity to reveal common trends in governance evolution. For this reason we propose in this chapter to review the functional and institutional paths of well established urban water services observed in Western Europe, Australia, and the United States of America. The chapter mainly focuses on drinking water services issues, but we call upon examples in waste water services as well, as there is a clear interaction between them in most of the cases studied.

We identify four major themes that have emerged in urban water sector governance when one looks at how territories and technologies in water services have evolved. These are, in turn, (i) the evolution of private sector participation in urban water services and the recurring public vs. private debate; (ii) the emergence of two opposite but complementary trends in territorial scales of infrastructure management and stakeholders' participation, from up-scaling to down-scaling; (iii) the potential alternative to these territorial reorganisations offered by various degrees of integration/unbundling of local public services; and (iv) lastly, the increasing reconsideration of the relationship between urban water services and water resources, as illustrated by the 'Water Framework Directive' set by European legislation (WFD 2000/60/EC, for integrated river basin management in Europe, adopted on 23 October 2000) the requirements of which imply that water policies must from now on start with the aquatic environment's recovery, i.e. improving *territorial* management before looking for *technological* solutions.

A lot of existing literature focuses only on the public vs. private debate (Bakker 2003; Bauby 2007). We argue that granting importance to this antagonism may result in overlooking other dimensions of the water services constitution. To give just one example, the well known and hotly debated divestiture and full privatisation of water services in England and Wales in the 1990s tends to obscure the much more important structural move – i.e. a progressive centralisation of water services towards the scale of supposedly 'optimal' water resources management (the scale of 10 regional water authorities) (Saunders 1983). We have long argued that the four themes of the evolution should be considered together, since it would better show the mutual influence of geographic, socio-economic, and political contexts on the scaling and technical–financial arrangements of water supply and sanitation (WSS) in various countries (Barraqué 1992; Pezon 2006). Additionally, we argued that resource management institutions and urban water services institutions should be studied separately, as their scales, laws, and economics differ in developed countries, and therefore discussions about privatisation of water resources and of urban water services should be held separately too (Barraqué 2003). It is all the more important that recent evolutions, in particular the E.U. WFD, tend to re-articulate the territories of services and those of resources, for the sake of reducing the ever increasing costs incurred by 'end-of-pipe' technologies.

The four themes of water service governance have followed different trends in the studied group of countries selected here. However, when studied collectively, they offer a heuristic approach to encapsulate the past and current trends in urban water services governance. The governance issues related to these four themes will be introduced first by a short insight into the ongoing institutional and functional evolutions of the urban water services governance in the Paris region, as we have studied it most thoroughly. It offers an interesting case of how complex these issues can be, even though Paris is neither a model nor an extraordinary case.

2 A Vast Array of Public–Private Arrangements

In the 1990s, the World Bank promoted private sector involvement aimed at improving urban infrastructure in developing countries, including WSS services. In less than a decade, many Third World NGOs organised a strong opposition to it, and their arguments against the social and political impacts of private sector participation on WSS services were rapidly taken up in OECD countries, where private participation was sometimes an ancient fact, as for example in France. Hence, a new phase of global debate opened up on the limits of private sector technical and financial involvement in WSS; it is now accepted as a conventional issue and discussed in the media.

However, history teaches that most countries developed their water services through a varying mix of public and private participation in the financial, technical, and institutional aspects, resulting in a wide range of hybrid governance models. A review of public–private partnerships in WSS might then focus on the implications of this hybridisation, rather than staying polarised in arguments supporting either public or private models.

2.1 Focus on the Paris Region

Historically, Paris's water services were developed with a mixed involvement of public and private partnerships which shared responsibilities, risks, ownership, etc. Since its creation in the early nineteenth century, the water service has shifted from fully public "in house" provision by a municipal department (but with several private supply systems for monasteries, hospitals, and palaces) to a growing but partial delegation to private companies. Typically, Baron Haussmann, appointed Prefect of the Seine County in 1853, set up a public monopoly for the production and distribution of water in Paris, but he contracted with a new private partner, Compagnie Générale des Eaux, to connect water customers, run the meters, and collect water bills.

More recently, a mixed economy company was set up in 1987 in order to take full charge of the transport and production of all the drinking and non-drinking water in Paris: the SAGEP was born. SAGEP shares were owned 70 % by the City of Paris and 14 % each by two private companies Générale des Eaux (now Veolia) and Eau & Force (Suez group), the rest, for legal reasons, being held by other public institutions. Furthermore, in 1985 these two private companies (Suez and Veolia) were entrusted with water distribution, billing, and network renewal in Paris through a 25-year management contract, with Générale des Eaux operating on the right bank and Eau & Force on the left bank of the Seine.

At the end of December 2009, the management contract signed with the private companies ended and all the water service activities reverted to fully local public management. Indeed, the left wing and green parties were in power – they had won the municipal elections in 2001 and again in 2007 – and Mayor Delanoë had officially announced his intention to reclaim the complete chain of potable water provision in Paris. Thus, on 1 January 2010, SAGEP (relabelled Eau de Paris) was changed into an EPIC (public department with separate budget and private law status) which was now in charge of all water services in Paris, from water resource protection to water production and distribution to 93,000 Parisian subscribers.[1] But it was impossible for the new EPIC to set up its own billing system rapidly enough, so Paris had to sign an additional 18-month service contract with its two former private operators in order to bill water services during the transition period to a fully public water service, which is now completed.

By way of comparison, in 1923 Générale des Eaux (Veolia) signed a management contract with SEDIF, the biggest joint board in France supplying 4 million inhabitants in 144 surburban communes around Paris. This contract finished at the end of 2010, but the majority of member communes of SEDIF voted to remain with the formula delegating water services, and the tendering for the services contract renewal was again won by Générale des Eaux. However, several communes tried to quit SEDIF in order to create their own independent joint board and public water service, following Paris's steps. On the south of SEDIF's territory, two communes reclaimed their water services into a new board named Communauté des lacs de l'Essonne, but at the time they could only change their water producer from Veolia to Suez, as their water supply depends on buying water from an existing water production plant, until they have access to one of Paris' long distance aqueducts, which runs trough their territory. On the east of SEDIF's territory, nine other communes, representing 10 % of SEDIF's total water consumption, also decided to reclaim the competence of water supply and waste water management under a joint institution named Est Ensemble. But the lack of alternative secure water supply forced the member communes of this new entity to stay within the SEDIF structure for another 2 years or more – until they could obtain the technical capacity needed to operate their own water service, purchase water from Paris or other private water producers, or to again delegate the water service management to a private operator.

[1] In 2009 2.2 million inhabitants lived in Paris but the water utility runs only 93,000 subscriptions as there is usually only one meter per building. The water bill is split between the households living in the building.

Thus, if we just consider the Ile de France region, there are now diverse formulas for the provision of water services, with different public–private combinations. The territories are modeled on the historical capital, keeping mostly public but leaving their own suburbs in the hands of the private sector for water supply. However, sewage treatment remains operated by a regional board set up back in the 1930s, which maintains this part of the water industry in public hands. Such complex public–private mix is quite common in France, but also in other countries.

2.2 *The Diversity of Water Supply and Sanitation (WSS) Models in Developed Countries*

A quick overview of WSS services in developed countries provides a wide range of public–private partnerships formulas. Instead of a clearly defined segmentation between public and private models, we consider there is a continuum ranging from completely public to completely private water services, with all sorts of hybrid arrangements in between.

In the USA, 85 % of water services are public commissions run directly by local governments and cities; the other 15 % are investor owned, i.e. fully private water services. In Germany and the Netherlands, the participation of private sector in water services remains quite low: WSS services are independent utilities (local and integrated in Germany, regional and unbundled in the Netherlands), operating under private law, but usually owned 100 % by local governments and counties. Countries such as France and Spain developed a wide array of public–private combinations resulting in variable local situations. The very common public–private mix is one where local public authorities keep the responsibility over the service and ownership of the assets, while a private company partly or completely operates it. At the private end of the public–private continuum, the UK decided in the 1980s to fully privatise WSS, selling all the assets of the 10 existing regional water authorities to private capital, and placing the resulting companies, plus the historically surviving smaller ‘water only’ companies, under the control of powerful national regulation agencies. Italy was attracted by the English model and decided – with the 1994 Galli Law – to ban local direct public management and to promote private participation in new ‘optimal territorial ambits’ set up at the level of provinces, in order to attract private sector funds and participation. Beyond such diversity, is there a possible common analysis?

2.3 *A Common Evolution Toward Hybrid Partnerships?*

A historical overview shows that there is a clear evolution of water services management toward hybrid public–private organisations. In most European countries, water and sanitation services in major cities were first developed by private companies in the nineteenth century, which were often replaced by municipal departments or transformed into public companies in the early twentieth century and managed under direct labour (Guerin Schneider et al. 2002). This happened first in Britain (Millward 2000; Saunders 1983), and the example was later followed on the European continent; except that in the continental cases the earliest water service companies were frequently held by foreign capital and attracted quicker criticism by the public (Calabi 1980). For instance, in Italy the 1903 law banned private management of water, gas, and transport utilities and replaced them with the (public) formula of *azienda municipalizzata* with financial autonomy, while in small towns water was run by a municipal department (Massarutto 2011). Yet, apart from Britain and Ireland, where costs were covered by rates paid according to the rent value of the house, drinking water coming from domestic taps was understood as a commercial service, so costs were covered by bills linked to volumes of water consumed. This was a powerful reason for cities to adopt private law institutional arrangements, even though they wanted to keep control of the capital invested. Additionally, under private law, accounting practices mandate that capital be depreciated and provision be made for renewal, while this is not always the case with public law management, so that, indirectly, local authorities are pushed to keep, or fetch, private partners as operators. So it was in France, where the central government maintained strict limits on the involvement of local authorities in the economy, and public accounting did not make it easy to depreciate and make forward provisions (Pezon 2000).

After WWII, legal provisions changed the status of sewers from an imposition to a service: advantage could be taken of the more widespread supply of sewerage to add sewer charges to drinking water bills. In some cases, this merger in billing resulted in the integration of waste water and water services into a single utility (as happened for the Portuguese and Italian reforms), while in Germany the municipal company (Stadtwerk) was typically put in charge of several utilities (water, gas, electricity, transport, district heating, etc.), but not sewage collection and treatment until recently.

The liberal-conservative revolution in the US and UK brought back the idea of water privatisation which took place in Britain in 1989. This new model strongly influenced the European Commission (and is especially supported by the Internal Market and Services Directorate General), which now favours unbundling and opening marketable services to competition and the private sector. From the 1970s to 1990s, several cities shifted from public, in-house water services to others with hybrid public–private management. In France, a growing number of middle size cities chose to delegate the management of their until-then public water services to private companies – for a large number of reasons, such as the need for higher technical capacities, the opportunity for mayors to transfer to the new private operator the blame for rising water prices, and to scale-up the water service by joining other neighbouring local water service for economies of scale (Pezon and Canneva 2009). In Australia, major water supply systems were reorganised in such a way as to separate production and transport on the one hand, and distribution on the other; in the last decade, the latter task was eventually reorganised into private companies owned 100 % by the state or by joint ventures formed with international private water companies (McKay 2005). However, for WSS services, this movement is being resisted in the northern part of Europe, which favours regional concentration and capital cross-holding with regional electricity or gas companies (e.g. in Germany and Switzerland; Pflieger and Paquier 2008). In Italy, the liberalisation movement was supported by the Berlusconi government with the 2009 Ronchi decree, restricting public ownership to a maximum 30 % of the shares in WSS utilities. But this reform was massively rejected by a referendum in 2011 (Majone and Veltri 2011). In other countries like France and Belgium, there is a growing coalition in favour of direct public management. Nevertheless, the commercial nature of WSS services is not really threatened, leaving open the possibility of benchmarking various mixed formulas.

The present evolution is a hybridisation of water service models, with various sorts of public–private partnerships rather than a total privatisation (with one exception: the UK divestiture). Thus it appears that there is no real general orientation towards fully public or fully private management. However, widespread networks of pipes and metering mean that most countries carry very large infrastructure. Eventually, this favours a combination of publicly owned capital (so as to benefit from cheaper money or remaining subsidies) and private or mixed economy operations and maintenance.

3 Upscaling/Downscaling WSS Management: The Right Mix?

In all of the countries we studied for this chapter, water services were initially run within the competence of the municipalities – or were organised at the scale of hamlets in rural areas, as is still the case in low density areas of Ireland (group water schemes: Brady and Gray 2013), Denmark, Portugal, etc. To face the challenges met in their development from the early nineteenth century, WSS services often changed territorial scales – due to resource problems, and also to the evolution of the local authorities and of their legal responsibilities. The general trend went towards larger territories to achieve economies of scale and a better management of risks (sanitary or financial). But in some cases, decentralisation and down-scaling took place to solve specific management issues. Today it seems that, regarding the sustainability of water services, debate on the appropriate territorial scale is more important than the private vs. public management issue. For what specific reasons, and how, did they move to present-day management territories?

3.1 The Parisian Case

Today around 15 plants produce drinking water for the Paris metropolitan area (10 million inhabitants), in addition to several long distance aqueducts (total length 470 km) conveying springwater and groundwater to Parisian reservoirs. The drinking water plants are either managed by Eau de Paris, Lyonnaise des Eaux (Suez), or Générale des Eaux (Veolia); the commune of Saint-Maur-des-Fossés also operates an additional drinking water plant for its own local water service under direct labour.

Back between World Wars I and II, regional planning studies considered the need for upstream reservoirs to regulate the flow of the Seine and its tributaries, with the official aim to avoid water scarcity problems, and allegedly reduce flood risks. These large dams were built from the 1960s to 1980s, with the financial support of the Agence de l'Eau Seine Normandie (river-basin financing institution). Furthermore, after the 1976 drought, the many water networks of the Paris metropolitan area were connected through additional security interconnections. These allowed water delivery to be maintained in the event of a pollution accident or a local water service failure, and catered for an ongoing but moderate growth in consumption.

There was, however, an unsuspected evolution: water consumption in Paris has been dropping since the early 1990s (at an average of −1.5 % per year). As a consequence, Eau de Paris decided to close one of its drinking water plants (Ivry-sur-Seine). In the same way, SEDIF's water sales decreased by 19 Mm^3 between 2005 and 2009. Consequently, mutualisation of the water production capacities in the Paris metropolitan area is now debated within broader discussions held about the "Greater Paris" project, which aims to scale up several urban infrastructures in the area. Common management of a limited number of drinking water plants in the area is acknowledged as an economic issue (Carroy 2010). But such an opportunity also raises the question about how to supply water services to territories larger than those operated by the current operators, as well as how to build a more appropriate new institutional framework.

As a point of comparison, the rationalisation of waste water treatment in the Paris region took place much earlier. At the beginning it was formed within the former Seine county.[2] The break up of this large county into four smaller ones in 1968 resulted in the creation in 1970 of a common board called SIAAP (Syndicat Interdépartemental d'Assainissement de l'Agglomération Parisienne) charged with treatment of the waste water from most of the central Paris area (almost nine million inhabitants). The upscaling of waste water treatment gave economies of scale which allowed SIAAP to invest in sophisticated technology and in measures for protecting the environment. Conversely, the sewerage systems stayed within the competence of the communes, while storm water drains and collectors are managed by the counties. This concentration and centralisation of both water and waste water services presupposes a complex organisation, in terms of both territorial boundaries and public–private ownership.

[2] The Seine county gathered together the four present-day départements – Paris, Hauts-de-Seine, Seine-Saint-Denis, and Val-de-Marne – until it was reformed in 1968.

3.2 *Concentration Process: Looking for an Improved Performance of Water Services*

The Parisian case illustrates how the concentration of water and sanitation services within a single entity, and the mutualisation of some equipment, can be justified on economic or technical grounds. This move towards the larger scale, with the creation of multi- or single-purpose joint boards, has taken place in all the countries we have studied. The Netherlands offers a remarkable illustration of this evolution, where today only 10 water companies are distributing water to 17 million inhabitants. Their territories partly match those of the Dutch provinces in size and they are run by mixed boards of provinces and communes (Loijenga 2009). On the other hand, sewerage remains a municipal competence. But sewage treatment was entrusted to the famous drainage and flood protection institutions, the 'Waterschappen'. These institutions were created in the Middle Ages and have intensely clustered in the last decades: there were more than 2,600 water boards in 1945; there are only 23 today (Kuks 2010). This concentration process went along with a decrease in the number of waste water treatment plants managed by the water boards: between 1980 and 2010, their number went down from 505 to 390. It appears that this concentration – based on a voluntary process – brought a solution to the need for rationalisation of water resources and sanitation management. Indeed many small Waterschappen went bankrupt after the deadly 1953 flood which killed 1830 people; moreover, water pollution issues also became more acute after the 1950s (Kuks 2010). To face these challenges the water boards had to make significant investments. They decided to merge into larger entities to better tackle the need for finance.

In England and Wales a similar concentration process of WSS services took place after WWII. It led, with the adoption of the Water Act in 1973, to a complete regionalisation of these services. At the beginning of the 1970s there were 198 water entities, of which 101 were joint boards, 64 were municipal services, and 33 small private companies (Barraqué 1995). The Water Act created 10 regional water authorities (RWAs), the borders of which matched hydrological basins (Saunders 1983). RWAs were put in charge of the whole drinking water service, waste water treatment, and resource management. Local sewer systems only stayed within the competence of the municipalities. This regionalisation illustrates how technical matters were meant to predominate over political or administrative considerations (Saunders 1983). Actually, the creation of the RWAs was the final outcome of a political process started in the 1940s: it eroded the local authorities' responsibilities and power over public services, and was a decisive step toward technical and industrialised management of water and sanitation services. The privatisation of WSS services in 1989 did not really impact the territorial organisation of the water services. It only resulted in a few more mergers of 'water-only' companies, and of planning and resource management authorities (from 10 to 8).

With the 1994 Galli law, Italy attempted to adopt the British model. The Italian water services needed huge investments for waste water collection and treatment so as to comply with European standards (Massarutto 2011). But the banks were unwilling to lend money to a multitude of small municipal entities. The concentration of 6,600 existing water services into larger entities, and their opening to the private sector, appeared as the solution for getting cheaper loans. Originally WSS services were supposed to be regrouped into local catchment districts called 'ambiti territoriali ottimali' (ATOs), but in most cases the regions – which were responsible for the concentration process – finally decided to merge water and waste water services together according to the administrative borders of the provinces (except in Sardinia and Puglia, where WSS were already regionalised). Thus, 91 ATOs were created to manage water and waste water services in order to produce economies of scale and economies of scope. Given the historical power of Italian mayors regarding all local public services – one used to talk about "the mayor's water" in Italy – this top-down reform met strong opposition among the local elected representatives. 15 years after the vote of the law, 19 ATOs have not yet appointed their water service operator (Mangano 2010), and commercial rationalisation is threatened by the success of an anti-privatisation referendum held in 2011. Italy's WSS services are in a stalemate situation.

In France, the large number of water services – 12,000 entities for 64 million inhabitants in 2009 – is a legacy of the extreme fragmentation of the 36,000 communes (Pezon and Canneva 2009). So far, no reform to enforce the merger of communes has been adopted by parliament. Inter-municipal cooperation has emerged as a practical alternative to rationalising the administration and territories of some public services. Single-purpose joint boards have come into existence since 1890; multi-purpose joint boards since 1959. The 1999 Chevènement law created three new kind of multi-municipal boards, one for the rural area (the Communauté de Communes), one for the urban communes (the Communauté d'Agglomération) and one for the metropolitan areas (the Communauté Urbaine). The Communautés Urbaines are responsible for water and sanitation services; the two other types of communautés can choose to provide these services to their citizens. So far it has led to a relative concentration of the water services, as shown in Table 2.1.

Still, the Chevènement law has been a turning point regarding the territorial evolution of water and waste water services: for the first time, the number of water services has been decreasing and has overturned the traditional organisation of municipal competence.

Table 2.1 Organisation of water services in France in 1988 and in 2007

	Communes	Water services	Water services managed by single-purpose boards	Water services managed by multi-purpose boards	Water services managed by the communautés created by the Chevènement law
1988	37,200	15,400	3,375	455	
2007	36,682	11,846	2,949	374	306

Source: Pezon and Canneva (2009)

3.3 *Centralisation of Water Production: A Strategic Path to Deal with Water Scarcity*

As mentioned, private companies initiated the first modern water supply systems. But when municipalities became convinced of the benefits of water for public health, they took over the services, following the "gas and water socialism" model developed in Great Britain (Barraqué 1992). Behind this action was the principle that Graham and Marvin (2001) called "the modern infrastructural ideal": every urbanite living on the service's territory should be connected to a centralised water supply system providing the same quality of service to everyone. This approach has been essential for the development of water services at the municipal scale. Today it seems that centralisation at a larger territorial scale could also help overcome some difficulties in water supply. In Germany, for example, infrastructure for water production in areas with low rainfall but major urbanisation has been managed at a supra-local scale since the early twentieth century. Thus, in 1912 in Land Baden-Württemberg the emperor set up a public enterprise, the 'Staatliche Landeswasserversorgung', to produce and transport drinking water to almost 100 different communes via a 750 km network.[3] In 2007, the 100 biggest water operators delivered half the total volume of drinking water produced in Germany, illustrating clear centralisation (Kraemer et al. 2007).

Centralisation as a solution to water scarcity has also been adopted in the United States of America – even if, given their remarkable diversity, it is difficult to present here an overview of American water services. In southern California, the risk of water shortage led utilities around Los Angeles to build a large aqueduct to bring water from the Colorado River. A "wholesaler", the Metropolitan Water District of Southern California (MWD), was put in charge to produce and transport water to 26 water distribution districts, supplying 19 million inhabitants. A similar organisation was developed in Boston but for different reasons: in 1985 the Environmental Protection Agency (EPA) required better treatment of waste water. Boston decided to create the Massachusetts Water Resources Authority (MWRA), a regional institution responsible for the treatment of effluents and the production of drinking water for 61 communities. Today, more than 2.5 million people from Boston and its suburbs get water from the MWRA. Member municipalities distribute water and collect effluent on their territories. The drinking water and sewage treatment plants, and the main sewers are managed centrally.

As for Australia, after the droughts of the early 2000s, the federal government and the water operators made huge investments in infrastructure to improve water security. Desalination and water recycling plants flourished, whereas local governments tried to rationalise and interconnect their networks. The situation in the south-east of Queensland, where 2.8 million inhabitants live around the fast-growing city of Brisbane, offers a good illustration of attempts to centralise water production.

[3] However, after WW2, this State enterprise was typically decentralised into a joint board of the municipalities served.

The South East Queensland Water Grid, initiated by the Queensland government, is one of the biggest Australian infrastructure projects.[4] AU$9 billion is being spent on an integrated water production system to secure water supply in this attractive area. The grid consists of a large network of drinking water plants and pipes which allow transfers from existing or new water sources. According to the designers, this system has the advantage of relying on water sources which depend in different ways upon the local climate – dams or rain water storage vs. desalination or recycling plants. Centralised management also allows better allocation of water resources between different users, including the aquatic environment itself.

In the end, the result of both concentration and centralisation processes is the rise of multi-level governance, whereby municipalities and their joint boards are not the only institutions providing WSS services. Frequently there is a multi-scale tier of water governance. Paradoxically, in one of the most centralised countries – France – municipalities still play an active role, while many reformers (e.g. the Cour des Comptes, the French financial revenue court in charge of overseeing public accountancy) call for increased concentration. The real political issue is whether to completely eliminate the role of local authorities, which have long carried the democratic and welfare ideals of public services, while remaining close to citizens. Typically in Australia, local stakeholders, who were once put aside, are now coming back to take part in the management of WSS services. First, local authorities are producing recycled water for their parks and sports grounds. Then citizens, individually or as part of local communities, are progressively turning into co-producers of the water service at the infra-municipal scale, since the use of rain water harvesting or greywater recycling has become compulsory for many new buildings. To what extent do these decentralised systems reveal a common trend in the functional evolution of water services in other developed countries?

3.4 Decentralised Systems: Emergency Patches or Next Step Towards Sustainability?

In contrast to the big cities, which developed centralised water supply systems in the nineteenth century, hamlets and scattered settlements in rural areas have always relied on self-sufficient water systems, traditionally managed at the neighbourhood scale. In France, where the communes are already very small entities, these decentralised infra-municipal systems are virtually absent. But in Denmark, Lower Saxony, rainy countries like Ireland, or the north of Portugal and Spain, this kind of neighbourhood association in charge of the local water supply is still common. Moreover, while they seemed bound to disappear in the short term (given industrialisation of water services), they are now being reconsidered. Indeed, the emergence

[4] See http://www.seqwater.com.au/

of new sustainable urban settlements encourages the development of decentralised water and waste water systems, which are considered closer to natural water cycles.

In Magdeburg, the capital city of the German Land Saxony-Anhalt, where the population has dropped by 20 % since reunification, the area supplied by the municipal water service is steadily losing consumers such that by 2050 there will be only 540,000 people compared to 800,000 today. The technical impacts of the subsequent decrease in water consumption are already serious. In some neighbourhoods, water stays for more than 25 days in the pipes before being used (Herdt 2009)! Today, as the public water service does not find enough industrial clients to compensate for the decrease in residential consumption, the local Stadtwerk plans on combining up-scaling and down-scaling solutions: a supra-municipal enterprise could produce bulk water for the whole region, while decentralised systems at neighbourhood or building scales could make it drinkable and deliver it to the end-consumers. This would minimise the time spent by the drinking water in the pipes. Many cities from the ex-East Germany are facing the same situation, so some researchers are questioning the traditional centralised water supply system and are now thinking about adapting the actual network to shrinking demand via well-designed alternative systems (Moss 2008; NetWORKS 2008).

In France, despite governmental incentives towards sustainable urban projects, decentralised water systems are still seen as technological innovations and remain scarce. But we could very well imagine a quick development of these systems in the near future. In a way it would be consistent with the five million septic tanks still functioning on French territory! The general issue at stake is the risk that decentralised solutions re-introduce strong differences in the quality of service provided to one social or territorial group and another. Graham and Marvin (2001) call attention to this potential 'splintering urbanism' which might bring rich countries' water services close to the fragmented situation existing in developing ones. Arizona offers a striking example of this fragmentation (Megdal 2012): there are very large public utilities (e.g. in Phenix and Tucson), but a significant fraction of the population depends on private community systems, sometimes with no or tiny networks, and very different degrees of technical sophistication. Individualism and parochialism makes it unlikely to see a development of public services using decentralised systems. Yet there are projects combining robust small scale systems with sophisticated NTIC systems to allow for remote control and reduced operating and maintenance costs.[5]

In the end, territorial scales of management for water services are moving in opposite but not incompatible directions, combining up-scaling and down-scaling, concentration and decentralisation. The municipal territory is no longer the only and most relevant level to deal with the concerns of WSS services – water shortages, lack of money to renew old and centralised assets, drop in water consumption, etc. The various examples given in this part of the chapter show that the evolution of a territorial scale of management often goes along with dismantling the integrated

---

[5] E.g. the work by Yoram Cohen on decentralised systems in the Institute of Environment and Sustainability at UCLA.

water service production chain. For instance, there may be a disconnect between water production at the regional scale and water distribution at the municipal or infra-municipal scale. Can disintegration and/or unbundling be part of the solution to some water services issues?

4 Integration Versus Unbundling of Water Services

WSS management encompasses all phases of the "small water cycle", starting from water resource abstraction and protection, drinking water production and distribution, wastewater and stormwater collection and treatment, and ending with disposal of treated water into the environment. These different steps have been either integrated or unbundled, depending on the time or place considered. So, in addition to the question of geographic scaling of water services, one can consider the degrees of integration of the water supply and sanitation services according to the two following concepts:

1. vertical and horizontal integration (or concentration): how far are the various tasks of water management, planning, and operation integrated together or separated (vertical integration vs. unbundling), or related to other sectors (single water service or multi-activity service, i.e. horizontal integration);
2. structural integration (or autonomy): to what extent is the water service autonomous and independent from other public administrations (in terms of legal status, budget, accounting, etc.).

4.1 Insight into Paris

The story of water and waste water management in the Paris metropolitan area can be considered from a different, but complementary, perspective to that of the upscaling/ downscaling trends studied above. During most of the nineteenth century, WSS services were provided by a unique municipal department under the direct responsibility of the prefect of the Seine county (Paris plus 80 communes). In 1860s, with Baron Haussmann and chief engineer Eugène Belgrand in charge, all water supply and sewerage systems in Paris were managed, in a fully integrated way, under a powerful technical and administrative *Direction* (Gaillard-Butruille 1995). For some years, this municipal *Direction* was in charge of the whole water and wastewater cycle, covering all the communes of the Seine county. But in 1867 and 1869, the suburban communes of the Seine county split from Paris and chose to delegate their water service to the 'Générale des Eaux' (now Veolia), first as individual communes and then within several larger joint boards, SEDIF being the chief one with 144 communes today (Defeuilley 2004). In Paris city, water supply and distribution continued to be run by a municipal *Direction* under the rule of the prefect, while water metering and billing were externalised and continued to be operated by Générale des Eaux until 1984, when a mixed economy company was created with both Générale and Eau et Force (Suez group) (Chatzis 2006). Conversely wastewater collection and treatment remained under state control at the Seine county level (Cebron de Lisle 1991), with all the wastewater collected ending up in a single sewerage treatment plant downstream of Paris.

This configuration lasted for many decades, until the 1960's, when the regionalisation reform of the Paris area led to the breaking up of the Seine county into four new counties (Paris, Hauts de Seine, Seine St Denis, and Val de Marne). In 1970, Paris and its new neighbouring counties again joined together in order to form SIAAP, so as to continue to operate a single downstream sewerage treatment plant – to which were progressively connected additional wastewater discharges produced by a few sewerage boards from the outer ring. Even though SIAAP remained the sole operator for wastewater treatment, it had to break up its linear upstream-to-downstream treatment arrangement in order to develop complementary new sewage treatment plants upstream of Paris. Conversely, after the creation of the four new counties, rainwater collection fell under the control of the counties and wastewater collection fell under that of the communes.

In the end, water resource management, water services, and waste water services in the Paris metropolitan area are now managed in a disintegrated way: they have separate budgets, are managed by different operators, are under the control of different public authorities, and involve various stakeholders. Nevertheless, the existing tariff system still provides a degree of integration between the three water-related services, as they all rely financially on the same bills.[6] Indeed French citizens cover the costs of their drinking water, waste water collection and treatment, and levies to the Agence de l'Eau through a single bill, the fee for which is calculated upon the volume of drinking water used, the quality of water resources accessed, and the amount of pollution generated. Thus, consumers usually perceive water and wastewater as an integrated service, which implies some interdependency between these activities. As an illustration of such, SIAAP recently asked Eau de Paris to stop encouraging "water efficiency" and the reduction of water losses, as the consequent decrease in water consumption (–30 % in 20 years) ends up impairing the operation of SIAAP's waste water plants, and automatically reduces the income received because of a decrease in the volume of water purchased, thereby risking financial imbalance (SIAAP 2010). At the opposite extreme, when Paris decided to reduce the drinking water price by 8 % (which was promised by mayor Delanoë before his re-election in 2007), SIAAP took advantage of this by increasing the unit price of sanitation by 6 % – and without attracting the consumer's wrath (SIAAP 2010); so too did Agence de l'Eau.

The history of water and wastewater services in the Paris region shows alternating movements between integration and disintegration, bundling and unbundling. Today's situation seems to favour a disintegrated and unbundled approach, but

[6] Storm water is managed separately from water services and waste water services, and its costs cannot be covered by water bills, only by local taxes.

recent tensions around water bills show that the trend is not as clear as it seems. Furthermore, all the different items included on a water bill received by the end user are difficult to understand and to that extent are unable to attract criticism. As in most very large metropolises, new projects pretend to bring more integration (e.g. setting up a regional board for bulk production of drinking water), but the complexity of the local reality makes change difficult, as detailed in the next section.

4.2 Vertical Integration Versus Unbundling

In some countries, WSS services are considered a single unified sector under a unified management (e.g. the privatised regional water companies in England and Wales, the 'consorzi idrici integrati' in Italy, the integrated water utilities in Australia before 1994, etc.). Alternatively, WSS services can be split into different sectors ruled by different organisations (e.g. in France, the Netherlands, etc.). Most of these institutional evolutions aim at improving the efficiency of the water and wastewater sectors, either separately or together. But it typically results in the unbundling of some activities which were previously integrated at the local level, so that now they need better coordination of activities/institutions.

An emerging model is the separation of production plants from distribution systems. This sometimes includes regionalisation of water resource abstraction, and eventually of drinking water production ("upstream" activities), together with waste water treatment ("downstream" activities). This model can be found in Australia since 1994 (e.g. Melbourne) and in Portugal (Correia 2011), where the government has supported the creation of public companies whose ownership is shared between an ad hoc national water company and several neighboring local municipal boards. The company manages drinking water production and sewage treatment at the regional level ('agua alta' as opposed to 'agua baixa', i.e. municipal management of water and sewer systems). Other countries present contrasting examples in which regional authorities provide bulk water and do sewage treatment for their member municipalities while other companies supply drinking water only: Metropolitan Water District of Southern California, USA; Canal de Provence for cities of the Mediterranean coast in France[7]; Aguas Ter Llobregat for Barcelona area, Spain; and regional water production and transport in Baden-Württemberg, Germany. Less frequently, the companies treat waste water only (but there is the above mentioned example of SIAAP in Paris). Of course, some other large cities retain an integration of bulk water production, transport over some distance, and distribution: there is New York and its Catskill Mountains and Upper Delaware; San Francisco with its Hetch Hetchy aqueduct; Paris with Belgrand's nineteenth century aqueducts; Madrid with Canal Isabel II; etc. These cities are large enough to fund the whole chain of water supply, while the dual set-up between regional and local levels arises

[7] In fact this company is a regional bulk water production and transfer company, supplying cities, industry, and agriculture with clean but untreated water.

in cases where local authorities need to band together to solve their supply problems. There is even the case in France where all the municipalities of Vendée county (excepting the three largest cities) joined into a single distribution system, but left the water production units in the hands of each local joint board. Mutualising water distribution (and pricing) harmonised the financing of a county-wide network, and secured the provision of water on the coast in summer.

4.3 *Horizontal Integration Versus Unbundling*

Water supply can be organised as a single-purpose activity or, conversely, be brought under a broader multi-utility company. In the beginning, water supply was initiated by private companies, which were single purpose. Sewer services were often linked with drainage and street maintenance, and were frequently publicly managed. When cities took over water supply from an inefficient private sector, they did not merge water and waste water; instead, they usually kept water supply separate as a commercial service managed by a single purpose public company. In Belgium and France, both activities long remained separate, and in fact are to some extent still separate today.

In Germany, conversely, water supplies were at first separate from the main municipal organisations and were later transformed into mixed public/private companies. As early as the inter-war period (1920s–1930s) they were frequently merged into multi-utility services called 'Queverbund' or 'Statdwerk' which included water, gas, electricity, district heating, transport, etc. This reorganisation allowed water services to access a broader range of financing resources (public loans, financial markets) and operate financial and operational cross-subsidies among different local public utilities (Herdt 2009).

This horizontal integration of water services into multi-utilities can also be found at a lesser degree in Austria, Switzerland, and Italy (with the well known example of Rome's ACEA, responsible for water supply chain, sewage treatment, and energy supply). Copenhagen also offers an example of the complex mix that can emerge: while the suburbs have opted for joint boards, the central city has an integrated water and energy company. This company is now a regional power supply company, which still provides water to the central city but also provides a supplementary volume to suburban water supplies (Barbier and Michon 2010).

4.4 *Water Services: Financial Autonomy or Dependence?*

A core factor in determining whether WSS services are integrated or unbundled is how their financial relations with other services and with the budgets of local authorities are arranged. The century-long transformation of water services into commercial utilities has turned them into autonomous entities with their own financial power and status. But the evolution was uneven. As mentioned above, waste water services were mostly developed by local city governments along with drainage, linked with street maintenance, and funded by taxes. For public health reasons they were usually imposed as administrative services rather than as commercial operations. Conversely, drinking water services were often created by private companies and, apart from the UK (and Ireland), their costs were covered by bills; even after water services were reclaimed by municipalities as public entities, their operating costs were frequently charged to customers. However, major amounts of public money were injected to extend water services to all inhabitants (Pezon 2000). Progressively, however, water utilities took advantage of metering and billing to get closer to full internal cost pricing, and an increase in tariffs is now supported by a general move to protect water resources and reduce the water footprint of cities (Walker 2009). Additionally, waste water services are increasingly considered as an amenity, just like drinking water, and their costs are incorporated into water bills. This has led to sharp price increases (doubling or more over time), and can cause secondary sustainability problems – like large consumers exiting the public utility, creating an imbalance in the water services budget, or like the 'water poor' phenomenon. Yet there is little chance for WSS to be reintegrated into the general budgets of local or regional authorities, and their financial autonomy will probably keep increasing.

The financial autonomy of WSS services is achieved in France and the UK through various forms of privatisation or delegation to private companies, which have always operated water services commercially. But it is also the case in other continental European countries, where cities or regional governments have broadly reorganised their WSS services but have kept them as public utilities. It is less the case in Mediterranean countries, where climate variability and periods of authoritarian governments have led to regional/national bulk water production and transport paid by public money. Waste water is still directly or indirectly subsidised, so that water prices paid by consumers remain low, and the push toward full cost pricing and commercialisation is resisted by local populations (e.g. the 2011 referendum on water in Italy).

WSS services end up being organised as a complex array of formulas involving direct public control, financial autonomy, delegation to the private sector, publicly owned private companies, multi-utility companies, and the like, all aimed at interfacing with other policies in an efficient and coordinated way. Economies of scale and scope seem to be the main driver for vertical or horizontal integration. But the lack of specific studies on this matter, and the preservation of unbundled and efficient water services, tend to suggest that no 'one solution fits all'.

5 Interactions Between WSS and Water Resources

In previous sections the change in management scales, or integration/unbundling of WSS services, were analysed chiefly in terms of economic rationalisation. We presented a wide array of institutional arrangements and governance, but did not take into account water resource allocation issues. Indeed, the merit of sanitary engineering, with the invention of water and sewage treatment, was to liberate cities and water services from water resources constraints, particularly in temperate areas where water quality was the chief problem (see Barraqué 2015, Chap. 9 this volume).

However, this technology-based model is increasingly being questioned. First, making drinking water production systems over-complex results in the following loop: criteria → monitoring → chemical treatment → new dangerous substances → more criteria... To avoid the costly race for quality, the water industry (e.g. in Germany) prefers to look for naturally clean water. Second, large cities impact water resources not only with waste water, but also with urban run off, which is polluted and can create problems like local flooding and the choking of rivers. Land-use based management and urban hydrology (reinflitration, storage, etc.) might not only help in reducing the reliance upon infrastructure, but also turn a nuisance into a resource. This new way of thinking also supports reconsideration of the limits of traditional WSS services based on networks, and their partial substitution with on-site solutions. It is obvious in the case of septic tanks, but there is also renewed attraction in developed countries towards rainwater harvesting and individual groundwater bores.

5.1 *Situation in Paris*

In Paris city, half of the potable water supplied originates from surface water (mostly from nearby waterworks located on the Marne and Seine rivers), and the other half comes from groundwater and springs located in four remote water catchments in the south-east and west. These springs have been tapped since 1870 and they flow under gravity to Paris (which is at a lower elevation) through several aqueducts. Such water remains cool in the summer and safeguards Paris's water supply should the Seine or Marne be accidentally polluted. Up until 2004, these water sources were naturally drinkable and chlorine was only added in order to secure its quality during transport to Paris's reservoirs.

However, back in 1971 nitrates began to be monitored in groundwater samples (Zakeossian 2011): the water catchments are located in semi-karstic rural areas and were sensitive to rising agricultural pollutants (nitrates and pesticides). In the early 1990s, SAGEP launched its first attempt to convert farmers to more eco-friendly fertilisation practices. But it soon appeared obvious that these programs were not enough. SAGEP therefore had to choose technological solutions in order to cope with rising groundwater pollution issues. In 2004 and 2005, it invested 18 M€ in two small treatment plants, with activated carbon filters, located at the intakes of the Voulzie and the Loing water catchments and aqueducts. This system was complemented in 2007 and 2008 with two other treatment units, at Saint-Cloud and l'Haÿ-les-Roses, at a cost of 40 M€ and 57 M€ respectively. In 2012, excluding depreciation and staffing costs, the annual operating costs of the last mentioned two plants is around 2.7 M€.

Facing these new investment and operating costs, SAGEP (now Eau de Paris) realised its actions were not enough to prevent further pollution problems. Consequently, Eau de Paris developed local 'territory projects' with the aim of protecting the local water resource while enhancing locally sustainable agriculture. To reach its goal, Eau de Paris now combines various measures such as regulatory requirements, voluntary measures, and training programs in order to reduce agricultural inputs in well-head catchments areas. Other actions include European agri-environmental measures, land acquisition, and support for agricultural transition to other production processes and methods.

Eau de Paris offers farmers stringent but well-rewarded commitments. The program is sufficiently attractive that an increasing number of farmers are being convinced they should engage in it and work towards leaving better quality water. For the Voulzie watershed (11,000 ha located east of Paris), 30 farmers representing 3,900 ha have already signed contracts with Eau de Paris to reduce the use of herbicides or switch to organic production. In 2007, only 1,000 ha were involved. However, at this point it is probably too soon to judge the success of this new approach of a water service toward its water resource territories. According to Eau de Paris, the main challenge today is to ensure the continuance of the financial engagements the farmers entered into, which still requires a sustained commitment from local authorities (Zakeossian 2011).

Compared to the well known 'payments for ecosystems services' which has already been experimented with in New York City, the difference here is that French farmers are more numerous and organised into powerful pro-intensive farming unions. The emerging issue is one of how effective will be efforts to reorganise drinking water production at the scale of the whole region – will it have much impact in protecting water resources? In any case, since Paris city water consumption has gone down by 30 % since 1990, if water from distant sources can be protected at lower cost than surface water treatment, some of it might be made available to suburban areas even if interconnection will raise both economic and political issues.

But the largest problem faced in the interplay between water resources and water services relates to waste water: the Paris region at large represents 40 % of French economic activity, and the Seine's flow is too low to cope. Typically, the 'linear' model used by engineers led to a concentration of all discharges downstream of Paris (first in Clichy, then in Colombes, and later in Achères). In 1942 a large sewage treatment plant began operation, and it soon became the largest in Europe. Despite its good performance, it lacks the physical space to treat wastewater so as to comply with the EU urban wastewater directive (UWWD, 91/271/EEC). So a huge project was undertaken, consisting of building a large new sewage treatment plant upstream of Paris on the Seine (in Valenton), and modernising another one located on the Marne River. This required the reshaping of part of the sewer system in order to redirect waste water to the new upstream water treatment plants. This new system permits full modernisation of the Achères treatment plant so as to meet the aquatic environment targets defined by the European water framework directive (WFD).

For both water and waste water, territorial reorganisation projects were discussed and then supported by the Agence de l'Eau Seine Normandie, one of six river-basin institutions created in 1964 to bring users of the same watershed together and co-ordinate their efforts to improve water quality through economic incentives. In its first decades of existence, the Agence chiefly funded the construction of sewage treatment plants and the extension of existing sewer systems, plus construction of some multipurpose water reservoirs. Yet, in the early 1980s it was thought that individual investments in point-source pollution controls would not suffice (Barraqué 2008). Together with the Water Directorate of the Ministry of the Environment, the Agence developed a new policy based on river contracts in which investments would be co-ordinated between users. But involvement in water resources planning is now essential: the WFD requires that water policy be based on recovery of aquatic environmental quality, which means that, before looking for technological solutions, water users are expected to manage water demands and water resources together as the long-term solution to reducing investments.

Some would argue that the European water Directives, and the WFD in particular, call for new forms of water policies based upon watershed partnerships – i.e. communities formed for an equitable and reasonable water allocation, together with a legal re-definition of water as a common pool resource. This is what the 1992 law reckoned, resulting in a new two-tier planning system: one master plan (SDAGE) for each of the six hydrographic districts (corresponding to the WFD river-basin approach) and as many SAGE (catchment plans) as possible. The system is not mandatory, but rather a bottom-up process: one cannot mandate stakeholder participation. In doing so the French have taken inspiration from other countries and regions where the link between the management of both WSS services and resources has been either maintained or developed.

5.2 *Water Services Manage Water Resources…*

In Germanic customary law tradition, water is a thing not to own but to use reasonably, and it is the community of users that sets the rules. Frequently the rules are quite restrictive, e.g. the riparian rights doctrine in English common law. But in some cases, the community is vital for managing water in times of scarcity or flood. In Europe, for example, such communities have existed for centuries and are essential for irrigating the 'huertas' in Spain, the mountain prairies in the Alps and other mountainous areas (still active in Switzerland, Aosta, and Roussillon), and in the lowlands along the North Sea (the 'Wasser- und Boden Verbände' in Germany). Usually they remained small and rural, but in the case of the Netherlands, the famous 'Waterschappen' (water boards) were not only task based, but also became fully functional institutions through a process of concentration and institutionalisation. Today they maintain dikes and canals, hasten the drainage of winter rain, treat waste water from cities, and even get involved in biodiversity recovery. Some Dutch water managers want to rename them regional water authorities. They offer a clear case of joint waste water and water resources management, even though in this area, apart from the Maastricht zone, there are no defined watersheds. Interestingly, they are funded not by water bills but through taxes on families (Mostert 2011).

In the Netherlands the common handling of water resources and services is traditional, rural, and based on water quantity policies. But in north-west Germany there is an outstanding example of WSS services taking over water resources management for water quality purposes. The example is in the Ruhr area (Ruhrgebiet) where cooperative boards were born at the beginning of the twentieth century. The 'Genossenschaften' manage rivers in their catchment so as to meet the needs of urban and industrial users. Their history goes back to a time of rapid industrialisation and urbanisation, which had catastrophic consequences on water quality and gave rise to epidemics. Local government and industry together decided to focus on three parallel rivers: the Emscher in the center was lined with concrete and became a collective sewer, with a primary treatment system at its mouth on the Rhine; the Lippe in the north was devoted to industrial and agribusiness uses; and the Ruhr in the south became the noble river with freshwater storage, leisure activities, and collective management of wastewater discharges through sewage treatment plants (Barraqué 1995; Raasch 2010; Bäumer 2010). Pre-existing water associations were authorised by the imperial government to become institutions, i.e. to make financial contributions mandatory and make their boards representative of various water users (not one man – one vote, but seats for each stakeholder type). This model was widely adopted locally, so that today there are 11 such institutions in North Rhine–Westphalia, but none in the other Länder.[8] Today their boards include representatives of territorial and local governments, industries, consumers, and environmental protection associations. This type of institution inspired the creation of the French 'agences de l'eau' 60 years later. Today, they do more than run water and waste water infrastructure: they undertake programs to recover aquatic environment quality (the Emscher 'Renaturierung' project), fight emerging micropollutants, and turn stormwater from a nuisance into a resource.

In Germany there are also many ongoing experiments of cooperative agreements between water suppliers and farmers, based on the notion of payments for ecosystem services. These projects are usually voluntary, and go further in terms of environmental recovery than the agri-environmental measures supported by the second pillar of the 'common agricultural policy' of the EU. But since they remain local, payments to farmers (compensations for loss of income) are not considered as government subsidies (which would contravene the equal opportunity principles of the European Common Market).

[8] For instance, the above-mentioned regional water supplies of Baden Württemberg are not based on hydrographic districts, and their boards are only composed of water suppliers.

5.3 ...and Water Resource Institutions Manage Water Services!

This dynamic is more or less the opposite of what is described above: river basin boards are entrusted with the provision of WSS services. It is not common, but it did take place in England and Wales between 1974 and 1989. The poor status of British rivers due to early industrialisation and urbanisation led to the development of 32 river boards in the 1950s, while in the same period WSS services were concentrating at county or subregional level. In 1974, water resources planning, water policing, and WSS were merged into 10 regional water authorities (Saunders 1983). This made sense, since in the UK WSS services are the first users of water resources. Half their board members were representatives from local authorities. But these were later replaced by consumer and other NGOs, within a general movement turning WSS into commercial services. In the end, the Thatcher government opted for privatisation of WSS, while planning and policing were reorganised through the creation of a National Rivers Authority; the latter was soon merged into the UK Environment Agency. Privatised water services remain broadly organised on a hydrographical district basis.

One is also reminded about the cases of Spain and Portugal, where river basin institutions were set up with extended responsibilities for water planning, design, and operation of large infrastructure. However, conversely to the British example, the Spanish 'confederaciones hidrograficas' did not focus on WSS, but rather on the joint management of hydroelectricity and irrigation. The centralisation approach was strong enough to allow for national water plans, whereby river basin institutions were supposed to store water and transfer it to drier watersheds, but not necessarily to secure water supplies of cities (such as Barcelona, which recently opted for desalination because there had been too many successive water plans that in vain promised water transfers from the Ebro basin; Sauri 2011).

To conclude this section, one could argue that it is centralisation that motivates water resources institutions to take over the provision of WSS. However, in countries with a tradition of decentralised policies, change towards the integration of water resources management and WSS is motivated by the principle of efficiency, rather than sovereignty, in a bottom-up process. Interestingly enough, the French 'agences de l'eau' are an interesting and unique case of the second approach in a centralised country where there has always been strong confrontation between central and local governments. With the 'agences', French water policy moved partially out of its centralising Napoleonic tradition towards the community type of water policy practiced in the Netherlands, Germany, Switzerland, and Northern Italy.

6 Conclusion

The aim of this chapter was to illustrate the complexity of the many arrangements found in several developed countries' cities to improve WSS sustainability. But maybe the most important is that such arrangements are now beyond the public vs. private debate that came up in the 1990s, opposing the World Bank and neo-liberal economists to the pro-public alter-globalist movements. In particular, the institutional diversity of water services in Europe not only resulted in a vast array of public–private arrangements, but the choices have usually been combined with other issues: local level provision vs. concentration/centralisation at upper scales, integration vs. unbundling, with diverse links between WSS services and water resources management.

In the end, privatisation and water marketing remains quite limited in the countries reviewed in this chapter. But this does not mean water is to be considered as a pure public good to be managed by administrations either. Water as a resource is still frequently considered a common pool resource, resulting in the survival of community-type institutions, functioning under equitable principles. Conversely, water as a service grew out of another form of impure public goods, defined by Samuelson (1954): club goods. As an invention of the Enlightenment, the club replaces community obligation and equity rules with new rules based on freedom and equality among its members.[9] Public services, in turn, are a special kind of club, since they are not closed but potentially open to all. They are also supposed to be funded by their users through some form of consumerisation (i.e. billing). This makes a serious difference in developing countries where water services are not yet complete: lack of financial capital delays the universalisation of good WSS services. For all those who are ill-connected or not connected, their relationship to water is to a resource, not to a public service, and in most cultures access to water resources for domestic uses should be free. This attitude explains the common low level of social acceptance to pay for water services, which in turn prevents improvements to them. In developed countries, a great deal of complexity, and sometimes misunderstanding, results from the coexistence of the two forms of impure public goods (common pool and clubs) in the distribution of water resources and the framing of WSS policies. Additionally, national governments, even in federal countries, still play a role in the regulation of water resources policies, and more recently in the protection of nature and of water as a natural capital.

As in other public policies, there is a growing need for co-ordination between various types of institutions related to water services – territorial communities, club

[9] Indeed, citizens are not obliged to subscribe to the public water service, and in some countries, there is in fact no obligation to extend the service to all citizens (such as hamlets far from the network). The tariff is the same for the entire category, and those who do not pay are temporarily disconnected. Until recently, few WSS operators considered the issue of the 'water poor', since water was affordable to virtually all the population. Major prices increases have given rise to a new issue that brings the developed countries closer to those of developing ones, and has created an unprecedented level of complexity.

goods, market institutions, and the state – resulting in better coordinated multi-level governance. Still, how to achieve such co-ordination is not obvious, as these policy mixes also call for a change in government culture. The failure of the socialist model, as developed in communist regimes, parallels the questioning of the national statism policies supported earlier by the World Bank, in which economic development is based upon the mobilisation of water resources in multipurpose projects. But the alternative is not just markets. Water management in particular illustrates the notion of multilevel governance. Here lies the opportunity for a water culture which would make room for all different kinds of water institutions, and the capacity to combine them, as a new form of social capital.

Acknowledgements This study has been supported by French National Research Agency (ANR) under the Sustainable Cities Research Program (project EAU&3E, coordinated by B. Barraqué, grant number ANR-08-VILL-0007). The authors wish to thank all the partners of the project who took part in interesting discussions on the issues addressed in this paper and who contributed to some of the case studies reported here.

References

Bakker, K. J. (2003). From public to private to … mutual? Restructuring water supply governance in England and Wales. *Geoforum, 34*(3), 359–374.

Barbier, R., & Michon, S. (2010). *Gestion de l'eau destinée à la consommation humaine: enquête sur l'action des départements*. Retrieved April 23, 2012, from http://aquadep.irstea.fr/wp-content/uploads/2012/04/Aquadep-Synthese-Enquete-CG.pdf

Barraqué, B. (1992). Water management in Europe: Beyond the privatization debate. *Flux, 7*, 7–26. doi:10.3406/flux.1992.1165.

Barraqué, B. (1995). *Les politiques de l'eau en Europe*. Paris: La Découverte.

Barraqué, B. (2003). Past and future sustainability of water policies in Europe. *Natural Resources Forum, 27*, 200–211.

Barraqué, B. (2008). Gestion intégrée et participative des ressources en eau: une perspective de sciences sociales. Keynote speech to the 9th Kovacs Colloquium, UNESCO, In S. Demuth, P. Hubert, & I. Tchiguirinskaia (Eds.), *River basins: From hydrological science to water management* (pp. 111–123). Wallingford: International Association for Hydrological Science.

Barraqué, B. (2015). Three engineering paradigms in the historical development of water services: more, better and cheaper water to European cities. In R. Q. Grafton, K. A. Daniell, C. Nauges, J.-D. Rinaudo, & N. W. W. Chan (Eds.), *Understanding and managing urban water in transition* Dordrecht: Springer. (pp. 201–216).

Bauby, P. (2007). Quels modèles de service public dans les pays de l'OCDE? *Cahiers français*, Les services publics, juillet-août 2007, La Documentation française, pp. 21–25.

Bäumer, A. (2010). *Ruhrverband: Current economic and sanitation issues*. Retrieved April 23, 2012, from http://f.hypotheses.org/wp-content/blogs.dir/146/files/2010/11/Athens_Bäumer_10_vuBB.pdf

Brady, J., & Gray, N. F. (2013). Analysis of water pricing in Ireland and recommendations towards a more efficient water sector. *Water Policy, 15*(3), 435–457.

Calabi, D. (1980). I servizi technici a rete e la questione della municipalizzazione nelle città italiane 1980–1910. In P. Morachiello & G. Teyssot (Eds.), *Le macchine imperfette. Architettura programma, institución nel XIX seculo*. Roma: Officina edizioni.

Carroy, C. (2010). En direct de… l'Île-de-France: vers une agglomération parisienne de l'eau? *Techniques Sciences Méthodes, 3*, 8–11.
Cebron de l'Isle, P. (1991). *L'eau à Paris au XIXe siècle*. Dissertation, University of Paris-Sorbonne.
Chatzis, K. (2006). Brève histoire du compteur d'eau à Paris, 1880–1930. *Terrains et Travaux, 2*(11), 159–178.
Correia, F. N. (2011). *Water and wastewater services in Portugal: An approach based on multilevel governance*. Retrieved April 23, 2012, from http://eau3e.hypotheses.org/files/2011/11/ATHENS_11_Correia.pdf
Defeuilley, C. (2004). Contrats et regulation des services d'eau, le Syndicat des eaux d'Ile de France. *Flux, 55*, Jan–Mar 2004.
Dupuy, G. (Ed.). (1984). *Réseaux techniques urbains: Les annales de la recherche urbaine*, nr 23–24, Dunod (ISBN/ISSN 0180930).
Gaillard-Butruille, S. (1995). *Evolution des traitements de potabilisation jusqu'en 1940, l'exemple de Paris*. Paris: SAGEP.
Graham, S., & Marvin, S. (2001). *Splintering urbanism. Networked infrastructures, technological mobilities and the urban condition*. London/New York: Routledge.
Guerin Schneider, L., Nakhla, M., Grand d'Esnon, A., & Baudot, B. (2002). Gestion et organisation des services publics d'eau en Europe. Management and organisation of public water services in Europe. *Cahiers de recherche – Ecole des mines de Paris, Centre de gestion scientifique*, 19. http://cat.inist.fr/?aModele=afficheN&cpsidt=14885350
Herdt, H. (2009). *Städtische Werke Magdeburg*. Retrieved February 13, 2012, from http://eau3e.hypotheses.org/files/2009/11/ATHENS_Helmut_Herdt.pdf
Kraemer, R. A., Pielen, B., & De Roo, C. (2007). Regulation of water supply in Germany. *CESifo DICE Report, 5*, 21–26.
Kuks, S. (2010, November). *Water management by the regional water authorities in the Netherlands*. Retrieved February 13, 2012, from http://eau3e.hypotheses.org/files/2010/11/Athens_Kuks_NL_10.pdf
Loijenga, H. (2009, November). *Water governance in the Netherlands*. Retrieved February 13, 2012, from http://eau3e.hypotheses.org/files/2009/11/ATHENS_Henk_Loijenga.pdf
Majone, U., & Veltri, M. (Eds.). (2011). Fra Referendum e norme: il futuro dell'acqua. *Rivista l'Acqua, 3*.
Mangano, A. (2010, November). *Water services in Italy*. Retrieved February 13, 2012, from http://eau3e.hypotheses.org/files/2010/11/ATHENS_Mangano_10.pdf
Massarutto, A. (2011). Urban water reform in Italy: A live bomb behind outward unanimity. In B. Barraqué (Ed.), *Urban water conflicts* (pp. 247–268). London: Taylor & Francis.
McKay, J. (2005). Water institutional reforms in Australia. *Water Policy, 7*, 35–52.
Megdal, S. B. (2012). The role of the public and private sectors in water provision in Arizona, USA. *WRRC. Water International, 37*(2), 156–168. Routledge.
Millward, R. (2000). The political economy of urban utilities. In M. J. Daunton (Ed.), *The Cambridge urban history of Britain, 1840–1950* (pp. 313–349). Cambridge: Cambridge University Press.
Moss, T. (2008). 'Cold spots' of urban infrastructure: Shrinking processes in Eastern Germany and the modern infrastructural ideal. *International Journal of Urban and Regional Research, 32*, 436–451.
Mostert, E. (2011, November). *Water and sanitation services in the Netherlands*. Retrieved April 23, 2012, from http://eau3e.hypotheses.org/files/2011/11/Athens_11_Mostert.pdf
Networks. (2008 September). *Minute*. Minute of the international workshop held in Berlin on "Sustainable Urban Water Infrastructure: Possibilities of Adaptation and Transformation".
Pezon, C., & Canneva, G. (2009). Petites communes et opérateurs privés: généalogie du modèle français de gestion des services d'eau potable. *Espaces et Sociétés, Usages et régulations des eaux urbaines, 4*, 21–38.

Pezon, Ch. (2000). *Le service d'eau potable en France de 1850 à 1995*. Paris: Presses du Centre de Recherche en Economie et Management, 441pp.

Pezon, Ch. (2006). *Intercommunalité et durabilité des services d'eau et d'assainissement. Etudes de cas français, italiens, portugais*. Report for the Programme Politiques Territoriales et Développement Durable, MEDD, 136pp.

Pflieger, G., & Paquier, S. (2008). L'eau et les services industriels de Genève : aux origines du modèle suisse des services urbains. *Entreprises et histoire, 50*, 36–51.

Putnam, R. (1993). *Making democracy work: Civic tradition in modern Italy*. Princeton: Princeton University Press.

Raasch, U. (2010, November). *Water management by waterboards: The Emschergenossenschaft example*. Retrieved April 23, 2012, from http://f.hypotheses.org/wp-content/blogs.dir/146/files/2010/11/athens_10_Rash_corr.pdf

Samuelson, P. A. (1954). The pure theory of public expenditure. *Review of Economics and Statistics, 36*(4), 387–389. doi:10.2307/1925895.

Saunders, P. (1983). *The "Regional State": A review of the literature and agenda for research*. Urban and regional studies paper, University of Sussex.

Sauri, D. (2011, November). *Challenges for sustainable water management: Scarcity, drought and debt in Barcelona*. Retrieved April 23, 2012, from http://eau3e.hypotheses.org/files/2011/11/ATHENS_11_Sauri1.pdf

SIAAP. (2010). Statement of the board meeting, 15 Dec 2010.

Walker, A. (2009). *The independent review of charging for household water and sewerage services*. Final report. Published by the Department for Environment, Food and Rural Affairs. Retrieved April 23, 2012, from http://www.defra.gov.uk/publications/2011/12/06/walker-review/

Zakeossian, M. (2011). Protection des eaux souterraines: exemples d'actions engages dans le domaine agricole par Eau de Paris. *Responsabilité & Environnement, 63*, 80–86.

Chapter 3
Coping with Extraneous Water in Sewerage Systems

Galina Dimova, Irina Ribarova, and Franz de Carné

1 Evolution of Urban Wastewater Management

1.1 Evolution in Thinking

People's need for potable water in daily life results in the production of wastewater, which then has to be carried away beyond the perimeter of human activities. The problem of how to *evacuate wastewater* has been recognised since civilization began. Evidence for the existence of drainage and sanitary sewer systems has been found for the Mesopotamian empire (4000–2500 BCE), the Indus civilization (3000–2000 BCE), the Aegean civilization (3000–100 BCE), Egypt (2000–500 BCE), and many other ancient places (Schladweiler 2002). With time, people found that the conveyance of wastewater can sometimes turn into a question of survival, not just an issue of getting rid of unpleasant sights and smells. In the early nineteenth century, after several severe disease outbreaks, scientific studies discovered that fatal diseases like typhoid and cholera are waterborne, with wastewater as their main carrier. Many towns then started to develop sanitary sewer systems, partly in response to diseases and in order to mitigate their spread. Furthermore, as a result of a combination of factors – the industrial revolution, the concentration of populations in towns and cities, the invention of continuous potable water supply to urban

G. Dimova (✉) • I. Ribarova
University of Architecture, Civil Engineering and Geodesy,
1, Christo Smirnenski, blvd., Sofia 1046, Bulgaria
e-mail: gdimova@hotmail.com; ribarova.irina@gmail.com

F. de Carné
Tailored Financial and Consulting Services S.a.r.l,
20, rue de l'Eau, Luxembourg City L-1449, Luxembourg
e-mail: tfcs@pt.lu

Q. Grafton et al. (eds.), *Understanding and Managing Urban Water in Transition*, Global Issues in Water Policy 15,
DOI 10.1007/978-94-017-9801-3_3

households, and the general increase of standards of living – the need was recognised for *the organised evacuation of urban wastewater through sewer systems.*

Social and economic developments in the twentieth century increased demands for better water supplies and wastewater evacuation. In the first half of the century the concept of *wastewater treatment* was introduced, because increased discharges of wastewater into receiving waters was causing severe pollution and detrimental effects on soils and groundwater. By the end of the 1980s, the concept of solving water pollution problems case by case, without consideration of follow-on effects, was replaced by a fundamentally different understanding of global socio-economic dynamics – *sustainable development.* The new concept meant meeting the needs of the present without compromising the ability of future generations to meet their own needs (United Nations 1987). The World Commission on Environment and Development first introduced the concept in the Bruntland Report of 1987. Subsequently, in 1992, Agenda 21 of the Rio Declaration on Environment and Development presented a comprehensive plan of action in every area in which human activities exert impacts on the environment (United Nations 1992). The key concept of sustainable development is that humans are part of nature; they depend strongly on nature and must preserve it. Thus, environmental issues become an integral part of all decisions concerning social and economic life.

Water is not like a commercial product; instead it is a heritage which must be protected, defended, and treated with respect (EU 2000). It is a part of our physical life, social well-being, economic progress, environment, and culture. Thus solving water problems in a sustainable manner affects all spheres of human life. The concept of sustainability has been implemented in the management of urban potable water, drainage, and wastewater systems, and is called *sustainable urban water management* (SUWM). However it is not an easy task to develop policies and to define sound principles in the design and operation of water supply and sewerage facilities. When trying to define what is appropriate for sustainable development of the water sector, SUWM has been applied using physical variables such as space and time (Larsen and Gujer 1997); other scientists have analysed SUWM in terms of environmental and cultural values (Harremoës 1997), while others have identified certain criteria upon which SUWM should be based: health and hygiene, environmental impact, functionality and related technical issues, economic criteria, and social and cultural criteria (Aspergen et al. 1997; Butler and Parkinson 1997; Hellström et al. 2000). Thus, SUWM has become a philosophy with various interpretations. It should be noted, however, that sustainability is not a steady state situation in a physical sense, since nature itself, as well as anthropogenic activity, is never constant or in a steady state (Aspergen et al. 1997). Therefore SUWM should also be regarded as a dynamic process, in the same way as all other global processes, and should be used to help the environment come to an acceptable balance.

The recent strategies (CEC 2001, 2010, 2011a, b) elaborated by the European Commission have made further progress in defining the concept of sustainable development, seeing it as promoting "resource efficiency" in moving towards sustainability. Resource efficiency means to "do more with less" and to decouple production from resource use and pollution. These documents call attention to restructuring or relinking production–consumption systems so as to use resources optimally. The Europe 2020 strategy promotes a "resource efficient Europe" and "resource efficient technologies" to help decouple economic growth from the use of resources (CEC 2010). In terms of SUWM it means producing less wastewater by using less water, adapting pipe diameters to use less energy, using fewer chemicals for treatment, and reusing natural resources after discharge.

1.2 Evolution of Technologies

There are two principal models of urban wastewater management: (1) decentralised, where wastewater is collected, treated, and/or disposed of within the place of origin; and (2) centralised, where all the waste water is conveyed to a specific location for treatment and disposal. The conveyance of storm water was introduced to centralised sewer systems later in the piece, since rapid urbanisation demanded an appropriate solution to this problem as well. From the beginning of the nineteenth century, both decentralised and centralised approaches have been used in parallel, and both systems have been improved as new technologies developed in response to increased concern for environmental protection. However, a dispute about the applicability of combined and separate sewer systems is still continuing.

Decentralised systems have evolved from simple cesspits to small-scale treatment plants which operate close to the pollution source; technologies like artificial wetlands, which emulate natural systems, have also recently been introduced. Small-scale systems can reuse wastewater, and use organic wastes as fertiliser, all without exporting them from where they originated. It is a common belief that decentralised systems do not require high investments and that their operation and maintenance are easier compared to centralised systems. It turns out, however, that the operation of some decentralised systems (e.g., artificial wetlands) is quite complicated and the treatment performance and final ecological effect can be questioned (Harremoës 1997; Orth 2007). In terms of cost, the general idea that decentralised systems are cheaper than centralised ones is also questionable (Orth 2007). Some authors conclude that decentralised systems suffer from lack of adaptability to many urban environments, lack of manageability and control, straying from environmental protection standards, and loss of economies of scale (Butler and Parkinson 1997). Engineering experience has proven that there is no universal technology. Certainly decentralised systems have their place in modern urban developments; however, their adequacy needs to be gauged case by case (Chung et al. 2008).

Centralised urban sewer systems appear preferable for densely populated urban areas, where space is limited. They provide an effective public health barrier when operated appropriately, prevent damage caused by storm water or high groundwater tables, and allow monitoring and control of pollutants in the wastewater before it is discharged to rivers.

Early centralised sewer systems were combined, i.e., urban wastewaters (including domestic sewage and trade effluents) were conveyed with storm water in a single pipe to the wastewater treatment plant (WWTP). Along the main trunk routes are installed combined sewer overflows (CSOs), which, during rainy weather, discharge a part of the mixed flow into the receiving water body. Over time, engineers have started to pay particular attention to the pollution of receiving water bodies caused by CSOs, and have recently developed new techniques (e.g., screens) for improving their performance (Butler and Davies 2000).

With growing urban populations and shifting development patterns, changing characteristics of wastewater quantity and quality, and rising concerns over the environment the interest in wastewater transport and treatment has shifted towards *separated sewer systems* (Burian et al. 2000). Some well-known advantages of separate sewer systems are: potentially less pollution of watercourses due to lack of CSOs, smaller diameters of the collectors, smaller wastewater treatment works, less variation in flow, and concentration of wastewater. However they also have some basic disadvantages like extra cost for two pipes, additional space occupied in narrow streets in built-up area, more house connections, and no flushing of deposited wastewater solids by storm water (Butler and Davies 2000).

Separated sewer systems are usually the preferred option for construction of new systems, since converting an old combined sewer system into a separate one seems to be expensive (US EPA 1999; Butler and Davies 2000). Some studies are strongly of the opinion that separate systems are not always the best solution because street runoff contains heavy metals and oils and its discharge directly into rivers may cause appreciable impact (Burian et al. 1999; Sieker 2003; De Toffol et al. 2007).

Increasing concern over the deterioration of receiving waters resulting from wastewater discharge has meant that wastewater treatment has become an integral part of sewer management. Initially, treatment consisted only of mechanical processes (screening and sedimentation) but later biological treatment was introduced. The first biological treatment process was intended to remove only organic carbon, but since the 1980s biological removal of nitrogen and phosphorus has become necessary, in line with growth of our understanding of how human activities impact on nature. Enhanced requirements for wastewater treatment mean higher investments and much higher operation and maintenance costs.

The centralised collection and treatment systems of urban wastewater require an increasingly sophisticated framework of procedures. They demand high energy consumption, expensive high-tech electronic monitoring, computerised decision-making tools, and skilled and motivated staff (Butler and Parkinson 1997). In this sense, their management turns out to be resource demanding rather than resource efficient. In addition, their environmental adequacy and sustainability are often questioned, since the environmental consequences of failure – leakages, inadequate treatment and discharge of effluent, or improper disposal of contaminated sludge – are much more severe than from that of a single decentralised system (Harremoës 1997; Beck 1996).

In spite of these disadvantages, centralised urban sewer systems, in particular combined ones, are still prevalent in most cities. They were built over decades and their upgrading to comply with more stringent requirements for wastewater quality and sludge quality is a major undertaking, and one that requires advanced engineering approaches and significant investments. This is now the challenge of many towns all over the world.

2 Extraneous Water: A Common Deficiency in Sewer Systems

The definition of extraneous water varies depending on the standards and regulations consulted and it changes with time. "Extraneous water" can be defined as ground water, storm water, or wrongly connected drainage entering the sewer system through defective pipes, joints, and manholes (DIN 1999). It may be described as "unwanted discharge in the sewer systems" (Pfeiff 1989; DIN 2008). The German Water Association gives the following definition: "extraneous water is water discharged to sewer systems which is neither qualitatively influenced by domestic, industrial, agricultural or other usage nor specifically collected and discharged during precipitation" (ATV-DVWK 2003). The phenomenon of extraneous water entering the sewer system is called "infiltration".

Thorsten (2006) undertook a major investigation of published data on the quantity of extraneous water. The results vary widely: in Germany, from 0 to 100 % surcharge to foul water, in UK from 15 to 50 %, and in Switzerland, 47 %. According to estimates made by the EPA, in the USA infiltration and inflow represent almost half of all flow at treatment plants nationwide (Vipulanandan and Ozgurel Gurkan 2004).

Regardless of its origin (groundwater, drainage, household leakages, spring water, cooling water, or wrongly connected storm water), extraneous water will increase hydraulic loads, resulting in higher operation and maintenance costs for pumping and treatment. Apart from purely operational problems, extraneous water can also directly impinge upon domestic customers, such as backflow into basements due to hydraulic overloading of sewer collectors (having highly unpleasant consequences for the residents and even generating political debate).

Ecological and sustainability issues also arise when the environmental effects of extraneous water are considered. Increased hydraulic flows often cause uncontrolled spills of untreated wastewater out of sewer overflows, causing pollution and risking public health. Inflow of ground water from a surrounding aquifer into sewer pipes can carry fine soil particles into the pipe, and these, when deposited, can result in reduced pipe capacity and can damage the ground structure surrounding the pipe.

The reduced effectiveness of the sewer system due to extraneous water can cause higher energy demands for wastewater pumping and, to some extent, treatment. Thus, in the wider picture, extraneous water is critical for SUWM, since it promotes inefficient use of resources.

The occurrence of extraneous water in sewerage systems depends on various factors that can be summarised in the following main categories:

- Natural characteristics, geology, and hydrology of the drainage area, in particular the aquifer level and the soil's ability to let storm water infiltrate. In addition, site topography and climatic conditions.

- Characteristics of the sewer network: the method of laying pipes, making joints, and installing manholes. Also the pipe material, and how connections are made to private houses.
- Characteristics of the water supply system, since potable water leakages often end up as inflow into the sewer system. Potable water pipes are usually laid above sewer pipes, following the same track along the street. Depending on the geology of the ground, leaking water can either drain to groundwater or, if there are damaged walls or decrepit pipe connections, seep into the sewer. The surrounding sand bed, in which the collector is laid during pipe-laying, facilitates this process.

On the other hand, wastewater from leaking sewer pipes contaminates groundwater, preventing it from being beneficially used for irrigation or potable water. This phenomenon is known as "exfiltration". It is estimated that in Germany several hundred million cubic meters of wastewater leak every year from partly damaged sewer systems into soil and reach groundwater (Eiswirth and Hoetzl 1997).

The phenomena of exfiltration and infiltration usually run in parallel, although usually in different parts of the sewer system and depend on the groundwater level in relation to the sewage collector. If groundwater levels change, one and the same section may, at different times of the year, exfiltrate sewage water and infiltrate extraneous water. The exfiltration/ infiltration along the sewer network dilutes the wastewaters to an extent which can negatively affect biological processes at the treatment plant.

The characteristics of the sewer network are probably the most important factor relating to infiltration/exfiltration. The most common defects that can cause infiltration are: cracks, fractures, joint displacement, root intrusion, deformation, collapse, poorly constructed connections, and abandoned laterals left unsealed (Vipulanandan and Ozgurel Gurkan 2004). With aging of the sewer network, the number of defects increases. The same defects also cause exfiltration as well.

The spatial extent of infiltration/exfiltration in sewer pipes and the flow rates of extraneous water are extremely difficult to determine. A well-known approach is direct flow measurement of the sewer system during the minimum night-time discharge, taking into account the diurnal variations in sewage discharge from users of the sewerage network. This method however demands extensive preparatory analytical and experimental work to determine an optimal measuring point within the sewer system (Thorsten 2006). The 'chemical method' can also be used, where the concentration of a specific added tracer is monitored and the amount of extraneous water is calculated based on its dilution. In cases where a WWTP exists, the rate of extraneous water can be determined by using different approaches such as: (1) comparing annual sewage discharge with drinking water consumption; (2) comparing measured minimum dry weather flow with expected daily discharge of sewage; and (3) analysing the multi-year discharge hydrograph curves (Bosseler et al. n.d.).

It is difficult to quantify extraneous water flow without undertaking several extended campaigns of flow measurement during dry weather. Because the amount of extraneous water in a sewerage system is not constant during the year, it is best to carry out measurement campaigns over at least 1 year so that the amount of extraneous water can be measured over different seasons. Long-term changes in climate may also affect the amount of extraneous water, since changes in rainfall and duration and height of snow cover will affect the level of the ground water table.

Performing prolonged direct measurement campaigns on a sewer network is laborious and costly, and such measurements are not within the common duties of a sewer operator. But since taking account of extraneous water is necessary for planning future investments in sewer systems, here a simplified methodology is suggested, one that was applied to two case studies in Bulgaria.

3 Simplified Methodology

To obtain general estimates of the availability and amount of extraneous water in an existing sewerage system, the following steps were considered as a minimum requirement.

Step 1 On-site survey along the river banks within the area served by the sewerage network, to identify:

- the number, type, and condition of sewer outlets (outfalls or CSOs);
- whether there was existing flow in dry weather.

Step 2 On-site manhole survey to:

- check and verify the course and diameter of sewers and the state of manholes;
- check and verify important levels (bottom level of the inlet and outlet pipes, level of the manhole lid);
- identify different sub-catchment areas in the sewerage network and select points for flow measurement.

Step 3 Inspect parts of the sewer network (around 1 km in length) with a closed circuit television (CCTV) camera in order to collect images on the state and functional conditions of the sewer pipes.

Step 4 Undertake a 24 h campaign of flow measurement in representative sub-catchment areas (selected based on the results from Steps 1 and 2), respecting the following criteria:

- covering a large, representative area of the town;
- returning a clear picture of the way in which the wastewater is conveyed through the catchment area;
- enabling the installation of a water measuring device at the end of the sub-catchment area.

Step 5 Measure wastewater quality at the same location where flow measurements are done in order to allow indirect assessment of the presence of extraneous water and comparison of the results with the existing data base (if any).

Estimation of the extraneous water flow rate is based on two approaches:

Approach 1 The difference between the measured average daily flow in the sewer and the analytically calculated domestic and industrial sewage flows.

Approach 2 Analysis of the average minimum flow measured during early morning hours when there is virtually no domestic or industrial activity in the sub-catchment area.

This methodology carries some uncertainty, since: (1) the period of investigation is short, i.e., 24 h; (2) the analytical calculation of the domestic and industrial wastewater flow is based on norms that are representative of the long term rate, but come from just a single day of measurement; and (3) only part and not the whole area of the town is encompassed. Nevertheless the results obtained can indicate major problems in the sewer system and can help engineers select technically and economically appropriate solutions.

In the study presented below, a flow meter, PCM4, manufactured by NIVUS, Germany, was used. It employs the doppler effect to measure velocity and a depth sensor to calculate free-surface flows in pipes or channels. During the campaign the depth and velocity were entirely within the ranges of accurate measurement: velocity between 0.1 m/s and 3 m/s, and depth greater than 30 mm. The flow rate was recorded at 5-min intervals, giving a total of 288 values/day at each site.

A grab sample of 150 ml was taken from the incoming waste water every 15 min during the flow measurements and every 3 h the samples were mixed to form a composite sample. The samples were analysed for 5-day biochemical oxygen demand (BOD_5), chemical oxygen demand (COD), suspended solids (SS), total nitrogen (TN), total phosphorus (TP), and pH.

4 Estimation of Extraneous Water in Sewer Systems of Two Towns in Bulgaria

4.1 Background

Bulgaria (110,000 km^2; population 7,365,000) is situated in Eastern Europe on the Balkan Peninsula and became a member of the EU in January 2007. Pursuant to the Accession Treaty, Bulgaria must meet the goals of Acquis Communautaire, stipulated in a number of EU legal documents. Under the sector "Environment", one of the main challenges to be faced by the country in 2007–13 is the fulfilment of EU Directive 91/271/EEC with regard to: (1) construction of sewerage networks and urban wastewater treatment plants in agglomerations with more than 10,000 PE (population equivalent); (2) construction of sewerage networks and urban wastewater treatment plants (WWTPs) in agglomerations with PEs between 2,000 and 10,000. Due to the extremely high investment costs and the time needed for the necessary technology, Bulgaria has requested a transitional period for item (1) until 2011.01.01 and for item (2) until 2015.01.01.

Table 3.1 Comparison between some key parameters connected with the operation and maintenance of the water supply and sewerage (WSS) infrastructure in Bulgaria

Key issue	Before 1989 (socialist period)	Today
Ownership of WSS assets	Only state	13 companies – 100 % (state share);
		21 companies – 100 % (municipal share)
		16 companies – 51 % (state share), 49 % (municipal share); 2 concession contracts
Price for WSS service	Equal throughout the state around 0.03 EUR/m³, relatively stable	Different for different WSS companies, continuously increasing
		In the region of case studies:
		0.59 to 0.69 EUR/m³, of which 0.18 EUR/m³ are for domestic sewage conveyance and treatment
Price for electricity	Equal throughout the state, relatively stable	Different for different electrical companies, continuously increasing
	Day tariff – around 0.023 EUR/kWh;	In the region of case studies:
		Day tariff – 0.092 EUR/kWh
	Night tariff – around 0.01EUR/kWh	Night tariff – 0.063 EUR/kWh

The last decade of the twentieth century is marked in the history of Bulgaria as a time of major transition from a socialist political system and planned economy to a parliamentary democracy and market economy. This change has affected all spheres of political, social, and economic life in the country. Table 3.1 marks some key issues of this transition relevant to the operation of water supply and sewerage infrastructure.

Before 1989 all decisions concerning water supply and sewer infrastructure were taken at a state level. The lack of public participation in decision making, combined with low prices for water and energy, encouraged customers to retain a low public awareness of the high cost for construction, operation, and maintenance of water and wastewater infrastructure and the relevant environmental issues.

Nowadays municipalities are taking a role in the decision making process, since they are the main beneficiaries of the infrastructure projects financed by the EU. Customers are facing the challenge of continuously increasing prices for water and energy services. At the same time, the water supply and sewerage infrastructure is very old and technically out of date.

According to data from the National Statistical Institute of Bulgaria, in 2010 the average percentage of non-revenue water (NRW)[1] in the supply systems was 57.6 %, while there are towns where the NRW is over 80 %. A big share of NRW is the physical loss of water through leakages. A large part of the leaking potable water is believed to seep into the sewerage system, which in general is in very poor condition.

[1] According to the International Water Association, non-revenue water is water that has been produced and "lost" before it reaches the customer.

There are no statistical data on the amount of extraneous water in sewerage; however wastewater quality entering some WWTPs proves that it is much diluted. The problem of extraneous water has so far been somewhat neglected, since the sewer systems are usually operated by gravity and in many towns there is no WWTP, resulting in low operation cost and generating little or no public attention. The design and construction of a WWTP, however, demands a technically and economically feasible approach, and one that is socially acceptable, for coping with extraneous water.

Below, two case studies in Bulgaria are presented – in the towns of Bansko and Gotse Delchev, which demonstrate how the problem of extraneous water was tackled. The work was in preparation for determining the priority measures needed for rehabilitation and/or reconstruction of the towns' water supply and sewerage infrastructure. The studies were carried out within the framework of the EU ISPA project "Technical assistance for project preparation in the water sector – Asenovgrad, Bansko and Gotse Delchev", EUROPEAID/124,486/D/SV/BG (MOEW 2011).

4.2 Case Study: Bansko

4.2.1 General Information

The town of Bansko is located in South West Bulgaria at the foot of the highest mountains on the Balkan Peninsula – Pirin (Vihren Peak, 2,914 m) and Rila (Musala Peak, 2,925 m). The Pirin mountain provide excellent opportunities for skiing and since 1989 the region has been turned into a dynamic winter tourist business. The municipal administration is endeavouring to turn the region into an attractive year-long tourist destination.

The number of permanent residents in Bansko is a little over 9,000, as registered in the 2011 national census. The intensive tourist activities have led to a significant increase in accommodation facilities over the last 10 years. In 2010, the number of registered hotel beds was more than 18,000. In addition, there were around 21,000 privately owned flats, among which only 16.5 % belonged to permanent residents. The rest are mostly occupied for 1–4 weeks during the winter season. Thus, the number of seasonal residents can increase the population by 3 times or more compared to the number of permanent residents.

The water supply system of Bansko, which is of the gravitational type, is fed by several mountain springs and provides water to 100 % of the residents. Some 56 % of the network was constructed using old (more than 40 years) asbestos cement (AC) and steel pipes, while the rest, constructed of high-density polyethylene (HDPE) pipes, has been laid during the last 10 years. The potable water network is characterised by frequent breakdowns and significant leakages. The average NRW in 2008 was estimated at 47 %.

The existing wastewater network, situated on both sides of the Glazne River, is completely gravitational. The old, central part is served by a combined sewer system, while the newly built resort area, on the south, is served by a separated sewer system of which part is still under construction. The connection rate of households to the sewer system is 100 %. The sewer network consists mostly of concrete pipes (77.4 %) more than 40 years old.

At present there is no WWTP in the town. A survey of the river banks found that the sewer water discharges into the Glazne River through 8 outfalls and 5 CSOs; however due to the significant amount of extraneous water the CSOs discharge also in dry weather. Five other outfalls discharge into small creeks on the outskirts of the town.

The Glazne River is a tributary of the Mesta River. In the river basin management plan for the Mesta, the Glazne is classified as a "water body in bad status" in the section below Bansko. Due to the mountainous character of the landscape, which provides good opportunities for natural self-purification, the whole catchment of the Mesta River is not identified as a "sensitive zone", as defined in Annex II of Council Directive 92/271/EEC concerning urban wastewater treatment (EU 1991).

4.2.2 The Problem of Extraneous Water

Bansko is situated in a natural kettle surrounded by high mountains with an inclination and natural outflow to the northeast (Fig. 3.1). The upper geological layer of the kettle consists mainly of terrace depositions from the Glazne River, represented by

Fig. 3.1 Aerial view of Bansko

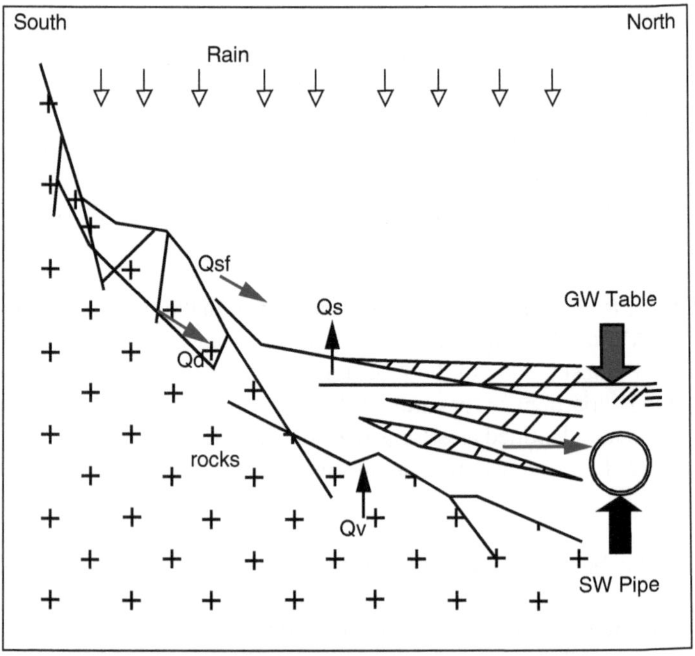

Fig. 3.2 Scheme of feeding of aquifer (Legend: *Qsf* – surface run off, *Qd* – mountain groundwater flow, *Qv* – groundwater flow inside of the kettle, *Qs* – groundwater overflow (springs), *GW* – ground water, *SW* – sewer)

boulders and pebbles with sandy-clayey filling. This context between the natural outflow in the northeast and the upper geological layer forms a natural aquifer.

The aquifer is recharged by melting snow and surface run-off in the kettle, and by subsurface run-off from the surrounding mountains that penetrates through the cracks in the rocks and causes underground flow toward the kettle, which laterally feeds the aquifer (Fig. 3.2). Depending on the groundwater level, the Glazne River acts as a natural receiver of groundwater or as a natural feeder of the aquifer. Consequently, the groundwater table inside the kettle is stable and relatively high during spring and summer, resulting in significant amounts of extraneous water in the sewer network.

The on-site manhole survey encompassed 308 manholes (25 % of the total) and together with verification of key parameters (levels, diameters) the flow of the combined wastewater system was estimated visually. The key results related to the problem of extraneous water are as follows:

- Small flow was observed in 103 manholes – less than 10 % of the pipe cross-section carried water; the flow ran slowly without evident turbulence;
- Medium flow was observed in 123 manholes – less than 50 % and more than 10 % of the pipe cross-section carried water, and the flow was turbulent;
- Strong flow was observed in 82 manholes – more than 50 % of the pipe cross-section carried water, and the flow was very turbulent.

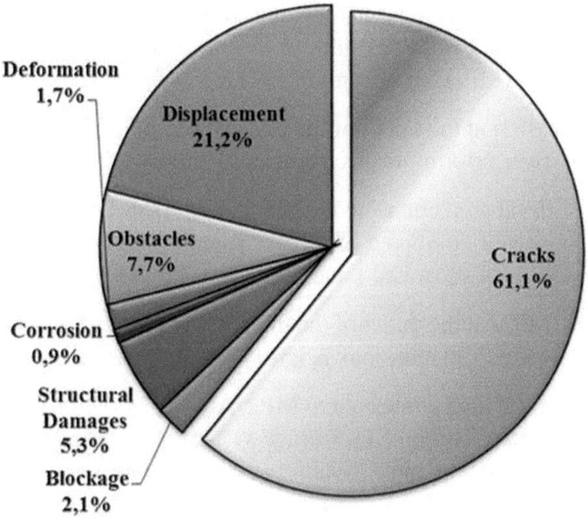

Fig. 3.3 Classification of the pipe damages in the sewer network in Bansko

The characteristic smell of sewage was not detected on opening most of the manholes and the appearance of the water looked clean and even transparent.

The CCTV inspection of 19 sewer collectors sections (in total 1 km length) and the manhole survey have shown that the sewer system is in very bad condition. Twelve sewer sections with length between 30 and 100 m were inspected completely, while in 8 sewer sections the CCTV camera couldn't reach the end of the section due to impassable obstacle. The average distance, within a section, at which the CCTV camera detected a problem was 1.60 m, while the longest distance without observable defects was 25 m. Figure 3.3 presents a summary of the observed problems.

The concrete collectors of the combined sewer system were partly broken, axially displaced, and had reduced hydraulic capacity (greater than 50 %) due to deposits of sand, stones, bricks, and other debris. In addition, some pipes of the relatively new separated network were inappropriately laid, and the cross-sections had undergone serious deformation, compromising the joints between the sections.

These figures suggest that extraneous water is a significant problem in the sewer network. If the problem is neglected, it will result in significant hydraulic loads that will affect the operation of both the new sewer network and the WWTP. Dilute wastewater will not provide the necessary organic substrate for stable biological treatment at the future WWTP. This will lead to higher operating and maintenance costs in the system.

4.2.3 Wastewater Flow Measurement Survey

Wastewater flow was measured for 24 h under dry weather conditions from Thursday 21 May to Friday 22 May 2009. During this time of year tourist activities are minimal. The measuring points were chosen to be places where good estimates of the largest possible sub-catchment areas could be achieved. The two measuring points (MPs), shown in Fig. 3.4, are as follows.

- MP1: at the outlet of the main sewer collector of DN 1000 on the east side of the river prior to its discharge to the Glazne River.

  Total area connected to MP1, $A_{MP1} = 188.19$ ha, of which,
  area of combined sewerage system, $A_{comb} = 88.91$ ha
  area of separated system (wastewater only), $A_{sep} = 99.28$ ha.

- MP2: at the outlet of the main sewer collector west of the river (oval profile, max width 700 mm; max height 800 mm) prior to the outfall into the Glazne River.

  Total area connected to MP2, $A_{MP2} = 131.03$ ha, of which,
  area of combined sewerage system, $A_{comb} = 75.96$ ha
  area of separated system (wastewater only), $A_{sep} = 55.07$ ha.

The total of catchment area in Bansko is 500.36 ha, therefore the two measuring points collect the wastewater discharge from 63.8 % of the town.

In parallel, a flow and pressure measurement campaign was carried out in the water supply network during the period between 20 and 27 May, 2009. Flow meters were installed at the outgoing pipes of the three water reservoirs that supply the whole town. Pressure measurement was carried out at 7 points of the water supply network. The results of the pressure measurements show that the average pressure was sufficient in all parts of the network. The maximum recommended pressure of 6.0 bars was exceeded at 43 % of the locations by up to 42 %. The minimum pressure did not drop below the recommended value of 1.0 bar.

Figure 3.5 presents a summary of the simultaneous measurement of wastewater flow and potable water flow. As mentioned, the water flow measurement encompassed the entire territory of the town, while the wastewater flow measurement covered only 63.8 % of it. Nevertheless the comparison between the two curves is stark. The waste water flow is about twice the potable water flow, and therefore, on an area basis, it is 3–4 times higher than the water supply flow.

At MP1 the typical morning and evening peak flows can not be seen; only a slight decrease of the relatively constant flows is noticeable between 1:00 and 6:00 a.m., when the average value was 127 l/s. At MP2 the flow was also relatively constant, with a slight increase between 10:00 and 14:00; the average flow during night-time, between 01:00 and 06:00, was 15.8 l/s. Table 3.2 presents the observed characteristic flows of wastewater during the recorded 24 h.

The infiltration flow (Q_{inf}) was calculated using the approaches described in Sect. 3.

Approach 1

The domestic wastewater flow was calculated based on potable water consumption per capita and the served population in the corresponding sub-catchment area. The last two parameters were calculated based on a study of water consumption, the average occupation rate of households, and the seasonal occupation rates of the

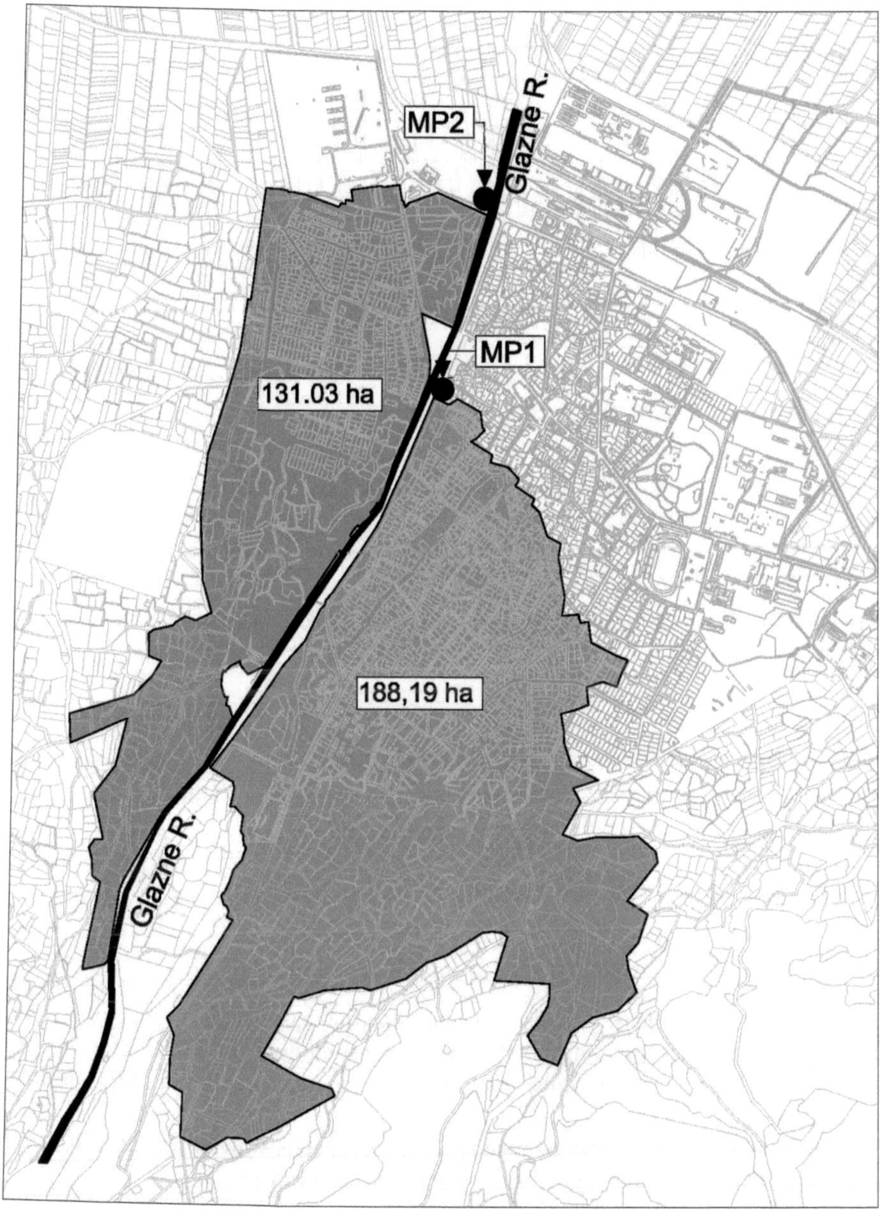

Fig. 3.4 Location of the sewer flow measuring points in Bansko

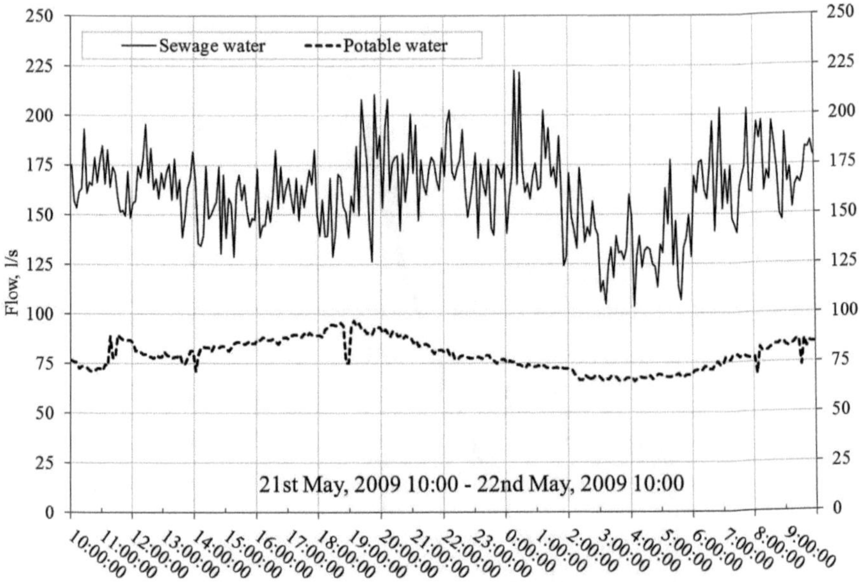

Fig. 3.5 Comparison between total measured wastewater and total measured potable water flow in Bansko

Table 3.2 Measured characteristic wastewater flows in Bansko

Parameter	MP1	MP2
Dry-weather peak flow	203.5 l/s at 00:25 a.m;	30.2 l/s at 12:14 a.m;
Minimum flow	90.0 l/s at 04:05 p.m	11.5 l/s at 16:25 p.m;
Average flow	142.5 l/s	18.6 l/s

holiday flats and hotels. It was also assumed that the wastewater discharge is equal to 90 % of the corresponding potable water consumption. Table 3.3 presents the results from the calculation of the domestic flows and the prediction of the infiltration flows.

Based on these figures the average specific extraneous water flow rates, R_{inf}, for the connected sub-catchment areas are calculated as follows:

- At MP1: $R_{inf(MP1)} = Q_{inf(MP1)} / A_{MP1} = 0.69$ l/(s · ha)
- At MP2: $R_{inf(MP2)} = Q_{inf(MP2)} / A_{MP2} = 0.079$ l/(s · ha).

Although both drainage areas are comparable, the calculated infiltration flows differ substantially because the natural drainage conditions and the structural condition of the collectors also differ. The aquifer gradient goes in a northeast direction, so on the left side of the river (the western part) groundwater can easily drain directly into the river, whereas on the right side (the eastern area) the ground water

Table 3.3 Determination of municipal flows and infiltration flows at MP1 and MP2 in Bansko

	Connected people	Potable water norm	WW return ratio	Wastewater flow
	nr	l/ca/d	%	m^3/d
MP1				
Form residential areas, incl. holiday flats	4,960	110	90	491
From hotels	2,912	200	90	524
Total calculated domestic waste water				**1,015**
Average measured waste water flow				**12,312**
Resulting infiltration flow				**11,297**
MP2				
Form residential areas, incl. holiday flats	5,195	110	90	514
From hotels	1,112	200	90	200
Total calculated domestic waste water				**714**
Average measured waste water flow				**1,607**
Resulting infiltration flow				**893**

table goes in direction away from the river (Fig. 3.4). With these factors in mind, it can be assumed that most of the sewer network connected to MP2 is above the groundwater table, whereas most of the collectors connected to MP1 are within the aquifer. It should also be noted that the sewer network in the western part was built about 10 years ago, while most of the network connected to MP1 is more than 40 years old.

Approach 2

The infiltration flow is calculated based on the minimum wastewater flow registered during the night. The measured minimum flows for both water supply and sewerage systems occurred between 3:10 and 5:35 a.m. The calculated average minimum potable water flow during this period was 68 l/s, while the measured instantaneous minimum flow of 66 l/s occurred at 4:05 a.m. If it is assumed that the water demand at this time should be minimal (close to 0 l/s), then the loss due to leaks in the water supply network for the whole town would be around 66 l/s (5,700 m^3/day).

The total average minimum wastewater flow across both drainage areas (MP1 + MP2) during the same period was calculated as 132 l/s, since the minimum instantaneous flow, approximately 103 l/s, occurred also at 4:05 a.m. (Fig. 3.5). With the same assumption that there is not likely to be any significant domestic activity at this time of day, this leads to the conclusion that the measured flows are mostly due to extraneous water. It is possible that part of the extraneous water is actually leaking potable water. Assuming that all the leaking potable water enters the sewer system and that the leakages are equally distributed throughout the town, it means that the contribution of leaked potable water to the sewer flow, in the measured sub-catchment areas, is around 30 %. It appears that ground water (including water from wells and springs within the town) makes the major contribution to the extraneous water flow. Based on both approaches presented in the methodology, the extraneous water was quantified as follows:

- 12,190 m^3/d of extraneous water, calculated on average daily measured flows and calculated domestic sewage flows;
- 11,405 m^3/d of extraneous water, calculated on average minimum flows during the night.

It should also be noted that in the sub-catchment area connected to MP2 (apart from the discharge where the flow meter was installed), there were two combined sewer overflows that have permanent flow to the Glazne River. These flows were measured for approximately 1 h during the measurement campaign. They were also analytically calculated using Poleni's formula (weir-height and length and estimated height of overflow) during the initial on-site survey along the river banks. The average total flow of both outfalls was determined to be 11.5 l/s. The water from the flows is most probably underground water, which adds 994 m^3/d more extraneous water to the analysed sub-catchment area.

4.2.4 Wastewater Quality Measurement Survey

The results from the wastewater quality analyses during the measurement campaign are presented in Fig. 3.6. For all 8 samples, the values for COD and BOD_5 over the 24 h were below the norm for treated water, and only one value of SS was above the normative value. The results for TN were all below 12 mg/l and for TP below 1 mg/l.

Wastewater quality at 4 sewer network outlets is also monitored within the regular state monitoring program, an activity performed by the Regional Inspectorate on Environment and Water (RIEW) once every 6 months. The outlets are located along the Glazne River in the central and northeast part of the town. The samples are grab samples taken randomly during the day from some of the outlets. In total, 14 samples were collected during the period 2004–08. Samples from all 4 outlets were taken on the same day only on 17 January 2007. Figure 3.7 presents the results of COD and SS.

As shown in Fig. 3.7, the concentrations of both COD and SS were low compared to those of typical municipal wastewater. Indeed, about half the samples show values even lower than the normative requirements for treated wastewater.

All the available quality analyses confirm the results from the wastewater flow measurements: that is, the sewer network acts mostly as a drainage system for groundwater. As a result, dilution is so high that during certain periods of the year there is no need for wastewater treatment. Such periods usually occur when there is no large tourist influx (e.g., during spring, summer, and autumn) and the ground water table is high enough to enter the sewer network (usually in spring and autumn). Nevertheless, infiltration of groundwater into the sewer network, or exfiltration of sewer water into ground water, cannot be considered a sustainable solution in line with the priorities of the EU Water Framework Directive 2000/60/EC for protection and management of groundwater (EU 2000).

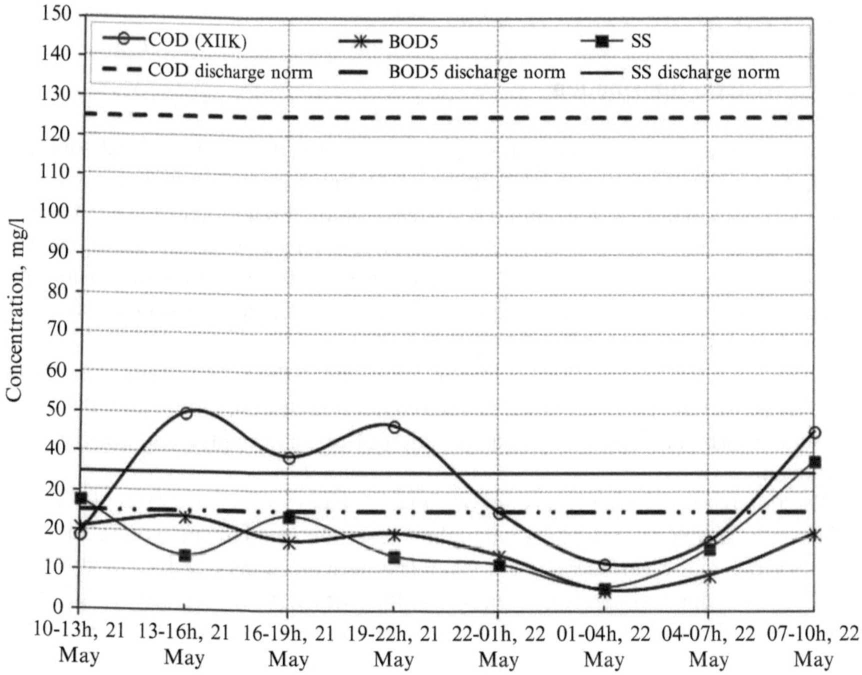

Fig. 3.6 Results of the wastewater quality monitoring during the flow measurement survey in Bansko

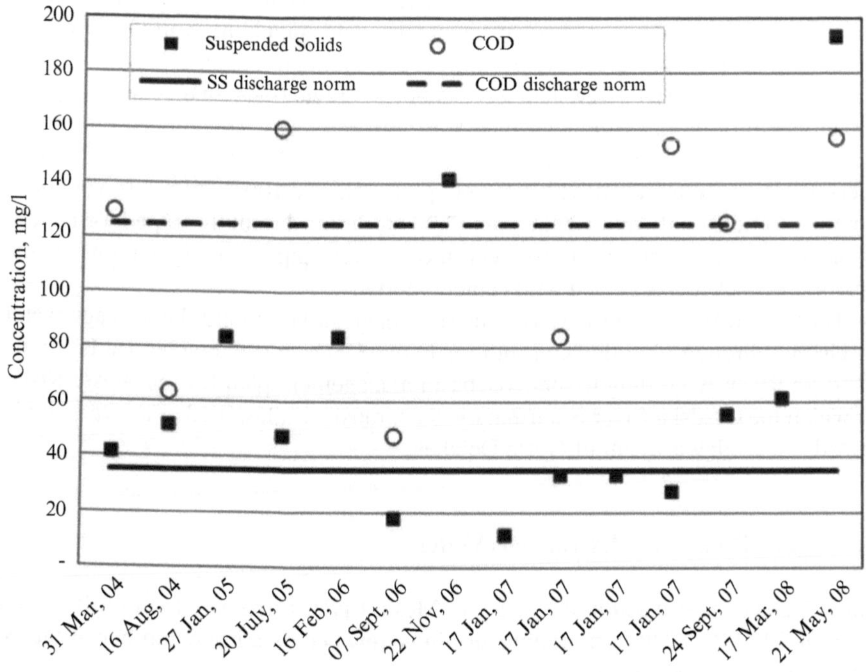

Fig. 3.7 Quality of sewer wastewater in Bansko (Data source: RIEW)

4.3 Case Study: Gotse Delchev

4.3.1 General Information

Gotse Delchev, with the Gradska River running through its western and southern parts, is also located in South West Bulgaria, 20 km from the frontier with Greece. Due to a combination of rich natural resources, a skilled workforce, and advances in restructuring the local economy it has good potential for economic development. The policies of municipal authorities have attracted foreign investors, who have created a growing business environment for industries and medium enterprises such as textiles, clothing, and shoe-making.

There are more than 19,000 permanent residents in the town as registered in the national census of 2011. The town is not a major tourist destination, although there are some tourist activities. Seasonal variation of resident numbers is therefore not significant.

The potable water supply system of Gotse Delchev, piped to 100 % of the residents, includes one treatment plant and three reservoirs, and the network is fed from three springs and a surface water intake. The network, operated by gravity, has a total length of approximately 63 km. About 45 % of the network piping is AC, 35 % steel, 4 % galvanised iron, and 16 % PE, with diameters varying between 60 and 400 mm. The PE pipes were installed in recent years to replace damaged AC or steel pipes. The age of the pipes can be divided into younger than 10 years (16.3 %) and older than 40 years (83.7 %). The many old pipes in the network results in abundant leakages and frequent incidents (Fig. 3.8). The replacement of some old distribution pipes has led to decreases in accidents along the water supply network; however the number of accidents in house connections is continuing to increase.

The existing wastewater system is a combined, gravitational one, with an approximate length of 66 km which was laid some 40 years ago. The connection rate of households to the sewerage system is presently 98 %. More than 90 % of the network has been constructed with concrete pipes. Nearly 100 % of the sewers have circular sections; only a few collectors (2.2 km) are oval, mouth-shaped, or rectangular in section. Connections between the concrete pipes are rigid joints of the spigot-and-socket type sealed with cement mortar.

There is no WWTP in the town. The river bank survey found that sewage water discharges through 12 outfalls, in addition to two CSOs, located on both sides of the Gradska River. According to the river basin management plan for the Mesta River, of which the Gradska River is a tributary, the Gradska is classified as a "water body in bad status" downstream of Gotse Delchev.

4.3.2 The Problem of Extraneous Water

Gotse Delchev is situated in the valley of Mesta River, surrounded with hills rising to around 1,000 m. The surrounding terrain is trenched by a large number of faults, where mineral springs occur.

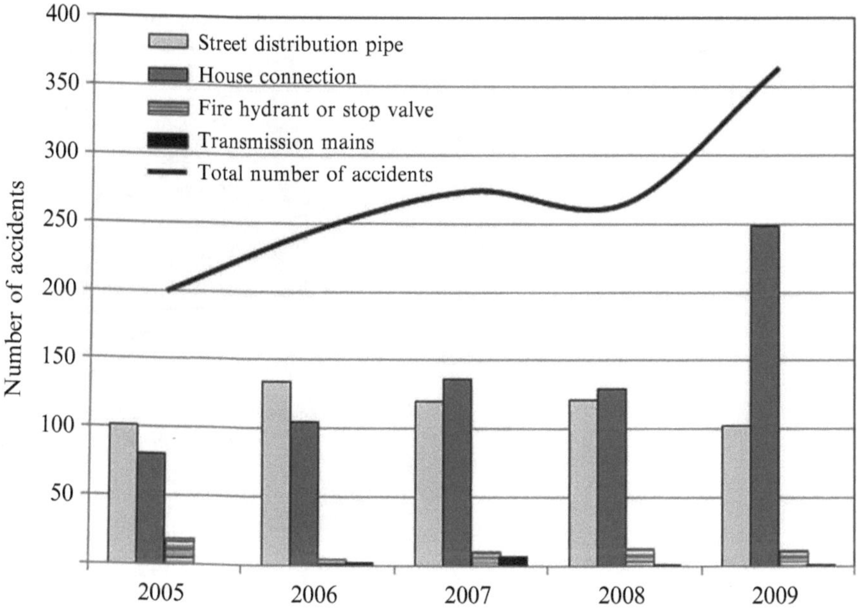

Fig. 3.8 Number of accidents along the water distribution network in Gotse Delchev

The manhole survey covered 271 manholes (30 % of the total). Evidence of significant amounts of clear, not foul-smelling wastewater was found in 77 manholes, mainly in the sewer collectors close to the river. 30 sewer sections with length between 30 and 110 m (in total 1 km length) were selected for inspection with CCTV camera. Only 9 sewer sections however were inspected completely, while in the rest 21 sections the camera couldn't reach the end due to impassable obstacle. The average distance, within a section, at which the CCTV camera detected a problem was 1.20 m, while the longest distance without observable defects was 17.40 m. The CCTV inspection showed that the sewer collectors were in very poor condition, resulting in partly broken or cracked cement pipes or offset joints, as well as significantly reduced hydraulic capacity (more than 50 %) due to deposits of sand, stones, bricks, etc. (Fig. 3.9). It was also found that, over time, the cement mortar that sealed the joints had washed off and there was actually no waterproof connection between the concrete pipes.

4.3.3 Wastewater Flow Measurement Survey

A wastewater flow measuring campaign was carried out, in dry weather conditions, from Wednesday 6 to Thursday 7 May 2009. The methodology and the flow measuring device were similar to the Bansko case study

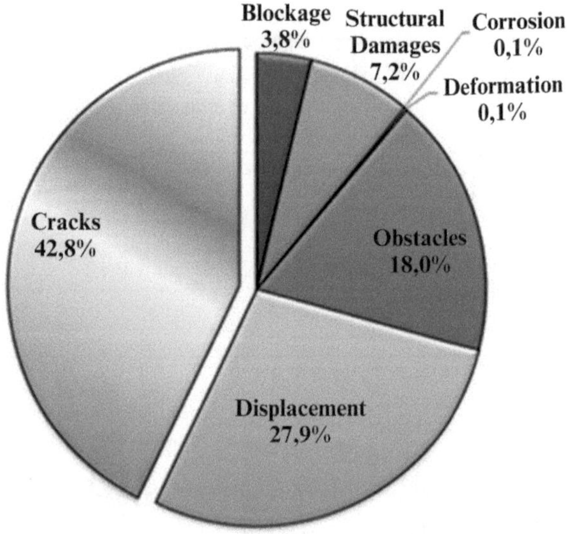

Fig. 3.9 Classification of the pipe damages in the sewer network in Gotse Delchev

A single measuring point was selected for the survey (Fig. 3.10). The flow meter was installed at the outlet of the main sewer collector on the left bank of the Gradska River (an oval pipe with dimensions of 1,800 by 1,200 mm). The total connected area at the measuring point (MP) was 184.12 ha; since the total area of Gotse Delchev is 301 ha, the measuring point represents 61 % of the town's drainage area. In parallel with the wastewater measurements, a flow and pressure measurement campaign was carried out in the water supply network from 30 April to 7 May 2009. Flow meters were installed at the outgoing pipes of the three water supply reservoirs. Pressure measurement was carried out in 8 points within the potable water network. The results show that the average pressure is adequate in all parts of the network. The maximum recommended pressure of 6.0 bars was exceeded (but by no more than 15 %) at 50 % of the locations. The minimum pressure never dropped below the recommended value of 1.0 bar.

Figure 3.11 presents a summary of the simultaneous measurement of wastewater flow and potable water flow. Here it should be noted that the potable water measurements represent flow to the whole town. The observed dry-weather peak flow was 192 l/s at 12:55 a.m., whereas the lowest observed flow was 80 l/s at 03:50 a.m. The calculated average daily flow was 107 l/s. A short rainfall event occurred on the morning of 6 May, which resulted in a peak in wastewater flow observed between 12:30 and 13:30. For the rest of the time the wastewater flow followed a stable trend. For the purposes of this investigation, the most important values are the average and the night-time minimum wastewater flow. Since the influence of the peak flow on the average values is not significant, the authors consider that the flow measurement results are representative. The extraneous water flow Q_{inf} was calculated using the same approach as was used for Bansko.

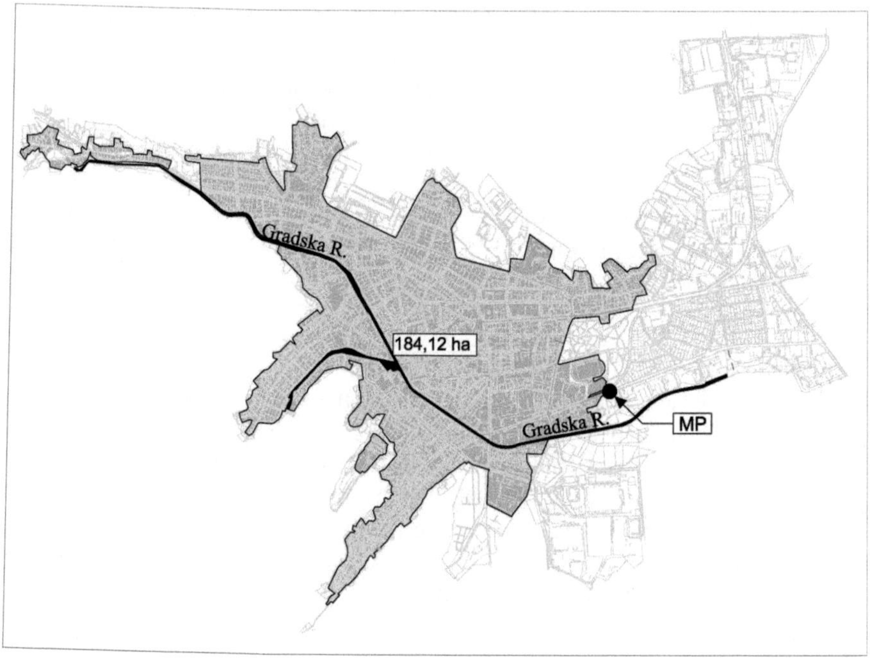

Fig. 3.10 Location of the sewer flow measuring point in Gotse Delchev

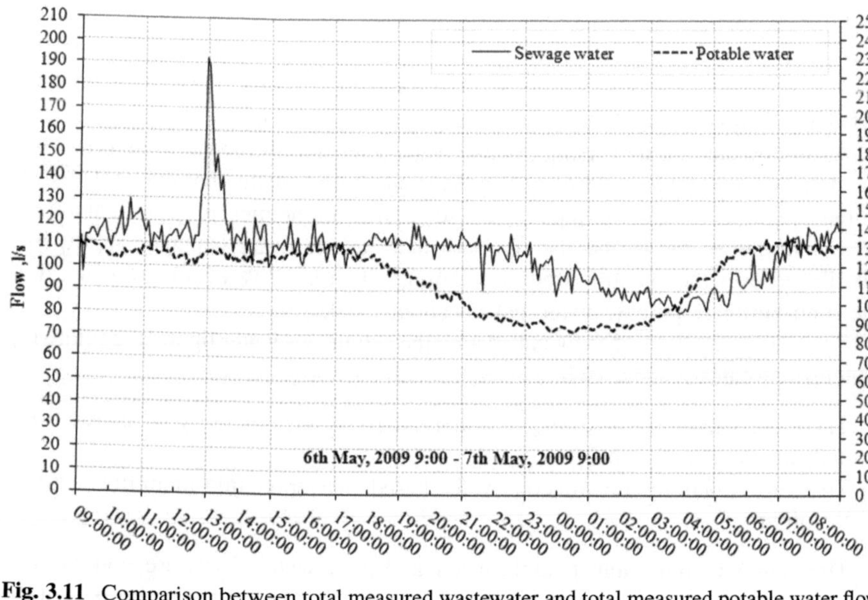

Fig. 3.11 Comparison between total measured wastewater and total measured potable water flow in Gotse Delchev

Table 3.4 Determination of municipal flows and infiltration flows at MP1 in Gotse Delchev

	Connected people	Potable water norm	WW return ratio	Wastewater flow
	Nr	l/ca/d	%	m^3/d
Form residential areas	14,780	110	90	**1,463**
From industrial units				**140**
Total calculated municipal waste water				**1,603**
Average measured waste water flow				**9,245**
Resulting infiltration flow				**7,642**

Approach 1

The domestic sewage flow was calculated using the same methodology described previously. The industrial effluent flows were calculated based on the water consumption of the corresponding plants in the sub-catchment area, assuming a 100 % discharge rate. Table 3.4 presents the results, at the measuring point, of the calculation of domestic sewage flows and estimates of infiltration flows in the sub-catchment area. Based on these figures, the average rate of extraneous water flow Q_{inf} for the measured area is 0.48 l/(s · ha).

Approach 2

The minimum potable water flow occurred between 23:35 and 2:30 with an average value of 88.1 l/s and a transitory minimum of 85.4 l/s at 00:30. If it is assumed that the water demand at this time is close to 0 l/s, this leads to the conclusion that the losses due to leaks in the water supply network for the whole town are around 88 l/s (7,603 m^3/day). The period of minimum sewage flow is typically delayed in time and occurs between 2:55 and 5:25, with an average value of 85.8 l/s (7,413 m^3/d) and instantaneous minimum of 80.5 l/s at 3:50. Using the same assumption that there is not likely to be any significant domestic or industrial activity at this time, the measured flows are therefore considered as extraneous water.

Assuming that all the leaking potable water goes into the sewer system and that the potable water leakages are equally distributed, this means that the contribution of the leaking potable water pipes to the sewer flow in the measured sub-catchment area is around 63 %. If one also assumes that potable water leakages are equally distributed over the encompassed population, which is 77 % of all residents, then the contribution of potable water leaks is about 79 %.

Based on both approaches for calculating the extraneous water figure, the following conclusions can be made.

- There is 7,642 m^3/d of extraneous water, based on average daily measured flows and calculated municipal flows;
- There is 7,413 m^3/d of extraneous water, based on average minimum flows during the night.

The values of minimum potable water and wastewater flows are comparable, which prompts the conclusion that potable water leakages form the major part of the extraneous water in the sewer system.

4.3.4 Wastewater Quality Measurement Survey

The results from the wastewater quality analyses during the measurement campaign are presented in Fig. 3.12. The values for TN ranged from 11 to 24 mg/l (average 16 mg/l), while the values for TP ranged from 0.9 to 2.4 mg/l (average 1.5 mg/l).

The wastewater quality analyses show that the concentrations of the mentioned parameters are relatively low. During the night hours the values were well below the discharge norms for treated wastewater. The discharge from one sewer collector in Gotse Delchev was monitored within the regular state monitoring program, which is performed by RIOSV Blagoevgrad once every 6 months. Figure 3.13 presents the results of BOD_5 and SS obtained from campaigns taking place during the period 2004–08. The sampling methodology is similar to the one described for Bansko. The values for COD and BOD_5 are comparable with the ones recorded during the 24-h measuring campaign (Fig. 3.12). The concentrations are lower than the values typical for a municipal sewerage system. The quality analyses confirm the conclusion that the wastewater is much diluted and not typical of municipal sewage.

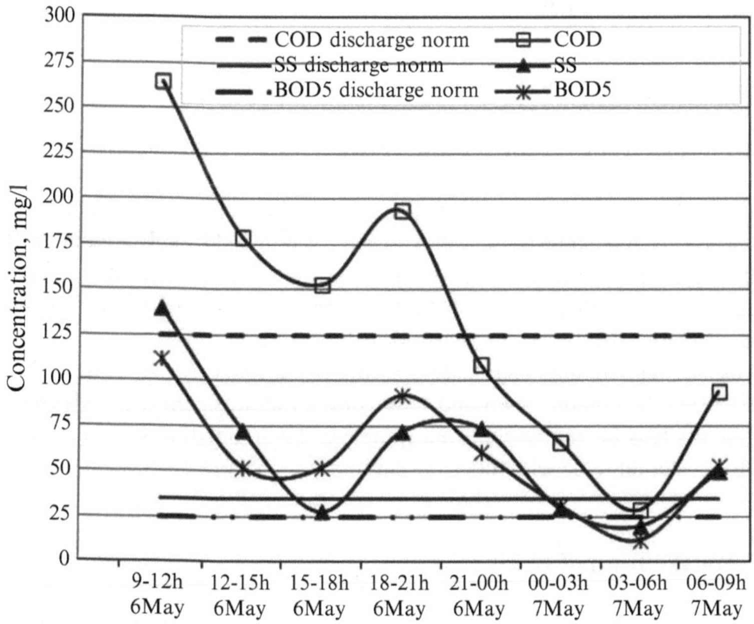

Fig. 3.12 Results of the wastewater quality monitoring during the flow measurement survey in Gotse Delchev

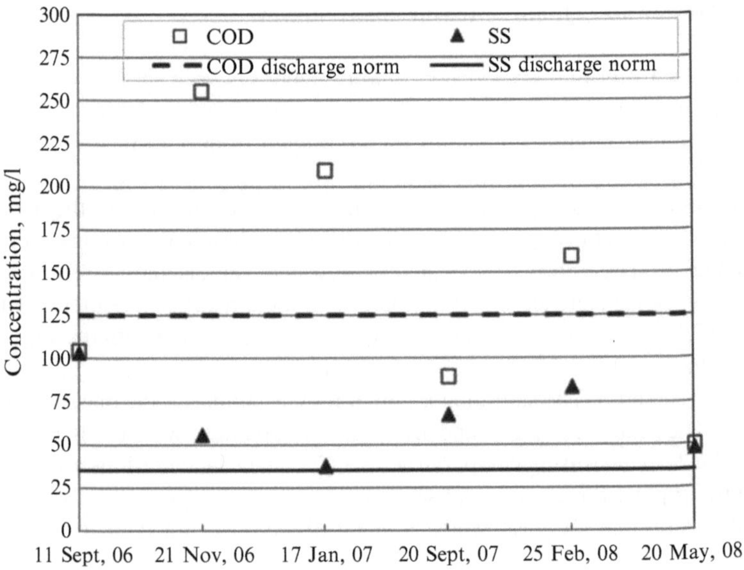

Fig. 3.13 Quality of sewer wastewater in Gotse Delchev (Data source: RIEW)

4.4 Recommendations for Sustainable Management in the Case Studies

Both case studies indicate that extraneous water represents a major part – 80 to 83 % – of the sewage flow, which means that the combined sewerage system is actually operating as a drainage system for extraneous water. Although both towns have a similar problem, in Bansko ground water infiltration is the major cause due to a naturally high aquifer level, while in Gotse Delchev the extraneous water appears to come mostly from leakage of the water supply system and its subsequent infiltration into sewers. The situation in both towns translates to unsustainable urban water management, seen either in environmental, economic, or social terms. The negative impacts of these aspects are summarised in Table 3.5.

This situation demands engineering solutions be found for rehabilitating and/or reconstructing the potable water and sewer infrastructures, solutions which are both economically feasible and socially acceptable and which will answer all the requirements for sustainable development.

4.4.1 For the Town of Bansko

Recommendations have been made to replace Bansko's 18.9 km of old water supply pipes with new ones in order to reduce potable water leakage from its current 47 % to 28 % in 2015. It has also been suggested that the existing

Table 3.5 Components of sustainable development that are negatively affected by the present state of potable water and sewer systems in the town of Bansko and Gotse Delchev

Component	Impact
Environmental component	Inefficient utilization of the natural water resource due to significant leakages in the potable water system;
	Pollution of groundwater due to high rate of infiltration/exfiltration in the sewer systems;
Economic component	Increased operation and maintenance costs for potable water treatment and transportation;
	Increased operation and maintenance costs for sewer system maintenance;
Social component	Continuously increasing price of water supply and sewage conveyance service

combined sewer network be transformed into a separated one by constructing a new separate network (25.2 km) for domestic and industrial urban wastewater. Another suggestion is to rehabilitate the existing combined sewer network (14.9 km, or around 22 % of the existing one) so that it can continue to operate as a drainage system for both ground water and storm water, since the groundwater needs to be drained anyway.

The suggested rehabilitation measures in the sewer network, combined with rehabilitation of the water supply network, is expected to reduce extraneous water flow to a target value of 0.05 l/(s·ha) in 2015. This value is recommended as a design parameter in the German standard DWA-A 118 (2006).

If the amount of extraneous water is decreased in this way, the design flow for the future WWTP would be reduced by about two-thirds. The concentration of organic matter in the wastewater will correspondingly increase, thus permitting the biological stage of the future municipal WWTP to operate effectively.

4.4.2 For the Town of Gotse Delchev

Rehabilitation of 22.5 km of water supply pipes has been suggested, a length that represents 36 % of the whole distribution network. This would reduce potable water losses from 73 % at the moment to 38 % in 2015. Due to financial limitations, further reduction of water losses will not be possible within the budget of the "Environment 2007–2013" operational program.

Concerning the sewer network, there have been suggestions for rehabilitation of the 11.3 km of existing collectors, construction of 11.4 km of new collectors, and construction of 7 CSOs. Such measures would affect 47 % of the whole network and, together with the rehabilitation measures in the potable water system, it is expected that they would reduce the current extraneous water rate to a target value of 0.05 l/(s·ha). In this way, the design flow for the future WWTP is reduced by almost 38 %. After implementation of these measures, all residents in both towns will be connected to a sewer system that includes wastewater treatment.

5 Conclusion

This chapter presents a practical, fast, and low cost approach for estimating extraneous water flows in existing combined municipal wastewater systems. The approach was applied in two case studies to determine the magnitude and origin of the extraneous water in the sewer system. The studies assisted the responsible authorities at municipal and state levels to select appropriate measures for solving the problem. Accordingly, the two towns will:

1. Reduce inefficiencies in water resources use through decreasing leakages in the potable water system;
2. Reduce the operation and maintenance costs of the water supply system;
3. Reduce significantly the hydraulic load in the sewerage system through preventing extraneous water infiltration;
4. Reduce underground water pollution through exfiltration from the sewerage system;
5. Reduce the capital cost for construction of future WWTPs;
6. Reduce the energy demand for pumping wastewater at the inlet to the WWTP;
7. Reduce the operational cost of the WWTPs and promote lower tariffs for wastewater treatment.

These achievements will contribute to a shift towards more sustainable urban water management in both towns.

Acknowledgements The authors would like to acknowledge the advice and guidance of the Bulgarian Ministry of Environment and Water (MoEW), the Regional Water Company (RWC) Blagoevgrad, the Municipality of Bansko, and the Municipality of Gotse Delchev. They thank these authorities for their valuable collaboration and support.

References

Aspergen, H., Hellström, B., & Olsson, G. (1997). The urban water system: a future Swedish perspective. *Water Science and Technology, 35*(9), 33–43.

ATV-DVWK. (2003). Fremdwassersituation in Deutschland (Situation of extraneous water in Germany). Report of working group ES-1.3, "Extraneous water". *KA Abwasser Abfall, 50*(1), 70–81.

Beck, M. B. (1996). Transient pollution events: Acute risks to the aquatic environment. *Water Science and Technology, 33*(2), 1–14.

Bosseler, B., Puhl, P., & Birkner, T. (n.d). *Reducing extraneous water: Detection and condition assessment of leaking private sewers*. Retrieved from http://www.ikt.de/down/english/ReducingExtraneousWater.pdf

Burian, S. J., Nix, S. J., Durrans, S. R., Pitt, R., Fan, C.-Y., & Field, R. (1999). *The historical development of wet-weather flow management*. EPA/600/JA-99/275.

Burian, S. J., Nix, S. J., Pitt, R. E., & Durrans, S. R. (2000). Urban wastewater management in the United States: Past, present, and future. *Journal of Urban Technology, 7*(3), 33–62.

Butler, D., & Davies, J. W. (2000). *Urban drainage*. London: E. & F. N. Spon.

Butler, D., & Parkinson, J. (1997). Towards sustainable urban drainage. *Water Science and Technology, 35*(9), 53–63.

CEC. (2001). *A sustainable Europe for a better world: A European Union strategy for sustainable development. COM (2001) 264*. Brussels: Commission of European Communities.

CEC. (2010). *Europe 2020: A strategy for smart, sustainable and inclusive growth. COM/2010/2020*. Brussels: Commission of European Communities.

CEC. (2011a). *A Resource-efficient Europe: Flagship initiative under the Europe 2020 Strategy*. Brussels: Commission of European Communities. Retrieved from http://ec.europa.eu/resource-efficient-europe/pdf/resource_efficient_europe_en.pdf

CEC. (2011b). *Innovation for a sustainable future: The eco-innovation action plan (Eco-AP)*. Brussels: Commission of European Communities. Retrieved from http://ec.europa.eu/environment/etap/inaction/pdfs/1_en_act_part1_v5.pdf

Chung, G., Lansey, K., Blowers, P., Brooks, P., Ela, W., Stewart, S., & Wilson, P. (2008). A general water supply planning model: Evaluation of decentralized treatment. *Environmental Modelling & Software, 23*(7), 893–905.

De Toffol, S., Engelhard, C., & Rauch, W. (2007). Combined sewer system versus separate system: A comparison of ecological and economic performance indicators. *Water Science and Technology, 55*(4), 255–264.

DIN. (1999). *DIN 4045 Abwassertechnik: Begriffe* (Wastewater engineering vocabulary). Berlin: Beuth Verlag GmbH.

DIN. (2008). *DIN EN 752:2008–04 Entwässerungssysteme außerhalb von Gebäuden*. Berlin: Beuth Verlag GmbH.

DWA-A 118. (2006). *Hydraulische Bemessung und Nachweis von Entwässerungssystemen*. Hennef: Deutsche Vereinigung für Wasserwirtschaft, Abwasser und Abfall.

Eiswirth, M., & Hoetzl, H. (1997). The impact of leaking sewers on urban groundwater. In: *Groundwater in the urban environment*. Proceedings of the 27th IAH congress, Nottingham, pp. 399–404.

EU. (1991). Council Directive of May 1991 concerning urban waste water treatment. *Official Journal of the European Communities*, L135/40.

EU. (2000). Directive 2000/60/EC of the European Parliament and of the Council of 23 October 2000 establishing a framework for community action in the field of water policy. *Official Journal of the European Communities*, L327/1.

Harremoës, P. (1997). Integrated water and waste water management. *Water Science and Technology, 35*(9), 11–20.

Hellström, D., Jeppson, U., & Kärrman, E. (2000). A framework for system analysis of sustainable urban water management. *Environmental Impact Assessment Review, 20*, 311–321.

Larsen, T., & Gujer, W. (1997). The concept of sustainable urban water management. *Water Science and Technology, 35*(9), 1–2.

MOEW. (2011). *Technical assistance for project preparation in the water sector: Asenovgrad, Bansko and Gotse Delchev*. EUROPEAID/124486/D/SV/BG, Feasibility Studies. Ministry of Environment and Water, Bulgaria.

Orth, H. (2007). Centralized versus decentralized wastewater systems? *Water Science and Technology, 56*(5), 259–266.

Pfeiff, H. S. (1989). Das problem "Fremdwasser" (the issue of external water). *Korrespondenz Abwasser, 36*(4), 471–481.

Schladweiler J. C. (2002). *Tracking down the roots of our sanitary sewers*. Retrieved from http://www.sewerhistory.org/chronos/roots.htm

Sieker, F. (2003). Regen(ab)wasser and Misch(ab)wasser, eine vernachlässigte Verschmutzquelle? *gwf Wasser Abwasser, 144*(9), 614–617.

Thorsten, F. (2006). *Spatial classification methods for efficient infiltration measurements and transfer of measuring results*. Dissertation, Technical University Dresden.

United Nations. (1987). *Report of the world commission on environment and development*. A/RES/42/187, 9th plenary meeting. Retrieved from http://www.un.org/documents/ga/res/42/ares42-187.htm

United Nations. (1992). *Report of the united nations conference on environment and development, Rio de Janeiro*. A/CONF.151/26 (Vol. 1). Retrieved from http://www.un.org/esa/dsd/agenda21/

US EPA. (1999). *Combined sewer overflow management fact sheet. Sewer separation.* EPA 832-F-99-041. Columbus. Retrieved from http://water.epa.gov/scitech/wastetech/upload/2002_06_28_mtb_sepa.pdf

Vipulanandan, C., & Ozgurel Gurkan, H. (2004). *Methods to control leaks in sewer collection systems.* White paper prepared by Center for Innovative Grouting Materials and Technology (CIGMAT), University of Houston, Texas.

Chapter 4
Challenges for Water Supply and Sanitation in Developing Countries: Case Studies from Zimbabwe

Innocent Nhapi

1 Introduction

Many developing countries are not on course to meet the Millennium Development Goals for water supply and sanitation (WHO/UNICEF 2012). There continues to be disparities in urban and rural services and also between water supply and sanitation (WSS). Water supply has continued to receive preferred attention compared to sanitation. WHO/UNICEF (2012) figures show that access to safe water supply in developing countries stands at about 86 % compared to 99 % in developed countries. Poor water quality has resulted in many waterborne disease outbreaks in developing countries, such as cholera, dysentery, and typhoid. For sanitation, the situation is critical, with coverage figures at 56 % for developing countries versus 95 % in developed countries (WHO/UNICEF 2012). Poor sanitation, unsafe water supplies, and unhygienic practices can cause the transmission of a wide range of diseases (especially diarrheal diseases, skin and eye diseases, and worm infestations). Poor water and sanitation also worsen malnutrition, which in turns leads to stunting, lower school and work productivity, and impaired cognitive function and learning capacity.

Using UNDP (2013) figures, the world population has increased by an average 1.3 % per annum since 1990 and currently stands at about 7 billion. The Sub-Saharan Africa (SSA) population has also increased by an average annual rate of

I. Nhapi (✉)
Department of Environmental Engineering, Chinhoyi University of Technology, Private Bag 7724, Chinhoyi, Zimbabwe
e-mail: I_nhapi@yahoo.com

Q. Grafton et al. (eds.), *Understanding and Managing Urban Water in Transition*, Global Issues in Water Policy 15,
DOI 10.1007/978-94-017-9801-3_4

2.6 % since 1990 and currently stands at about 860 million. Global urbanisation has increased from 43 % in 1990 to 51 % in 2010 and the rising trend is expected to continue as poor countries industrialise and floods and droughts affect rural communities. The SSA region is urbanising at a rate of 4.0 % per annum. This urbanisation, which is highest in developing countries, has led to the mushrooming of informal settlements where water supply and sanitation services are almost non-existent. It has also placed serious pressures on ageing water and sanitation infrastructure in existing high density and low income areas.

This chapter looks at the broader challenges in water supply and sanitation in developing countries, focusing particularly on the SSA experience and using Zimbabwe as a case study. The reason for choosing SSA instead of the whole of Africa is that the other North African countries (Egypt, Libya, Morocco, Tunisia, Algeria) generally show a substantially different picture of development indicators (WHO/UNICEF 2013; UNDP 2013). Including these countries would distort the indicators used in this chapter. The distinction also enables regional comparisons, as most UN and other multilateral reports always distinguish between North Africa and SSA. The chapter starts by outlining the general water supply and sanitation issues, covering both domestic and productive water, and how this relates to environmental and health issues in SSA. It then uses Zimbabwe as a case study for going into further details and introducing the institutional responses. It also looks at the urban and rural aspects of water supply and sanitation services. This work is based on a review of the literature, practical experience, and personal knowledge of the water and sanitation sector in Zimbabwe.

Figure 4.1 shows a map of the Southern Africa Development Community (SADC) countries. The SADC is a Regional Economic Community in Africa comprising 15 member states: Angola, Botswana, Democratic Republic of Congo, Lesotho, Madagascar, Malawi, Mauritius, Mozambique, Namibia, Seychelles, South Africa, Swaziland, Tanzania, Zambia, and Zimbabwe. Established in 1992, SADC focuses on regional integration and poverty eradication within Southern Africa through economic development and ensuring peace and security. Zimbabwe was once a success story in SADC in terms of water supply and sanitation, with almost complete coverage in urban areas and above 60 % coverage in rural areas (WHO/UNICEF 2013). This scenario changed at the turn of the century when a political crisis affected the economy resulting in annual inflation soaring to billion percentage levels and economic collapse. A great cholera outbreak in 2008/09 attracted the attention of international donors, leading to a current recovery. This chapter discusses the water and sanitation sector in Zimbabwe in more detail, including a background to the cholera outbreak, water supply and sanitation challenges in rural and urban areas, institutional responses, and financial issues.

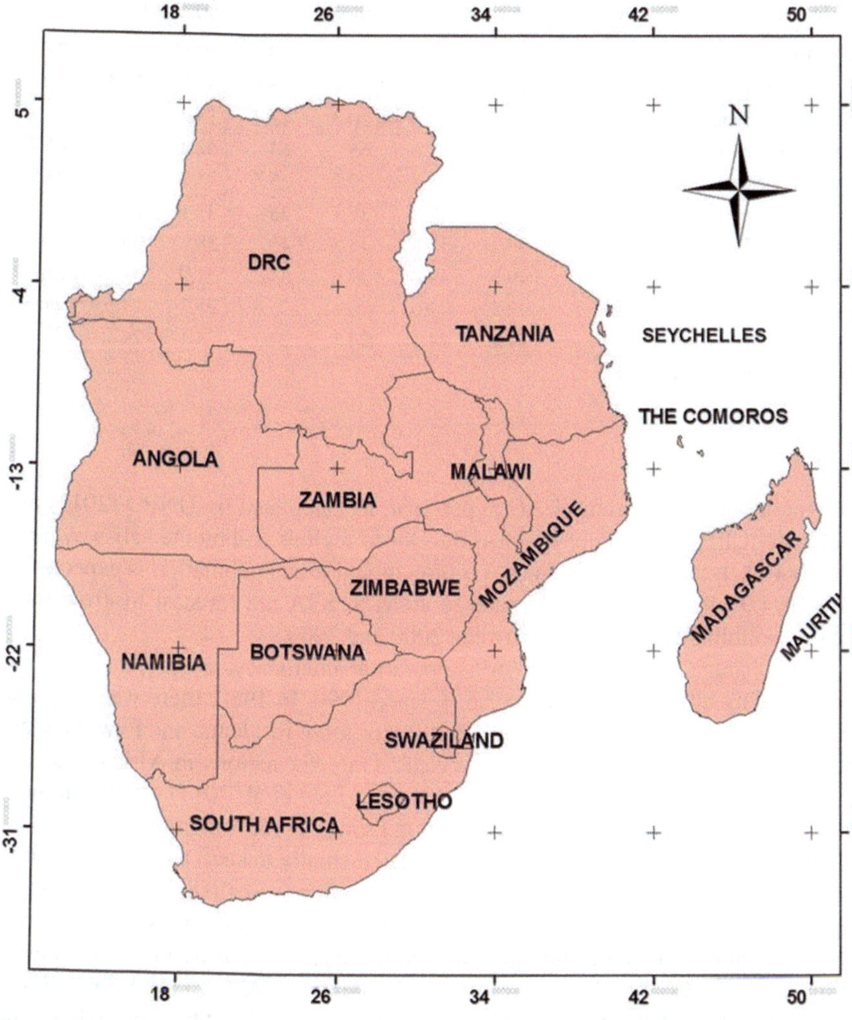

Fig. 4.1 Location of Zimbabwe within the Southern Africa Development Community (Map by W. Gumindoga)

2 Challenges to Water Supply and Sanitation in Sub-Saharan Africa

2.1 *The Challenge of Urbanisation and Slum Developments*

Like other developing countries, African countries are urbanising very quickly. Some 49 % of the world's population is estimated to be living in rural areas; in SSA, it is estimated that about 37 % of the population live in cities and these cities are

Table 4.1 Percentage of slum dwellers in urban population by region, 1990–2012

Major region or area	1990	1995	2000	2005	2007	2010	2012
Developing regions	46.2	42.9	39.4	35.6	34.3	32.6	32.7
Northern Africa	34.4	28.3	20.3	13.4	13.4	13.3	13.3
Sub-Saharan Africa	**70**	**67.6**	**65**	**63**	**62.4**	**61.7**	**61.7**
Latin America and the Caribbean	33.7	31.5	29.2	25.5	24.7	23.5	23.5
Eastern Asia	43.7	40.6	37.4	33	31.1	28.2	28.2
Southern Asia	57.2	51.6	45.8	40	38	35	35
South-eastern Asia	49.5	44.8	39.6	34.2	31.9	31	31
Western Asia	22.5	21.6	20.6	25.8	25.2	24.6	24.6
Oceania	24.1	24.1	24.1	24.1	24.1	24.1	24.1

Source: UN-HABITAT (2012)

growing at an average rate of 2.6 % per year, as calculated by UNPD (2013) and WHO/UNICEF (2013). Globally, less developed regions will hit the half-way point later, but likely before 2020. The 3.3 billion global urban population is expected to grow to 4.9 billion by 2030. Urbanised areas in SSA are forecast to grow most rapidly, doubling in population between 2000 and 2030.

Sub-Saharan Africa is headed for a population emergency, as the population has been growing at an annual rate of 2.6 % since 1990. In 1960, there was only one SSA city (Johannesburg) which had 1 million or more residents, and now there are over 30 megacities, straining urban services. Only six nations in Africa had economic growth rates above 7 % in 2010 (Chuhan-Pole et al. 2011) – the minimum rate believed necessary to support population growth of this magnitude.

The rapid growth of African cities has failed dismally to keep pace with required urban shelter. This has given rise to the development of so-called 'peri-urban housing' or slums. From Table 4.1, it can be seen that urbanisation in SSA has become virtually synonymous with slum growth, and there is a similar trend in most developing countries. Common examples of slum settlements are found in Kenya, South Africa, Tanzania, Uganda, Zambia, and Zimbabwe (UN-HABITAT 2012). The battle to achieve the MDGs will therefore have to be waged in urban slums. This is due mainly to the failure of governments to cater for the influx of people from rural areas, amid growing poverty levels in many SSA countries. In other cases, wars and the HIV/Aids pandemic have displaced many people who have been forced to seek shelter and better opportunities in urban areas. The majority of slum dwellers in African cities are between the ages of 15 to 24 (Zulu et al. 2002). Table 4.1 shows a positive trend in the reduction of urban slum dwellers from 70 % in 1990 to about 62 % in 2012. However, the figures for SSA remain the highest compared to other regions and the downward trend could be reversed due to recent droughts and an economic recession mainly affecting rural communities.

2.2 Water Resources Challenge

There are basically two main challenges related to water resources management for urban areas. The first one is how to supply enough water of adequate quality to all, at all places, at all times, and at an affordable cost. The second and emerging challenge is related to adaptation and sufficient resilience against the extreme climatic conditions of floods and droughts. These challenges are compounded by rapid urbanisation and massive land use changes. Often, the floods that kill people in urban areas are strongly concentrated in areas with high population growth. The global energy crisis has resulted in many urban areas resorting to firewood for fuel, a situation that has resulted in massive deforestation in urban areas and surroundings. The increase in the use of firewood has resulted in some cities experiencing problems with sand and silt in sewers since households resort to sand for removing soot from their pots. Harare in Zimbabwe is typical of such a problem (see Fig. 4.2). Global warming and climate change are also altering temperatures, affecting evaporation and rainfall, and ultimately increasing the frequency and severity of floods and droughts.

The quest for better opportunities and jobs in towns and cities has, no doubt, exacerbated rural-to-urban migration, but global warming and climate change might also have been a factor. A lack of developed water supply infrastructure, such as dams and boreholes, in rural areas could motivate people to migrate from drought-prone areas into cities.

2.3 Water and Health

Water is responsible for spreading 80 % of diseases in Africa. The main causes of death in children under five are diarrhoea, malaria, and measles (Lopez and Mathers 2006). Malnutrition accounts for about one-third of the disease burden in low- and middle-income countries. The most effective management intervention is providing

Fig. 4.2 High levels of sand in sewers (*left*) can cause downstream problems at sewage treatment plants (*right*) (Photographs: City of Harare)

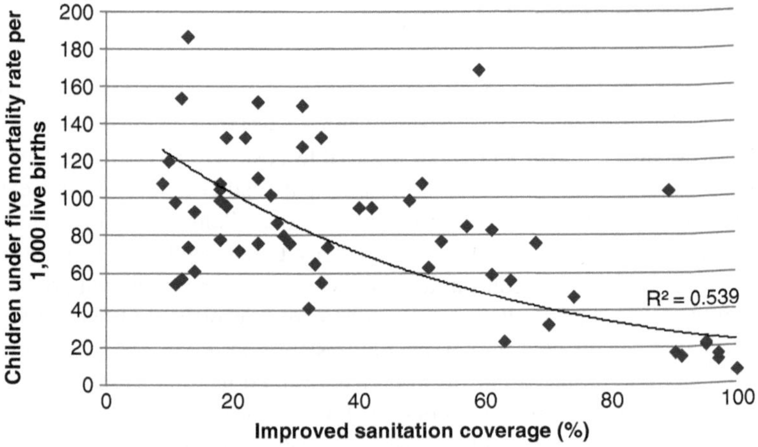

Fig. 4.3 Under-5 child mortality versus sanitation coverage for African countries. Each *dot* represents data from one country (Data source: WHO/UNICEF 2013)

safe drinking water and proper disposal of human waste (UNEP 2010). Figure 4.3 shows that there is a relationship between child mortality and access to improved sanitation ($R^2 = 0.54$). Although the graph does not show other factors that affect child mortality, such as general nutritional standards, access to health care and medical treatment, and so on, it is clear that there is a causal connection between access to sanitation and health. Additionally, improved sanitation will provide multiple other benefits: reducing the incidence of diarrhoeal diseases is important, especially for young children.

Poor sanitation, unsafe water supplies, and unhygienic practices lead to the transmission of a wide range of diseases, especially diarrheal diseases, skin and eye diseases, and worm infestations. Poor water and sanitation also leads to malnutrition, resulting in stunted growth, lower school and work productivity, and impaired cognitive function and learning capacity. The cost of not investing in water and sanitation also affects education, water resources, the environment, poverty reduction, tourism, and economic development. The impact of water and sanitation is often underestimated: the cost of not investing can be a cost to society of 2–5 % of GDP (Rees et al. 2012). Sanitation has therefore been described as "the greatest medical milestone since 1840" (British Medical Journal 2007).

3 Water Management and Challenges in Zimbabwe

3.1 Cholera Outbreaks in Zimbabwe

Zimbabwe has been sporadically affected by cholera outbreaks on an annual basis since 1998 (Red Cross 2010). The protracted 2008/09 cholera crisis was attributed to weakened public health and municipal services, with local authorities unable to provide adequate access to safe water, waste disposal, and sanitation. By the end of June 2009, 98,702 cases of cholera had been reported, with 4,282 deaths and a cumulative case fatality rate (CFR) of 4.3 %. According to WHO (2004), a CFR of <1 % from cholera is acceptable. Harare alone reported a total of 15,773 cases and 485 deaths, giving a CFR of 3.1 % (UNICEF 2010). Actually, it is possible that many more people might have died but were not recorded, especially in peri-urban and rural areas. The incidence of cholera in Zimbabwe during 2008/09 is shown in Fig. 4.4.

Partners in the Water, Sanitation and Hygiene (WASH) cluster estimated that, just before the cholera outbreak, 6 million people in Zimbabwe had limited or no access to safe water (UNICEF 2010). This was largely due to the unavailability of water treatment chemicals, irregular refuse collection, and inadequate sanitation facilities, as well as a lack of resources to repair damaged infrastructure. Leaking sewers in some urban residential areas and insufficient number of latrines in many rural areas resulted in unhygienic conditions and practices which led to the contamination of water sources, contributing significantly to the outbreak (Gwinji 2010). The situation was further exacerbated by the deteriorating socio-economic condi-

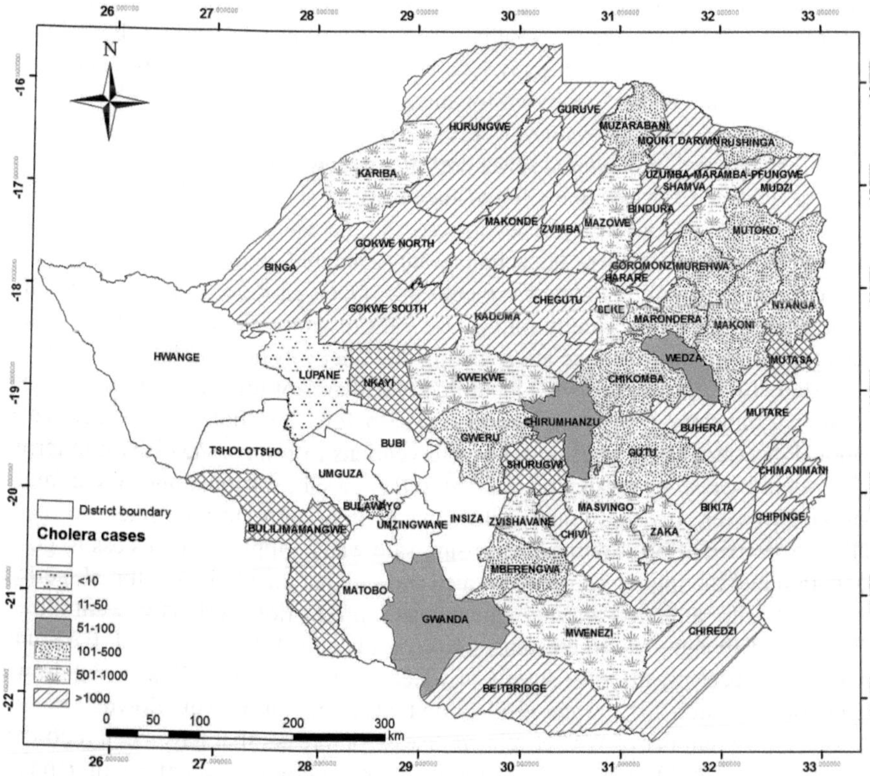

Fig. 4.4 Overview of the cholera outbreak in Zimbabwe, August 2008 to June 2009 (Map by W. Gumindoga based on data from UNICEF Zimbabwe)

Fig. 4.5 Use of unprotected shallow wells (*left*) and protracted sewer overflows (*right*) became very common, especially in high density suburbs, during the 2008 cholera outbreak in Zimbabwe (Photographs: I. Nhapi)

tions and food security in the country, a dire situation which called for major humanitarian intervention. The cholera outbreak in 2008 coincided with a period of breakdown of many systems, including the national financial system, hyperinflation, and little or no municipal service delivery (Red Cross 2010). Most residential areas in urban areas had to survive for many days or even months without water supply, and most sewerage systems were choked by reduced water inflows (UNICEF 2010). The blocked or collapsed sewers resulted in sewage flowing all over residential areas and the sewage flows found their way to shallow wells and boreholes (Fig. 4.5). Most people were forced to resort to the use of river water or shallow unprotected wells. With less household water available, personal hygiene greatly suffered and most public toilets were unusable. Open-air defecation became common, increasing the opportunity for disease transmission since even the basic practice of handwashing with soap after visiting the toilet was not possible.

A study by Kumwenda (2012) identified two phases of the 2008/09 cholera outbreak in Zimbabwe (Fig. 4.6). Period A was before the cholera was declared a national disaster and the start of donor interventions to control the disease in terms of preventative actions and treatment. Period B shows the response and containment period. However, this study makes clear that cholera outbreaks are likely to recur unless concerted efforts are made to ensure safe water supplies and access to good sanitation into the future. The analysis also showed that the cholera outbreak period coincided with a period during which water production seriously went down (Fig. 4.7). Although the circumstances have not been fully documented, the study noted that apparently after the supply of water treatment chemicals was withdrawn by donors in March 2012, water production by City of Harare went down.

There are a number of WASH-related disease outbreaks that have been recorded in Harare from 2000 to 2011, the main ones being cholera, dysentery, diarrhoea, and typhoid (Mokitimi 2012). These are waterborne diseases that are caused by

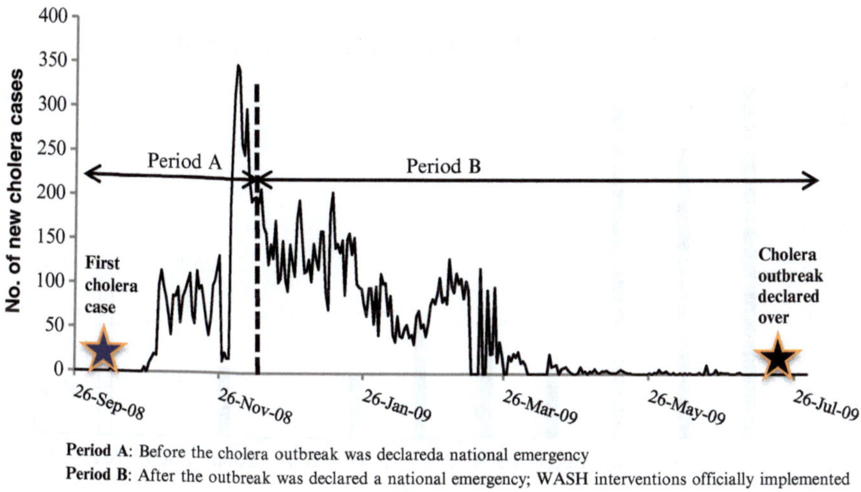

Fig. 4.6 Cholera cases from October 2008 to July 2009 (Source: Adapted from Kumwenda 2012)

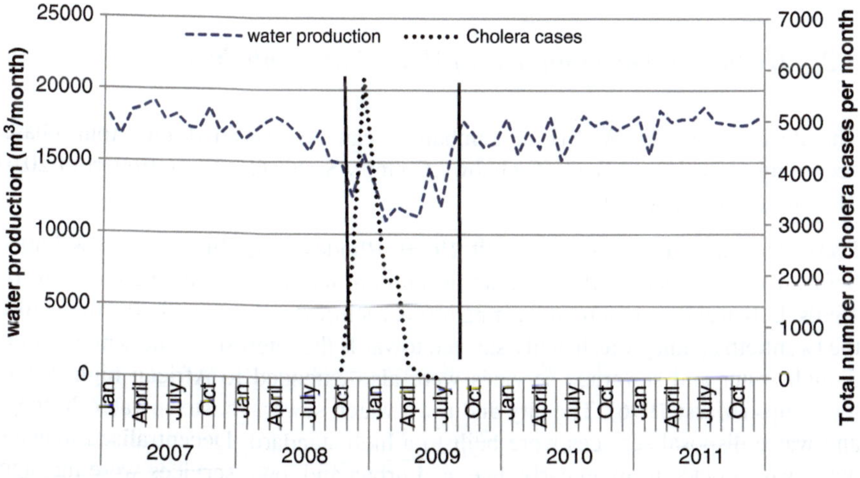

Fig. 4.7 Monthly water production (2007–11) and monthly cholera cases (2008–09) (Source: Adapted from Kumwenda 2012)

pathogenic microbes that can be directly spread through contaminated water. Dysentery rates appear to have been fairly consistent over the years, resulting in a number of fatalities since 2000 (Fig. 4.8). However, the cholera outbreak of 2008/09, discussed earlier, stands out as the most severe and well-known outbreak of waterborne disease in the country.

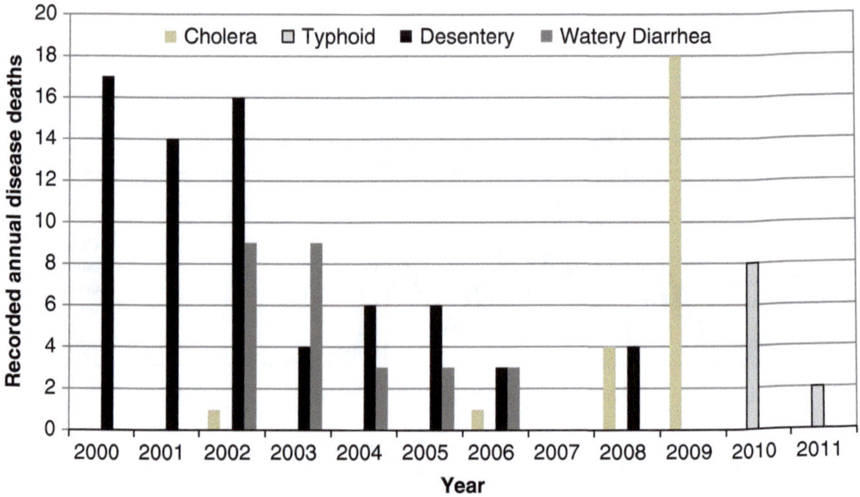

Fig. 4.8 WASH diseases annual total cases in Harare from 2000–11 (Source: Own elaboration based on data from Mokitimi 2012)

3.2 Historical Development of WASH in Zimbabwe

The development of WASH in Zimbabwe can be classified into four phases according to the World Bank Zimbabwe Country Status Overview Report of 2010. The phases are elaborated below.

Phase I: Rhodesian Legacy (Until 1979) Rhodesia, as Zimbabwe was known before independence in 1980, began developing its urban infrastructure from the 1890s. Urban and commercial farming infrastructure development in the first half of the twentieth century was heavily skewed towards the interests of the white settlers, though water and sewerage services were also extended to African high density townships (Masst 1996). The city and town water, sewerage, stormwater drainage, and waste disposal services were built to a high standard. Decentralised management was in place from an early date, and urban and town services were managed through water and sewerage departments in local authorities (AMCOW 2010). Zimbabwe's urban infrastructure during the latter part of the twentieth century was built on the revenue from urban consumers. Water supply and sanitation services in rural areas, where the majority of the indigenous population lived, were typically neglected (AMCOW 2010).

Phase II: Independence (1980–1999) At independence in 1980, an international conference on reconstruction and development (ZIMCORD) sought to rebuild the country's infrastructure after the devastating effects of the war of liberation. The main water supply and sanitation challenge was to extend services to rural areas. The 20 years ending at the millennium saw a near doubling of coverage (WHO/

UNICEF 2013). During the early years of independence, the government developed an Integrated Rural Water Supply and Sanitation Program (IRWSSP), based on government-subsidised, low-cost, locally developed technologies (Bush pumps, VIP latrines) (AMCOW 2010). The IRWSSP was built on external development assistance grants to the Government of Zimbabwe (GoZ). By contrast, the growing urban population continued to be served by decentralised municipal authorities and coverage levels of nearly 100 % were maintained for urban water supply and sewerage services. Milestones in sector development were: adoption of a National Master Water Plan for Rural Water Supply and Sanitation (NMWP) in the mid 1980s; the establishment of a National Action Committee (NAC) and its secretariat (the National Coordination Unit, NCU) to coordinate the IRWSSP; revision of the Water Act in 1998 to establish catchment management councils; and the creation of a central parastatal, the Zimbabwe National Water Authority (ZINWA) for commercial management of the country's water resources (AMCOW 2010).

Phase III: Collapse (2000–2008) The decline of the economy (due to donor confrontation over the land reform program), the collapse of government funding, and the pull-out of donors from 2000 saw reduced investments in service delivery for nearly a decade, characterised by failure to repair or maintain an already aging infrastructure. The collapse in public sector salaries led to a significant exodus of skilled manpower. Capacity shortages developed in the public and private sector along the entire value chain: local manufacturing and supply of essential equipment, spares, chemicals and commodities, management of water treatment and wastewater plants, engineering supervision, finance, administration, project design, contract management, policy guidance, and necessary skills at provincial, district and village levels (AMCOW 2010). A nationwide cholera epidemic started in August 2008 and spread to 60 out of 62 districts, as well as into neighbouring countries. The outbreak was an indicator of the state of national neglect of the water and sanitation sector.

Phase IV: Emerging Recovery (2009 Onwards) The cholera epidemic generated a considerable humanitarian response from all parties. This, together with the introduction of a multi-currency system and the establishment of a Government of National Unity (GNU), led to important sector developments. The cholera emergency generated a strong response from donors and NGOs with UNICEF playing a leading role in coordinating the humanitarian response (IWSD 2009). A WASH cluster mobilised and coordinated the rural sector and an Urban Emergency Rehabilitation Program focused on supplying chemicals to urban local authorities, drilling boreholes and implementing cholera educational programs (GoZ and WHO 2011). The definition of humanitarian assistance was expanded to include "quick win" repairs to essential components of the water infrastructure. An analytical Multi-Donor Trust Fund (MDTF) was established to support critical studies and capacity building. A programmatic MDTF was also set up by the World Bank. In February 2010 the Ministers from the four leading water-related ministries met and agreed on plans to restructure the water and sanitation sector leadership, building momentum for a new era in sector development (see Sect. 14.3.7).

3.3 *Zimbabwe WSS Sector Challenges Versus MDG Targets*

3.3.1 Zimbabwe MDG Targets

The aftermath of the cholera outbreak saw a lot of investment in the supply of water treatment chemicals and the rehabilitation of critical water and sanitation infrastructure. Other areas targeted included the improvement of solid waste management and improved health and hygiene awareness. It was later realised that all these investments were going to count for nothing if they were only supported by donors, with councils collecting very little in terms of revenue (GoZ and WHO 2011). A World Bank scoping report in April 2009 estimated that less than 20 % of urban water and wastewater expenses were being recovered from revenue collections. A lot of effort thus began to be focused on tariff setting, management, and revenue collection. High water losses, up to 50 %, also meant reduced revenues for councils, so funding was therefore directed at further rehabilitation.

The Zimbabwe water and sanitation MDG targets are shown in Table 4.2. However, Zimbabwe also developed a draft water policy in 2004, a policy that was never implemented, and this had ambitious targets of almost universal coverage by 2015. The target was set despite the fact that in 2004 the Zimbabwean economy was already on an uncontrolled decline. The country's inflation environment, shown in Fig. 4.9, derailed all investment efforts to meet the MDG targets.

Table 4.2 Zimbabwe MDG targets and achievements (Source: AMCOW 2010; WHO/UNICEF 2013)

	MDG targets (2015)	2004 Draft policy targets (2015)	Latest JMP figures (2011)
Water coverage	90 %	100 %	80 %
Rural	85 %	100 %	69 %
Urban	99 %	100 %	97 %
Sanitation coverage	79 %	85 %	40 %
Rural	68 %	80 %	33 %
Urban	100 %	100 %	52 %

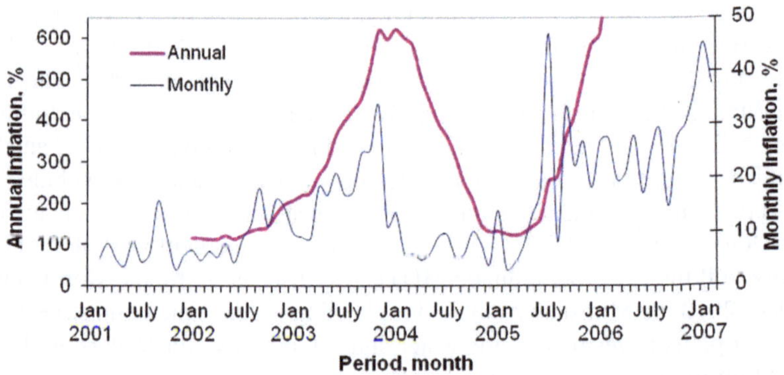

Fig. 4.9 Inflation in Zimbabwe from Jan 2001 to Jan 2007 before it spiralled to billions in 2008, signalling the collapse of the water and sanitation sector, among other calamities (Own elaboration using data from www.rbz.co.zw)

Zimbabwe has numerous national water supply and sanitation targets. Estimates of coverage and investment requirements also vary considerably. Two scenarios of the current status can be developed which represent the range of sector status:

Scenario 1 – an optimistic scenario is presented which uses the lower targets, is based on WHO/UNICEF Joint Monitoring Programme (JMP) figures, and assumes sufficient and sustained budget allocations from 2010 onwards with supplementary funding from the Multi-Donor Trust Fund. In 2008, JMP suggested that 82 % of Zimbabweans had access to improved drinking water and only 68 % had reasonable access to an improved toilet.

Scenario 2 – a more pessimistic scenario which uses the higher targets, Zimbabwean sector agencies' own figures, and assumes a scenario where, without a political settlement, external resources are more limited and budget allocations cannot be sustained. This scenario uses the lower estimates that in 2008, 46 % of Zimbabweans had access to improved drinking water and 30 % to improved sanitation facilities.

3.3.2 JMP Drinking Water and Sanitation Coverage Trends in Zimbabwe, 1990–2010

The water supply coverage in urban areas of Zimbabwe shows very high coverage of piped water (Fig. 4.10) but this coverage has been declining over recent years. This is attributed to a number of new housing plots under development by cooperatives where plot owners have moved onto their plots without first finishing the house to an acceptable state. Performance issues include high unaccounted-for water, sub-economic tariffs, and reluctance to pay for water due to erratic and poor service delivery (Manzungu and Chioreso 2012; Ndebele 2012).

Rural water supply coverage is, surprisingly, showing limited movement (71–69 %) despite huge investments into the rural water sector at the end of the last millennium. The land reform of 2000 also resulted in people moving from communal areas to formerly commercial farming areas (Scoones et al. 2011). Although no scientific studies have been conducted to ascertain the water and sanitation situation in resettled areas, it is most likely that very few resettled farmers have reasonable toilet facilities, and heavy soils in resettled areas have poor well water yields. With a declining economy, very few households can afford to drill and equip boreholes which cost from USD4,000 upwards.

Compared to drinking water coverage, national improved sanitation coverage is virtually stagnant at 40 %, despite large investment programs in rural sanitation. This situation (Table 4.3) suggests there is still a large number of people who must use open-air defecation in Zimbabwe and points to a significant failure of the rural sanitation policy in the country. Although urban sanitation shows extremely high coverage, it is now declining. The general conclusion is that either new investment or a new policy addressing current urban sanitation challenges is required. The current high standards of urban sanitation (sewerage provision) might be difficult to sustain and justify in the long-term.

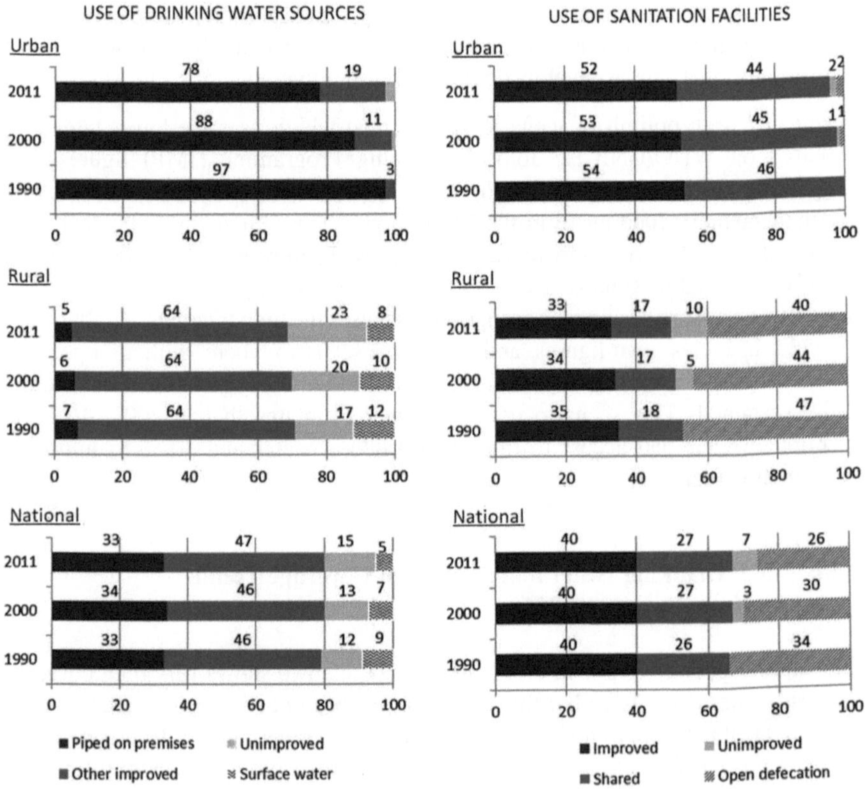

Fig. 4.10 JMP-estimated trends of drinking water and sanitation coverage in Zimbabwe to 2011 (Source: WHO/UNICEF 2013)

3.4 *Key Rural WASH Challenges in Zimbabwe*

Zimbabwe's rural water and sanitation services, once a source of national pride, have suffered a major collapse in the past decade due to, among other reasons, the flight of external support agencies in 2000, persistent droughts that have resulted in severe stress in both surface and underground water, general economic decline, eroded institutional and community capacity, and the effects of the HIV/Aids pandemic. In the new resettlement areas there are very few water and sanitation facilities and people rely on unsafe sources of water and use bushes for disposal of excreta. From a ministerial meeting on water sector coordination held in Nyanga in February 2010, key challenges in the rural WASH sector were identified. For ease of discussion here, they are grouped into the following broad categories: inadequate organisation/coordination of water and sanitation utilities at all levels; the economic situation (an external factor); and deficiencies in public action at the national level.

Table 4.3 Modified wastewater production scheme for medium- and low-density residential areas of Harare

Wastewater source	Medium-density areas L/cap/d	Low-density areas L/cap/d	Actions required to reduce wastewater production
Bath	50	50	Reduce larger water heater sizes to 50 L
WC	50	50	Replace larger size cisterns with 5/10 cisterns: 5 L for flushing urine and 10 L for flushing faeces (or smaller)
Laundry	8	11	Laundry once a week. 25 L for washing and 25 L for rinsing
Kitchen	9	13	Based on three washings (dishes or hands) and cooking routines using a double-bowl sink of 15 L/bowl. Seven and 10 people per household for low- and medium-density respectively
Other	12	12	10 % contingencies to cater for leakages and incidental washings
Total	129	137	
Target figure	135	150	Amount to be targeted in wastewater reduction efforts
Current figures	210	315	Source: JICA (1997)
% reduction in consumption that can be reached with the actions described	36	52	
Cost savings, USD/household/year	**162**	**227**	

- *Inadequate organisation/coordination of water and sanitation utilities at all levels* – There has been a lack of clarity about overall leadership in rural water supply and sanitation (RWSS). Coordination structures for the water and sanitation sector at all levels are currently weak and thus require urgent rescue and rejuvenation. The District Development Fund (DDF), under the Ministry of Transport, Communications and Infrastructural Development (MoTCID), has been leading the drilling of boreholes and a piped water scheme, while the development of toilets and shallow wells is being managed by the Ministry of Health and Child Welfare (MoHCW). The Ministry of Local Government, Public Works and National Housing (MoLGPW&NH), through the district councils, is responsible for rural development. This has led to confusion on who is in charge, with the Ministry of Environment, Water and Climate (MoEnvWC) largely left out of local initiatives. In addition, responsibilities for rural sanitation and hygiene have not been clear because several ministries are involved.

  Borehole drilling has not been the sole responsibility of any agency and no government department claims sole authority over it. The DDF and the Zimbabwe

National Water Authority (ZINWA), for example, have always drilled boreholes on a service provider basis. However, the sector recognises the current limited capacity within government departments to provide drilling services, and thus encourages the participation of the private sector. Responsibilities for financing, coordinating, drilling, and maintenance therefore need review and clarification.

- *Economic situation* – Over the last decade the public sector budgetary allocation to the rural WASH sector has been on the decline, largely due to the economic challenges facing the country after the debilitating economic sanctions imposed by mainly western donors since 2002. In particular, budgets for rural WASH since 2002 have declined to negligible amounts, with several NGOs largely taking up the responsibilities. The Ministry of Health's budget for environmental health development has largely been inadequate for its responsibilities. The Rural Capital Development Fund (RCDF), established under MoTCID's Department of Infrastructure Development, was the main instrument through which the National Rural Water Supply and Sanitation Programme was financed from Treasury. These allocations have never been adequate given the increased demand and nationwide need for water and sanitation services. The RCDF does not fund MoHCW for key promotion, control, and hygiene education functions. A national financing strategy for the RWSS sector, which takes into account the current financial resource base and implementation capacities at all levels, is essential to mobilise resources.

  It is estimated that more than 65 % of water points in rural areas are non-functional at any given time, owing to the many challenges bedevilling the existing DDF-led three-tier maintenance system (Gwinji 2010). In response to this challenge, NAC adopted (though with some challenges) the Community Based Management System (CBM), aimed at empowering communities to operate and maintain their water facilities with limited outside support. The main impact of the economic situation, however, is that government agencies are massively under-resourced to undertake needed functions in the RWSS sector. To compound the situation there is high staff turnover and shortages of vehicles, equipment, and communication material to support operational activities. Conditions of service have declined, staff loss and rates of turnover have been high, together with inadequate budgets for specialised expertise, training, and operational activities. It will take time for Zimbabwe to recover economically using its own resources. Therefore the reengagement of the donor community to complement government efforts to rebuild the sector has become a priority.
- *Deficiencies in public action at the national level* – There was no national policy in place until March 2013 when a new National Water Policy was launched. A national Sanitation and Hygiene Strategy was also developed and launched in 2011, but the official document was only released and circulated in early 2013. To better coordinate government and donors, a national investment framework is required for specific sectors. In 2012 the government, with assistance from the World Bank, has been developing a national water sector investment framework and, as it is not sector-specific, it is likely to reduce the focus on specific sectors like rural WASH.

From the foregoing, there is the major issue of resilience that should be addressed: due to unfavorable economic conditions the system has collapsed and it is failing to recover from the shock. There is therefore a need to increase resilience (or decrease vulnerability), such as increasing revenues from water users (to become more independent from donors) and use of appropriate technologies. Technologies implemented should be screened for appropriateness and those who use them should be trained as necessary. Water and sanitation coverage continues to decline, rendering the meeting of MDG targets a challenge to the sector (WHO/UNICEF 2013). Under the current situation, the Zimbabwe Government provides rural water supply services with limited community participation and ownership. A sustainable option that encourages community-managed rural water services development, with user communities contributing to capital development and operation and maintenance costs, could be adopted. Of priority in the sector is the rehabilitation of all non-functional infrastructure, establishing a sustainable community-managed operation and maintenance system and implementing community participatory approaches to hygiene and sanitation improvement. A tremendous and immediate turn-around in sector investment and performance is a priority if sector MDGs targets of 100 % for water supply and 80 % for safe sanitation are to be achieved. The government needs to create a favourable environment for private sector participation and direct foreign investment. This, among others, involves fostering confidence and certainty on tariffs, ownership, and profit remittances.

Following the cholera outbreak, NGOs and the donor community (e.g., EU, DFID, USAID, AusAid, SIDA, CIDA, ECHO, GIZ) have played a crucial role in the recovery of rural water supply and sanitation in Zimbabwe. These have continued to provide infrastructure, and have also enabled communities to manage their own water infrastructure under the Community Based Management Programme. Organisations such as World Vision, Practical Action, and PumpAid have done a lot of work in this area. For example, PumpAid (http://pumpaid.org) introduced a simple water abstraction technology called the elephant pump. The technology is not as complicated as the bush pump which needs a specially trained person to maintain and repair. The majority of villagers can be trained in its maintenance, although there are questions on its life span. Also notable is the financial assistance provided by organisations such as the EU and DfID under the Protracted Relief Programme.

3.5 *Urban Domestic Water Supply and Sanitation Challenges in Zimbabwe*

One of the major challenges in urban water supply in Zimbabwe is the current high levels of water consumption for domestic purposes. Current life styles can result in an enormous thirst for both energy and water, with the high usage directly related to the type of water installation in the home. The use of large water heaters (or geysers), jacuzzis, and large cisterns needs revisiting. In Zimbabwe water use varies from 80 to about 600 L/cap/d for poor to affluent residents, respectively (JICA

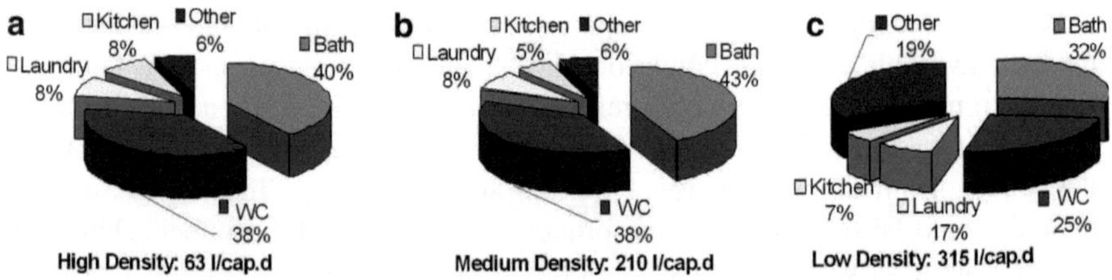

Fig. 4.11 Household water use and wastewater production in Harare (Source: Nhapi et al. 2002). Figures below each pie chart relate to wastewater production, not water consumption

1997, Nhapi and Gijzen 2004). However, under the post-2008 water rationing regime, demand has generally been reduced to about 20 to 100 L/cap/d (Chisango 2012). A high water demand results in the generation of large volumes of wastewater which requires treatment, and the capital and running costs for wastewater treatment plants are directly related to the volume treated. Thus the generation of huge volumes of wastewater has a direct bearing on the cost of treatment. High volumes of dilute wastewater also reduce options for wastewater treatment, making technologies such as anaerobic treatment unattractive.

A combination of end-use efficiency, system efficiency, stormwater harvesting, storage innovations, and reuse strategies would greatly reduce current municipal water demand. The estimated proportion of household water demand required for potable uses (drinking and cooking) is only 5–8 %, whereas 25–38 % is used for toilet flushing in Harare (Fig. 4.11). Greywater constitutes about 56 % of water used in all housing categories. The high water consumption for bathing and toilet flushing can be explained by the sort of water systems installed in homes. A large reduction in consumption is possible if water saving measures and devices were to be used. Education and community awareness campaigns could greatly reduce water consumption as reported in Bulawayo, Zimbabwe's second largest city, where reductions of more than 50 % were reported (BCC 2002). Water-saving devices are useful for water conservation in Zimbabwe and these include specific types of showerheads (3–5 L/min), smaller or dual-flush cisterns (6 or 3 L/flush), specialised toilet pans, and low water-usage faucets (2 L/min) (Chaplin 1998). In low-density areas, bore water could be used for supplementing municipal supplies. However, this alternative needs further investigation because extracting ground water can have negative ecological or other effects. More efficient methods and devices for watering gardens can also be used, e.g. drip irrigation, depending on cost. The possible methods and their wastewater volume reduction potentials are given in Table 4.3.

Table 4.3 shows that a reduction in wastewater production is feasible without greatly compromising present comfort, convenience, dignity, or quality of life. The high wastewater volumes produced in medium and low-density areas are a direct result of installed systems in homes. For example, water heaters (geysers) in most affluent homes are as large as 200 L and there is a tendency to empty the geyser each time a person takes a bath (Nhapi 2004). A 50-litre geyser will still be adequate. Most of the toilet cisterns being sold in Zimbabwe have capacities of around 15 L with no provision for separately controlling flushing volumes for urine or faeces, although the flushing of each requires different water volumes. A '3/6' cistern is used in Australia and other countries (Chaplin 1998), with the 3 L button used for flushing urine and the 6 L for flushing faeces. Because of different development densities, a '5/10' cistern will still do for Zimbabwe and would realise substantial savings. There is a tendency for housemaids to do laundry more frequently, and washing could be reduced to once per week without inconvenience so that water usage could be reduced. In kitchens, double-bowl sinks of 15 L each could replace washing under a running tap.

Applying the above-mentioned changes and adding a 10 % contingency level for leaks and other incidental washings show that it would be feasible to reduce wastewater production from the current 210 L/cap/d to about 135 L/cap/d (a 36 % reduction) for medium-density residential areas. For low-density, the reduction would be from 315 L/cap/d to 150 L/cap/d, implying a 52 % reduction in wastewater production. It is therefore concluded that reduction of wastewater production is technically feasible in Harare. The cost savings per household per year based on the average cost of treated water of USD0.5/m^3 are USD162 for medium and USD227 for low density areas. Some of the measures described here for households could also be applied for commerce and industry.

The reuse of wastewater, especially greywater from laundry, kitchen, and bathing, and the harvesting of rainwater could potentially reduce water consumption by more than 50 % (Chaplin 1998). Rainwater harvesting is a possible substitute to a treated water supply. With an annual rainfall of 830 mm/annum in Harare (ZINWA records), and assuming 50 % losses, about 0.4 m^3/m^2/year could be collected. For an average built housing area of 120–300 m^2, this means 4–17 m^3/cap/year (depending on the housing category, water consumption is 30–230 m^3/cap/year). In certain parts of Hawaii, the Virgin Islands, Australia, and Texas, rainwater collection systems provide the entire water supply for many homes (Chaplin 1998). Progressively high water tariffs could also be used as a demand management tool since almost all properties in Zimbabwean towns have water meters. An increasing block tariff structure applies punitively high charges for high water consumption.

3.6 *Water Tariffs and Funding of Urban WSS in Zimbabwe*

3.6.1 Water and Sewerage Tariffs

In 2011, the Government of Zimbabwe, with assistance from the World Bank, commissioned a study on water and sewerage tariffs. The study, among others, aimed at developing a methodology for establishing cost recovery tariff levels and an appropriate tariff and subsidy structure to simultaneously meet a number of national objectives and to develop financial tariff models for each of the selected seven municipalities (Harare, Bulawayo, Chitungwiza, Mutare, Kwekwe, Masvingo, and Chegutu). The study also looked at options for an institutional regulatory framework to manage the tariff review process, monitor and benchmark performance, and provide incentives for continuous performance improvement. Only two of the seven municipalities, Mutare and Kwekwe, were more-or-less meeting the demand for potable water. The main problems faced by water and sewerage (W&S) departments were high population growth, high losses (technical and commercial), a high fraction of non-functioning meters, and low recovery ratios on billed water (World Bank 2011). People without adequate municipal supplies of water were surviving through access to boreholes (using hand pumps in the high density areas), shallow wells, surface water, and water purchased from unlicensed vendors.

The state of the infrastructure in each town varies, but one of the common problems is lack of functioning meters. People getting water are using more than the volume they are metered and charged for. People not getting water are unwilling to pay monthly fixed charges (Ndebele 2012). On the sewerage side, nearly all of the biological nutrient removal (BNR) plants are not working. The ponds are still working, but overall there is a high degree of pollution of streams due to spillage from sewers, failure of sewage pumping stations, discharge of raw/partially treated sewage from treatment works, and failure of effluent pumps (Dlamini 2012). Pollution has had the most impact in Harare and Chitungwiza, where the sewage discharge is upstream of water intakes (Kibena et al. 2013).

The financial position of W&S departments is generally precarious due to the following factors (World Bank 2011; Ndebele 2012):

- Collection efficiency is low at about 30–60 % (customers are unwilling to pay when service is poor).
- Non-revenue water (NRW) is extremely high at about 40–60 % (leaking old pipes and lack of working meters, as already noted, contribute in a big way).
- Spending on essential fixed costs is low (salaries, repairs and maintenance, capital).
- Transfer of funds from the water account to other municipal uses, such as social services (schools, clinics, amenities), is significant (Harare is an exception).

The municipalities expect to make significant improvements in service delivery and revenue generation in the next couple of years. Cumulative benefits are expected as non-revenue water declines, access to water improves, and willingness to pay increases. The resulting downward pressure on tariffs is offset by the need to increase infrastructure spending (maintenance, rehabilitation, and investment). Zimbabwe's urban areas have some of the lowest cost of treated water in the region (the lowest being USD0.05/m^3 in Mutare). Tariffs are also very low by regional standards (van den Berg and Danilenko, 2011).

3.6.2 Setting of Water Tariffs in Zimbabwe

Setting of urban water tariffs in Zimbabwe is premised on a number of legal instruments, Acts of Parliament, subsidiary legislation or regulations in place, and directives from respective Ministers for Water and Local Government. Section 8 of the ZINWA Act (Chapter 20:25) allocates the function of water pricing to the Authority (GoZ 1998a). It also allocates the particular function of encouraging and assisting local authorities to discharge their obligations with regard to the development and management of water resources in areas under their jurisdiction – in particular, the provision of potable water and the disposal of wastewater. Section 30 of the ZINWA Act sets water and other charges. ZINWA is empowered to fix charges for the sale of raw or treated water from waterworks operated or controlled by the Authority. The charges by ZINWA must be approved by the Minister of Water after considering various factors. There is no reference to consultations with users of water, or setting water costs according to full cost recovery for the dams from which the water is extracted. There is no consideration at this level as to the ultimate cost of water for the users, especially the urban vulnerable groups.

The Water Act (Chapter 20:24), Section 6, under the General Functions of the Minister, states that it "shall be the duty of the Minister of Water Resources to secure the provision of affordable water to consumers in under-privileged communities" (GoZ 1998b). Under the Urban Councils Act (as amended 2008), (Chapter 29:15, Part XIII), the powers of the council in relation to water include entering into agreements for the purchase and sale of water (GoZ 2008). It is not clear whether there is any agreement between the urban councils and ZINWA on the pricing of raw water and the method of calculating the price of water to consumers in order to satisfy the legal requirements of the Minister of Water Resources to meet the provision of affordable water to consumers in under-privileged communities (World Bank 2011).

The post-independence Urban Councils Act has progressively been altered, in part to deal with challenges from inefficient and increasingly ineffective urban councils, as well as with their reduced accountability/transparency. In short, councils have over time been perceived as not representing the interests of their electors. The Minister has increased his oversight powers to intervene. These changes have usually been promulgated as regulations and not amendments to the Act, using especially Part XXI of the Act under "General" (World Bank 2011). There are provisions that could inadvertently undermine the powers and duties of councils through many open powers being delegated to the Minister of Local Government. The Minister, for political or other reasons, may scale down justifiable tariffs through cost recovery objectives, thereby undermining service delivery, maintenance, and water production capacity.

The Urban Councils Act does not mention systems for fixing water tariffs. Municipal councils are required to balance their budgets with the general consent of their residents. However, since independence a number of regulations and directives by successive Ministers of Local Government have defined priority areas for high density areas on issues of affordability, especially the minimum charge (World Bank 2011). For example, in April 2011 the Minister of Local Government, with the concurrence of the Minister of Water Resources, slashed the price of water for Harare (World Bank 2011). Harare's high-density fixed charge was reduced by the Minister's directive from USD7 to USD5 and for low density from USD13 to USD11. This was done after the City budget had been approved and 3 months into the financial year. There is presently no independent regulator to whom both the municipal council and the Minister can refer any contentious issues.

3.6.3 Affordability

In recent years, there have been few studies on household income and expenditure. One exception is the Zimbabwe Vulnerability Assessment Committee (ZimVAC) report of April 2011, the Urban Livelihoods Assessment Report. This is based on studies of vulnerable households in a sample of urban areas which is shown in Table 4.4, which also gives the expenditure on water as a proportion of income. The average household expenditure on water and sanitation is 5.2 % of monthly household income. If the reference point is that expenditure on water by vulnerable households should be less than 5 % of their income (Franceys et al. 1992) then there were, in 2011, four towns in which this affordability threshold is breached (Harare, Bulawayo, Chitungwiza, and Mutare). When expenditure is higher than affordability, payment for services becomes an issue.

Reasons for the high cost of water and sanitation can be attributed to inefficiencies and losses in the treatment and distribution systems, as summarised in Table 4.5. Consumers are in a way compelled to cover the costs of water which does not reach them or is not metered. In Table 4.5, the first panel has estimated demographic information. The results of the 2012 Zimbabwe national census were released as preliminary results in December 2012 but these have been disputed by almost all councils according to media reports. The next panel records the number of connections, the number of staff, current water treatment works (WTW) capacity, and current production. Only in Mutare is WTW capacity fully utilised at present (in fact over-utilised, 70 ML/d being produced from a plant with nameplate capacity of 64 ML/d). The amount of water that Chitungwiza purchases from Harare is uncertain because it is not metered. The city is charged for 27 ML/d, but the indications are that the volume is larger.

Table 4.4 Affordability of water and sanitation in selected towns of Zimbabwe. (Source: World Bank 2011)

Town	Average household income, USD/month	Household expenditure on water and sanitation, USD/month	Household expenditure on water and sanitation, % of monthly income
Harare	457	26.10	5.5
Bulawayo	311	19.86	6.4
Chitungwiza	475	26.10	5.5
Mutare	293	17.35	5.9
KweKwe	359	17.53	4.9
Masvingo	331	12.08	3.6
Chegutu	308	13.37	4.3
Average	**362**	**18.91**	**5.2**

Water losses are high in all the towns covered, ranging from 30 % in Chitungwiza to 57 % in Harare, but there seems to be considerable uncertainty about the accuracy of the figures kept by councils (World Bank 2011; Ndebele 2012). The cost of losses reported in Table 4.5 represents the revenue lost when supplying treated water that never gets sold, compared to benchmark losses of 20 % – the figure for Harare is USD35 million per year. The costs are incurred whether the water is sold or not – the opportunity cost that is involved in non-revenue water is the effective tariff, which takes the collection ratio into account (actual collected revenue divided by billed volume), multiplied by the excess losses (difference between actual losses and the benchmark level of 20 %).

The next panel deals with present levels of supply of piped water and of consumption of water (piped plus self-supply water). Only Mutare and Kwekwe supply between two-thirds and three-quarters of unconstrained demand with piped water (Table 4.5). The lowest performer is Chitungwiza, which only supplies 15 % of unconstrained demand, which is itself based on a lower assumed consumption per capita than other councils. Self-supply is thought to be important in Harare, Bulawayo, Chitungwiza, and Chegutu, with the actual values on a ML/d basis being as shown in the table. The relative contribution of self-supply is highest in Chegutu (12 %) and Chitungwiza (9 %). Total consumption of water is closest to unconstrained demand in Mutare (80 %) and Kwekwe (66 %). The unconstrained figure for Harare is high because it includes provision for bulk sales to Norton and Chitungwiza as well as a large industrial demand.

3.7 Institutional Structure and Coordination Mechanisms

The cholera outbreak of 2008/09 exposed the bottlenecks in the institutional system in Zimbabwe, and both donors and the government realised the need for better coordination and accountability in the sector. A ministerial meeting was therefore organised. The Nyanga Ministerial Retreat held in February 2010 was attended by ministers and senior government officials from the key ministries involved in the WASH sector. The following are the major outcomes from the Nyanga Retreat:

- Development of a common understanding of issues affecting the WASH sector in Zimbabwe
- Development of a comprehensive framework clearly showing leadership, sub-sectoral roles and responsibilities, and key coordination and financing mechanisms for the sector
- Identification of interventions for sector recovery, including development of policy, legislation, institutional development, and capacity building, and
- Agreement on important follow-up steps, including relevant interventions, timeframes, and responsibilities.

The institutional structure for the water sector in Zimbabwe, after extensive discussions, was subsequently changed – see Fig. 4.12. The Government of

Table 4.5 Basic information on water supply, demand, and wastewater management in selected towns of Zimbabwe, based on available 2011 data (Source: World Bank 2011)

Demand and supply	Units	Harare	Bulawayo	Chitungwiza	Mutare	KweKwe	Masvingo	Chegutu
Stands	#	177,000	160,000	67,000	34,000	17,500	14,300	12,100
Population	#	2,500,000	985,000	400,000	200,000	155,000	120,000	100,000
of which HDA	#	2,200,000	757,000	395,000	281,000	143,000	115,000	90,000
prop. of total	%	88 %	77 %	99 %	141 %	92 %	96 %	90 %
Unconstrained water demand	ML/d	1,200	221	136	45	54	50	21
Water connections	#	177,000	160,000	67,000	34,000	17,500	14,300	12,100
Staff involved in water services	#	2,480	530	365	155	112	77	42
Water treatment works capacity	ML/d	704	260	0	64	90	30	12
Current water production	ML/d	570	140	30	70	60	24	8
Daily water losses	ML/d	325	49	9	36	24	10	3
Losses as % of daily. production	%	57 %	35 %	30 %	52 %	40 %	41 %	40 %
Opportunity cost of losses	$M p.a.	35.3	2.8	0.3	0.6	0.8	0.6	0.1
Net current piped supply	ML/d	244	91	21	34	36	14	5
Prop. unconstrained demand	%	20 %	41 %	15 %	75 %	66 %	28 %	22 %

Self supply	ML/d	53	11	12	2	0	0	3
Prop. unconstrained demand	%	4 %	5 %	9 %	5 %	0 %	0 %	12 %
Total consumption	ML/d	298	102	33	36	36	14	7
Prop. unconstrained demand	%	25 %	46 %	24 %	80 %	66 %	29 %	34 %
Sewage treatment works operational capacity	ML/d	220	89	56	34	35	23	12
Current volumes treated	ML/d	275	91	29	43	23	14	7
Capacity utilisation	%	125 %	103 %	52 %	128 %	66 %	60 %	59 %

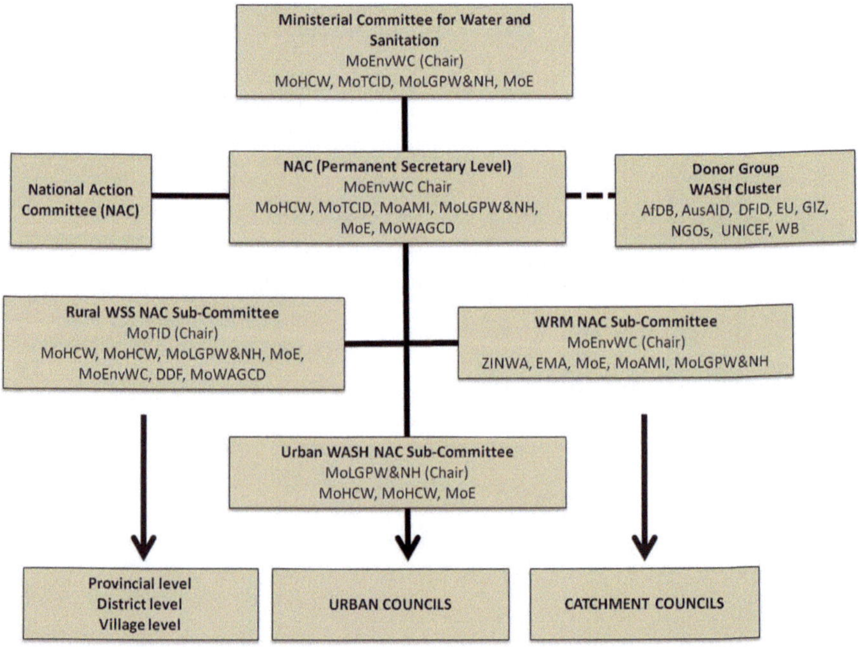

Fig. 4.12 Zimbabwe water sector institutional structure as established in June 2010 (Updated from AMCOW 2010)

Zimbabwe agreed to consolidate the water sector from June 2010 through a National Action Committee on water, looking at all aspects of water in both rural and urban areas as well as water resources management issues. The Government decided to transform and enhance the old National Action Committee on rural water supply and sanitation which had hitherto focused on rural areas only. The new NAC has three sub-committees, each looking at specific issues:

I. Rural Water Supply and Sanitation Sub-Committee,
II. Urban Water Supply and Sanitation Sub-Committee, and
III. Water Resources Management Sub-Committee.

The outcomes of the Nyanga retreat have far-reaching implications for the recovery of the WASH sector. Sector steering will now be the responsibility of the Cabinet Steering Committee for Water and Sanitation chaired by the Minister of Environment, Water and Climate (MoEnvWC). Included in this Cabinet Steering Committee are Ministers for Agriculture, Health and Child Welfare, Transport Communications and Infrastructure Development, Energy, and Local Government. The national coordination and management of the sector will now be through a National Action Committee for water and sanitation chaired by the Permanent Secretary for the ministry responsible for water. Members of this committee are Permanent Secretaries for the Ministries of: Health and Child Welfare, Transport Communication and

Infrastructure Development; Agriculture, Local Government; Women Affairs Gender and Community Development; Energy; and Finance. Each Permanent Secretary can nominate an alternate whose level shall not be below that of a Director. The new NAC is required to work closely with the Donor Coordination Group and the WASH cluster.

4 Conclusions

Rapid population growth and urbanisation are key drivers in water supply and sanitation in developing countries. These have resulted in pressures in the provision of key rural and urban infrastructure such as housing, safe water supplies, and adequate sanitation. In some cases, these pressures have resulted in the mushrooming of slum settlements and the prevalence of waterborne diseases such as cholera, dysentery, typhoid, and diarrhoea, among others. The impact of poor water supply and sanitation services could potentially affect other key development indicators since many lives and livelihoods are affected. The problem in most developing countries is that they lack the financial muscle to deal with a myriad of problems facing them, meaning they cannot reach the targets of the Millennium Development Goals. In this context, urban water supply and sanitation should aim for reduced water usage to cut conveyancing and treatment costs and to widen management choices. Rural sanitation is not being given the attention it deserves, and particular attention needs to be paid to coordination, appropriate technology, and financing mechanisms.

In this analysis, Zimbabwe has been used as a case study to show the key challenges in water supply and sanitation in developing countries. The challenges include lack of investment in the water and sanitation sector, inappropriate technologies, ill-defined institutional frameworks, capacity limitations, and neglect of rural infrastructure. These challenges should be factored in at the planning and organisation of the sector to ensure a win–win situation under various environmental and economic situations. This calls for innovative thinking, judicious choice of technologies, robust institutional structures, a responsive legal framework, and political will. Poor water supply and sanitation in Zimbabwe is typified by the cholera outbreak of 2008/09 which killed nearly 4,300 out of the 99,000 that were affected. The cholera epidemic caught the attention of the international community and attracted donor funding. Responses included a review of the policy environment, appropriateness of the institutional structure, accountability, and choices of technologies. The emphasis now is shifting to appropriate technologies centred on waterless toilets and natural sewage treatment systems. For water supply, the focus is now on demand management, reduction of unaccounted-for water, and innovative methods for enhancing revenue collection. Two important lessons for other developing countries are that the water sector must be streamlined to give efficient service delivery at all times, and that the technologies chosen should be robust, resilient, and appropriate for the situation.

References

AMCOW. (2010). *Water supply and sanitation in Zimbabwe: Turning finance into services for 2015 and beyond*. Nairobi: World Bank.

BCC (Bulawayo City Council). (2002). *Study on water use, water conservation and affordability, May 2002*. Bulawayo: Department of Works.

British Medical Journal. (2007). Sanitation is greatest medical milestone since 1840. Press release, 18 January 2007. http://group.bmj.com/group/media/press-release-archive-files/BMJ/bmj-2007/BMJ-18Jan-2007.pdf. Accessed 23 Feb 2014.

Chaplin, S. (1998). *Water efficiency: The next generation*. Draft report 5–9, Rocky Mountain Institute, Colorado. http://www.rmi.org/Knowledge-Center/Library/W98-07_WaterEfficiencyNextGeneration. Accessed 23 Feb 2014.

Chisango, H. (2012). *Investigating opportunities for improving service through water demand management in Harare*. MSc Thesis, University of Zimbabwe, Harare.

Chuhan-Pole, P., Korman, V., Angwafo, M., & Buitano, M. (2011). *Africa's pulse: An analysis of issues shaping Africa's economic future* (Vol. 3). Nairobi: World Bank.

Dlamini, S. (2012). *Evaluation of the water quality status in Lake Chivero, Zimbabwe*. MSc Thesis, University of Zimbabwe, Harare.

Franceys, R., Pickford, J., & Reed, R. (1992). *A guide to the development of on-site sanitation*. Geneva: WHO.

GoZ (Government of Zimbabwe). (1998a). *ZINWA Act (Chapter 20:25)*. Harare: Government Printers.

GoZ (Government of Zimbabwe). (1998b). *Water Act (Chapter 20:24)*. Harare: Government Printers.

GoZ (Government of Zimbabwe). (2008). *Urban Councils Act (Chapter 29:15, amended)*. Harare: Government Printers.

GoZ (Government of Zimbabwe) & World Health Organization. (2011). *Intersectoral actions in response to cholera in Zimbabwe: From emergency response to institution building*. Paper presented at the World Conference on Social Determinants of Health, 19–21 Oct 2011, Rio de Janeiro.

Gwinji, G. (2010). *The key rural water, sanitation and hygiene issues in Zimbabwe*. Paper presented at the Inter-ministerial Conference on Water, Nyanga, 10–11 Feb 2010.

IWSD (Institute for Water and Sanitation Development). (2009). *Evaluation of the WASH response to the 2008–2009 Zimbabwe cholera epidemic and preparedness planning for future outbreaks. Zimbabwe WASH Cluster Report*. Harare: Institute for Water and Sanitation Development.

JICA. (1997). *The study of water pollution control in Upper Manyame river basin in the Republic of Zimbabwe*. Ministry of Local Government, Public Works and National Housing, Harare, Zimbabwe.

Kibena, J., Nhapi, I., & Gumindoga, W. (2013). Assessing the relationship between water quality parameters and changes in landuse patterns in the Upper Manyame River, Zimbabwe. Physics and Chemistry of the Earth, Parts A/B/C (in press). doi:10.1016/j.pce.2013.09.017.

Kumwenda, V. (2012). *Evaluating the impacts of WASH interventions on the 2008–09 cholera outbreak in Harare, Zimbabwe*. MSc Thesis, University of Zimbabwe, Harare.

Lopez, A. D., & Mathers, C. D. (2006). Measuring the global burden of disease and epidemiological transitions: 2002–2030. *Annals of Tropical Medicine & Parasitology, 100*(5/6), 481–499.

Manzungu, E., & Chioreso, R. (2012). Internalising a crisis? Household level resistance to water scarcity in the city of Harare, Zimbabwe. *Journal of Social Development, 17*(1), 111–136.

Masst, M. (1996). *The harvest of independence: Commodity boom and socio-economic differentiation among peasants in Zimbabwe*. PhD Dissertation, Roskilde University, Roskilde.

Mokitimi, L. (2012). *Linking institutional and legal framework with accountability and possible implications on urban water sub-sector performance in Zimbabwe*. MSc Thesis, University of Zimbabwe, Harare.

Ndebele, P. (2012). *Assessment of water revenue collection strategies in selected urban local authorities in Zimbabwe*. MSc Thesis, University of Zimbabwe, Harare.

Nhapi, I. (2004). *Options for wastewater management in Harare, Zimbabwe*. PhD Thesis, Wageningen University, The Netherlands. Taylor and Francis, London, UK (ISBN 9058096971). http://edepot.wur.nl/193572. Accessed 23 Feb 2014.

Nhapi, I., & Gijzen, H. J. (2004). Wastewater management in Zimbabwe in the context of sustainability. *IWA Water Policy Journal, 6*(6), 115–120.

Nhapi, I., Siebel, M., & Gijzen, H. J. (2002). *Options for the decentralised management of wastewater in Harare Zimbabwe*. IWA Water and Environmental Management Series, Dec. 2002, pp. 955–964.

Red Cross. (2010). Final report, MDRZW004 – Zimbabwe cholera. Harare, Zimbabwe: International federation of red cross and red crescent societies.

Rees, N., Chai, J., & Anthony, D. (2012). *Right in principle and in practice: A review of the social and economic returns to investing in children*. New York: United Nations Children's Fund (UNICEF).

Scoones, I., Marongwe, N., Mavedzenge, B., Murimbarimba, F., Mahenehene, J., & Sukume, C. (2011). *Zimbabwe's land reform: A summary of findings*. Brighton: Institute of Development Studies, University of Sussex.

UNDP. (2013). *Human development report 2013. The rise of the south: Human progress in a diverse world*. New York: United Nations Development Programme. http://hdr.undp.org/sites/default/files/reports/14/hdr2013_en_complete.pdf. Accessed 23 Feb 2014.

UNEP. (2010). *Assessing the environmental impacts of consumption and production: Priority products and materials*. Report of the Working Group on the Environmental Impacts of Products and Materials to the International Panel for Sustainable Resource Management. http://www.unep.fr/shared/publications/pdf/DTIx1262xPA-PriorityProductsAndMaterials_Report.pdf. Accessed 31 Dec 2012.

UN-HABITAT. (2012). *State of the world's cities 2010/2011: Bridging the urban divide*. Nairobi: United Nations Human Settlements Programme (UN-HABITAT).

UNICEF. (2010). The 2008–2009 cholera epidemic in Zimbabwe. United Nations Children's Emergency Fund, unpublished report, Harare.

UNPD (United Nations Department of Economic and Social Affairs, Population Division). (2013). *World population prospects: The 2012 revision, key findings and advance tables*. Working paper No. ESA/P/WP.227, New York.

van den Berg, C., & Danilenko, A. (2011). *IBNET water supply and sanitation performance Blue Book: The international benchmarking network for water and sanitation utilities data book* (p. 176). Washington DC: World Bank.

WHO. (2004). *Assessing the outbreak response and improving preparedness*. Geneva: World Health Organization.

WHO/UNICEF. (2012). *Progress on drinking water and sanitation: 2012 update*. New York: UNICEF/World Health Organization.

WHO/UNICEF. (2013). *Progress on sanitation and drinking-water, 2013 update*. Geneva: WHO.

World Bank. (2011). *Zimbabwe urban water tariff study: Final report*. Report prepared for the World Bank and the Government of Zimbabwe. Harare: World Bank.

Zulu, E. M., Dodoo, F. N., & Chika-Ezee, A. (2002). Sexual risk-taking in the slums of Nairobi, Kenya, 1993–8. *Population Studies, 56*, 311–323.

Chapter 5
The Challenges of Providing Water and Sanitation to Urban Slum Settlements in South Africa

Ephias M. Makaudze and Gregory M. Gelles

1 Introduction

Like many other developing countries, South Africa has committed itself to the United Nations Millennium Development Goals (MDG) of 8 goals and 18 targets. One of the key goals is MDG 7 and Target 10 which calls to halve the proportion of people without sustainable access to safe drinking water and basic sanitation by 2015. According to the South African Department of Water Affairs (DWA 2013), the country has surpassed this goal including its targets and is moving towards achieving 100 % water and sanitation provision to its citizens by 2014. The DWA also developed and implemented the Free Basic Water Policy (2000) so as to ensure that the poor are not denied access to water on the basis of financial inability.

However, despite this remarkable achievement, the country faces daunting challenges meeting Target 11 (MDG 7) which calls for a significant improvement in the lives of at least 100 million slum dwellers by 2020. As acknowledged by DWA, to provide improved water and sanitation services to an ever-growing slum population, estimated as 2.2 million households (Rwida 2013), is indeed the greatest challenge facing the government. The task becomes more cumbersome due to the influx of illegal immigrants (from neighboring countries) who gravitate to slum settlements.

E.M. Makaudze (✉)
Department of Economics, University of Western Cape, Cape Town, South Africa
e-mail: emakaudze@uwc.ac.za

G.M. Gelles
Department of Economics, Missouri University of Science and Technology, Rolla, MO, USA

Q. Grafton et al. (eds.), *Understanding and Managing Urban Water in Transition*, Global Issues in Water Policy 15,
DOI 10.1007/978-94-017-9801-3_5

2 The Early Rise of Slum Settlements in Urban Areas

The major cities across South Africa continue to experience the proliferation of 'slum' or 'squatter' settlements, predominantly on their outskirts. In Cape Town, for example, there are more than 300 slum settlements with a population of at least one million people. It is estimated that, by 2015, approximately five million households will be living in slum settlements in South Africa (UN-Habitat 2006).

The problem of slum settlements in South Africa originated during the apartheid era. Beginning with the Pass Act (1977), stringent laws were introduced so as to curb the local movement of migrants from homelands or townships to urban centres. Only those gainfully employed were allowed into urban areas, and even their families were not allowed. In 1986, following intense international lobbying and stiff internal political resistance, the Pass law was repealed. Following its repeal, the immediate effect was a drastic rise in rural-to-urban migration by millions of job-seekers who flocked to cities in search of job opportunities and better living conditions. This sudden rise in urban population created greater problems for city authorities than anticipated. What perhaps was initially viewed by city authorities as a few temporary slum settlements has instead grown into large settlements composed of millions of haphazardly scattered and crowded dwelling units. The dwelling units are constructed and/or improvised from a collection of all sort of materials, including rusty corrugated iron bars, metal sheets, plastic sheets, wooden planks and bricks.

The problem associated with the growth in urban slums has illustrated a little known aspect of the period of white minority rule in South Africa. The policy of apartheid was not just about political repression and racial discrimination; it also involved depriving (or ill-providing) the black majority with basic services like water and sanitation. Following the 'balkanisation' of the country into townships and homeland territories, the responsibility for water resource management and water services was fragmented and scattered. No single national government agency was responsible for water supply and services. According to Lefebvre et al. (2005), water management was the responsibility of 11 departments of Water Affairs and 14 departments of Agriculture, each having a specific organisational structure. Besides these departments, there were other institutions charged with the responsibility to provide water services. For instance, there were regional councils which also shouldered a responsibility to provide water services on a regional basis. The councils consisted of representatives of local authorities and provided services like bulk water supply, bulk electricity supply, and sewerage works. Under such an institutional arrangement it was not surprising that duplication, conflicts, disparities, and inefficiencies in water provision were common. Most of the disparities were manifested in terms of a highly skewed water resource distribution between white areas versus townships/homelands, as the bulk of financial resources were channelled towards development of white-dominated areas.

It is important to emphasise that during apartheid it was not the absence of water institutions per se but rather the existence of a multitude of water-related institutions, many with overlapping and conflicting missions that contributed to inefficiency in water provision. A similar conclusion is reached by the Department of Water Affairs and Forestry (DWA) which asserted that the proliferation of water institutions during apartheid largely contributed to duplication of functions with no clear responsibilities. There was no central department of government to coordinate the various water institutions and water resource management in South Africa (Nnadozie 2011). Muller (2000) provides a comprehensive review of the institutional setting of water supply and sanitation in South Africa under apartheid, and makes the following observations: (i) institutional fragmentation worsened due to the proliferation of autonomous and semi-autonomous government structures under policies of territorial and political segregation; (ii) communities with the greatest water needs had the weakest institutional structures; (iii) all levels of services were characterised by low levels of accountability because of institutional fragmentation; and (iv) financial inequity and economic inefficiency in the allocation of water and sanitation services increased during apartheid.

3 An Inherited Problem

As a result of apartheid policies the new democratic government inherited huge backlogs in the provision of basic services – housing, electricity, waste disposal, water, and sanitation. It has been estimated that by 1993 about 16 million people lacked access to potable clean water and 20 million had no access to safe sanitation (DWA 1994). Faced with this problem, the new government instituted fundamental policy changes, as reflected in a number of policy documents such as the White Paper on Water and Sanitation (1995), the Water Services Act (1997), the White Paper on National Water Policy for South Africa (1997), and the National Water Act (1998). These sweeping water policies and legislations were introduced so as to quickly address severe water allocation inequalities and imbalances.

The first major policy drive started with the government's approval of the Water Law Principles in 1996. These principles were constructed along the lines of the Dublin Principles on Water Management, and contained the following salient points (Earle et al. 2005): (i) the abolition of water rights and the discontinuation of private ownership of water; (ii) the establishment of an environmentally sustainable and economically efficient measure of water benefits as a criterion for water resource management and allocation; (iii) provide economic instruments for the management and control of pollution; and (iv) the beneficiaries of water use should contribute to the cost of its establishment and maintenance.

Following the crafting of the Water Law Principles, two additional pieces of legislation were passed: the Water Services Act (Act 108 of 1997) and the National Water Act (Act 36 of 1998). The Water Services Act 1997 prescribes the legislative duty of municipalities as water service authorities to supply water and sanitation according to national norms and standards (DWA 2013). Hence the Water Service Act 1997 sets the legislative ground for water and sanitation in terms of:

- guaranteeing everyone the "right of access to basic water supply and basic sanitation"
- pre-setting national standards and norms for water prices
- providing a regulatory framework for water services institutions and water service intermediaries
- establishing water boards and water service committees, making transparent their powers and duties, and
- offering financial assistance/subsidies to water institutions.

Another important piece of legislation was the National Water Act (Act 36 of 1998). The National Water Act 1998 sought to ensure that the country's water resources are protected, used, developed, conserved, managed, and controlled in a sustainable and equitable manner for the benefit of all people. The Act assigns the national government as the public trustee of all water resources. Because of the highly inequitable distribution of water due to past discriminatory laws, under the National Water Act the government assumed direct control, ownership, and overall responsibility for water resource management, including equitable distribution to all citizens. In 2000 the government adopted the free basic services policy encompassing all basic services – water supply, sanitation, solid waste collection, and electricity. In particular, the government instituted the popular 'free basic water' (FBW) policy. The FBW supply is defined as the "minimum quantity of potable water of 25 litres per person per day or 6 kilolitres per household per month". In the case of a common stand water tap, this must be placed within 200 m of a household.

By declaring water and sanitation provision a 'human right' for its citizens, the South African government has advanced beyond many countries in Africa, as only a few have taken such a bold stance. This policy has been lauded internationally and credited as the main policy force behind South Africa's phenomenal achievement in water provision within its short history of democracy. The Bill of Rights (Chapter 2 of the Constitution of Republic of South Africa) affirms the "democratic values of human dignity, equality and freedom" (Earle et al. 2005). Section 27 of the Constitution declares that everyone has the "right of access to basic services including sufficient food."

4 Water and Sanitation in Post-apartheid South Africa

As discussed earlier, the government has exceeded MDGs in water and sanitation provision. Current statistics (Statistics SA 2012) indicate that at least 89.5 % of the population in South Africa have access to clean water – a substantial increase from 56 % in 1993. Disaggregating this figure, 43.3 % have piped water in their homes, 28.6 % have access to water in their yards, 2.7 % use a neighbour's tap, and 14.9 % use communal taps. For sanitation, 78.2 % have access to safe sanitation and this represents a significant improvement from 43.4 % in 1993. Table 5.1 provides an

Table 5.1 Water and sanitation backlogs 1994–2012

Province	Backlog 1994	Backlog 2001	Backlog 2006	Backlog 2008	Backlog 2012	% backlog (2012)	% no access (2012)
(a) Water backlogs (people)							
Eastern Cape	3,689,500	3,117,600	1,579,000	1,097,700	549,839	32.1	22.2
Free State	624,800	404,900	207,400	39,300	39,165	4.8	2.2
Gauteng	1,235,500	898,200	683,000	352,800	180,473	4.6	1.8
KwaZulu Natal	3,863,600	4,047,900	2,584,200	2,128,800	549,774	21.6	14.1
Limpopo	2,405,600	2,172,300	1,484,700	1,081,200	386,445	27.3	14.0
Mpumalanga	1,221,200	1,016,600	731,800	474,400	205,911	19.1	12.6
Northern Cape	1,030,800	909,800	612,900	381,400	173,857	16.4	8.4
North West	392,700	170,800	117,200	78,800	27,769	9.2	2.6
Western Cape	1,426,600	364,500	217,700	125,500	53,320	3.3	0.9
National	15,890,200	13,102,600	8,217,800	5,759,900	2,157,600	8.8	8.8
(b) Sanitation backlogs[a] (households)							
Eastern Cape	1,001,500	899,600	1,658,900	514,800	677,263	40.1	12.7
Free State	399,800	348,400	266,300	265,500	194,177	23.6	3.5
Gauteng	612,100	511,600	427,100	425,600	433,309	11.1	1.1
KwaZulu Natal	939,900	1,063,600	672,700	604,200	820,899	32.3	6.3
Limpopo	800,400	840,300	711,700	665,700	881,665	62.2	7.2
Mpumalanga	421,800	434,300	369,500	359,600	460,041	42.8	6.3
Northern Cape	377,500	423,000	343,500	324,400	450,552	42.4	5.8
North West	143,100	85,600	61,500	59,000	73,269	24.3	8.0
Western Cape	388,700	153,200	94,700	92,700	146,776	9.0	3.1
National	5,084,900	4,759,700	3,525,800	3,311,500	4,137,951	28.8	5.2

Source: DWA (2013)

[a]Figures indicate number of households

overview of government achievement in reducing water and sanitation backlogs across provinces. In 1993 a total of 15.9 million people were without access to water and 20.6 million lacked access to safe sanitation. In 2012, 2.2 million households were still without access to clean water, but this translates to a reduction of more than 80 % in water backlogs. Across most provinces (except Eastern Cape, KwaZulu Natal, Limpopo, and Mpumalanga) water backlogs have been reduced to less than 10 %. There are a number of reasons for this achievement:

- Coordinated and revitalised water institutions with strong linkages at local, provincial, and national level of governance have been developed.
- Public–private partnerships in water resource management have increased, including allocation, distribution, operations, and maintenance.
- An increase in national budget allocations and a policy commitment by government to water infrastructure reconstruction and development.
- Enactment of supportive pro-poor water policies, particularly the free basic water supply policy.

For sanitation provisions, while 20.6 million people lacked safe sanitation in 1993, this had been reduced to 16.4 million people (5.2 million households) by 2012, a 20.4 % reduction (Table 5.1). Except for Gauteng and Western Cape, all other provinces are struggling to clear sanitation backlogs. Worst affected provinces are Eastern Cape, North West, Limpopo and KwaZulu Natal. Thus, compared to water, the government lags behind in sanitation provision as it is still grappling with significant backlogs.

5 Challenges of Water and Sanitation Provision

Despite the remarkable achievements in water and sanitation provision over the nearly two decades of democratic transition, a number of challenges still remain, and these are summarised in Table 5.2. The challenges can be classified into five broad categories: governance, institutional, financial, community, and water resources. Below we discuss these challenges.

5.1 Governance/Policy Challenges

5.1.1 Urban Sprawl and the Proliferation of Slum Settlements

The proliferation of slum settlements poses the greatest challenge for many urban municipalities. Statistics SA (2012) estimates that approximately 14.1 % of South Africa's 50 million people currently live in slum settlements located predominantly on the periphery of urban areas. UN-Habitat (2006) predicts that the urban population in South Africa will grow at 5.8 % per annum. A majority of the new urbanites

Table 5.2 Main challenges affecting water provision in urban areas of South Africa

Category	Nature of challenge
Governance/policy	No strict enforcement of regulations
	Inadequate by-laws/regulations
	No will to control sprawling of slum settlements
	Unsustainable policy
Institutional	Lack of human capacity
	Conditions and age of infrastructure
	Lack of maintenance
	Badly designed infrastructure
	High levels of leakage
	Corruption
Financial	No revenue collection
	Tariffs not affordable
	Cost of supply greater than income
	Equitable Share Grant funds less than operation and maintenance costs
Community	Excessive demand and expectation crisis
	Culture of non-payment
	Vandalism, theft, and illegal connections
	Inadequate education
Water resources	Insufficient water supply
	Drought impact
	Pollution and deterioration of water quality
	Threat of waterborne diseases

Source: Adapted from DWA (2013)

flocking to cities are poor and live in insecure, squalid, and over-crowded conditions (Makaudze et al. 2011). Many municipalities have demonstrated lack of political will to deal with slum settlements; enforcement of bylaws and regulations to govern and promote orderly settlements is lacking, so the current haphazard type of settlement is presently the norm.

The situation is worsened by a high influx of illegal immigrants, estimated to be over six million people flocking from neighbouring countries like Zimbabwe, Malawi, and Mozambique. This creates further problems for municipalities in several ways: first, informal settlements grow haphazardly, including the invasion of preserved land often not suitable for human settlement (e.g. flood-prone areas); second, the logistics of providing water and sanitation services under such conditions are difficult to implement, monitor, and enforce; third, with a high influx of internal and external migrants moving into cities each year, it is difficult for municipalities to plan, manage, and budget for water and sanitation provision in slum settlements; fourth (as discussed later), most of these households are poor and cannot afford to pay for water services; and fifth, many of these settlements have become enclaves of socio-economic problems such as HIV/AIDS, prostitution, rape, and drug-related crime. It is estimated that about 16.2 % of residents in slum settlements in South

Africa suffer from HIV/AIDS, requiring special consideration for water and sanitation provision (Makaudze et al. 2011).

5.2 Institutional Challenges

5.2.1 Lack of Skilled Manpower

Lack of skilled manpower has badly affected the ability of many urban municipalities to deliver efficient water services. In particular, the shortage of water engineers is critical. A research study conducted in 2005 found that on average there are currently three engineers per 100,000 people manning civil water-related works within cities – a drop from 20 engineers per 100,000 during apartheid. Such a ratio is too low to deliver, operate, and maintain water in cities in a sustainable and efficient manner.

5.2.2 Deteriorating Water Quality

According to household surveys conducted by Stat SA (2012), only 62.1 % of those surveyed rated the quality of water-related services as 'good' compared to 76.3 % in 2005. About 7.5 % of the respondents declared that their water was not safe to drink, 8 % indicated that their water was not clear, 8.9 % said it tasted bad, and 11.1 % said their water was not free from bad smells. Thus overall perception is that water quality in urban cities has been deteriorating. The increasing child mortality for children under 5 years of age is evidence of deteriorating water quality in South Africa. The major cause of these deaths is diarrhea, which is prevalent in slum settlements. Diarrhea is a symptom of an illness that can be produced by high levels of *E. coli* in drinking water – an indication of poor water quality.

5.2.3 Severe Water Losses and Leakages

Many cities suffer severe water losses. The Water Research Commission reports that 36.8 % of South Africa's municipal water is lost before it reaches customers. Most of the losses occur as a result of leaks and deteriorating water supply systems. As a result, the government incurs huge losses, estimated as 1,580 million liters of water each year, the equivalent of ZAR7billions in revenue. This does not augur well for a country like South Africa which is regarded as 'water stressed'. Table 5.3 shows losses based on three indicators: non-revenue water (NRW) loss, water loss, and infrastructure leakage index. Across all metropolitan cities, the indicators show significant water losses. For instance, cities like Buffalo City and Nelson Mandela Bay record the highest losses with NRW greater than 40 %. Johannesburg has the

Table 5.3 Percentage of NRW and water losses across major metro cities

Metro area	NRW (%)	Water losses (%)	ILI[a]
Johannesburg	37.8	36.5	8.3
Tshwane	23.8	22.9	5.6
Ekurhuleni	40.8	31.8	5.0
eThekwini	35.6	35.3	6.8
Cape Town	20.7	15.2	2.1
Nelson Mandela Bay	43.1	32.4	4.4
Buffalo City	47.7	28.9	3.5
Mangaung	39.5	35.7	6.0
Average	**33.7**	**29.7**	**5.7**

Source: DWA (2013)

[a]Infrastructure Leakage Index (*ILI*) – values range from 1 (best practice), to 10 (implying high leakages and bad practice)

highest ILI (8.3), symptomatic of serious water leakage losses. Cape Town recorded the lowest scores across the three indicators. Overall, across all cities, more than one-third of water is lost as NRW or pure water losses and high ILIs (≈6) point to high leakages.

5.2.4 Aging Water Infrastructure

Not many water services authority have managed to maintain proper management of water service infrastructure, with some water infrastructure having outlived its life span. As a result many municipalities are experiencing frequent water service failures resulting from non-functionality of their regulatory plans. DWA (2013) reported that in 2012, 71 water-related protests occurred largely as a result of non-functional water infrastructure.

5.3 *Financial Challenges*

5.3.1 Huge Municipal Water Deficits Are Unsustainable

At the national level, municipalities face massive water debts estimated to total ZAR870 billion (US$87 billion) due to non-payment for water, predominantly by households (Department of Cooperative Governance and Traditional Affairs 2013). This is of serious concern as it inhibits the ability of municipalities to sustain efficient delivery of water and sanitation services. While high levels of poverty and unemployment are largely to blame, as they render many people unable to afford water, weak revenue collection systems by municipalities has also contributed to the problem.

At the municipal level, although the free basic water policy is mainly aimed at benefiting poor households, the poor have not always been the primary beneficiaries. The policy has entailed huge costs, estimated at ZAR1.5 billion or 0.15 % of GDP. The deficit is projected to increase further to 0.25 % of GDP by 2015. The policy has largely been financed either through subsidies from the national government via such sources as the Equitable Share grant, Municipal Infrastructure grant (MIG), or through cross-subsidies from other users or local taxes.

A fragmented institutional service delivery approach has also contributed to inefficiencies in water provision, resulting in waste of resources. For instance, responsibilities for water services have been delegated to a number of government departments such as DWA, Department of Human Settlement, and the Water Services Authorities. From statistics provided by DWA (1994), out of 169 service providers, 29 provide free basic water to all their residents, 136 to some, and 4 (very small municipalities) to none.

This highlights three related issues. First, unless there is careful planning and implementation of the free basic water policy, the policy can actually be a source of inequality across municipalities, with some delivering and others not. Second, it can be argued that free basic water largely benefits the people who can afford to pay for it. This follows the widespread sentiment that the free basic water policy has been implemented more successfully in wealthier municipalities than poor ones. Recently, the government suggested that it will review the implementation strategy for free basic water and possibly revisit the need for developing registers of poor users, with the overall thrust being to improve subsidy-targeting. Third, the salient issue of treating water and sanitation as a basic human right has also introduced its own challenges. In essence, this has transformed water and sanitation into highly politicised commodities.

The number of cases of water-related conflicts between residents and municipalities ending up as Constitutional Court hearings is increasing. A few illustrative examples have been the Gauteng Constitutional Court case between Johannesburg Water versus Phiri residents (2008) who resisted the installation of prepaid water meters (referred to as the Mazumbuko case); the case of City of Cape Town versus Makhaza slum settlements, with residents resisting an "open toilets saga" (2009); and the recent "feces or toilet war" (a war cry in 2013 against the bucket toilet system in Cape Town). The "toilet war" is not only whether or not sanitation is provided but also about the standard of provision (Penner 2010). Residents voice strong preference for "flush toilets", which in their views, imbues dignity and respect and seen as being at par with counterparts in formal settlements. With water and sanitation being transformed into highly sensitive politicised commodities, many municipalities have found themselves in a difficult financial situation as they are not only unable, but also afraid to implement effective cost recovery strategies lest they face unpopular uprising and unrest by local residents.

5.4 Community Challenges

5.4.1 Pervasive Culture of Non-payment for Water Services

A more complex problem is that there exists a culture of non-payment for basic services including water and sanitation. In this regard Indigent policy sends a wrong signal. Although the policy is morally and socially justified (as it addresses equity and poverty concerns for poor households), it perpetuates the 'free lunch' mentality that makes provision of basic services unsustainable. To explain non-payment of services, McDonald and Pape (2002) offered two explanations: first a 'culture of non-payment' and second the issue of 'non-affordability'.

Historically, during apartheid, people (especially blacks) boycotted payment for services on political grounds. Unfortunately, this culture of non-payment persists, even under the new democratic dispensation. Several arguments have been advanced to explain this behaviour: some argue that, as enshrined in the Bill of Rights (1996, s27.1.b), citizens have every right to continue receiving free services; others argue that it is a 'crisis of expectation' as people become increasingly frustrated with corruption within the public sector and the government's failure to deliver on provision of key basic services, water and sanitation in particular. For instance, the fact that 20 % of households still use the bucket system of sanitation two decades after democracy began underlies a deep problem.

The second explanation relating to non-payment for service is simply the issue of non-affordability. A significant proportion of households living in slum settlements cannot, because of high levels of poverty, afford to pay for municipal services. Close to two-thirds of households living in slum settlements survive on incomes of less than ZAR2,000 per month and many depend on the social grant system. A few of the employed are burdened with other pressing financial obligations (e.g. school fees, health care, funeral) and hence severely constrained in their ability to pay water bills.

5.4.2 Expectation Crisis

The end of apartheid and the ushering in of a new democratic government in 1993 raised high hopes among many people in South Africa, particularly blacks. Many expected increases in job opportunities and employment, a better quality of life, improved delivery on basic services (especially water, sanitation, electricity, and housing), political freedoms, and human rights. Two decades later, many people appear disillusioned and their spirits broken. For many people despair has given rise to anger, and anger has led to violence. Despite scoring commendable progress, government policies have fallen short in meeting the expectations of the general population. In recent years we have seen growing social protests across the entire country as people vent their anger and frustration against poor service delivery, water and sanitation included. Today, it is the provision of basic services – water and sanitation – that constitute the most formidable challenge to post-apartheid South Africa.

5.4.3 Theft and Vandalism

Acts of theft and vandalism are rife in slum settlements and these impose huge and often unplanned cost on maintenance of water and sanitation infrastructure by municipalities. For example in Cape Town, every ZAR1.00 invested in water and sanitation in Khayelitsha, between 10 and 20 cents goes towards repairs and replacements due to theft and vandalism (Makaudze et al. 2011). For instance, taking the total budget for water and sanitation in the city of Cape Town during the 2010/2011 financial year of ZAR19.6 billion, it follows that approximately ZAR1.8 billion had to be expended on repairs and replacements due to theft and vandalism. If extrapolated to the entire country, this runs into billions worth of water and sanitation infrastructure lost due to theft and vandalism.

5.5 Water Resource Challenges

5.5.1 Unprecedented Increase in Water Demand

South Africa is a water-stressed country with limited freshwater resources. Otieno and Ochieng (2004) predict that the country is likely to experience physical water scarcity by 2025. Many cities in South Africa are not only experiencing an explosion in population growth, but also are undergoing rapid growth in industries such as agriculture, manufacturing, and mining. With such a rapid expansion of the economy, the demand for water will grow. While new infrastructure for water will be necessary, water supply resources are limited and hence the key to sustenance of water resources is effective water conservation and water demand management strategies (including better water cost recovery strategies). Failures in water supply and demand management would entail huge costs that could hurt the entire economy.

6 Conclusion

Unless improved policy measures and planning are put in place at both the national and local government levels to address the challenges of providing water and sanitation in urban slum settlements, then South Africa could face social unrest. Already there are ominous signs that the country could experience violence and social protests as people vent their anger and frustration against poor service delivery – water and sanitation included. The xenophobic attacks (2009/10), the 'open toilet' saga (2010/11), and the current 'feces protests' are all evidence of swelling discontent by the general population against poor service delivery in urban areas. This paper has highlighted the challenges, and key amongst them include a continued rise in water and sanitation backlogs in slum settlements, poor cost-recovery systems, a pervasive culture of non-payment for water services, and huge and unsustainable water deficits. These challenges are hindering the ability of municipalities to provide sustainable and efficient services. They demand urgent attention.

References

Department of Cooperative Governance and Traditional Affairs. (2013). *South Africa's municipalities owed billions of rands due non-payment for water by residents*. Mail and Guardian press statement.

Department of Water Affairs and Forestry. (1994). White paper on the *water supply and Sanitation*. Pretoria.

Department of Water Affairs. (2013). *Strategic overview of the water sector in South Africa*. Pretoria: Department of Water Affairs.

Earle, A., Goldin, J., & Kgomotso, P. (2005). *Domestic water provision in the democratic South Africa: Changes and challenges*. Paper for Nordic Africa Institute's conflicting forms of citizenship program, University of Pretoria.

Lefebvre, M., Morardet, S., Montginoul, M., & Farolfi, S. (2005). *How to finance multiple use water systems for the rural poor? Lessons learnt from the domestic water sector in the Oliphant River Basin, South Africa*. Paper to an international workshop on "Water poverty and social crisis", Agadir.

Makaudze, E., du Preez, M., & Potgieter, N. (2011). *How does the HIV and AIDS epidemic in South Africa impact on the water, sanitation and hygiene sectors?* Pretoria: Water Research Commission.

McDonald, D. A., & Pape, J. (2002). *Cost recovery and the crisis of service delivery in South Africa*. Cape Town: Human Science Research Council.

Muller, M. (2000). *Transforming water law to achieve South Africa's development vision: A case study in national law*. Second World Water Forum, The Hague.

Nnadozie, R. C. (2011). Access to adequate water in post-apartheid South African provinces: An overview of numerical trends. *Water SA, 37*(3), 339–347.

Otieno, F. A. O., & Ochieng, G. M. M. (2004). Water management tools as a means of averting a possible water scarcity in South Africa by the year 2025. *Water SA, 30*, 120–124.

Penner, B. (2010). *Flush with inequality: Sanitation in South Africa*. http://places.designobserver.com. Accessed 10 Dec 2013.

Republic of South Africa. (1997). *Water services act*. Government Gazette of the Republic of South Africa, Cape Town, 390 (18522).

Rwida, U. (2013). *Mortgage default insurance: Workshop with PC: Human settlements*. Johannesburg: National Treasury.

Statistics South Africa. (2012). *General household survey*. Pretoria: Statistics South Africa.

Tshikotshi, V. (2009). *The challenges of eradicating informal settlements in South Africa by 2014: The case of Seraleng sustainable human settlement, Rustenburg local municipality, North West Province*. MSc thesis, University of Witwatersrand, Johannesburg.

UN-Habitat. (2006). State of the world's cities 2006/07. Nairobi. Kenya.

Chapter 6
Integrating Water Quality into Urban Water Management and Planning While Addressing the Challenge of Water Security

Françoise Bichai and Patrick W.M.H. Smeets

1 Introduction

Public water services, including clean water supply and sewerage services, have been crucial in improving public health and hygiene, and centralisation of drinking water supply systems has contributed to eradicating a wide range of waterborne diseases. However, in the face of growing demand pressures, urbanisation, and climate change, freshwater resources are becoming more scarce and supply planners are turning to less traditional water sources, such as treated wastewater and urban run-off (stormwater), sources which may pose health risks to consumers. At the same time, traditional surface and groundwater resources are being subject to increased contamination. Pollution of freshwater resources diminishes water security, so proper treatment and recycling of polluted run-off and wastewater could achieve the twofold benefit of increasing water quantity and improving water quality.

In rural and agricultural landscapes, wastewater and surface water of poor quality have long been used for low-grade applications such as crop irrigation. In the urban context, highly centralised systems usually provide potable quality water for all kinds of uses. People use 45–455 l of drinking water daily (IWA 2012), of which only 0.1–1.55 l (0.02–3.4 %) is used for drinking (Mons et al. 2007). However, its high quality also prevents any risk to health through showering or food preparation,

F. Bichai, PhD (✉)
Harvard Water Security Initiative, School of Engineering and Applied Sciences,
Harvard University, 29, Oxford St., Pierce Hall, Cambridge, MA 02138, USA
e-mail: francoisebichai@gmail.com

P.W.M.H. Smeets, PhD
Water and Health, KWR Watercycle Research Institute, Groningenhaven 7, P.O. Box 1072,
3430 BB Nieuwegein, The Netherlands
e-mail: Patrick.smeets@kwrwater.nl

Q. Grafton et al. (eds.), *Understanding and Managing Urban Water in Transition*, Global Issues in Water Policy 15,
DOI 10.1007/978-94-017-9801-3_6

or by accidental cross-connection of a dual-pipe system. This traditional approach of providing water as one product of uniform quality for all purposes is only possible when sufficient water is available. With global population growth and expected full exploitation of water sources, the annual per capita availability of water is expected to drop by 30 % over the next 20 years (WBCSD 2006). Today over 430 million people live in countries considered water-stressed. Population Action International (PAI) projects that, by 2050, the percentage of the world's population living in water-stressed countries will increase at least threefold. With less water available to meet consumption needs, a 'fit-for-purpose' water quality approach may need to be introduced to the water supply sustainability culture so as to effectively harness the potential of alternative water sources.

With the introduction of alternative sources, and the delivery of fit-for-purpose water quality, the level of public health safety and protection needs to be both maintained and demonstrated. A systematic but flexible approach is needed to manage public health risks, either by framing and guiding the development of new supply schemes, or by assessing and validating the safety of existing schemes from contaminated sources. Water Safety Plans (WSPs) provide a risk management framework around water intakes and supply systems. Within WSPs, questions arise such as the tolerable health-risk target to be achieved, the efficacy of contaminant barriers, and the effect of system failures on health. In WSPs, a semi-quantitative approach is taken based on experience and WSP-workers' knowledge. However, with new water supply schemes, experience is lacking. Being overly conservative towards public health protection may increase costs and environmental impacts, or discourage the industry from innovating with alternative water supply schemes. Therefore, objective, science-based risk quantification is needed to support a decision to implement any risk-management measure, with the aim, ultimately, of raising public acceptance of alternative schemes.

Quantitative Microbial Risk Assessment (QMRA) is a method that allows quantitative estimates to be made of microbial risks related to exposure of humans to water, either through drinking or other uses. Water-related outbreaks of illness demonstrate that the risk from waterborne pathogens remains the most acute and prominent risk associated with the human use of water (Hrudey and Hrudey 2004). Although novel contaminants are of increasing concern to public health authorities, their impact can usually be determined by water quality monitoring, whereas microbial risks are a real threat to public health at contamination levels far below pathogen detection limits. In addition, microbial contamination of natural water sources can vary on timescales of just hours. This chapter focuses on controlling microbial risks in urban water systems. Nevertheless, the QMRA approach is transferable to chemical contaminants, given that there is sufficient scientific data to characterise their health impacts.

In some cases QMRA can help support decisions on water supply schemes. In this chapter, the role of a systematic quantitative risk management approach, such as QMRA, is described as an important contribution to urban water security and water quality management. This role answers a fourfold need: (i) for technical guidance in the design of alternative supply schemes; (ii) for regulation to protect public health, both in traditional and alternative supply sources; (iii) for regulatory frameworks and institutions to enable innovation and development; and (iv) to assess new risks from innovative supply schemes and compare them to traditional water supply or other public health risks.

More broadly, consideration of water quality and public health risks can point towards smart decisions on water supply schemes, as certain 'green' approaches may not always be effective solutions if they allow public safety to be compromised.

2 Water Quality Challenges of Water Quantity Solutions

2.1 Alternative Water Supply Sources

When traditional surface and groundwater resources are depleted, polluted, or insufficient to meet increasing water demand, strategic measures on both the demand and supply sides are needed to achieve water security. In order to augment available water supplies in water-stressed areas, alternative water supply sources, as listed in Table 6.1 below, are often considered. Planning of such schemes must ensure there are no potential routes by which public safety is compromised.

Among the alternative water sources listed in Table 6.1, some, like rainwater and stormwater harvesting, are highly rainfall-dependent. Variability and uncertainty of rainfall patterns is one of the major challenges of climate change. Rainwater harvesting usually involves rain collection on clean rooftops. Stormwater is collected through an urban drainage system (separate from sewage collection), and is therefore subject to contamination by urban pollution from animals, cars, rubbish, etc. In certain locations, retaining stormwater in small storage facilities can have minor flood-mitigation benefits, and its treatment and reuse can help protect aquatic ecosystems by reducing pollution from urban run-off and erosion due to peak flows.

During prolonged droughts, domestic wastewater, or 'used water', can become an important resource. Being less affected by rainfall, it is more reliable year-round. Table 6.1 distinguishes between greywater, blackwater, sewer mining, and centralised municipal wastewater as sources. Greywater is generally considered at the household level, and includes all used water except human sewage (from toilets), i.e. water from the shower, sink, laundry, and dishwasher. Blackwater and sewer-mining options rely on domestic wastewater (including human sewage) upstream from the municipal wastewater treatment plant. They are small to medium-scale, on-site (or near-site) reuse options. Blackwater can typically be reused for non-drinking uses in large buildings. Sewer-mining schemes can be of larger scale, and depend on the available flow to be extracted from the sewer and on the demand nearby, typically from municipal green space irrigation, parks, sports fields, or golf courses. On-site treatment is usually involved. Table 6.1 provides an indication, based on *Cryptosporidium* concentrations, of the contamination level of each source.

Table 6.1 Alternative water supply sources and typical intended scales and uses

Sources	*Cryptosporidium* (oocysts/L)	Scales	End uses[a]
Rainwater	0.01–0.1 (Ahmed et al. 2010)	Small *(household/building)*	Residential non-drinking[b]
			Shower[c]
			Drinking[d]
			Green space irrigation
Greywater	1 (Birks et al. 2004)	Small *(household/building)*	Residential non-drinking[b]
Blackwater	2000 (NRMMC–EPHC–AHMC 2006)[f]	Small–medium *(building)*	Residential non-drinking[b]
Sewer mining	2000 (NRMMC–EPHC–AHMC 2006)[f]	Medium	Residential non-drinking[b]
			Green space irrigation[e]
Stormwater	1.8 (NRMMC–EPHC–NHMRC 2009b)[g]	Medium	Residential non-drinking[b]
			Green space irrigation[e]
			Industrial/commercial demand
			Fire fighting?
			Potable demand
Treated municipal wastewater (centralised)	600 (NRMMC–EPHC–AHMC 2006[h]	Medium–large	Residential non-drinking[b]
			Green space irrigation[e]
			Industrial/commercial
			Fire-fighting
			Agriculture irrigation
			Potable demand

[a]Based partly on an approach developed by the Australian water industry, described in Bichai et al. (2013)
[b]Residential non-drinking uses include: outdoors (e.g. garden watering), toilet flushing, and laundry
[c]Potential application for hot water systems
[d]Potable uses occur, although not under regulated control (e.g. Australia)
[e]Including sports fields, golf courses
[f]Raw sewage value used in the AGWR (NRMMC-EPHC-AHMC 2006); the WHO Guidelines for Drinking Water Quality (2011) provide a range of 1–10,000 oocysts/L
[g]Based on the AGWR–Phase 2 on Stormwater Harvesting and Reuse (NRMMC-EPHC-NHMRC 2009b)
[h]Approximate concentration in secondary treated effluent, assuming a minimal 0.5-log removal from raw sewage based on the indicative AGWR values (NRMMC-EPHC-AHMC 2006)

Centralised municipal wastewater reuse schemes require collection of domestic wastewater (with some contribution from industrial discharge) and treatment at the municipal plant, often followed by additional advanced treatment to allow recycled water uses. They can supply fit-for-purpose water to nearby irrigation or industrial schemes, or provide, via dual-pipe systems, recycled water for non-drinking residential use in new developments. Ultimately, advanced treatment can allow the re-injection of recycled water into the main water supply reticulation system. It permits potable reuse, either directly (as done in Windhoek, Namibia, for nearly 50 years) or through environmental buffers – such as surface water supply reservoirs (e.g. Singapore) or aquifer storage (e.g. California or, more recently, Perth, Australia). Such schemes, however, still face major public reluctance in many parts of the world.

2.2 *Alternative Water Supply: Context and Challenges*

The rise of the fit-for-purpose water quality approach, which is often viewed as resource-efficient and part of a 'green' paradigm, entails, in an urban context, a duplication or even multiplication of water reticulation systems, since separate pipes are needed to supply water of different quality. One or more alternative water pipes are installed next to the drinking water supply pipes. Using a centralised approach, the building of dual-pipe systems in new developments introduces extra up-front costs, while retrofitting such a system in existing residential or commercial areas is generally even less cost-efficient. Alternatively, a multitude of decentralised systems can be used, which involves a larger paradigm shift in water and sewerage infrastructure management. Decentralisation also means a lower level of control over public health protection and risk management related to water quality. Even though the effect of one decentralised system is smaller, the number of such systems needs to be higher in order to serve the total population, making the task effectively larger. To prevent increasing health risks, the level of safety of each individual system needs to be of just as high a standard as a centralised system. In Europe, centralised water supplies are almost 100 % compliant to drinking water standards, whereas only about 60 % of the small supply systems comply, and microbial problems prevail (Hulsmann and Smeets 2011). This suggests that, overall, quality and risk management of multiple small, decentralised systems can become more challenging than with one large centralised system.

In general, the centralised approach offers an economy of scale, while small-scale systems are more costly. Extra pipe placement and maintenance costs can be balanced, to some extent, by savings on treatment due to providing a lower quality water, by preventing transport of water over large distances, and ultimately by avoiding or deferring costs of large system augmentations. Nevertheless, the actual production costs for drinking water are very low, even when a contaminated surface source is used: in Amsterdam, for instance, the life-cycle costs for water production are only €0.10/m^3 (Barrios et al. 2008), while the tariff to consumers is around €1.76/m^3 (VEWIN 2007). In addition to treatment costs, the tariff covers costs of distribution, water quality monitoring, institutional costs, etc., which will also be incurred for alternative water supply. Hence, the benefits of producing lower quality water are, in terms of overall cost, likely to remain small. When combining financial and environmental costs and benefits of some dual-pipe systems, Chen and Wang (2009) concluded that a system involving dual-pipe collection and greywater treatment and reuse could only possibly reach a net benefit value (NBV) greater than zero if all greywater is reused and environmental benefits are included in the NBV.

In terms of public health, dual-pipe water distribution systems generate concern due to well-recognised health risks of cross-connections between the fit-for-purpose pipes and the drinking water pipes, which can lead to disease outbreaks (Anonymous 2003; Storey et al. 2007). The costs and environmental impacts of disease outbreaks, or even incidental illness, generally outweigh the benefits of alternative supplies. In the Dutch example of Leidsche Rijn, over a 1-year pilot period ending in December 1999, several dozen people fell ill due to inadvertent cross-connections in the dual-pipe system, a system which was intended to provide pre-treated surface water for household non-drinking uses only; however, intrusion of lower-quality water into the drinking water system occurred (Anonymous 2003). Since there is no shortage of water in the Netherlands, and economic and environmental benefits are marginal if any, this serious outcome led to a ban of large-scale dual-pipe systems. The historical and contemporary challenges of the Dutch water supply industry do not revolve around insufficient water, but lie in supplying safe services from highly contaminated water resources.

In Australia, the water availability challenge is real: pressure on water resources is increasing in urban areas, climate variability is creating high water stress, and water systems need to adapt accordingly. In various arid and semi-arid countries, urban centers face similar conditions. To prevent or limit the serious consequences of water insecurity, some cities may not have the luxury of ignoring alternative supply options, but instead must initiate risk management strategies for extracting the most from all available water resources and produce safe water. QMRA can play an important role in developing efficient risk-management frameworks for water supplies facing the challenges of urban water security.

3 QMRA for Safety Management in Urban Water Services

3.1 What Is QMRA?

QMRA is a method that allows estimating the risk, in terms of probability of infection, based on two primary factors: the human dose–response parameter for a given pathogenic microorganism and the ingested pathogen dose from exposure to contaminated water. Typically, exposure to water is assessed through the drinking route, i.e. based on the average volume of water that people drink daily. Exposure may also be estimated, for example, as the volume of water accidentally ingested during showering or bathing, or as the volume of water that could be ingested from irrigated crops when eaten raw. The main equations involved in QMRA calculations for water safety assessment (Haas et al. 1999) are summarised in Table 6.2.

The daily pathogen dose to which humans are exposed (through various water uses) depends firstly on the volume of water used, and secondly on the water quality in terms of pathogen concentration. The health risk is then determined based on the ingested pathogen dose and the capability of the pathogen to create an infection in

Table 6.2 Main QMRA equations

Output variable	Equation	Input variables
Probability of infection	$P_{\text{inf (annual)}} = 1 - (1 - R)^n$	R = daily risk of infection
		n = number of exposure days (365)
Daily risk of infection	$R = 1 - \exp(-r \times D)$	r = infectivity parameter describing the host–pathogen interaction (dose–response)
		D = dose (number) of pathogens ingested daily
Daily dose	$D = C_{pathogen} \times V_d$	$C_{pathogen}$ = concentration of pathogens in the supplied water
		V_d = volume of water consumed daily (exposure)

Table 6.3 Risk of infection (per person per day) at an average concentration of one organism per volume of daily water consumption, based on the dose–response relations in the Dutch legislative model *QMRAspot* (Schijven et al. 2011)

	Original publication	Dose–response model	Org/223 mL	P_{inf} (pppd)
Maximum risk[a]		Exponential	1	63 %
Campylobacter	Teunis et al. 2005	Hypergeometric	1	28 %
Cryptosporidium	Teunis et al. 2002a, b	Hypergeometric	1	14 %
Giardia	Teunis et al. 2002a, b	Exponential	1	2 %
Rotavirus	Teunis et al. 1996	Hypergeometric	1	27 %

[a]Maximum risk is the assumption that ingesting a single organism has 100 % probability of causing an infection. At an average concentration of one organism per volume of water. Consumed daily, there is a 63 % chance that one or more organisms are actually present in the consumed water

the host's body (usually via the gastrointestinal tract). The infection parameter varies for each pathogen and each host (due to age, immunity, etc.). The daily risk of infection is expressed in Table 6.2 by using an exponential model to represent the dose–response relationship of the pathogen infecting the human body (e.g. *Cryptosporidium*, *Giardia*). Alternative models that are used to describe this relationship include the beta-Poisson and the hypergeometric model. Table 6.3 provides an indication of the infectivity of some waterborne pathogens (e.g. using *Giardia* as the least infectious of those listed). Table 6.3 shows that ingestion of one single pathogenic organism by a human is likely to create an infection. The reader is referred to Petterson et al. (2006) for further details on dose–response relationships and for an exhaustive description of QMRA calculations.

The annual probability of infection is then calculated for a whole population covered by water services. Some QMRA models (e.g. Australian Guidelines for Water Recycling) use a deterministic approach, where each of the variables included in the equations of Table 6.2 is expressed as the likeliest or standard value (e.g. mean value, 95th percentile, etc.), and risk is calculated as a point estimate. Other models (e.g. the Dutch legislative model) use more complex calculations involving a stochastic approach, where each of the variables is described by a probability distribution function (PDF), and the output probability of infection is expressed as a PDF as well.

An extension of QMRA calculations allows translating the probability of infection (P_{inf}) to the Disability-Adjusted Life Year (DALY) unit, which accounts for the severity of the consequences of an infection on a human life (burden). This procedure allows waterborne infection risks to be compared, from a policy and risk management standpoint, with other environmental and health risks (WHO 2011).

3.2 QMRA: The Regulatory Approach

The absence of an indicator microorganism (such as enterococci, fecal coliforms, or *E. coli*) in a routine monitoring sample (typically 100 mL) has commonly been interpreted and communicated as the absence of risk from consuming the supplied water. The traditional form of the European legislation is framed on that basis (European Union Drinking Water Directive 1998), and so is the current form of the Australian Drinking Water Guidelines (NHMRC–NRMMC 2011). Over the past decades, large disease outbreaks have occurred through the drinking of water in which there were no detectable levels of indicator organisms (Hrudey and Hrudey 2004), a fact which illustrates one of the shortcomings of the indicator concept. QMRA forms the basis of a scientifically structured approach, and emerges as an alternative to indicator monitoring for the regulation of water quality.

QMRA sets a quantitative target to describe the tolerable health risk from water uses within a population, and a systematic approach is taken to verify compliance with that target. Figure 6.1 illustrates how QMRA-based frameworks regulate water quality. The sequence involves (i) source characterisation to express the initial concentration of a selection of reference pathogens in the raw water; (ii) process performance monitoring to calculate the capacity of the multiple-barrier treatment system put in place to efficiently remove such reference pathogens; (iii) calculation of the resulting concentration of reference pathogens in the distributed water; and (iv) calculation of the probability of infection from all reference pathogens in the finished water to demonstrate compliance with the health-risk target. The focus is on demonstrating the safe use of water, from a more or less contaminated source, by achieving appropriate pathogen removal through treatment.

The QMRA approach as a form of regulation has been proven to effectively support water safety management and practice in the Netherlands, a country that has long had to rely on poor-quality surface water sources to supply drinking water to its population. Australia is transitioning towards an increased use of the QMRA approach in water quality regulation, especially at the moment with the Australian Guidelines for Water Recycling (NRMMC-EPHC-AHMC 2006, 2008, 2009a, b). Experiences and challenges of both of these two very contrasting countries in terms of water supply landscape and history are reported in Bichai and Smeets (2013). The WHO also supports the QMRA approach in its recent water quality guidelines (2011). Some countries base their water quality regulations on the same logic as the QMRA approach,

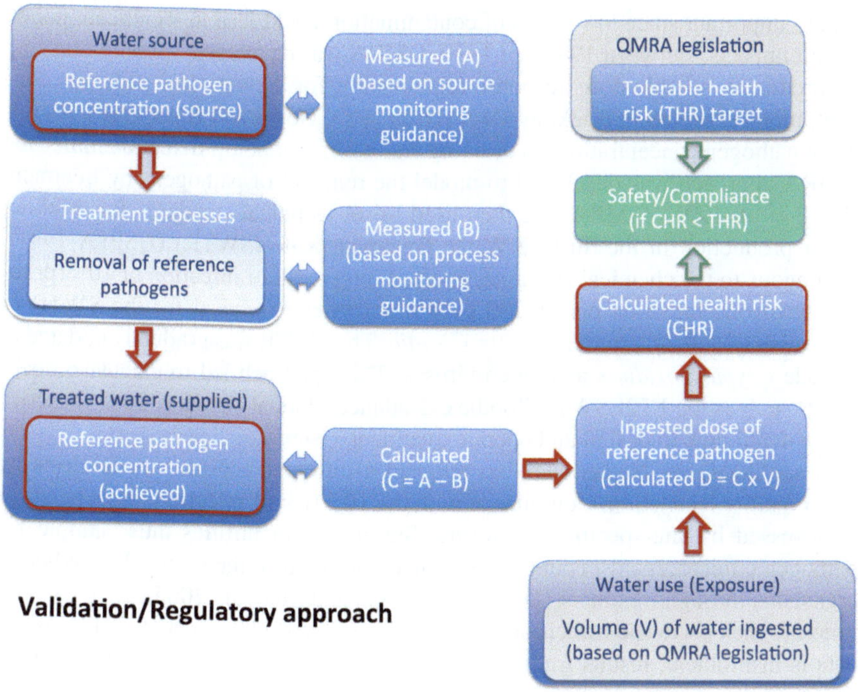

Fig. 6.1 How to use QMRA to validate and regulate safety of water supply

without requiring utilities to perform on-site source and process monitoring, and QMRA calculations to demonstrate compliance with a health risk target. For example in the United States and Canada, regulations provide treatment performance requirements based on a classification of source water quality (USEPA 2006).

In general, regulating water quality in water supply services effectively means deciding on a tolerable health risk associated with water uses and exposure in a serviced population, whether or not this choice is made implicitly or explicitly. The use of a quantitative health-based target in the QMRA approach makes this public health choice explicit, which therefore allows addressing the balance between between water safety, engineering costs, and environmental costs.

3.3 QMRA Context and Uses to Date

The environment has always been a source of health risk, and managing these risks has been a long learning process through history. By the 1970s, it became clear that risks could not be absolutely eradicated, but needed to be quantified and compared in order to find a balance between the costs and benefits of risk mitigation measures. The risk from pathogenic microorganisms in water and food was recognised as highly important since low levels of contamination could lead to significant public health impacts. Between 1980 and 1990, QMRA was developed, initially focusing on producing reliable dose–response relationships for the very low pathogen concentrations expected in drinking water (Haas 1983; Gerba and Haas 1988). As relevant pathogen concentrations in drinking water were far below detection limits, the QMRA approach was developed to model the removal of pathogens by treatment (Regli et al. 1991). QMRA was then used to set technical standards for drinking water production in the Surface Water Treatment Rule (SWTR) (USEPA 1991). Analogous to the chemical risk assessment, a target risk of infection of 10^{-4} (probability of one infection per 10,000 people per year) was used for the SWTR to determine treatment requirements for *Giardia*. The SWTR was gradually updated to include *Cryptosporidium* and enteroviruses. This approach led to extensive documents such as the LT2ESWTR Toolbox Guidance Manual (USEPA 2003) describing in detail how to design and operate treatment systems.

Since 2001, the Netherlands has adopted a different approach in its drinking water quality legislation. Compliance with the 10^{-4} risk of infection target needs to be assessed by site-specific monitoring. Surface water utilities must sample for pathogenic microorganisms in the raw water, and monitor the removal of indicator organisms through treatment (De Roda Husman and Medema 2005). Risk calculations are then performed stochastically using this data (Schijven et al. 2011). When data is insufficient, results from pilot-scale treatment tests or knowledge from the literature are used.

The World Health Organization (WHO) first mentioned the use of QMRA for setting health-based targets and system assessment in the 2004 Drinking Water Guidelines (WHO 2004). A risk target of 10^{-6} DALY (Disability Adjusted Life Years) is generally applied (although there are debates around the need to adapt the target to local circumstances). Using the WHO reference pathogens, this DALY target translates to a $2.5–16 \times 10^{-4}$ risk of infection, depending on the pathogen. The health target can then be assessed directly (following the approach in Fig. 6.1), or translated into technical requirements.

Countries that formerly relied on the traditional absence of fecal indicators to assess water safety are now starting to implement a risk assessment approach. With an increasing need to use more contaminated water sources, the balance between risks, costs, and environmental impact is sought. Australia has formally taken the QMRA turn in water quality regulation with the recent Australian Guidelines for Water Recycling (NRMMC-EPHC-AHMC 2006; 2008; 2009a, b), based on the deterministic QMRA approach. Health targets were set based on the 10^{-6} DALY endpoint (as with the WHO). National guidelines in Australia are (voluntarily) adopted by the States, and integrated in their own regulatory framework, which has progressively happened in all States (Bichai and Smeets 2013). Health Canada has included the QMRA approach in recent Canadian Guidelines for Domestic Reclaimed Water for Use in Toilet and Urinal Flushing (Health Canada 2010) and also developed a QMRA-based tool for utilities to use freely and voluntarily (Health Canada 2011).

QMRA is also applied in other water-related fields. The WHO guidelines for the safe use of wastewater and excreta in agriculture (WHO 2006) promote the use of QMRA to assess risks to farmers and crop consumers, and provide basic data and examples to guide quality control measures. For instance, agriculture is by far the most widespread form of reuse worldwide, with uncontrolled health risks in many developing countries where regulatory frameworks and structures are lacking. Such applications often involve more variables and are typically less controlled than centralised water supply systems, and standardising quality and risk assessment is a complex process. Recent USEPA guidelines for water reuse (USEPA 2012) and for recreational water (USEPA 2012) also refer to the use of QMRA to assess risks for various water reuse schemes. QMRA can help improve insight into the factors that contribute to microbial health risk in bathing water, especially in situations where sewage impact is not the main source of contamination. The use of QMRA in that context is explored in various studies (Ashbolt et al. 2010; Tseng and Jiang 2012).

The aforementioned QMRA applications focus on fecal contamination of water and exposure through ingestion. However, opportunistic pathogens can, under favorable conditions, grow via other routes to numbers that can cause illness. QMRA of *Legionella*, an opportunistic pathogen that causes lung infection after inhalation, has recently been attempted, but lack of knowledge about exposure through inhalation and dose–response relationship prevent real quantitative estimates (Storey et al. 2004; Armstrong and Haas 2008; Schoen and Ashbolt 2011; Thomas 2012; Buse et al. 2012).

In summary, QMRA has been used to gain an understanding of microbial risks both from conventional and new water uses. This insight is used to set health-based targets, to provide technical guidelines, or as a direct site-specific measure of compliance to water quality regulation. Experience has shown the added value of QMRA (Bichai and Smeets 2013), but it has also identified challenges such as gaps in scientific knowledge and data, monitoring and analytical costs, modeling issues, and regulatory implementation challenges, especially for smaller water supply systems. Addressing such challenges can help make QMRA an effective and efficient tool to manage microbial risks, from both new and traditional water sources and uses, while meeting urban water security challenges (Bichai and Smeets 2013).

4 Using QMRA in Alternative Water Supply Planning

4.1 *The Planning and Technical Guidance Approach*

Various water-stressed and expanding cities are experimenting with novel water management options to improve long-term security and sustainability of their water supplies. In wealthier countries especially, there is a growing focus on addressing environmental costs and benefits, as well as the livability of their cities, through water resources planning. The challenge of water availability is therefore leading to an increasingly integrated approach to urban water management. Such models often provide novel planning options to deal with water quantity concerns, taking account of environmental and social impacts. Implicitly, water quality requirements are assumed to be met, or may be left undetermined. To determine water quality requirements while ensuring public health safety, the only systematic approach, and one now widely recognised by scientists and the water industry, is that based on QMRA (Bichai and Smeets 2013).

The QMRA planning approach requires: (i) setting a health risk target; (ii) characterising human exposure (ingested water volume) to microbial hazards (reference pathogens) from specific water uses; (iii) determining the permitted concentration of reference pathogens at end use, based on the health risk target; (iv) characterising source water quality (reference pathogens concentration); and (v) calculating treatment requirements to meet required quality at end use (Fig. 6.2). From that end output (treatment performance requirements), an estimate of treatment costs can be calculated by selecting preliminary treatment processes that meet pathogen removal requirements. Hence, QMRA can help in developing models that comply with integrated urban water management principles, but in which crucial public health concerns are also addressed by including technical water quality requirements in the planning and evaluation stages of supply options.

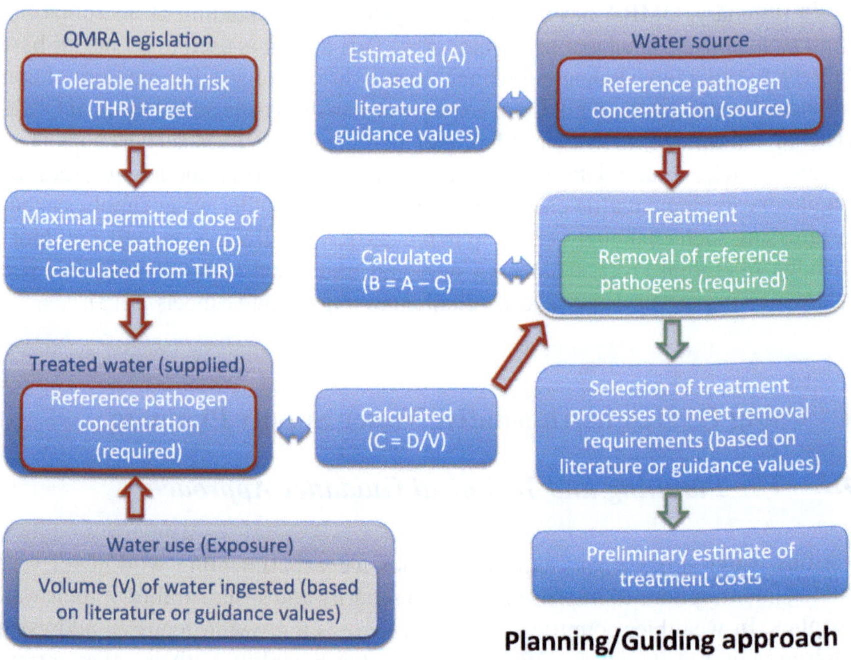

Fig. 6.2 How QMRA can be used to integrate water quality requirements into water supply planning

4.2 QMRA to Improve Understanding of Innovative Water Supply Scheme Risks

Potable reuse, especially from wastewater, is by far the most controversial example of alternative water supply. Because drinking water involves the most direct route for waterborne pathogens to infect humans and the highest level of exposure, health risk concerns are paramount when it comes to distributing recycled wastewater, even highly treated, through the main drinking water supply pipes. In cases of dual-pipe systems, which allow recycled wastewater to be distributed for non-drinking uses, may be too costly or bring excessive cross-connection risks, and here a viable option for water-scarce cities may be centralised potable reuse – if public reluctance can be overcome.

Potable reuse can be viewed as the ultimate reuse scheme, allowing large-scale reuse of wastewater for all uses, and being compatible with the traditional centralised water services approach. A long residence time of treated wastewater in soils prior to distribution may improve public acceptance, and potable reuse through aquifer recharge is becoming common in arid parts of the United States, for example. In Singapore, augmentation of the water supply includes the dilution of highly treated wastewater effluents into the main water supply reservoirs. Although such a scheme is innovative, it can be compared to the typical scenario of several cities in the world which are supplied by a surface water source that is fed by other cities' wastewater discharges some distance upstream. Such traditional surface water supply can be viewed as 'de facto' potable reuse, albeit avoiding controversy. The case of Windhoek, Namibia, is the only known example today of direct potable reuse, i.e. where the treated wastewater is injected back into the drinking water distribution network without prior surface or underground storage, and has done so since 1969. The debate around potable reuse in water-stressed countries like Australia usually excludes direct potable reuse, which is less likely (than indirect reuse) to overcome risk perception barriers.

QMRA can be used as a quantitative tool to calculate the relative risks of possible water supply schemes. For example, QMRA can numerically assess the contribution from environmental barriers – which are important for public acceptance of (indirect) potable reuse schemes – in reducing microbial risks. A scientific, systematic QMRA demonstration that potable reuse schemes do not introduce additional risks compared to a traditional surface water supply may help garner general acceptance. An example is provided in a recent US National Research Council report on water reuse (NAP 2012), where hypothetical but realistic scenarios were compared. Using typical pathogen concentrations in wastewater secondary-treated effluents and typical removal capacity of common treatment processes, QMRA calculations showed that the risk associated with potable reuse, through aquifer recharge, was significantly lower than with a traditional surface water supply scheme (NAP 2012). The example used, as a point of reference, a typical surface water supply scenario in which surface water, impacted by upstream effluents ('de facto' reuse), was treated by conventional treatment (coagulation, flocculation, filtration), UV, and chlorination. Following advanced wastewater treatment (including reverse osmosis and advanced oxidation) and aquifer recharge, reference pathogen concentrations in the potable water recycled from wastewater were at least 7 orders of magnitude lower. The approach was extended to a number of chemical contaminants (including disinfection byproducts, hormones, pharmaceuticals, antimicrobials, etc.) and showed that potable reuse systems are expected to provide barriers to trace organic contaminants at levels comparable to traditional surface water supply systems.

4.3 Examples of Water Quality Challenges in Urban Water Supply Innovation

QMRA can be used to evaluate the health risk of alternative water supply systems and compare them to conventional systems. Two examples illustrate, through simplified calculations, possible use of QMRA to improve understanding of alternative supply risks. Other examples can be found in the AGWR (NRMMC-EPHC-AHMC 2006).

4.3.1 Example 1: Rainwater for Watering the Garden

Concept Description Water from rooftops is collected through gutters and pipes into a storage tank. The water is used for watering the garden with a pump and spray hose.

Health Risk Assessment Water quality assessments of roof-harvested rainwater are scarce and very site-specific. All studies clearly show some level of contamination for a proportion of the investigated systems. Abbott et al. (2007) found that 46 % of New Zealand roof harvesting systems were heavily fecally contaminated. Ahmed et al. (2011) showed that fecal indicator bacteria were found in 36–100 % of systems worldwide. In addition, various pathogens such as *Giardia*, *Cryptosporidium*, *Salmonella*, *Campylobacter*, *Legionella*, and *Aeromonas* were detected. This example focuses on the risk from *Cryptosporidium*, which has been detected in concentrations of 1–10 oocysts/100 L (Ahmed et al. 2011).

In the Australian Guidelines for Water Recycling (AGWR), ingestion of water while watering the garden has been estimated as 0.1 mL per event, with a yearly occurrence of 90 events per year (NRMMC-EPHC-AHMC 2006). Roughly, in a million garden watering events in which accidental ingestion of water occurs, only one leads to ingestion of an oocyst, equal to a 14 % increase in infection risk. The risk per event is 0.14×10^{-6}, and with 90 events per year the annual risk is 0.13×10^{-5} infections per person per year, or 1.3×10^{-8} DALY. Hence the health risk from this application would be considered acceptable according to international guidelines (10^{-4} infections or 10^{-6} DALY).

Another route of exposure listed in the AGWR is a more extreme "accidental ingestion" scenario of 100 mL occurring once a year. Since the ingested volume is 1,000 times higher and the frequency is 90 lower, the risk from accidental ingestion is about 10 times higher than for watering the garden. The 10^{-4} infection risk target might not be met, but the 10^{-6} DALY target would not be exceeded. Such extreme accidental ingestion is most likely among children, since they cannot read warnings and are less aware of quality issues. Children are more susceptible to *Cryptosporidium* than adults. Assuming maximum infectivity (63 %), the risk would further increase by 4.5 times, to an infection risk of 6×10^{-4}, or 0.7×10^{-6} DALY, approaching a relevant risk.

This example only assesses risk from *Cryptosporidium*, a fecal pathogen. Opportunistic pathogens such as *Legionella* may pose a greater risk as they can multiply in rainwater systems. Simmons et al. (2008) reported a *Legionella* outbreak in which *Legionella* strains were also found in the rainwater storage tanks. Abbott et al. (2007) stress the need to properly design and maintain roof water collection systems to reduce risk, although some risk will always persist when used for drinking. Lye (2009) provided several examples of outbreaks caused by drinking roof runoff.

Quantitative risk assessment shows that the risk from *Cryptosporidium* is small when roof-harvested rainwater is used to water the garden. The highest risk is from significant (but infrequent) accidental ingestion, especially by children. Hence, such water should not be used by children. Installation of pumps to spray the water (via sprinklers) is undesirable due to the increased health risk, as well as the environmental impact (energy use) and the inefficient application of water.

4.3.2 Example 2: Greywater for Food Crop Irrigation

Concept Description Greywater (from showers, sinks, laundry) is collected at the household level and used to irrigate garden crops that are eaten raw.

Health Risk Assessment Various studies collected by Winward et al. (2008) have shown that greywater consistently contains indicators of fecal contamination. O'Toole et al. (2012) detected pathogenic bacteria and viruses in 21 % of greywaters collected at homes, but presence of *Cryptosporidium* and *Giardia* was not tested. Birks et al. (2004) found both pathogenic protozoa in greywater collected in London. Clearly the level of contamination of greywater varies largely from site to site depending on the types of water collected and the health status of the people that have used it: laundry water from a family with small children is more likely to contain fecal pathogens than from a household without children. Greywater often provides favorable conditions for growth of opportunistic pathogens such as *Legionella*, *Aeromonas*, and *Pseudomonas* (Winward et al. 2008): temperatures are relatively high since some of the water has been heated (shower, laundry, dishwasher) and there are enough nutrients for organisms to grow.

Actual pathogen concentrations have rarely been determined in greywater. For this example, a concentration of 1 *Cryptosporidium* oocyst per liter is used (Birks et al. 2004). The AGWR suggest that a person could ingest 5 mL of greywater from one serving of irrigated lettuce seven times per year, and 1 mL of greywater from other produce 50 times per year. The risk of ingesting an oocyst from lettuce is 0.5 % per event and 0.1 % from other produce, resulting in 0.07 % and 0.014 % risk of infection respectively. The combined annual risk of infection is then 5.5×10^{-2} (or 5.8×10^{-5} DALY), well above the health risk target. Ceasing irrigation several days before harvesting could reduce the risk since pathogens can potentially be inactivated by the UV in sunlight. However, pathogens may be shaded by crop leaves, and persistent pathogens like *Cryptosporidium* could accumulate on and in the produce during repeated irrigation with contaminated water. When the produce is cooked, pathogens are inactivated; hence prohibiting the use of greywater to irrigate crops eaten raw may be a safer approach.

5 Discussion

5.1 Context of Alternative Water Supply Options

Worldwide, reusing wastewater for irrigation has been spontaneously done for centuries (Dreschel et al. 2010). With the agriculture sector being the major global water consumer, irrigation remains a prominent use of reused water. Various post-irrigation measures –especially cooking – can help mitigate contamination risks prior to crop consumption. The major concern with water use in irrigation is when there is an absence of risk-management and regulatory frameworks, as is the case in many developing countries. In the urban context, landscape irrigation with recycled wastewater remains a relatively low-risk opportunity to save water in periods of drought while improving cities' livability by providing community green spaces.

In an urban context, some countries (like the Netherlands) may not need to invest in alternative water supplies at all, because the challenge is not really one related to quantity and availability of the resource. Where water is abundant, water-recycling schemes involving dual-pipe distribution systems introduce unnecessary risk.

In other countries, the only alternative to water recycling for cities in need of water supply augmentation will often be seawater desalination. This may be the case in several cities in arid or semi-arid regions (e.g. various Middle-Eastern countries, Western and Southern parts of the USA, Australia) or in heavily populated islands and cities, like Singapore, which struggle to develop water autonomy despite abundant, year-round rain because of a lack of storage capacity. In such cases of severe water scarcity, water recycling may be defended as more environmentally friendly and cheaper than desalination (although several reasons may explain the quicker and more widespread expansion of desalination compared to water reuse). Also, contrary to desalination, wastewater reuse does not require proximity to the coast.

In terms of water service management, decentralised options generally involve greater institutional challenges, as institutions are typically centrally based, and decentralised control of public health matters is more complex. But given an appropriate framework for risk assessment and risk management, it becomes possible to guide and regulate alternative water supply options, either centralised or decentralised, in a way that ensures a level of safety comparable to the typical water supply scenario.

QMRA is gaining prominence in the water industry, receiving strong support from water and public health scientists and responsible agencies. It is the only approach to determining water safety requirements in a systematic and quantitative way, covering scheme design, validation, and regulation (O'Toole 2011; Bichai and Smeets 2013). With urban water systems evolving in response to the complex challenges of water scarcity, expanding populations, environmental sustainability, and urban livability, the QMRA approach has much to offer urban water managers and planners. QMRA lets them innovate and improve water security while dealing with the vital public health component of water services.

5.2 *The Role of QMRA in Designing Water Quality Regulation*

When innovation is called for, for example to develop a novel supply scheme to harness stormwater and wastewater in a water-scarce city, proponents often need guidance, especially when design experience is lacking. Regulatory and public health agencies need adequate scientific support to ensure that innovation is safe. In several parts of the world, a lack of existing water quality regulations for recycling schemes is a barrier to development. When implementation is essential, as it was the case in the city of Windhoek when it developed its direct potable reuse scheme, then creating purpose-made water quality regulations may be necessary. In Australia, even though the (national) AGWR framework intends to cover a wide range of recycling schemes, there remains a lack of regulation for stormwater reuse at the State level (e.g. in Victoria), a situation which leaves potential developers with a shortage of technical support.

In designing recycled water quality regulations, there is a tendency to be overly conservative on the public health protection side. This can lead to unnecessary cost in terms of infrastructure, energy, and the environment, and may ultimately discourage the water industry from innovating with a full range of alternative schemes. It is important that modern regulations be sensible and not prescribe that all water systems be over-engineered. At the same time, the regulations must maintain confidence in the safety of water, both in terms of the public and of developers. In this role, QMRA can provide a sound basis for assessment and demonstration of water safety. To build and refine QMRA models, there is a crucial role for science in improving the accuracy of basic estimates and assumptions, particularly in collecting more data. Effective implementation of QMRA-based regulations also calls for training and institutional support (Bichai and Smeets 2013), since, in the end, risk assessment requires professional interpretation and judgement.

6 Conclusion

Water scarcity and water security are at the core of important policy concerns in many water-stressed areas today, especially in urban areas with expanding populations. As water quality (at the point of delivery) is regarded as a public health policy matter, it is often left out of mainstream water services planning debates, decisions, and policy.

QMRA is a systematic and flexible approach based on science. Its prime advantage is that it integrates multiple water quality considerations with diverse urban water planning approaches. Crucially, it brings public health concerns back into the decision-making process, and balances them with water availability issues and environmental, social, and economic factors.

References

Abbott, S., Caughley, B., Douwes, J. (2007). The microbiological quality of roof-collected rainwater of private dwellings in New Zealand. In 13th international conference on rainwater catchment systems. Sidney, Australia 2007.

Ahmed, W., Vieritz, A., Goonetilleke, A., & Gardner, T. (2010). Health risk from the use of roof-harvested rainwater in Southeast Queensland, Australia, as potable or non-potable water, determined using quantitative microbial risk assessment. *Applied and Environmental Microbiology, 76*(22), 7382–7391.

Ahmed, W., Gardner, T., & Toze, S. (2011). Microbiological quality of roof-harvested rainwater and health risks: A review. *Journal of Environmental Quality, 40*(1), 13–21.

Anonymous. (2003). *Contamination of drinking-water in Leidsche Rijn* (in Dutch: Verontreiniging drinkwater Leidsche Rijn). The Hague: Raad voor de Transportveiligheid.

Armstrong, T. W., & Haas, C. N. (2008). Legionnaires' disease: Evaluation of a quantitative microbial risk assessment model. *Journal of Water and Health, 6*(2), 149–166.

Ashbolt, N. J., Schoen, M. E., Soller, J. A., & Roser, D. J. (2010). Predicting pathogen risks to aid beach management: The real value of quantitative microbial risk assessment (QMRA). *Water Research, 44*(16), 4692–4703.

Barrios, R., Siebel, M., van der Helm, A. W., Bosklopper, K., & Gijzen, H. (2008). Environmental and financial life cycle impact assessment of drinking water production at waternet. *Journal of Cleaner Production, 16*, 471–476.

Bichai, F., & Smeets, P. (2013). Using QMRA-based regulation as a water quality management tool in the water security challenge: Experience from the Netherlands and Australia. *Water Research.* http://dx.doi.org/10.1016/j.watres.2013.09.062

Bichai, F., Komatsu, R., Ryan, H., Fitzgerald, C., Abdelmoteleb, A., Williams, K., & Briscoe, J. (2013). *Understanding the role of water recycling in an urban water security strategy: An analytical framework for decision-making.* Second Water Research Conference, 20–23 Jan 2013, Singapore.

Birks, R., Colbourne, J., Hills, S., & Hobson, R. (2004). Microbiological water quality in a large, in-building, water recycling facility. *Water Science and Technology, 50*(2), 165–172.

Buse, H. Y., Schoen, M. E., & Ashbolt, N. J. (2012). Legionellae in engineered systems and use of quantitative microbial risk assessment to predict exposure. *Water Research, 46*(4), 921–933.

Chen, R., & Wang, X. C. (2009). Cost–benefit evaluation of a decentralized water system for wastewater reuse and environmental protection. *Water Science and Technology, 59*(8), 1515–1522.

De Roda Husman, A. M., & Medema, G. (2005). Inspectierichtlijn Analyse microbiologische veiligheid drinkwater Artikelcode: 5318. In VROM-inspectie (Ed.), VROM-inspectie. Haarlem, the Netherlands.

Dreschel, P., Scott, C. A., Raschid-Sally, L., Redwood, M., & Bahri, A. (2010). *Wastewater irrigation and health: Assessing and mitigating risk in low-income countries.* London: IDRC/IWMI.

European Union Drinking Water Directive, Council Directive 98/83/EC of 3 Nov 1998 on the quality of water intended for human consumption.

Gerba, C. P., & Haas, C. N. (1988). Assessment of risks associated with enteric viruses in contaminated drinking water. *ASTM Special Technical Publication, 976*, 489–494.

Haas, C. N. (1983). Estimation of risk due to low doses of microorganisms: A comparison of alternative methodologies. *American Journal of Epidemiology, 118*(4), 573–582.

Haas, C. N., Rose, J. B., & Gerba, C. P. (1999). *Quantitative microbial risk assessment*. New York: Wiley.

Health Canada. (2010). *Canadian guidelines for domestic reclaimed water for use in toilet and urinal flushing*. Ottawa: Health Canada.

Health Canada. (2011). *Guidelines for Canadian drinking water quality: Guideline technical document: Enteric viruses*. Ottawa: Water, Air and Climate Change Bureau/Healthy Environments and Consumer Safety Branch/Health Canada. 70pp.

Hrudey, S. E., & Hrudey, E. J. (2004). *Safe drinking water: Lessons learned from recent outbreaks in affluent nations*. London: IWA Publishing.

Hulsmann, A., & Smeets, P. (2011). *Towards a guidance document for the implementation of a risk-assessment for small water supplies in the European Union, Overview of best practices*. Nieuwegein: KWR.

IWA. (2012). *International statistics for water services: Information every water manager should know*. Specialist Group Statistics and Economics, the Hague, the Netherlands, 16pp.

Lye, D. J. (2009). Rooftop runoff as a source of contamination: A review. *Science of the Total Environment, 407*, 5429–5434.

Mons, M. N., van der Wielen, J. M., Blokker, E. J., Sinclair, M. I., Hulshof, K. F., Dangendorf, F., Hunter, P. R., & Medema, G. J. (2007). Estimation of the consumption of cold tap water for microbiological risk assessment: An overview of studies and statistical analysis of data. *Journal of Water and Health, 5*(Suppl 1), 151–170.

NAP (National Academies Press). (2012). *Water reuse: Potential for expanding the nation's water supply through reuse of municipal wastewater*. Committee on the Assessment of Water Reuse as an approach to meeting future water supply needs. Washington, DC: National Research Council.

NHMRC–NRMMC. (2011). *Australian drinking water guidelines*. Canberra: National Health and Medical Research Council and Natural Resource Management Ministerial Council, Australian Government.

NRMMC–EPHC–AHMC. (2006). *Australian guidelines for water recycling: Managing health and environmental risks. Phase 1*. Canberra: Natural Resource Management Ministerial Council/Environment Protection and Heritage Council/Australian Health Minister's Conference/Australian Government.

NRMMC–EPHC–NHMRC. (2008). *Australian guidelines for water recycling: Managing health and environmental risks. Phase 2: Augmentation of drinking water supplies*. Canberra: Natural Resource Management Ministerial Council/Environment Protection and Heritage Council/National Health and Medical Research Council/Australian Government.

NRMMC–EPHC–NHMRC. (2009a). *Australian guidelines for water recycling: Managing health and environmental risks. Phase 2: Managed aquifer recharge*. Canberra: Australian Government.

NRMMC–EPHC–NHMRC. (2009b). *Australian guidelines for water recycling: Managing health and environmental risks. Phase 2: Stormwater harvesting and reuse*. Canberra: Australian Government.

O'Toole, J. E. (2011). *Identifying data gaps and refining estimates of pathogen health risks for alternative water sources*. Waterlines report series no. 50. Canberra: National Water Commission. ISBN 978 1 921853 23 4.

O'Toole, J., Sinclair, M., Malawaraarachchi, M., Hamilton, A., Barker, S. F., & Leder, K. (2012). Characterising microbial quality of household greywater. *Water Research, 46*(13), 4301–4313.

Petterson, S., Signor, R., Ashbolt, N., & Roser, D. (2006). QMRA methodology. In G. J. Medema, J. F. Loret, T. A. Stenstrom, & N. Ashbolt (Eds.), *Quantitative microbial risk assessment in the water safety plan, MicroRisk final report*. Nieuwegein: Kiwa Water Research. Available at http://www.microrisk.com/uploads/microrisk_qmra_methodology.pdf. Accessed 31 Oct 2013.

Regli, S., Rose, J. B., Haas, C. N., & Gerba, C. P. (1991). Modeling the risk from Giardia and viruses in drinking water. *Journal of the American Water Works Association, 83*(11), 76–84.

Schijven, J. F., Teunis, P. F., Rutjes, S. A., Bouwknegt, M., & Husman, A. M. (2011). QMRAspot: A tool for quantitative microbial risk assessment from surface water to potable water. *Water Research, 45*(17), 5564–5576.

Schoen, M. E., & Ashbolt, N. J. (2011). An in-premise model for Legionella exposure during showering events. *Water Research, 45*(18), 5826–5836.

Simmons, G., Jury, S., Thornley, C., Harte, D., Mohiuddin, J., & Taylor, M. (2008). A legionnaires' disease outbreak: A water blaster and roof-collected rainwater systems. *Water Research, 42* (6–7), 1449–1458.

Storey, M. V., Langmark, J., Ashbolt, N. J., & Stenstrom, T. A. (2004). The fate of legionellae within distribution pipe biofilms: Measurement of their persistence, inactivation and detachment. *Water Science and Technology, 49*(11–12), 269–275.

Storey, M. V., Deere, D., Davison, A., Tam, T., & Lovell, A. J. (2007). *Risk management and cross-connection detection of a dual reticulation system*. In S. J. Khan, R. M. Stuetz, & J. M. Anderson (Eds.), Proceedings of the 3rd AWA Water Reuse and Recycling Conference, UNSW, Sydney, July 16–18 2007.

Teunis, P. F. M., Van der Heijden, O. G., Van de Giessen, J. W. B., Havelaar, A. H. (1996). *The dose–response relation in human volunteers for gastro-intestinal pathogens*. RIVM Report 284550002.

Teunis, P. F. M., Chappell, C. L., & Okhuysen, P. C. (2002a). Cryptosporidium dose response studies: Variation between isolates. *Risk Analysis, 22*(1), 175–183.

Teunis, P. F. M., Chappell, C. L., & Okhuysen, P. C. (2002b). Cryptosporidium dose–response studies: Variation between hosts. *Risk Analysis, 22*(3), 475–485.

Teunis, P. F. M., Van den Brandhof, W., Nauta, M., Wagenaar, J., Van den Kerkhof, H., & Van Pelt, W. (2005). A reconsideration of the *Campylobacter* dose-response relation. *Epidemiology and Infections, 133*(4), 583–592.

Thomas, J. M. (2012). *The risk to human health from free-living amoebae interaction with Legionella in drinking and recycled water systems*. Sydney: University of New South Wales.

Tseng, L. Y., & Jiang, S. C. (2012). Comparison of recreational health risks associated with surfing and swimming in dry weather and post-storm conditions at Southern California beaches using quantitative microbial risk assessment (QMRA). *Marine Pollution Bulletin, 64*(5), 912–918.

USEPA. (1991). *Guidance manual for compliance with the filtration and disinfection requirements for public water systems using surface water source*. Washington, DC: Environmental Protection Agency.

USEPA. (2003). *Long term 2 enhanced surface water treatment rule toolbox guidance manual*. Washington, DC: Environmental Protection Agency.

USEPA. (2006). *LT2ESWTR, Long term second enhanced surface water treatment rule*. Washington, DC: Environmental Protection Agency.

USEPA. (2012). *Guidelines for water reuse*. EPA/600/R-12/618. Washington, DC: Environmental Protection Agency.

VEWIN and Inspectie. (2007). *De kwaliteit van het drinkwater in Nederland*.

WBCSD. (2006). *Water: Facts and trends*. World Business Council for Sustainable Development. http://www.unwater.org/downloads/Water_facts_and_trends.pdf. Accessed 31 Oct 2013.

WHO. (2004). *Guidelines for drinking water quality* (3rd ed.). Geneva: World Health Organization.

WHO. (2006). *Guidelines for the safe use of wastewater and excreta in agriculture*. Geneva: World Health Organization.

WHO. (2011). *Guidelines for drinking water quality* (4th ed.). Geneva: World Health Organization.

Winward, G. P., Avery, L. M., Frazer-Williams, R., Pidou, M., Jeffrey, P., Stephenson, T., & Jefferson, B. (2008). A study of the microbial quality of grey water and an evaluation of treatment technologies for reuse. *Ecological Engineering, 32*(2), 187–197.

Chapter 7
The Development of Private Bore-Wells as Independent Water Supplies: Challenges for Water Utilities in France and Australia

Jean-Daniel Rinaudo, Marielle Montginoul, and Jean-François Desprats

1 Introduction

From the 1930s to the 1960s, significant financial and technical resources were devoted to the development and geographical extension of public water distribution networks. In most developed countries, governments promoted the development of centralised technical systems allowing economies of scale. Water and sanitation services were delivered within private or public monopolies at municipal, metropolitan, or regional scales as part of a broader extension of welfare-state principles (Graham 2000). These services were then considered a public or quasi-public commodity, i.e. freely available to all individuals at equal cost and with standard quality within specific urban areas. For millions of inhabitants, access to public water supplies increased the availability, reliability, and quality of the water supply. This was clearly perceived as progress by households who previously relied on independent water supplies such as shallow wells, rain water cisterns, or even purchased water. These independent water supplies were then often abandoned or only kept operational for non-domestic uses, particularly in rural areas.

A half-century later, access to an abundant, reliable, and safe water supply distributed by public water utilities is taken for granted by populations in developed countries. Water services can be relied on more or less implicitly and tend to become

J.-D. Rinaudo (✉)
Water Department, BRGM, French Geological Survey, Montpellier, France
e-mail: jd.rinaudo@brgm.fr

M. Montginoul
IRSTEA – UMR G-EAU, 361 rue JF Breton, BP5095, 34196 Montpellier, Cedex 5, France
e-mail: Marielle.montginoul@irstea.fr

J.-F. Desprats
BRGM, 1039 rue de Pinville, 34000 Montpellier, France
e-mail: Jf.desprats@brgm.fr

Q. Grafton et al. (eds.), *Understanding and Managing Urban Water in Transition*, Global Issues in Water Policy 15,
DOI 10.1007/978-94-017-9801-3_7

'invisible' to users who know very little about how the service is produced. However, a number of factors are leading growing numbers of households to seek cheaper alternative water-supply sources as a substitute for scheme water. Unlike 50 years ago, there are now several efficient and affordable technological solutions available which households can use to create their own independent water supply. Examples include rain-water harvesting systems (with filtering and pressurising systems), grey water recycling systems, and private bore-wells – the focus of this chapter. Often, the substitution is only partial, and households continue to use small volumes of tap water for purposes requiring high-quality water, while they use their own supply system for others (such as garden irrigation and toilet flushing). Total substitution is possible if a household decides to invest in a water purification device (e.g. a reverse osmosis set up), several sorts of which are now available for domestic use at reasonable prices.

Worldwide, the development of domestic groundwater self-supply as a substitute for public water supply has occurred in very different institutional, economic, and climatic contexts. In some countries considered a source of problems (e.g. France, Belgium), in others it has been promoted (e.g. Western Australia). This chapter analyses this emerging trend based on examination of three situations in Southern France (Sect. 2) and one in Western Australia (Sect. 3). We then investigate households' motivations to develop self-supply and bypass collective scheme-water systems (Sect. 4) before highlighting challenges water utilities will need to overcome when confronted with this new social phenomenon.

2 Southern France

2.1 Context

In France, approximately 98 % of the population use water delivered to their homes by public water purveyors. Users generally benefit from high-quality water services at prices that are relatively low compared to other European countries.[1] However, the rapid increase in water rates that occurred in the 1990s has provided households in detached or semidetached housing units with incentives to drill private bore-wells. Montginoul and Rinaudo (2011) describe this evolution in regions that have very different climates, economics, and demographics. The trend took place relatively unnoticed by water managers and government authorities because households typically failed to declare their bore-well despite a legal obligation to do so.

[1]According to NUS consulting European water price barometer (2009), the average water price in France is 10 % lower than the European average (3.09 against 3.44 €/m^3) and far below German and Danish prices (respectively 5.29 and 6.42 €/m^3). The study is based on an analysis of the 5 largest cities in 10 EU countries, covering a total population of about 40 million inhabitants (Germany, Belgium, Denmark, Spain, Finland, France, Italy, the Netherlands, the United Kingdom, and Sweden).

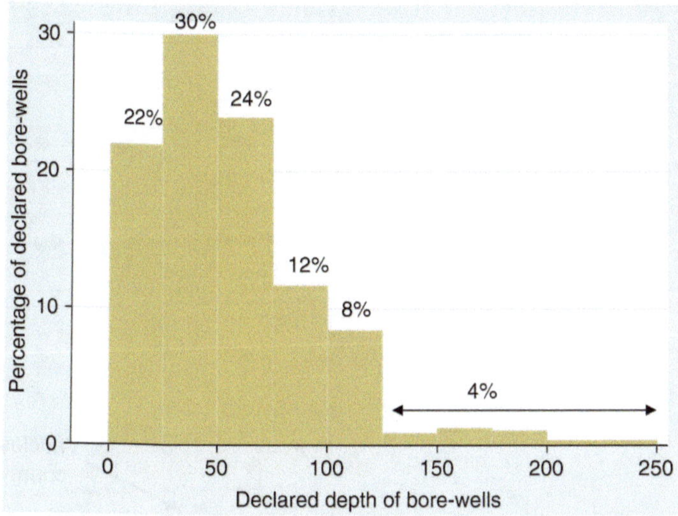

Fig. 7.1 Declared depth of domestic bore-wells drilled in the Languedoc–Roussillon region. N=547, 2008–2013 (Source: own elaboration from BRGM groundwater database (Banque du Sous-Sol), Languedoc–Roussillon region)

Although relatively uncommon, official bore-well declarations can be informative as to the type of bore-wells drilled by households. Figure 7.1 presents an analysis of the depth of bore-wells declared between 2008 and 2013 (July) in the Languedoc–Roussillon region of Southern France. The figure shows that a majority, 52 %, attained depths of 50 m or less, while 36 % drilled to depths of 50–100 m. A minority, 12 %, went deeper than 100 m. Note that the cost of a bore-well is not directly proportional to its depth: it also depends on the physical properties of the geological layers involved. Drilling in a sand aquifer or in soft alluvial materials requires expensive drilling techniques, while cheaper techniques can be used to drill through Jurassic limestone for instance.

2.2 *The Intensity of the Phenomenon*

A series of case studies were recently conducted by the authors with a view to assessing the frequency of the phenomenon at a regional scale. Insofar as very few domestic bore-wells were declared to government and/or municipal authorities, an indirect method was developed to estimate bore-well density at a regional level. The methodology consists in calculating the profitability for the owner of drilling a bore-well, defined as the savings made on their water bill from substituting cheap groundwater for more expensive scheme water, minus the cost of drilling the well. Benefits were expected to be proportional to the price of scheme water, while the cost of the bore-well itself was taken to be proportional to the groundwater depth,

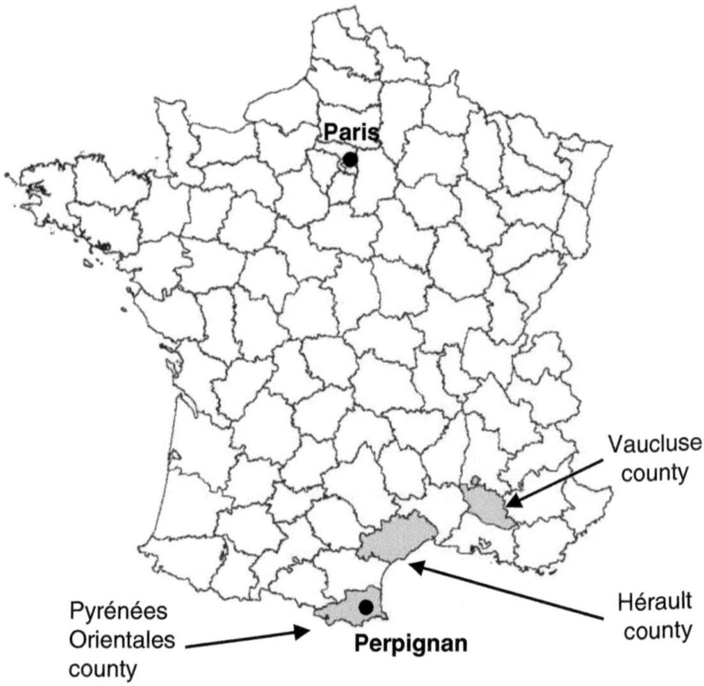

Fig. 7.2 Locations of the three regional case studies

adjusted to take into account the type of geological layer and uncertainty related to groundwater availability (for a detailed description, see Montginoul and Rinaudo 2011).

The methodology was applied in three counties in Southern France: Hérault, Pyrénées–Orientales, and Vaucluse (see Fig. 7.2).

Detailed results obtained in Vaucluse are presented in Fig. 7.3. The upper left map shows that the cost of drilling a well is relatively low in some areas (less than €5,000). The cost rises significantly in other zones where the geology is less suitable, water is at greater depths, and there is a higher risk of failure (a 'dry' well). The upper right map illustrates the variability of scheme-water prices, which exceed €4 per cubic meter (water, sanitation, and wastewater treatment included) in areas colored black. The lower left map depicts variations in the bore-well profitability threshold, defined as the minimum amount of groundwater that a household must substitute for scheme water in order to repay its investment. In areas shown in light grey, it is worth investing in a bore-well if the household uses at least 120 m^3 per year. Conversely, for areas colored in dark grey, constructing a bore-well is only profitable for households using more than 400 m^3 per year. The lower right figure presents an estimate of bore-well density, based on socio-economic information and the estimated profitability threshold (see Montginoul and Rinaudo 2011 for details).

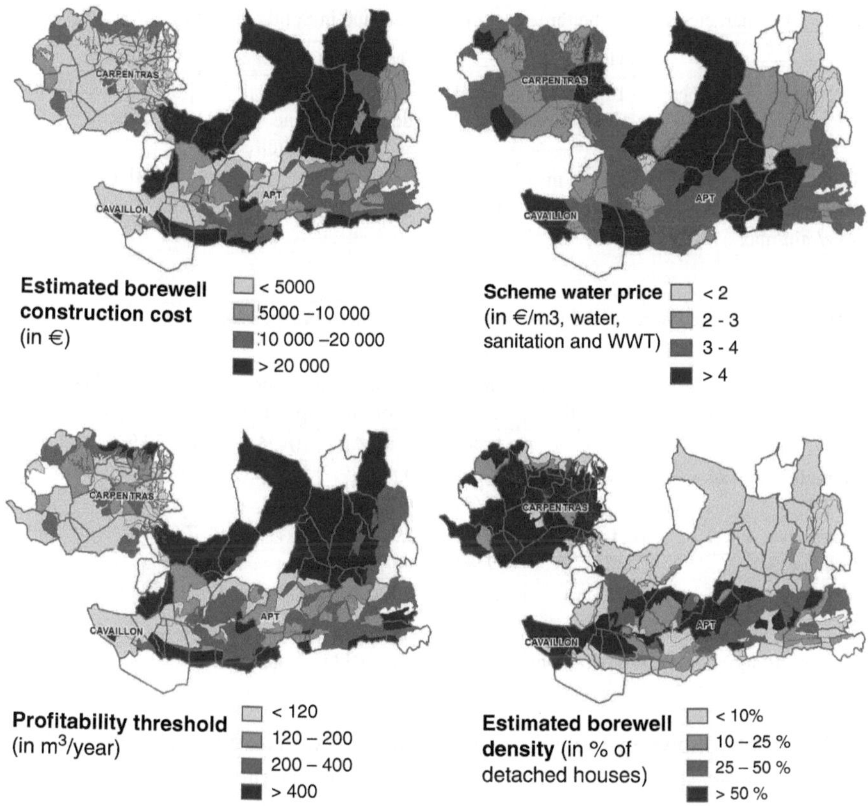

Fig. 7.3 Bore-well construction price, scheme-water price, profitability threshold, and bore-well density in the Calavon river basin and the Miocene aquifer basin, Vaucluse county, Southern France (Source: Desprats et al. (2012))

Density is expected to be very high (more than 50 % of single-family homes equipped) along the Calavon Valley and in the Miocene groundwater basin (a region of Carpentras in the northwest of the map). Estimated densities were cross-checked with the staff of municipal water utilities, who confirmed the range of values. Based on this methodology, the estimated number of bore-wells in this area ranges from 14,000 to 21,000, corresponding to a density of 32–48 % (calculated as the fraction of households living in detached houses).

The same method was applied in two other counties in Southern France (Table 7.1). These two case studies correspond to larger river basins, comprising respectively 186 and 229 municipalities. The geology is highly varied within these territories: in alluvial valleys, groundwater can easily be reached at depths of 5–10 m, while in karstified limestone areas, households may need to sink wells to depths of up to 150 m.

Table 7.1 Characteristics of the three case studies selected in Southern France

	Hérault	Pyrénées–Orientales	Vaucluse
Groundwater geology	Limestone, alluvial deposits, sands	Quaternary alluvial deposit, Eocene sedimentary aquifer	Limestone, alluvial deposits,
Depth of bore-wells	5–175 m	5–150 m	5–150 m
Water price (€/m³) **Avg**/min/max	**2.60**/0.53/4.23	**2.60**/0.24/3.71	**3.32**/0.80/4.50

Table 7.2 Estimated bore-well density in the three Southern France case studies

County	No. munici-palities	Estim. no. of bore-wells	Number of municipalities per class of bore-well density (density in % of municipal housing stock)				
			<1 %	1–10 %	10–25 %	25–50 %	>50 %
Hérault	186	8,800	108 (58 %)	47 (15 %)	24 (13 %)	18 (10 %)	7 (4 %)
Pyrénées–Orientales	229	16,400 to 22,000	135 (59 %)	39 (17 %)	46 (20 %)	7 (3 %)	2 (1 %)
Vaucluse	66	14,000 to 21,000	5 (8 %)	7 (10 %)	9 (14 %)	26 (39 %)	19 (29 %)

Source: Montginoul (2008), Montginoul and Rinaudo (2011), Desprats et al. (2012)

The results shown in Table 7.2 suggest that bore-wells are less common in these two other case studies. In Pyrénées–Orientales and Hérault, there are far fewer situations which combine high scheme-water price and easily accessible groundwater. Consequently, the number of municipalities where bore-well density exceeds 25 % remains low (14 % in Hérault, 4 % in Pyrénées–Orientales).

These three case studies conducted at the county level were supplemented by local investigations conducted in the Perpignan Méditerranée Urban Community, the metropolitan area of the Pyrénées–Orientales county. Our investigations relied on the use of various sources of information including: (i) the municipal register of bore-well declarations; (ii) an internet survey conducted on a sample of 204 households occupying single-family houses[2]; and (iii) an analysis of water-billing records over a 5-year period.

According to municipal records, there were only 351 domestic bore-wells officially registered in the community, which comprises more than 52,000 single family dwellings. This very low rate of bore-well ownership (0.6 %) was not consistent with the results of the study presented in Table 7.3, which estimated the total number of bore-wells in the Perpignan metropolitan area as 17,400, i.e., a 34 %

[2]The survey targeted a sample of 2,778 households living in detached houses. They were selected in 100 neighborhoods representative of the diversity of the metropolitan area in terms of income, housing characteristics, groundwater characteristics, and water tariffs. The response rate was just below 10 % (227 answers with 200 fully exploitable).

Table 7.3 Estimated rate of bore-well ownership from three distinct sources (the Perpignan Méditerranée Urban Community)

Estimation method	Estimated rate of bore-well ownership (in % of single family houses)
Official bore-well declaration registers	0.6 %
Economic modeling (all municipalities)	34 %
Internet survey (*N*=200)	24 %

rate of ownership (households living in single family units only). The internet survey conducted with 200 households confirmed a higher value, with 24 % of respondents declaring using a private bore-well.[3]

2.3 *How Do Households Use Their Bore-Wells?*

Previous studies suggested that most households only use bore-well water for irrigating gardens and filling swimming pools, and only a minority use it indoors for washing machines, toilet flushes, and sometimes personal hygiene (showers and baths) or even cooking and drinking (Montginoul et al. 2005). In the Perpignan Méditerranée Urban Community, we found that half the households equipped with a declared bore-well almost totally substituted untreated groundwater for municipal supply (Table 7.4). These households use less than 5 m^3 per year of scheme water, mainly for drinking and cooking. A second group of bore-well owners had drinking-water consumption ranging from 5 to 60 m^3 per year (average 32 m^3), indicating that they use scheme water for drinking, cooking, and showers but use bore-well water for washing machines and toilet flushes. The third group consumes more than 60 m^3 per year (average 124 m^3) which corresponds to the average water use in Perpignan, implying that the bore-well only supplies outdoor uses.

2.4 *Issues for Water Utilities*

The development of groundwater self-supply by households is generally perceived as a threat by water utilities, who cite three types of negative impacts (Montginoul et al. 2005). The first impact is on groundwater resources. The risk of groundwater contamination is increased in areas characterised by high bore-well density. Because

[3] The estimated rate of bore-well ownership in this metropolitan area is high compared to what has been found in other cities. In the nearby Montpellier area for instance, a similar internet survey conducted with 347 households showed a borehole equipment rate of 9 %. The difference is mainly explained by geological conditions.

Table 7.4 Scheme-water consumption for three types of bore-well owners in Perpignan Méditerranée Urban Community (based on municipal records, 351 households)

Type of bore-well water use	Average scheme water use (m^3/year/household)	Sample % (*N*=351)
Type 1: full indoor and outdoor substitution	Range=[<5]	49 %
	Average=0	
Type 2: partial indoor and full outdoor substitution	Range=[5, 60]	27 %
	Average=32	
Type 3: no indoor and full outdoor substitution	Range=[>60]	24 %
	Average=124	

they are often poorly constructed, private bore-wells often connect previously distinct hydrogeological layers and become multiple contamination vectors for groundwater resources. The development of bore-wells also increases total water abstraction, as households having free access to cheap groundwater will use more than when they fully rely on municipal supply. This is more of a problem when private bore-wells and public supplies tap the same aquifer.

The second impact is on public health. Untreated bore-well water is sometimes used indoors for personal hygiene, cooking, and even drinking without ensuring that its chemical and bacteriological quality complies with drinking water standards. Montginoul et al. (2005) also report cases where, due to improper construction of the dual-pipe system inside the house, high-pressure contaminated bore-well water flows back from a private bore-well into the municipal drinking water network.

The third impact is financial. The development of private bore-wells may reduce water sales and generate cost-recovery problems for public water suppliers. Moreover, where sanitation and waste-water treatment are charged proportionally to drinking water use (as is the case in France), the development of groundwater self-supply generates additional financial problems for utilities in charge of waste-water treatment and sanitation. This is particularly a problem where water demand is already in decline for demographic or economic reasons.

Legislative authorities recently reacted to these problems by allowing water utility staff to enter private properties in order to record the existence of bore-wells and to control the bore-well itself as well as the water pipe network inside the house.[4] Utilities are authorised to cut off the connection to scheme water if the installation does not comply with regulatory requirements. Domestic bore-well owners are also under the legal obligation to install a meter which can be used by utilities to charge sanitation and waste-water treatment services.[5]

However, enforcement of this new legislation remains highly problematic in France. Despite information campaigns, households rarely declare their bore-wells

[4]The 2006 water law, modifying article L 2224-12 of the Code Général des Collectivités Territoriales (CGCT) and application decrees of 2/07/2008, 17/12/2008.

[5]Article 2224-12-5 of CGCT.

to the municipality for fear of being charged for groundwater abstraction in the future (currently exempt for less than 1,000 m^3 per year) or for sanitation and wastewater treatment by the utility in charge of the service. Municipal water utilities are often unable to identify bore-well owners, for utilities frequently lack the technical capability to analyse water consumption data and cross-check it with other sources of information such as aerial photographs and household demographic data. The situation is exacerbated by a lack of political will: the mayor generally tends to refrain from sending staff to conduct property inspections lest he or she loses support for the next election.

3 Western Australia

By contrast with the French situation, the development of private bore-wells in Perth, Western Australia, was considered by water authorities as one way to reduce the growing pressure on drinking water supplies. Groundwater self-supply reduces the demand on scheme-water supplies, delaying the time when new resources (dams and well fields) will be needed.

With a population of 1.5 million (2006), Perth is the fourth largest Australian city. For the last 30 years, rapid population growth and economic development have been accompanied by an impressive urban sprawl. The urban landscape is characterised by very low-density suburban estates with detached housing (Kennewell and Shaw 2008). More than 75 % of the population lives in single-family detached dwellings. The fondness of Perth households for their garden (Syme et al. 2004) and the desire to replicate the English country garden in a semi-arid climate (Kennewell and Shaw 2008), has resulted in very high water use. Average water consumption was around 500 m^3 per household in 1975–1976. It has remained relatively high over time, as shown by Loh and Coghlan (2003), who estimate single residential use at 460 m^3 per year, with 56 % of that volume being used outdoors.[6]

This substantial urban growth and rising water demand has coincided with a declining trend in rainfall, which has diminished inflows to reservoirs and lowered groundwater levels, at a time when groundwater is making increasingly larger contributions to Perth's water supply. In the late 1970s, restrictions were imposed on outdoor water use due to a long spell of dry weather. Volumetric pricing was also introduced at that time, and campaigns were conducted to promote water-conservation practices. Many households did not reduce their usage of water, but responded by seeking alternative private sources of supply. Given that the Perth metropolitan area is underlain by extensive shallow aquifers, many unlicensed private bore-wells were drilled to substitute groundwater self-supply for scheme water. Since the aquifer is easily accessible and productive in most locations, private

[6] Based on a sample of 720 households living in detached houses and monitored during 1998–2000.

bore-wells can be constructed at reasonable cost, ranging from A$3,500 to 5,000.[7] The development of private bore-wells was facilitated by the regulatory framework, which does not require any licence for bore-wells tapping shallow groundwater and used to irrigate less than 0.2 ha. Bore-owners also do not pay any groundwater extraction fees and the amount of groundwater they use is not metered.

The development of private bore-wells was investigated in the late 1980s. A study published by the Metropolitan Water Authority (1985) found that 24 % of households owned a bore-well, with a further 3 % having access to one. According to Thomas et al. (1987, quoted in Thomas and Syme 1988), bore-well ownership increased from 11 % of households in 1976 to 27 % in 1982. More than two-thirds of the 64,000 bore-wells constructed at the end of the 1980s had been installed after the onset of drought and its consequent publicity, restrictions, and changes in price regime (Syme et al. 2000). Bore-well popularity continued to increase due to more stringent restrictions on the central water supply and mounting water prices.

The number of households that use bore-well water has risen from 64,000 in 1987; 99,600 in 1992; 135,000 in 2001; 150,900 in 2006; to 167,000 in 2010 (Department of Water 2011). The total amount of water pumped from these bore-wells is estimated at 75–120 million m³ depending on the source, corresponding to 30–50 % of the potable water supply of the metropolitan area (235 million m³/year, 1994–2005). Since 2003, bore-well installation has been encouraged by the government, which offers a rebate[8] to people in areas where bore-wells are considered suitable according to the Perth Groundwater Atlas. About 5,000 subsidies are granted each year (Smith et al. 2005). Bore-well density varies significantly from one area to another depending on the depth of the water table (the cost can be dissuasive when the water table is more than 10 m deep), the type of aquifer (drilling in limestone is more expensive), and the chances of success (bore-wells are fewer in areas of clay where bore-well yields are generally low).

Most studies and surveys report that self-supply groundwater is only used outdoors. Estimated volumes are relatively high by international standards (see Table 7.5), even in view of Perth's dry climate. This suggests that Perth households

Table 7.5 Estimated bore-well density and use as a function on property size

Property size (in m²)	Indicative groundwater use (m³/year)	Average bore-well installation rate (% of lots)
Less than 500	400	5
500–999	800	30
1,000–5,000	1,000	30

Source: Department of Water (2009)

[7]In 2013 A$, based on quotes provided by drilling contractors in the Perth metropolitan area; equivalent to €2,500–3,500.

[8]Up to A$300 (capped at 50 % of the installation cost) offered by Western Australian government as part of the 'Water Wise' rebate program.

are overusing water compared to what they would use if they relied solely on more expensive scheme water.

Concerns over increasing bore-well water use in Perth have been raised recently in relation to its possible impact on groundwater levels and quality. The response from the authorities was to impose partial groundwater use restrictions on bore-wells: watering is banned between 9 a.m. and 6 p.m. to reduce losses by evaporation; it is only allowed three days per week; and it is banned in winter. These restrictions do not apply to highly efficient irrigation methods such as sub-surface trickle irrigation. In the future, these restrictions are likely to be increased, if future climate and rainfall patterns persist (Australian Bureau of Statistics 2007). However, enforcement problems are reported by the Department of Water.[9]

4 Understanding Households' Motivations to Invest in Private Supply

Understanding households' motivations to drill is essential for predicting the future evolution of self-supply development and the resulting demands for scheme water. The analysis that follows attempts to identify these motives, based on the few studies that address this issue in France (Montginoul et al. 2005; Montginoul and Rinaudo 2011) and in Australia (Thomas et al. 1987; Thomas and Syme 1988). We also use the results of surveys of households' perception of alternative water sources (Hurlimann 2011), shortages, and conservation (Roseth 2006).

4.1 Maximisation of Utility Derived from Water Use

Households' decision to drill can first be analysed from a utilitarian perspective, assuming they seek to maximise the benefit they derive from water use. Self-supply can be a strategy to maximise utility, in particular where the performance of public schemes is mediocre. Independent access to groundwater self-supply can help cope with limited water availability, poor water quality, low pressure, intermittent supply, and temporary restrictions. This is clearly the main motive prompting the use of wells in developing countries, as illustrated in India (Raju et al. 2008), Sri Lanka (Nauges and van den Berg 2009), Nepal (Pattanayak et al. 2005), Pakistan (Madanat and Humplick 1993) and Kenya (Mu et al. 1990), where self-supply enables a wide range of costs to be avoided, including those entailed in health, water purchase from vendors, labor for collecting water from other sources, and tank construction to cope with intermittent water supply. A similar argument also applies to developed countries, and was reported in the two case studies described above. In France,

[9] In October and November 2010, almost a thousand garden bore-well users in the Perth metropolitan area were caught breaching restrictions. News release by Department of Water, 9 December 2010. Available at http://www.water.wa.gov.au/News+and+events/News+archive/2010/1711.aspx

some households interviewed by Montginoul et al. (2005) say that investing in a bore-well benefits them because of the higher pressure available than with scheme water, so they can irrigate large gardens more rapidly. In Perth, the development of private domestic bore-wells was clearly prompted by bans on sprinklers imposed during long droughts. By drilling bore-wells, households were both securing their water supply and ensuring that their gardens would not suffer losses. Such losses would have had a financial impact (reduced property value), but also a psychological one, given the importance of home gardens for a variety of quality-of-life variables such as avoidance of stress, recreation, and personal and social identity (Syme et al. 2004). The demand for garden bore-wells is largely related to the belief that a garden must be green to be healthy (Roseth 2006).

4.2 Cost Minimisation

Groundwater self-supply can also be a strategy to minimise the cost of water supply. Realising that they do not need high-quality water to irrigate lawns and fill swimming pools, some households decide to drill a bore-well in order to substitute cheap untreated groundwater for costly tap water in all outdoor uses. This strategy differs notably from the previous one in that it aims to bypass a public service that performs well simply to reduce the total cost of water. This decision is based on a simple cost–benefit analysis (CBA) of comparing investment cost with water bill savings. A majority of the French households interviewed by Montginoul et al. (2005) and Desprats et al. (2012) took the decision to drill a bore-well based on this type of CBA. The development of domestic bore-wells in France was apparently prompted by the drastic increase in scheme-water price in the 1990s and the prospect of a continuation of this trend.[10] The introduction of volumetric scheme water pricing in the late 1970s in Perth could also have played a role, albeit ranking behind sprinkler bans. This was shown by Thomas and Syme (1988), who conducted an econometric analysis to estimate a demand function for bore-wells. They identified three main significant explanatory variables: water price, household income, and the number of days of restriction. Estimated coefficients suggest that the demand for a bore-well was highly responsive to the level of restriction on water use.

4.3 Ethical and Political Motivations

The utilitarian framework presented above cannot fully account for the complexity of households' decisions to invest in self-supply. Utility maximisation interacts with ethical and political values and beliefs.

[10] In the medium term, and as energy prices rise, there could be cross-elasticity because of the cost of electricity or other fuel to provide energy for running the bore.

4.3.1 Green Households

Some households may view drilling as a commendable action from an environmental and social standpoint. A frequently cited argument is that it avoids wasting precious treated drinking water for watering lawns. A second one is that it alleviates technical and financial pressures on water utilities faced with rapid population growth and corresponding demand. Investing in groundwater self-supply is thought to be consistent with public policies which incentivise the installation of solar hot water systems or solar panels for electricity production. With this perspective, investing in a bore-well reinforces the idea of being environmentally neutral.

4.3.2 Libertarian Households

A number of citizens may also consider water-use restrictions as inherently limiting personal freedom and rights. Owning a private bore-well allows this freedom to be regained and provides the moral satisfaction of being independent. In her study of community views on water shortages in six Australian metropolitan areas, Roseth (2006) finds a minority of households (7–11 %) who believe it is their right to use as much water as they want when they want, given that they pay for it. However, while the majority accepts restrictions in principle, 50 % wouldn't accept having their garden be impacted, insofar as they have the *right* (our emphasis) to keep their garden looking green and healthy.

4.3.3 The Emergence of Infrastructural Consumerism

The development of private water supplies also reflects a more profound evolution of society. Increasingly, privileged users tend to reject the redistributive role of public water infrastructure monopolies. They expect to get what they pay for rather than cross-subsidising low-income users or other territories through a set of uniform rates. By bypassing the monopolistic water-supply system, they reclaim their right to freely choose their water supplier, somehow asserting "the moral superiority of individual choices over the tyranny of collective decision-making" (Leonard 1997, 4; quoted in Graham 2000). This societal evolution is exacerbated by what Graham (2000) calls "the emergence of infrastructural consumerism", referring to the fact that every household in one street can now sign up with a different electricity, gas, internet, or telephone company. This choice, however, is restricted to certain social groups, as the ability to access alternative water supply systems is dependent on income, housing type, and location.

5 Challenges for Water Utilities

The development of private wells raises a number of challenges for water utilities and calls for changes in water-management practices. Utilities need to develop strategies to monitor the development of private supply. They also need to control and possibly restrict private drilling where increased bore-well density poses environmental threats on the groundwater resource and dependent ecosystems. A third challenge consists in integrating the self-supply option into water-demand models used to predict future demand or simulate the impact of evolving water rates. Last but not least, utilities need to manage the social impacts associated with the development of self-supply.

5.1 *Challenge 1: Monitoring the Development of Garden Bore-Wells*

Obtaining accurate information on garden bore-well density is essential for operating and planning urban water management. The French and Australian examples show that very different strategies can be implemented. The first one consists in obliging bore-well owners to declare their wells. The French example shows that this strategy may not work well unless significant financial resources are put in to implement and enforce the system. In a context where the state restricts financial and human resources devoted to water and environmental protection, the spontaneous bore-well declaration rate is likely to remain very low.

An alternative approach to obtaining up-to-date information consists in conducting household surveys. The government of Western Australia has adopted this strategy, and it commissions the Australian Bureau of Statistics to conduct surveys at regular intervals (2003, 2006, 2009). This is considered to be a cost-effective approach, whereas the cost of licensing large numbers (160,000) of low-yield bore-wells is thought to be prohibitive and not an efficient use of public resources.

Surveys, however, do not always provide reliable information. In the French context, because bore-well owners fear they may incur new administrative costs and taxes if they declare their wells, non-participation or falsified response rates are high. An alternative solution, then, consists in developing an indirect bore-well detection method based on a consistency check between households and housing characteristics on the one hand and metered water consumption on the other. The objective is to identify abnormally low water consumption, revealing the likely presence of a bore-well.

Such a method was developed and tested in one of the French case studies described above, the Perpignan Mediterranée Urban Community. The method comprised five steps: (1) we first developed a geographic information system integrating a digitised cadaster, a high-resolution aerial photograph, and an address database; (2) the aerial photograph was then analysed using a supervised classification method

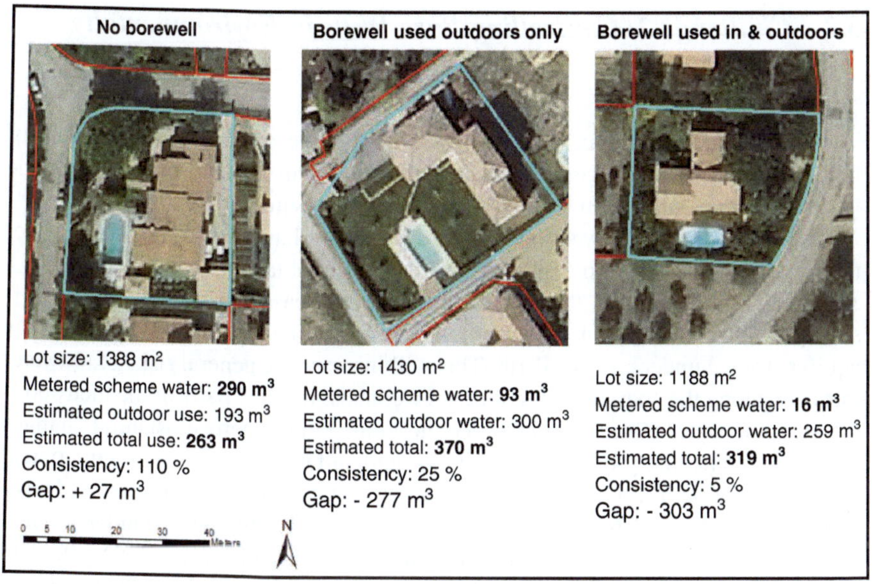

Fig. 7.4 Comparison of estimated water use and metered scheme-water consumption for three houses of Perpignan Méditerranée urban community

to assess each plot according to the area coming under major land-use classes, including built areas, swimming pools, irrigated lawns, tree plantations, and other non-irrigated areas; (3) next we used a simple agro-climatic model to estimate theoretical outdoor water use corresponding to irrigation requirements for irrigated lawns and trees, and evaporation for swimming pools; (4) we then cross-checked the estimated outdoor water use against metered scheme-water consumption; (5) finally, discrepancies were identified and a physical inspection by utility staff was performed to check the existence of a bore-well.

Figure 7.4 illustrates the results obtained with this methodology. The house on the left has an estimated outdoor water use of 263 m^3 consistent with a metered scheme-water consumption of 290 m^3. The house in the center has a metered consumption of 93 m^3, which roughly corresponds to indoor water use, while the total water use is estimated at 370 m^3. The difference suggests the presence of a bore-well which is only used outdoors (lawn and swimming pool). The difference between metered and estimated consumption is even greater for the house on the right: use was estimated at 319 m^3 but only 16 m^3 are withdrawn from the scheme, suggesting that the bore-well is not only used in the garden but also indoors (washing machine, toilet flushes, etc.).[11]

[11] This approach might not work for second homes which are only occupied during summer months. Errors are thus expected to be greater along the coast where seasonal occupancy is more frequent.

5.2 Challenge 2: Controlling Bore-Wells in Environmentally Sensitive Areas

The French and Australian case studies show that the continued development of groundwater self-supply could ultimately lead to environmental problems for the groundwater resource itself and for dependent ecosystems.

In Perth, bore-well development and dry climate have led to a lowered water table (1995–2004) over more than 40 % of the metropolitan area (Smith et al. 2005). Assuming continuation of the climate trend over the past 10 years, additional self-supply of groundwater from the superficial aquifer can be expected to further reduce aquifer storage under parts of Perth. This will increase the general risks of seawater intrusion along the coast and loss of urban wetlands. In some parts of the metropolitan area, the water table is liable to decline to the point where substantial changes occur in the chemistry of groundwater. As acid sulfate sediments come in contact with atmospheric oxygen, acidification takes place, leading to the release of heavy metals and arsenic. These contaminants are leached into the groundwater and eventually find their way into nearby wetlands or rivers, causing environmental and economic harm (Appleyard et al. 2006; Department of Water 2009). The risk is not only environmental but also economic. If shallow aquifers deteriorated to the point where they can no longer be used for garden irrigation (for example, following sea water intrusion), the demand for public supplies would increase tremendously, imposing very high costs on water utilities to develop new resources and rebuild the distribution network, which would then be undersized to meet this increased demand.

To cope with that risk, water policy in Perth has gradually shifted from an unconditional support of bore-wells for gardens to a spatially-differentiated policy which only supports the development of groundwater self-supply in well-defined zones (Department of Water 2011). In areas deemed unsuitable, the department does not support the establishment of new domestic garden bore-wells, but existing ones can still be used. Areas are deemed unsuitable if: (i) water quality is not suitable for irrigation (salinity, contaminants); (ii) the area is close to the ocean and vulnerable to seawater intrusion; (iii) it is near an important conservation wetland or groundwater-dependent ecosystem which could be adversely affected by bore-well use; or (iv) the area is over-allocated to existing users and future development of garden bore-wells is liable to generate damage to third parties or existing users. Areas suitable and unsuitable for garden bore-wells have been defined by the Department of Environment and published in a Groundwater Atlas which is available in an interactive format on the Internet (Department of Environment 2004).

In France, the major problem stems from the impossibility for utilities to ensure that bore-wells are properly built and designed. As mentioned above, drilling contractors do not always install a properly cemented bore-well casing. Accordingly, each bore-well becomes a potential contamination path for groundwater, with serious consequences when the resource is exploited for municipal use. In an area of the Perpignan coast, we found a situation where the third layer of the coastal confined aquifer became brackish in the space of a few years due to the improper construction of a dozen bore-wells, so that hydraulic communication between the brackish upper layer and the confined aquifer they were exploiting was established. The challenge for water utilities is to find ways of controlling the quality of wells (through video technologies for instance) and to impose, at least in sensitive areas, credible threats of sanctions (e.g. charging owners for the cost of refilling poorly built bore-wells). However, this would presuppose that utilities already had access to reliable information on bore-well location, which is obviously not necessarily the case.

5.3 Challenge 3: Including Bore-Well Development in Scheme-Water Demand Modeling

Having access to up-to-date information on bore-well density is also a prerequisite to any accurate modeling of scheme-water demand. This is of utmost importance when forecasting long-term demand, since the unit consumption ratio differs significantly depending on the presence or absence of a bore-well on the property. An underestimation of bore-well density will lead to an overestimated demand for scheme water and consequently a costly over-sizing of storage and conveyance infrastructures. Conversely, overestimating future bore-well density might result in shortages, as infrastructure will prove to be undersized in times of peak demand.

Knowledge of bore-well density is also crucial when simulating the impact of rate changes on demand and utility revenues. Several studies have shown that price elasticity differs between households who have access to an alternative water supply and those who do not (Nauges and van den Berg 2009). In Perth, Thomas and Syme (1988) showed that bore-well owners have very low price elasticity. They also indicated that cross-elasticity of bore-well installation rates and the price of public scheme water was also significant: a 32 % increase in water price would result in a doubling of the demand (+100 %) for bore-wells. In France, Montginoul and Rinaudo (2011) showed that an increase in scheme-water price would significantly raise the return on investment for bore-well construction and provide incentives for households to drill. The same impact is expected from increasing block or seasonal rates. Utilities need to take into account this difference in household price sensitivity when they calculate the impact of proposed rate changes.

5.4 Challenge 4: Maintaining Equity Between Bore-Well Owners and Other Customers

The development of bore-wells also raises an issue of consumer equity. In France for instance, it is most likely that bore-wells will be drilled by higher income households who own a detached house in a suburban estate with a garden and swimming pool. Their scheme-water consumption will decrease to a very low level, especially if they install a dual pipe system in their house to supply washing machines and toilets. Since water, and sanitation and wastewater treatment (S&WWT), are paid for on the basis of metered scheme-water use, their financial contribution to the service will decrease. To compensate for declining revenues and ensure full cost recovery, utilities will need to increase their rates. This will ultimately result in raising the bills paid by less wealthy consumers, increasing consumer inequity. In such situations, water utilities will have to develop innovative cost-recovery policies. A key issue is recovering S&WWT costs from bore-well owners who discharge wastewater into the public sewer without paying for it. This might be done by metering bore-well water entering the house, or by charging for S&WWT services using a flat rate.

The question arises differently in Perth. In 2006, a desalination plant was built to meet the demands of the growing population, and its cost will be paid for by scheme-water consumers in proportion to their level of water consumption. At the same time, authorities are contemplating a groundwater replenishment program in which the shallow aquifer is recharged with recycled waste water (Li et al. 2006). Clearly, sharing these costs on the basis of scheme-water consumption alone would be unfair from a consumer justice point of view. It would also probably be unfair in terms of social justice, as bore-well owners are probably among the better-off households. To maintain equity, utilities will have to design new cost-recovery schemes, for instance through charging bore-well owners a water extraction fee, as practiced in Orange County, California. Such a change which would probably trigger strong political opposition, in France and Australia alike.

6 Conclusion

For decades, centralised, standardised and universal water services were developed as part of a wider elaboration of the welfare state. Technologies and centralised management allowed public or private utilities to achieve high levels of service quality and reliability. We are now starting to see users who look beyond the tap, paying attention to how the service is operated, how the infrastructure is configured, and how costs are shared among users. Some of these users have started to question the fundamental assumptions – the universality of the service, social and territorial financial solidarity – and have resorted to alternative individual water-supply systems such as rainwater harvesting, grey water recycling, and groundwater self-supply. In this chapter, we have focused on one of these strategies which consists in using a private bore-well as a partial substitute for scheme water. The development of such self-supply systems merits attention from water utilities and policy-makers alike, for a number of technical, economic, and social reasons expanded upon in this chapter. We documented two case studies in Southern France and Western Australia, but it is worth mentioning that the issue has also been reported elsewhere in the world.

Acknowledgments This chapter was prepared as part of the EAU&3E project funded by the French Research Agency (ANR), grant VD-08-321989. We thank Céline Nauges and Geoffrey Syme for very useful comments on an earlier version of this paper. The usual disclaimers apply.

References

Appleyard, S. J., Angeloni, J., & Watkins, R. (2006). Arsenic-rich groundwater in an urban area experiencing drought and increasing population density, Perth, Australia. *Applied Geochemistry, 21*, 83–97.

Australian Bureau of Statistics. (2007). *Changing water and energy use in Perth homes*. Western Australian Statistical Indicators. Issue 1367.5, pp. 29–34.

Department of Environment. (2004). *Perth groundwater atlas* (2nd ed.). Perth: Government of Western Australia.

Department of Water. (2009). *Strategic policy 2.03: Managing unlicenced groundwater*. Perth: Government of Western Australia.

Department of Water. (2011). *Operational policy 5.17: Metropolitan domestic garden bores*. Perth: Government of Western Australia.

Desprats, J. F., Rinaudo, J. D., Moulin, M., & Graveline, N. (2012). *Evaluation des prélèvements d'eau issus de forages domestiques dans les bassins du Coulon-Calavon et de Carpentras (Vaucluse)*. Report 61565-FR, BRGM (French Geological Survey), Orléans, 61pp.

Graham, S. (2000). Constructing premium network spaces: Reflections on infrastructure networks and contemporary urban development. *International Journal of Urban and Regional Research, 24*, 183–200.

Hurlimann, A. (2011). Household use of and satisfaction with alternative water sources in Victoria Australia. *Journal of Environmental Management, 92*, 2691–2697.

Kennewell, C., & Shaw, B. J. (2008). Perth, Western Australia. *Cities, 25*, 243–255.

Li, Q., Harris, B., Aydogan, C., Ang, M., & Tade, M. (2006). Feasibility of recharging reclaimed wastewater to the coastal aquifers of Perth, Western Australia. *Process Safety and Environmental Protection, 84*, 237–246.

Loh, M., & Coghlan, P. (2003). *Domestic water use study in Perth, Western Australia 1998–2001*. Perth: Water Corporation.

Madanat, S., & Humplick, F. (1993). A model of household choice of water supply systems in developing countries. *Water Resources Research, 29*, 1353–1358.

Montginoul, M. (2008). *Estimation du nombre de forages domestiques – Application au cas des Pyrénées Orientales (66)*. Report for the Ministry of Ecology. Cemagref – UMR G-Eau, Montpellier, 51pp.

Montginoul, M., & Rinaudo, J.-D. (2011). Controlling households' drilling fever in France: An economic modeling approach. *Ecological Economics, 71*, 140–150.

Montginoul, M., Rinaudo, J.-D., de Lajonquière, Y. L., Garin, P., & Marchal, J.-P. (2005). Simulating the impact of water pricing on households behaviour: The temptation of using untreated water. *Water Policy, 7*, 523–541.

Mu, X., Whittington, D., & Briscoe, J. (1990). Modeling village water demand behavior: A discrete choice approach. *Water Resources Research, 26*, 521–529.

Nauges, C., & van den Berg, C. (2009). Demand for piped and non-piped water supply services: Evidence from southwest Sri Lanka. *Environmental and Resource Economics, 42*, 535–549.

Pattanayak, S. K., Yang, J.-C., Whittington, D., & Bal Kumar, K. C. (2005). Coping with unreliable public water supplies: Averting expenditures by households in Kathmandu, Nepal. *Water Resources Research, 41*, W02012.

Raju, K. V., Manasi, S., & Latha, N. (2008). *Emerging ground water crisis in urban areas. A case study of ward no. 39, Bangalore city*. Bangalore: Institute for Social and Economic Change.

Roseth, N. (2006). *Community views of water shortages and conservation*. Salisbury: Cooperative Research Center for Water Quality and Treatment.

Smith, A., Pollock, D., & McFarlane, D. (2005). *Opportunity for additional self supply of groundwater from superficial aquifer beneath Metropolitan Perth*. Canberra: CSIRO Water for a Healthy Country National Research Flagship.

Syme, G. J., Nancarrow, B. E., & Seligman, C. (2000). The evaluation of information campaigns to promote voluntary household water conservation. *Evaluation Review, 24*, 539–578.

Syme, G. J., Shao, Q., Po, M., & Campbell, E. (2004). Predicting and understanding home garden water use. *Landscape and Urban Planning, 68*, 121–128.

Thomas, J. F., & Syme, G. J. (1988). Estimating residential price elasticity of demand for water: A contingent valuation approach. *Water Resources Research, 24*, 1847–1857.

Thomas, J. F., Syme, G. J., & Salerian, S. N. (1987). Factors influencing bore ownership of residential households in Perth. In *Domestic water use study Perth, Western Australia* (pp. 72–124). Perth: Water Authority of Western Australia.

Chapter 8
Inter-Basin Transfers as a Supply Option: The End of an Era?

Jean-Daniel Rinaudo and Bernard Barraqué

1 Introduction

In the past, when water demand in large metropolitan areas outstripped local supply, additional water was often secured by tapping into other river basins. As a reflection of the enormous interests at stake, such economic centres devoted correspondingly impressive financial resources and technical ingenuity to the design and construction of inter-basin projects. Examples of projects transferring hundreds of millions of cubic meters per year over hundreds of kilometers abound in different parts of the world, including Australia, Europe, Northern and Southern America, and Asia. In the USA for instance, the largest metropolitan areas – including New York, Chicago, Los Angeles, Oakland and San Francisco, Houston, Dallas, and Miami – rely on large scale inter-basin transfers (IBTs). Most were built during the nineteenth and twentieth centuries, reflecting a phase that Barraqué calls the civil engineering paradigm (see Chap. 9, this volume).

The issue became more complex after WW2. On the one hand, under American influence, international finance institutions supported new-born nation-states, which, emerging from colonisation, endeavoured to develop their economies via hydraulic projects. On the other hand, in the US (and soon after in other developed countries), economic tools like cost–benefit analysis were increasingly used to legitimise projects funded from government and other sources. The idea that hydraulic projects should finance themselves via their beneficiaries slowly eroded the older 'pork barrel' philosophy. From about 1965, the environmental protection movement

J.-D. Rinaudo (✉)
Water Department, BRGM, French Geological Survey, Montpellier, France
e-mail: jd.rinaudo@brgm.fr

B. Barraqué
Centre International de Recherches sur l'Environnement et le Développement, Paris, France
e-mail: bernard.barraque@agroparistech.fr

Q. Grafton et al. (eds.), *Understanding and Managing Urban Water in Transition*, Global Issues in Water Policy 15,
DOI 10.1007/978-94-017-9801-3_8

brought additional arguments against large hydraulic projects, and plans for many yet to be built began to be shelved since they seemed to perform poorly on both economic and environmental grounds. However, even today some projects still proceed, and indeed some large ones have emerged, as in China.

This chapter is devoted to discussing the evolving role of IBT in urban water management. We argue that the era of large IBT development is about to end, at least in democratic developed countries. The chapter starts with providing an historical overview of IBT development, distinguishing ancient drinking water aqueducts and multipurpose urban water projects. The next section describes how IBTs are challenged by a change in the technological and economic context. It shows how the emergence of alternative technologies, such as desalination, wastewater reclamation and reuse, or managed artificial groundwater recharge is reducing the attractiveness of IBTs. Water utilities are also becoming increasingly aware that water conservation programs can save volumes of water at a much cheaper cost than IBT. In Sect. 4, IBTs are discussed within the changing socio-political context. We show, through various international examples, that IBTs trigger many questions and concerns from communities involved or affected, questions such as the environmental impact on donor and receiving river basins, the economic impact on donor regions, the impact on local cultures and livelihoods, how costs and benefits are distributed (social justice), and issues related to public participation. In a final section, we look ahead at new and more efficient uses of existing IBTs. We suggest they can be used for increasing the flexibility of how water is allocated over space and time. As conjunctive use management approaches gain support, IBTs will be operated in conjunction with aquifer storage and recovery schemes, and they will probably support the development of emerging water markets, in particular during drought years.

2 IBTs in Contemporary History

Cities, and urban civilizations, have always depended upon fresh water. There are many examples of once glorious civilizations which disappeared due to changes in hydrologic conditions or extreme events (which they themselves sometimes provoked): Fatehpur Sikri, Angkor Wat, Mayan cities. Other cities were defeated when their conquerors destroyed their water supplies.

In most of these cases, water was required as a vital resource, but somewhat indirectly for irrigation and other human uses, including for drinking. More directly, with the Roman Empire for example, long distance aqueducts were developed primarily for urban comfort and drinking, since the amount of (pure) water carried was insufficient for irrigating crops. Most aqueducts ceased to function after the fall of the Roman Empire, and only a few small ones were built during the Middle Ages under feudal rule (monasteries). The one exception was Rome where a single aqueduct survived (*Acqua Marcia*, 91 km).

The idea of acquiring clean, fresh water from a distance was picked up again by Italian engineers during the Renaissance. It was very much needed because at the time a proto-industry was being developed based, in part, on putrefaction techniques, the effect of which was to cause local urban water to become increasingly polluted (Guillerme 1983). Nevertheless, aqueducts were costly and reserved only for kings, their castles, and patrician palaces, although they would eventually come to serve public fountains. Most cities relied on local water sources, either wells or river water served by vendors.

During the eighteenth century water was increasingly considered a potential source of disease and even of epidemics (given the miasmatic theory). Experts of the time either favoured filtration, which was already experimented with, or distant sources. Before the discovery of microbes in the second half of the nineteenth century, there was no way to ascertain which of these two technologies was better. The choice was between small quantities of pure water from distant sources, or larger quantities of unsafe local water, which could be pumped thanks to the rapid spread of the steam engine at the turn of the nineteenth century. Fire-fighting and parks watering also required water, and urbanization led to a dramatic increase in solid waste accumulation in the streets; many local authorities imagined they should 'wash the city clean', an option that required yet more water.

2.1 *Potable Water Aqueducts*

Some of the aqueducts bringing water to palaces or monasteries, or even to public fountains, have long histories: the New River project in London dates to the sixteenth century and was 68 km long; the Medici aqueduct feeding the (now) Luxembourg palace in Paris from Rungis springs dates from 1623 and ran for 13 km. In turn, progressive improvement in the technologies of pipe welding and faucet design allowed the development of water supplies under pressure, extending to within residential buildings. Enlarged water capacity was also needed to deal with waste evacuation, notably once London decided to turn the drainage system to the discharge of human waste. The combined sewer was born, and this innovation needed flushing if the slope was insufficient or the rain too infrequent.

In Britain, however, taking freshwater from distant sources was not easy for both hydrological and legal reasons: not only were there no large mountains or aquifers to draw water from, but the riparian rights, which translated the common law to the water domain, meant that cities needed an act of Parliament to catch distant water. Even though some cities like Glasgow managed to draw clean water from up to 55 km away (and as early as 1859), others preferred to take water from a nearby river and filter it. In London, filtration became mandatory in 1852 after a major cholera outbreak. Birmingham, the champion city for public water (and gas) procurement, had to wait until 1902 to get fresh water from a reservoir built in Wales 100 km away (Barraqué and Andreas Kraemer 2014).

As early as 1837, New York decided to fetch water from Croton Lake, located north of Manhattan in a rural area about 60 km east of the Hudson river. A dam was built to store more water in the lake, and a 66 km aqueduct took water to a fountain in Manhattan in 1842. For centuries, Madrid in Spain had been served by groundwater, relying on the ancient Persian technology of the *Qanats* (horizontal galleries draining dripping groundwater). But in the nineteenth century, supplies became inadequate, and in 1851 a new aqueduct was built from mountains north-west of the capital, about 60 km away.[1] In Paris in 1860, Baron Haussmann, a powerful Prefect of the time, found an engineer of the corps, Eugene Belgrand, who knew the nature of all the springs around Paris, and this engineer drew up a plan to divert several springs to Paris via a system of aqueducts that would take water from around 100 km away. It took almost a century for the plan to be fully implemented, but today these aqueducts provide half of the water Parisians drink, while Paris suburbs depend either on surface water from the Seine and its tributaries or from shallow aquifers.

Before the First World War, New York City extended its long distance aqueducts up to the Catskill mountains 250 km to the north. Other cities in the eastern part of the country, like Boston, also built in-stream storage facilities further and further away from urban areas. The first reservoir, Cochituate, built between 1845 and 1848, was only 27 km away, but the most recent, Quabbin, built in the 1930s, is 105 km away. The longest transfers can be found in California, most likely due to the Mediterranean-type climate. Around WW1, Los Angeles managed to appropriate freshwater from the Owens Valley and transferred it across a mountain range (375+220 km), while San Francisco built the Hetch Hetchy aqueduct from the Yosemite Mountains (269 km), and Oakland another line from the Moktelumne river (153 km). Some time later, Southern California even took water from the Colorado River 389 km away across a range of mountains.

In Europe, construction of long distance transfers usually depended upon authorisations, and they required subsidies or low interest loans by national or central governments, the reason they were frequently reserved for capital cities. Lyon, the second largest city in France, always took its water from an alluvial aquifer in a natural area upstream of the Rhone. Conversely, Lisbon received water from a reservoir upstream through a magnificent aqueduct (*Aguas Livres*) built by the government in the nineteenth century, an aqueduct which incidentally provides little water compared to the needs of modern hygiene (Luisa et al. 2012). As soon as water treatment could take advantage of discoveries in microbiology, it became cheaper to provide safe water pumped from rivers close to cities, and many aqueduct projects became financial white elephants.

But of course, water transfer projects were still being built throughout the Mediterranean and in other areas of water scarcity. In France, the Canal de Marseille is an 80 km transfer from the Durance, providing water to Marseille as early as 1849, but which was constantly improved until 1970, when it was supplemented with a regional water scheme called Canal de Provence.[2] Another example is the *Landeswasserversorgung* in Baden Württemberg, a complex system of aqueducts from the mountains in the south to the urbanised north (in the Stuttgart area). It was

[1] See "The Croton Aqueduct", New York Historical Society at https://www.nyhistory.org/seneca/croton.html

[2] See the Canal de Provence website at http://www.canal-de-provence.com

a national project funded and owned by the German government in the early twentieth century, and it was only after WW2 that it was transformed into an inter-municipal joint board.[3] The most impressive example in Europe between the two World Wars remains the Italian Aquedotto Pugliese,[4] a regional scheme to bring water to the driest part of the country (the heel of the boot), and which has long been the largest regional water supply scheme in terms of population served (4.5 million).

2.2 Multipurpose Projects and Urban Interbasin Transfers

In the early twentieth century, the increasing involvement of national states in economic matters led to an unprecedented rise of large hydraulic projects, including dams and transfers, and this time it was not only to bring water to cities. Projects usually started with hydroelectricity generation, requiring the construction of reservoirs. But then it also became conceivable to regulate river flows so as to enlarge navigability periods, provide water for irrigation, and for industry. Eventually such projects also included water for cities, but it was not their main goal.

The United States played a major role in this development, the reason being that they could mobilise important major amounts of capital and direct them to areas that were wild or even desert. One of the first projects was the construction of a canal from the Colorado to the Imperial Irrigation District, a desert located between San Diego and the Mexican border (1898). This project was followed by many other state or federal projects, resulting in the interlinking of Californian watersheds and valleys from north to south.[5] Today, more than three-quarters of the water is used for irrigation, and progress in water efficiency has created the opportunity to re-allocate water for other higher value uses.

Similar projects were constructed in several western states, in particular in Colorado and Arizona. The city of Denver, Colorado, which is located on the eastern slopes of the Rocky Mountains, put in place a long distance infrastructure to convey water from the western side of the range. As a compensatory measure for people in areas from where the water came, reservoirs reserved for their sole use were also constructed (Blomquist et al. 2004).

In Arizona, the development of the Colorado Aqueduct Project was mainly intended to supply the rapidly growing areas of Tucson and Phoenix, which had outgrown local surface water and groundwater resources. As is frequently the case with multipurpose projects, the transfers were mainly paid for by urban users, with additional water sold to farmers at a much lower price.

---

[3] See website of Landeswasserversorgung – Trinkwasser für Baden-Württemberg at http://www.lw-online.de/

[4] See the Aquedotto Pugliese website at www.aqp.it

[5] See the Californian Department of Water Resources at www.water.ca.gov

However, the most representative example of this multipurpose project policy is to be found in the eastern United States: the Tennessee Valley Authority. This was a planned project to revitalise a depressed economic area after the 1929 crisis. It was also a federal government project to create a dream in which water mobilisation could create a new and modern way of life (with electricity for industry and homes). At the time, there was a similar project planned on the Rhone in France (CNR), but engineers and planners met strong resistance from local communities and the project remained limited to electricity generation on the Rhone between Switzerland and Lyon (the Génissiat dam). Similarly, the soviet regime built a similar project on the Dniepr under Stalin (the Goelro plan, with the Dnieprostroi hydroelectricity farm and education centre).

The allied victory in 1945 led to the setting up of a new world order based upon the creation of new nation states. These new states were encouraged to promote economic development through the development of hydraulic infrastructure. American cooperation was influential, and the TVA is still considered to as an inspiring model, as this quote from the TVA website indicates:

> Under the leadership of David Lilienthal ("Mr. TVA"), the Authority became a model for America's governmental efforts to modernize Third World agrarian societies.[6]

Many countries launched hydraulic projects combining electricity generation, irrigation, and urban water supply, as did Spain with the financial support of the US (Swyngedouw 2007). During the Franco dictatorship (1940–1975), Spain became, in terms of number of dams, the third highest-ranking country in the world. Brazil also developed complex water and electricity projects in the metropolitan areas of Sao Paulo and Rio de Janeiro, and later with huge dams on the Parana. Even small countries like Israel[7] or Tunisia[8] built national water carriers for the sake of developing arid southern regions. Socialist countries also invested in large projects, with the USSR developing some in central Asia for cotton growing (the Ferghana valley in Uzbekistan). This project led to the drying up of the Aral Sea, because the part of the scheme which involved changing the course of a Siberian river southwards to feed the dying sea could not be built. In many developing countries, however, multipurpose schemes were legitimised by appeals to the need for urban water, even if this need remained minor.

Despite growing doubts about the positive balance of pros and cons, and the retreat of the World Bank on several financing schemes, many of these projects went ahead: GAP (the Greater Anatolia project) in Turkey, Nile regulation with Toshka transfer in Egypt, the Three Gorges and South–North transfers in China, and the San Francisco–Ceara in Brazil. India is even now considering the construction of project

[6] Tennessee Valley Authority website at www.tva.gov

[7] See presentation of the National Water Carrier at http://www.mekorot.co.il/Eng/Mekorot/Pages/IsraelsWaterSupplySystem.aspx

[8] See website of the Tunisian National Water distribution society at www.sonede.com.tn/index.php?id=44

linking rivers in the Himalayas to Kerala and Tamil Nadu, with eastern and western branches.

After the rise of the environmental movement in the 1970s, multipurpose hydraulics started to receive criticism, beginning in the US, from ecologists and also economists. In France, the retreat of hydraulic engineers from the colonies gave rise in the 1960s to the development of new multipurpose projects at regional level. Three major regional planning companies were set up: Société du Canal de Provence, Compagnie du Bas-Rhone Languedoc, and Compagnie d'Aménagement des Côteaux de Gascogne. All of these started with irrigation as a major component, but Canal de Provence was more economically successful because it found urban and industrial customers able to pay close to the full cost of regional water.

It was the rise of alternative solutions for urban water provision which undermined the economic rationale of IBTs.

3 IBTs in a Changing Technological Context

Until the late 1980s, water transfers were often the sole technological option cities could rely upon to supplement local resources and meet fast-growing demand. The emergence of alternative technologies, such as desalination, wastewater reclamation and reuse, or managed artificial groundwater recharge – and the progressive decrease of their cost – strongly influenced the perception of IBTs by water managers and planners. At the same time, water utilities increasingly became aware that the demand for water could be curbed through proactive water conservation policies. Because significant volumes can be saved and devoted to new uses, water conservation has increasingly been considered as an alternative to water supply options – IBTs in particular. This section aims to illustrate how this change has happened through examples selected from different parts of the world.

3.1 *The Development of Non-conventional Supply Options*

3.1.1 IBT Versus Desalination

Desalination represents a serious alternative to inter-basin transfer for a number of reasons. The first one is the declining cost of this water supply option. The most common technology chosen for new desalination plants today is reverse osmosis (60 % of existing capacity), in which salt water is filtered under high pressure through semipermeable membranes. The cost of RO varies significantly depending of the quality of raw water used, the size of the plant, the site conditions, and the cost of energy (Zhou and Tol 2005; Ghaffour et al. 2013). Desalination of brackish groundwater, estuarine water, or contaminated fresh water is often preferred to seawater desalination due to reduced energy cost. Recent case studies have shown the full cost of such projects ranges between €0.3–0.7/m³ (Vedachalam and Riha 2012). A Spanish study estimated the cost of seawater desalination at about €0.7/m³ in the Segura basin, when facilities there operate at full capacity (Lapuente 2012). These costs are likely to continue to decrease if new low-energy technologies become successful (Ghaffour et al. 2013).

There are other reasons why desalination can be more attractive than IBTs for water utility managers. The prime advantage of desalination is that it is a secure and reliable source of supply, independent of climatic conditions, unlike IBTs which can be affected by drought. Desalination also produces very high water quality that can be blended with lower quality water supplies if necessary. For local politicians, desalination offers greater control of municipal water supplies, allowing them to freely consider a wide range of economic developments – without depending on water supplies from outside their jurisdiction. Last but not least, the mobilisation of financial resources is easier in the case of desalination as the private sector often gets involved through 'build–own–operate–transfer' (BOOT) contracts (Ghaffour et al. 2013). Since the associated costs are operational and maintenance costs rather than initial investment, they are easier to recover from customers, even if they are high.

Several large metropolitan areas confronted with mounting water scarcity (and even shortages during droughts) have preferred the desalination option rather than the construction of an inter-basin transfer. The case of Barcelona provides a good illustration. In the 1990s, Barcelona investigated the possibility of importing water from the Rhône river, through an extension of the canal system that diverted water to the Montpellier region. The project consisted in constructing a 330 km long pipeline intended to transfer up to 400 million m³/year to the metropolitan area. For reasons that will be further exposed below, the metropolitan area instead constructed a 200,000 m³/day desalination plant in 2009 which can satisfy 20 % of Barcelona's drinking water needs (Saurí et al. 2014).

Another example is the Kimberley pipeline scheme designed to supply Perth in Western Australia (Ghassemi and White 2007). Metropolitan Perth has a sustained population growth of over 1.5 % per year and it is expected that demand could increase by 600 million m³/year by 2050. In that context, numerous proposals have been developed since the late 1980s to transfer water from the wet tropical region of the Kimberleys to Perth and even Adelaide (op. cit. pp. 169–176). The preferred project consisted in constructing a 1,400 mm diameter pipeline 1,840–2,100 km long (depending on route). Investment cost was estimated at A$8–11 billion (1990 dollars) and operating and energy costs were more than A$2 billion/year. Overall, the total cost of imported water was estimated at A$5–6/m³, which was five times above the cost of desalinated water (A$1.1/m³) and other alternative options such as groundwater extraction (A$0.3–0.8)). After several studies, the Kimberley scheme was finally abandoned and a 45 million m³/year desalination project was approved in 2003 at a cost of A$350 million. The Western Australian government also adopted a strategy of improving water use efficiency in all sectors and promoting water reuse.

3.1.2 IBT Versus Wastewater Reclamation and Reuse

Another alternative to IBTs consists in using poorer quality water supplies and developing wastewater treatment (reclamation) for subsequent use as a water supply (reuse). Water reuse is growing in importance in the US, Australia, some parts of Asia (Singapore), and in Western Europe (Spain). Recycled water is most commonly used for groundwater replenishment or for landscape irrigation; much less frequently is it used for direct domestic reuse.

Reclamation and reuse can be either centralised or decentralised. In decentralised systems, water reclamation facilities are located close to areas where commercial or residential demand exists for reclaimed water. Only a fraction of the available wastewater is treated and distributed. This approach, which is sometimes referred to as "sewer mining" only requires the development of a small local secondary distribution network. Investment saving are likely to compensate the cost of no economy of scale.

In a centralised system, wastewater is collected and treated (possibly up to drinking water standards) before being introduced into a water supply source (river, reservoir, or aquifer). A combination of three purification technologies is frequently used: micro-filtration, reverse osmosis, and UV treatment (Zekri et al. 2013). A well-known example is the groundwater replenishment program of Orange County Water District (OCWD) in California. OCWD has recently completed a new recycling facility that provides a new resource for the region. Wastewater is treated using reverse osmosis technology. It is then injected into the aquifer through a series of shallow and deep wells. The system is used both as a seawater intrusion barrier and a source of water. Groundwater is then pumped and used for drinking supply (indirect potable reuse) and distributed to consumers using the existing reticulation system. A variant, observed in other Californian water districts (see Chap. 11, this volume), consists in directly distributing reclaimed water to customers through a dedicated reticulation network ('purple pipe') for non-potable use. The cost of this option is much higher due to the need to construct a dual distribution network. In Europe, the Dutch have long replenished the sand-dune aquifer between the Ranstad and the North Sea with monitored freshwater from the Rhine (when the quality is acceptable). In Barcelona, the AGBAR company also treats waste water through RO technology to recharge the alluvial aquifer of the Llobregat 15 km upstream from the drinking water plant of Sant Joan Despi.

A number of studies have shown that the costs of managed artificial recharge and water reclamation compare favorably with those of alternative sources. In Southern California, the total capital, operating, and maintenance costs were estimated at US\$1.34/m^3 for the West basin and US\$0.77 for Orange County Water District, which is cheaper than imported water supplied by the Metropolitan Water District of Southern California (National Academy of Sciences 2012). In San Diego, the cost of recycled water is estimated at about €0.35/m^3 (2001 value) which is equivalent to 90 % of the potable water cost. The same results were found in Florida where the cost of recycled water was estimated at €0.3–0.37/m^3 (Miller 2006). Other comparable cost figures are reported for several projects worldwide (Zekri et al. 2013).

Recycled wastewater can also be introduced into surface water resources. Singapore opted for this option with the well-known NEWater project in the early 2000s. Wastewater is treated and blended with surface water resources in reservoirs. Recycled water will represent 15 % of daily urban consumption once the planned treatment capacity is fully deployed. This technological change was implemented with the aim of reducing reliance on importing water from Malaysia. The cost of NEWater is estimated at less than €0.2/m^3 (Tortajada 2006).

3.2 IBTs Versus Water Conservation

Although urban water utilities are still inclined to consider importing remote new resources to satisfy the needs of growing populations and economies, they are increasingly aware that the demand for water can be curbed through proactive water conservation policies, which often turn out cheaper than new water supply schemes. A projected transfer from Sweden to Denmark across the Strait of Sund was abandoned after Copenhagen drastically reduced its domestic water consumption. In the UK, the regulator Ofwat banned water transfers from the north to the London area until water conservation measures were implemented to reduce leakage. In southern Italy, the Aquedotto Pugliese looked at the option of importing water from Albania across the Otrante Strait. But water conservation policies made this project unnecessary, as well as other, smaller and IBTs.

There are a number of benefits associated with water conservation: containing demand avoids or postpones huge capital costs associated with the expansion of water supply and waste water treatment infrastructure. A number of studies have shown that reducing water demand through active water conservation measures can be much cheaper than mobilising new conventional or non-conventional resources. In California, a study by the Pacific Institute estimated that more than 25 % of existing water use could be saved at a cost below the marginal cost of water supply (Gleick et al. 2003). Similar results were found in southern France (Rinaudo et al. 2010).

Three main control levers can be used to promote urban water conservation. The first one consists of imposing minimum technical (or economic) water efficiency standards that water utilities need to comply with. This particularly applies to water distribution networks, which are still often characterised by high rates of leakage. Leakage can be reduced by setting up a technical and managerial organisation that allows for real-time detection of pipe bursts or leakage and rapid repair. It also requires increasing the replacement rate of old pipes, which can entail huge costs, especially if maintenance has been deferred for many years. Different regulatory approaches can be selected. France for instance has opted for the definition of maximum admissible rates of leakage based on technical criteria such as the number of customers per pipe kilometre (regulation of 27 Jan 2012). A different approach is implemented in the UK where water companies are required to calculate an economically optimal level of leakage (ELL), which is such that the marginal cost of reducing leakages equals the marginal cost of mobilising new resources.

The second tool consists of economic incentives to promote the adoption of water-efficient technologies. Two types of incentives have been widely used in the USA and more recently in Australia: price rebates, and water conservation oriented water rates. Rebate policies aim at encouraging the purchase of water-efficient appliances (e.g. AAA clothes washing machines, timers for garden sprinklers) or retrofitting of old inefficient appliances (toilets). Rebates can target domestic, commercial, or industrial users. European countries are making tentative attempts at using them. Pilot projects have been implemented in the UK (Waterwise 2009), Spain (Shirley-Smith et al. 2008), and in some French counties (Corrèze, Gironde) through distribution of free or subsidised faucet aerators or 'water hippos' (bag-like devices that fit in the water cistern and reduce the volume of water flushed).

Water conservation oriented rates have also been used to promote the wise use of water. Pricing tools are intended to simultaneously change water use behaviour (reduction of wastage) and promote the adoption of water-efficient appliances, fixtures, and fittings. The most frequently used tariffs are increasing block rates and seasonal rates. A number of studies have demonstrated the potential savings that can be achieved through such tariffs (Rinaudo et al. 2012). More complex budget-based rates are also becoming more popular in south-western states of the US. This approach consists of rates that are tailored to the situation of individual customers.[9] In Europe, this rate budgeting is not frequent, but some water companies in England propose variable tariffs, with either high fixed parts and low volumetric prices (for customers who have regular water uses) or low fixed parts and high volumetric ones (for people who may have low consumption but some peaks).

The third lever consists of raising customer awareness of water scarcity through information campaigns. This lever is often considered as a catalyst for and used in conjunction with technical and incentive-based measures. A few econometric studies have shown that they can, for instance, significantly enhance the effectiveness of water pricing measures (Kenney et al. 2008).

Water conservation may also generate significant benefits related to energy savings. In the residential sector, these benefits may outweigh the cost of installed devices such as low flow shower-heads or high efficiency washing machines and dishwashers.

[9] A water budget rate is an increasing block rate structure in which the block definition is different for each customer, based on an efficient level of water use by that customer (Mayer et al. 2008, quoted in Beecher 2012).

3.3 Economic Comparison of IBTs with Alternative Solutions

In developed countries, water utilities are increasingly compelled by regulators to justify that the investment decisions they make are economically sound. This often implies conducting cost-effectiveness or benefit–cost analyses[10] to compare various water supply and demand management options. Inter-basin transfer projects must be compared not only to 'soft' water conservation measures (Gleick 2003) but also to alternative 'hard' water supply options, including desalination and waste water reuse. The economic comparison of alternative water management options is current practice in the US, where least-cost and benefit–cost analyses generally make up a full chapter of urban water management plans. It is also increasingly used in Europe, partly because the Water Framework directive requires it, but also because national regulators like Ofwat and the Environmental Agency in the UK (EA 2011) have made it compulsory. Many case studies have shown that water conservation measures and the use of non-conventional resources can be much cheaper than long-distance and large-scale inter-basin transfers. They might not, however, always be sufficient to bridge the increasing gap between demand and supply in regions confronted with rapid demographic and economic growth. This is now illustrated through references to several international situations.

3.3.1 Results of Economic Assessment in Selected Examples

In the UK, a systematic cost-effectiveness analysis of water supply and water demand management options is carried out for all water resources management plans (WRMPs) prepared by water companies, according to guidelines developed by Ofwat and the Environmental Agency. The EA states that the WRMP assessments have demonstrated that large-scale transfers of water to the south of England from Wales or the north of England would be more expensive than other options available to meet current forecast demand (Environment Agency of England and Wales 2011).

In the Spanish Catalonia region, Pouget et al. (2012) conducted a cost-effectiveness analysis to compare various adaptation strategies. Their study shows

[10] The cost-effectiveness approach consists in calculating the average annual cost of each cubic meter saved (or mobilised) with a set of water supply and water conservation measures. Annual costs are calculated by spreading investment costs over a duration corresponding to the technical lifespan of the project and adding recurring operational and maintenance costs. Diminishing effectiveness (in terms of volumes saved) can be accounted for in the calculation if relevant. The benefit–cost approach consists in estimating the costs and benefits of a set of water conservation measures that can be ranked and prioritised on the basis of a benefit–cost ratio. This approach has been widely used in the US, in particular because it was incorporated in software packages such as IWR-MAIN developed by the Army Corps of Engineers, or the Least Cost Planning Demand Management Decision Support System developed by Maddaus (Maddaus and Maddaus 2004).

Table 8.1 Levelised costs by water management options for the Canberra area (Turner and White 2003)

Option	Savings in 2023 (10^3 m^3/a)	Present value cost (M$)	Levelised cost ($/$m^3$)
Water conservation	12,500	45.2	0.30
Source substitution	6,000	308.5	4.50
Source augmentation:			1.45
New Cotter Dam	3,700	55.6	1.34
Tennent	3,700	77.4	1.87
Tantagara	3,700	47	1.14

that long-distance inter-basin transfer (bringing water from the Rhône in France to Barcelona) would cost up to five times more (€0.5/m^3) than water conservation in agriculture (modernisation of irrigation systems) or implementing seasonal tariffs (€0.1/m^3). Other water supply measures were also found to be less expensive than large inter-basin transfer projects, e.g. managed aquifer recharge (€0.25/m^3), improvements in the quality of currently polluted water resources (€0.25/m^3); and water reuse projects (€0.3/m^3). However, in the long term these measures would not be sufficient to bridge the gap between future water demand and available resources, as the latter would decline due to climate change. The mobilisation of new resources, through a transfer from the Rhône or desalination (the option which Barcelona ultimately chose), would thus be necessary. In our opinion, many studies of water demands are based on unquestioned optimistic demographic and economic growth forecasts, and this was indeed the case for the Barcelona water transfer justification.

Examples can also be found in Australia. Table 8.1 below emphasises the relevance of water saving measures in the Australian Canberra area.

3.4 *IBTs Versus More Gradualist Adaptable Solutions*

Due to the one-off nature and very long lifespan of IBT infrastructure, its design calls for a forecast of the long-term demand in the receiving basin and the long-term water resources availability in the donor basin. However, over the last decade we have seen how large are the uncertainties attached to demand and supply figures, and how difficult it is to forecast them accurately. Climate change in particular makes it difficult to predict the amount of water which will be available in both the receiving basin (local resources) and the basin of origin. New water infrastructure must be able to meet a large range of changing conditions. In that context, some authors recommend avoiding the use of large inflexible infrastructure which can't adapt to changing climatic or economic conditions (Hallegate 2009). Instead, they recommend adopting a gradualist strategy, reducing the time horizon for decisions and adapting management strategies as the situation unfolds. Such flexible approaches necessarily involve a combination of various infrastructure solutions (storage, transfer, recycling) and soft adaptation strategies (demand management).

In some countries, these recommendations have already been transferred into regulations. These impose an approach which combines infrastructure development with water conservation and recycling. This is the case for instance in California, England, and Wales. Moreover, cities which depend heavily on imported water supplies tend to invest in multiple alternatives so as to reduce their vulnerability to interruption of their major resource. This is true for cities subject to earthquakes, like Marseille in France or in California. The solution consists in developing local resources (groundwater in Marseille, the Encina desalination plant in San Diego) and implementing water demand management approaches (including setting the water price at the potential marginal cost in case a new supply is built).

3.5 Energy Considerations

Since IBT projects often require moving large quantities of water over long distances and significant elevation differences, they are high energy-consuming water supply options. For instance, the energy used by the California State Water Project (SWP) to transfer water from north to south California represents about 2–3 % of all electricity consumed in the state. This is equivalent to approximately one-third of the total average household energy use in the region (Cohen et al. 2004). With this perspective, non-conventional water supply options such as desalination of brackish water or reuse may be competitive in terms of energy requirements. For example, the water recycling system constructed in Orange County, southern California, uses only half the energy required to import water from Northern California with the SWP (Cohen et al. 2004). In a context of increasing concern over energy issues, the high energy requirements of IBT are likely to reduce their attractiveness against other technological solutions.

4 IBTs in a Changing Socio-Political Context

IBTs trigger many questions from communities affected. These questions involve a variety of issues including environmental impacts on donor and receiving river basins, the economic impact on donor regions, impact on local cultures and livelihoods, how costs and benefits are distributed (social justice), and issues related to public participation (Gupta and van der Zaag 2008).

4.1 A Growing Concern for Environmental and Socio-Economic Impacts of IBTs

4.1.1 Environmental Concerns

There is a growing awareness that IBTs can generate significant environmental impacts on donor and receiving basins. In the donor region, reduction of in-stream flow may endanger aquatic ecosystems through reduced flows, which results in increased temperature and reduced dilution of pollutant loads discharged to rivers. Receiving regions can also be impacted by the inflow of imported water. Ecosystems or species can be perturbed by changes in the hydrological regime of the receiving river, or by its quality. There are examples where an IBT has favored undesirable migration of species from one basin to another and even of invasive alien species. IBTs also modify hydrological regimes, reduce flow variability, and can cause a seasonal river to become a perennial river. Several examples in South Africa and the USA are quoted in Gupta and van der Zaag (2008).

Fears of environmental impacts do not always rely on strong scientific evidence. For instance, the Aqua Domitia IBT project (southern France) is strongly opposed by citizens and environmentalists who fear that receiving rivers could be contaminated by PCBs (polychlorobyphenyls) and even radioactive waste leaking into the donor river (the Rhône). Although water quality analyses apparently show that there is no such risk, opponents clearly do not trust the information provided to them and instead advocate the precautionary principle (Commission Nationale du Débat Public 2012).

The environmental impact of IBTs not only restricts the possibilities of constructing new infrastructure, it may also threaten existing ones. In California, the capacity of the Central Valley Project, which transfers water from the Sacramento River to the San Joachim Valley, was reduced by 11 % as a consequence of the Central Valley Improvement Act of 1992. The re-allocation of 985 million m^3 to in-stream flow was intended to regenerate salmon runs. Also in California, the Hetch Hetchy aqueduct to San Francisco was strongly contested at the time of its construction by John Muir, a prominent environmental activist. The pipeline, which captures water from the reservoir on the Tuolumne River in Yosemite National Park, is still a source of controversy today and environmentalists continue to request its removal (Ghassemi and White 2007, pp. 219–220).

The environmental constraint on IBTs has also been growing due to the recent evolution of water laws and regulations. In the California Central Valley example, the decision to reallocate water to the aquatic environment was a direct consequence of the Endangered Species Act (1973). In Europe, the Water Framework Directive is also likely to restrict IBT possibilities, as it strengthens requirements for minimum in-stream flows.

4.1.2 Increased Hostility from Donor Regions

In the basin of origin, a loss of water can be perceived as a risk for long-term economic prosperity and quality of life. Communities located in donor basins feel that IBTs unfairly direct economic development away from them, for the benefit of predatory metropolitan areas. As a US stakeholder put it: "when you ship water across the state, you ship jobs as well, for economic development cannot occur without sufficient water supplies".[11]

This potential harm has caused most inter-basin transfer proposals to be controversial, often pitting cities benefiting from the transfer against donor rural areas. Conflicts over IBTs are increasingly reported in both developed and developing countries (Cox 1999). For instance, in 1966 the city of Mexico started to extract and transfer groundwater from the Ixtlahuaca valley, 50 km outside the city. Groundwater tables rapidly declined, depriving the local population from access to a water resource used for drinking water and agricultural irrigation. This generated a strong conflict in the 1970s and 1980s, even threatening the physical integrity of the aqueduct (water robberies and direct attacks) and ultimately led to army intervention (Dyrnes and Vatn 2005).

Conflicts over water transfers may oppose counties, regions, or states. For instance, the Atlanta metropolitan area (Georgia, USA) planned to purchase and transfer water from Tennessee to meet increasing demand. The State of Tennessee opposed the transfer by passing an Interbasin Transfer Act of 2000 (Carver et al. 2011). Atlanta considered other long-distance transfers from two lakes in northern Georgia and from southern Georgia aquifers, but these projects were abandoned as too politically charged. Currently, the city of Atlanta is investigating more local solutions that only involve transfers from one district to another. But these projects too are highly controversial.

This led certain countries or states to tightly regulate transfers to avoid third-party effects. Indeed, in 1996 the water code of California included five articles on 'wheeling', which, among others, protects the rights of third parties impaired by transfers.

4.1.3 The Political Dimension of IBTs

Due to mounting opposition from parts of society, water transfers are increasingly debated in regional, state, or federal political arenas and can even influence the outcome of elections. There are a number of examples showing that political parties can place an IBT as a key issue on their political agenda prior to elections. In Western Australia for instance, the construction of the Kimberley scheme designed to supply the Perth metropolitan area was strongly supported by the Liberal–National coalition

[11] Comment by Ron Cross, posted on 21 March 2010 at: http://onlineathens.com/stories/032110/opi_593391532.shtml

during the 2005 election campaign, whereas the Labor Party favored desalination (Ghassemi and White 2007). In Spain, the conservative Aznar government actively supported a National Hydrological Plan during the 2004 elections. At the same time, the European Commission issued two 'reasoned opinions' criticising the lack of appropriate environmental impact and economic evaluation studies. The surprising victory of Aznar's socialist opponent Zapatero allowed the new premier to shelve the huge IBT and launch a desalination plan instead.

As a result, in many non-authoritarian countries only intra-regional transfers can still be implemented, using the regional solidarity argument. In France for instance, a few transfers have still been developed: in the Reunion Island (Indian Ocean) between west and east, in Provence Verdon-St Cassien (SCP); in Languedoc a regional water transfer called Aqua Domitia is being considered by the regional bulk water supply company BRL, but it is hotly debated (see below).

4.2 *Increased Public Participation Favors Opposition to IBTs*

Legislation in many countries now requires that consultation with people likely to be affected is carried out before a decision is taken on an IBT. Public consultation may lead to significant changes in the design of the IBT project (its route or size) or in the amount of compensation granted to donor regions; it may even lead to a different option or cancellation of the project.

In France for instance, all IBTs above 1 m^3/s are subject to a public hearing process. The consultation is organised by an independent state agency, the Commission du Débat Public, which nominates a magistrate in charge of organising the public consultation. The objective of the Commission is to ensure a fair debate, during which the voices and arguments of proponents and opponents to the project can be equally heard. At the end of the consultation, the Commission does not provide recommendations related to the project, it only summarises in the most impartial way the views expressed during the consultation. The contracting authority in charge of the IBT project must then prepare a report explaining how it has considered the arguments raised in the debate and present the decision it has made. The whole process ensures maximum transparency in how the decision was taken. Public consultation can represent a significant financial and political cost for the promoters of IBTs.

We illustrate this with the case of the proposed Aqua Domitia IBT project, in southern France. This project consists of extending an existing canal – the Lamour Canal – which diverts water from the Rhône River to supply the Languedoc coastal area. Designed and constructed in the 1960s, the Lamour Canal's main function is to promote development of a diversified irrigated agriculture and the supply of water to coastal tourist resorts. Later on, cities such as Nimes and Montpellier hooked into the pipeline to substitute or complement local resources. The canal is owned by the regional government and operated by BRL, a large semi-public engineering company. The Aqua Domitia Project aims to extend the canal to the west, with the construction of a 2.5 m^3/s capacity pipeline 130 km long at a total cost of €280 million. The pipeline would interlink three main coastal basins (Hérault, Orb, Aude) and supply cities, tourist resorts, and about 7,000 ha of irrigated agriculture. Although the agricultural component of the project is marginal (in terms of expected revenues from water sales), the promoters of the project have strongly emphasised this aspect in order to gain political support from rural areas and the agricultural community.

The public consultation consisted of three main phases: the Commission first heard from 92 experts and stakeholders to identify the main issues to be debated; it then organised a series of 13 public meetings in different sub-basins impacted by the project. Five of these meetings focused on thematic issues[12] and six meetings were held to allow all inhabitants to express local views of the problems and identify how an IBT might address these problems. All meetings were filmed and could be followed through the internet. A web site was established to collect comments and suggestions. Stakeholders were offered the option to write position papers which were disseminated by the Commission (Commission Nationale du Débat Public 2012). The debate mobilised experts from water management institutions of the territories concerned by the project and from associations representing environmental protection. Very few lay citizens participated to the meetings, in spite of an intensive communication campaign organised through various media.

The main arguments raised by opponents were the following. The assumptions underlying future water demand forecasts were contested and considered to be biased in favor of the project. The IBT would encourage the development of unsustainable water demand from business activities, housing development, and high living standards. The quality of Rhône water was considered as a source of risk for receiving basins, in particular due to the presence of heavy industries (chemical, nuclear) upstream of the IBT intake. Alternative cheaper solutions, such as using local groundwater resources and implementing water conservation measures, were considered to be not sufficiently well investigated.

The public consultation is complicating the decision making process for regional elected politicians and engineers of the water company. First, it adds a significant cost (about €1.2 million). Second, it offers a platform for opponent to express their views. Third, it weakens the legitimacy of the project if they decide to implement at the end of the consultation; project leaders had to admit that the lack of sufficient existing demand implied the need for public subsidies to fund what would become a regional water grid, which would in turn 'reveal unspoken demands'.

[12]The five thematic issues were the following: (1) the IBT project and urban/economic water demand; (2) the IBT project and irrigated agriculture; (3) IBT and climate change; (4) IBT and alternative solutions; (5) the cost of the project and its impact on water pricing.

4.3 Reduced Public Funding

IBTs are very costly projects which cannot be fully financed by urban water consumers, unless the water price is raised above what engineers and politicians consider to be the maximum acceptable in social and political terms. Their construction thus relies on subsidies. A strategy to attract subsidies has often been to develop multipurpose IBT projects, supplying urban areas, agriculture, industries, and occasionally providing power generation. In Australia, water transfers supply cities and the mining industry. In France, many recent projects have been marketed as combined urban and agricultural projects – in order to obtain European subsidies from the EU agricultural policy and to obtain political support from the agricultural lobby (which is much more proactive than urban water users). In the present economic crisis, financing projects through an increase in consumer water bills might be considered neither socially acceptable nor politically feasible.

4.4 Why Do IBTs Remain So Attractive?

In the western US, the development of new IBT projects is no longer favored because of their high cost compared to alternative solutions and their environmental and social impacts. Water planners increasingly rely on water conservation programs, water trading (purchase from irrigation districts), desalination, and waste water recycling. Conversely, IBTs are still considered by engineers and politicians in a number of European countries (e.g. France, Spain) but also in developing countries. What can explain the preference for multi-million euro water schemes when cheaper water demand management alternatives exist? Analysing the interests at stake for the different actors involved in such decisions can help answer this question. A preference for large scale expensive infrastructure projects can be said to reflect the existence of powerful coalitions between politicians, engineers, and financial actors who wish to protect political and economic interests.

From a political perspective, the choice of an inter-basin transfer solution is likely to yield significant benefits for locally elected representatives. When supporting an IBT project, they typically appear as "providers of a life-giving resource that enhances health, security and prosperity" (Gumbo and van der Zaag 2002). Since such large projects are often subsided by government agencies or international donors in developed countries, they can also claim to have attracted external funds to support the development of the regional economy, in particular in the case of multi-purpose projects supplying agriculture, industry, and the population. Leaving their personal mark in local historical records can be an additional motive: there are plenty of canals and pipelines around the world that bear the name of the local politician who initiated the project. By contrast, water demand management programs are perceived as constraints on the population, so there is little benefit for politicians to support this type of action. It is much more comfortable for a politician to be seen as finding ways to meet people's water demand rather than trying to change their behavior and infringe on their freedom to consume what they consider a basic good. Moreover, because the money spent on water conservation programs does not materialise as a tangible asset, it may be perceived as a less efficient use of public funds than the construction of a pipeline. However, in the end, large hydraulic schemes bring more benefits to governments in search of legitimacy than to their countries' economies (Allan 2002; Molle and Berkoff 2009).

From a technical perspective, engineers in charge of urban water schemes may prefer to build a water supply infrastructure than implement demand management because the infrastructure gives a prompt and long lasting solution to the water deficit problem. By contrast, investing in water conservation is considered more risky since the behavior of water users might change unexpectedly in the future. Technical staff are thus reluctant to engage in a strategy which does not allow them to certify that demand will be satisfied under all future circumstances. They may feel that accepting this risk undermines their professional legitimacy. Moreover, engineers tend to prefer large one-shot projects that they eventually can claim prowess for than immerse themselves in demand management projects, which may require many small and technically diverse engineering inputs, the involvement of non-engineering technicians such as economists (design of water conservation oriented rates), and communication specialists (awareness raising campaigns). The latter somehow implies losing control over a water system that they have been accustomed to operate by themselves.

From an economic perspective, politicians and managers alike generally prefer projects entailing high initial investment costs; these can be subsidised, whereas operation and maintenance costs are not eligible for public subsidies. This is a strategy that allows water prices to be kept at a lower level, with obvious political benefits. Also, engineering firms have a strong interest in supply-oriented options such as IBTs, which generate significant work and can be added to their list of technical references. Large scale projects may also be favored by financial actors, such as banks and international donors, because it allows them to meet their lending objectives. In some nations, individuals can obtain personal benefits (bribes) by supporting the construction of infrastructure (Davis 2004).

The existence of a powerful coalition supporting IBTs against a preferable alternative option is nicely illustrated in a case study conducted in the city of Mutare, Zimbabwe (Gumbo and van der Zaag 2002). The study showed that the choice of the pipeline option allowed the political and economic interests of the three main actors to align. Similar conclusions can be reached in a case study of Mexico City's water supply (Dyrnes and Vatn 2005). These authors conclude that the IBT solution was preferred to unpopular regulation and water conservation policies because it represented a relatively quick solution, popular with public opinion and Mexico City authorities.

5 Looking Ahead: Using Existing IBTs for Reallocating Water

For the different reasons set out in the previous section, fewer large IBT schemes are expected to be constructed in the twenty-first century than in the previous one. Moreover, it is likely that in the future existing IBT schemes will be used to increase flexibility in water allocation over space and time. As conjunctive use management approaches gain support, IBTs will be operated in conjunction with aquifer storage and recovery schemes. It is also likely that IBTs will support the development of emerging water markets, particularly during drought years.

5.1 IBT and Aquifer Storage and Recovery (ASR)

Because most IBTs have been sized to meet demand during drought years, their capacity is not fully used in normal and wet years, producing a significant cost burden. One solution to improve the financial sustainability of these schemes consists in using the excess capacity in normal years to store water in depleted aquifers. Artificial recharge can be performed during wet years, using infiltration ponds or deep injection wells. Stored water is then extracted in dry years when surface water is scarce and the full IBT capacity used. Conjunctive management improves water supply reliability. This strategy, referred to as conjunctive management, combines the use of a transport and a storage infrastructure to bridge the gap between water demands and available resources over time. Such conjunctive management is already implemented in Arizona, where Colorado water transferred with the Central Arizona Project is used for recharging groundwater in the region of Phoenix and Tucson.[13] Similarly, water from the California State Water Project is also used to recharge groundwater in the central valley (Kern county,[14] Santa Clara valley). The combine use of IBTs and ASR however requires that a specific institutional framework – water banking – be developed for tracking water rights and ensuring that users storing excess imported water during some years can retrieve it during drought years.

5.2 From Pipeline to Grid

While engineers and financiers will probably move away from large IBTs in the coming decades, existing projects are likely to continue to develop into water grids at the regional level. Here again, the rationale leading to the development of grids

---

[13] See Central Arizona Project website at: http://www.cap-az.com/index.php/departments/recharge-program

[14] See Water Association Kern County website at http://www.wakc.com/index.php/water-overview/sources-of-water/87-water-banking

is to increase the flexibility of supply over time and space through physically linking production and storage infrastructures. Grids enable water to be moved in several directions – from where it is available to where it is required. The best historical case of a regional water grid is California, where water can be moved from north to south through the State Water Project in the central valley, and via other IBTs from east to west, including the Colorado transfer. It is only after a sufficient number of links came into being that it was possible to imagine water trading. We have already pointed out that the potential third-party impacts, which add to the cost of transfers, usually limit the volumes exchanged (e.g. the IID in San Diego, see below).

Australian cities responded to the 10-year long recent drought by developing such grids. The Queensland government is spending A$ 9 billion on a South East Queensland water grid, consisting of 450 km of pipelines, two new dams, a desalination plant, and three advanced water treatment plants to deliver 350 million m^3/year additional water. It also includes greater reliance on recycled water. A similar project in the State of Victoria consists of creating another such grid, with six major links under construction. The project includes new water transfers such as the Sugarloaf Interconnector that would take water to Melbourne from the Goulburn River, a tributary of the water-stressed River Murray (Pittock et al. 2009).

In the UK, the Environmental Agency of England and Wales recognises that there may be further opportunities for developing greater linkages between existing water company systems, and for sharing of water resources to gain some of the same benefits expected of large-scale transfers (Environment Agency England and Wales 2011). However, a projected national water grid was shelved by the national regulator which instead pushed for conservation and leak control.

5.3 *Water Transfer Infrastructures and Trading in a Mature Water Economy*

In recent years, large-scale water transfer infrastructure has increasingly been used for reallocating water from some water right holders to others, based on voluntary agreements. IBTs can therefore support the development of water trading, which can in theory ensure improved water efficiency, particularly during drought conditions. Examples come mainly from the western states of the USA, Australia, and, more marginally, Spain.

In Southern California, the Imperial Irrigation District (IID) is the 'first in time' and largest water right holder in the Colorado basin (3.2 billion m^3 out of the 5.4 billion allocated to California). IID receives its water from the All American Canal (constructed in 1942) and distributes it to farmers through unlined distribution canals, where significant water losses occur. In the late 1980s, IID was compelled by the US Bureau of Reclamation to improve the technical efficiency of water use through lining distribution canals. Water conservation potential was estimated at about 400 million m^3. IID was then approached by the Metropolitan Water District of Southern California (MWD) who was interested in providing support for the lining, and recovering and transferring part of the water conserved. An agreement was signed between MWD and IID in 1989 for a 35-year term. After implementing conservation projects, IID reduced its water abstraction from the Colorado River at Imperial Dam, while MWD increased its pumping from the Colorado at Parker Dam and transferred the water to the Los Angeles area, using the 380 km long Colorado River Aqueduct. Later on, San Diego came into the re-allocation game, and requested a direct transfer of IID water for its own use (Pincetl 2012).

In 2003, after being accused of not using water efficiently, and under threat of the California Water Resources Board to get its huge share reduced, IID signed another transfer agreement with the San Diego County Water Authority (SDCWA), by which they agreed to transfer annually 247 million m^3 of conserved water for a term of 75 years. IID implemented additional water conservation measures at the farm level, and with other projects the volume of transferred water progressively increased so as to reach a maximum after 10 years. However, even though the SDCWA had gone through this lengthy process to acquire more independence vis-à-vis MWD (whose rival Los Angeles interests superseded), it turned out that the best economic solution was to use existing MWD aqueducts to re-allocate regional water and divert it to San Diego and to another irrigation district (which held senior rights on the water). Moreover, the environmental movement was also able to make sure that some of the water would help maintain the level of the Salton Sea (a saline lake between IID and San Diego which is an important stop on bird migration routes) (Pincetl 2012).

In Spain, two large IBTs, which were primarily constructed to meet growing urban and agricultural water demands in coastal areas, are now being used for transferring water based on voluntary exchanges between right holders. The first case is the Tagus–Segura IBT through which intensive fruit and vegetable growers in the Segura basin purchase water from more extensive farmers in the Tagus basin. The second is the Negratin–Almanzora connecting the upper Guadalquivir basin to the Andalucian coastal basins. This 120 km long pipeline was recently used to transfer water purchased by an Almanzora drinking water company in Andalucía to rice growers in the Guadalquivir delta (Garrido et al. 2013).

IBT infrastructure can also facilitate the development of option contracts between agricultural users and urban areas. In a typical option contract, an urban agency pays an option premium for the right to purchase water at some point in the future, if climate turns out to be dry. Option contracts do not imply the transfer of ownership, so right-holders retain access to the water allotment when the option is not exercised. This type of agreement was established in California in about 1994. In 2003, the MWD and Paolo Verde irrigation district signed an option agreement with a 35-year term by which MWD can call on from 25,000 to 110,000 acre-feet (31–135 million m^3) each year. The irrigation district receives a one-time fee of $3,170 for each acre participating in the program and $600 for each acre that is fallowed during the program. There were 15 such contracts signed between 2003 and 2008 in California (Hansen et al. 2008; Tomkins and Weber 2010). Overall, option contracts represent efficient risk management instruments and they offer new perspectives for using existing IBT infrastructure.

6 Conclusion

Large hydraulic projects have underlain the success of ancient civilizations, and they are part of humankind's water culture. However, most of them have been confined to a single river-basin. The aqueducts of the Roman Empire are a notable exception, and it remains fascinating that the Romans spent so much effort to build such impressive and long-lasting hydraulic architecture, but then, in the end, to transfer so little water. It would be interesting to link this with the economy of the time, e.g. looking at the slavery factor. In our capitalist economy, the value of labor and the depreciation of capital over time make such projects impossible to finance. And because water is heavy, hydraulic infrastructure usually comprises specific assets with high investment costs and slow technical depreciation.

Yet in order to provide clean water to cities, the idea resurfaced of transferring water over long distances, an idea that returned in the classical period and rose during the nineteenth and twentieth centuries. Additionally, the invention of the steam engine, of sealed pipe junctions, and also of tight-closing faucets, made it possible to raise water or build siphons and develop IBTs at an unprecedented scale.

However, at the turn of the twentieth century, water treatment was invented. This allowed a reduction in the initial investment cost by increasing operation and maintenance costs. In turn, this new engineering based on water treatment and sanitation made it possible to extract water from nearby surface water; it made urban water much more a local affair, and shifted the burden of finance from the state or rich entrepreneurs to the customers. From that point on, control of water volumes became essential, and this did not favor large IBTs, where the bigger the flow, the smaller the marginal cost.

Paradoxically, IBTs still fared very well in the colonies, and later in developing nation-states, for intensive agriculture: the motto after WW2 was multipurpose water projects for global economic development, meaning that industrial and urban water users would indirectly subsidise agriculture. But as soon as environmental impacts were translated into dollars, and the global economy became less confident in the future (thus increasing interest rates), many IBTs were weakened. Their future would have been better assured if they had been re-integrated into some regional water resources management scheme, perhaps associated with the conjunctive use of ground water and surface water. Large IBTs have been frequently associated with the power of upper layers of government, especially centralised ones. At a time when multi-level governance is on the global agenda, it becomes difficult to believe that transferring large quantities of water suffices to improve humankind's welfare: it has to be combined with less conventional technologies and other institutional designs.

Acknowledgements J. D. Rinaudo would like to thank Bruno De Grissac for his contribution to this paper through several informal discussions on water conservation related issues. He also acknowledges support from BRGM (scientific program 30) for preparing this chapter.

References

Allan, J. A. (2002). *The Middle East water question: Hydropolitics and the global economy*. London/New York: I. B. Tauris.

Barraqué, B., & Andreas Kraemer, R. (2014). Les services publics d'eau en Grande Bretagne et en Allemagne: origine commune, trajectoires différentes. In B. Barraqué (dir.), *in L'eau urbaine en Europe et en Amérique du Nord: origines et développements* (No. 97–98, pp. 16–29). Flux, Metropolis, 2014–3.

Beecher, J. (2012). The ironic economics and equity of water budget rates. *Journal of the American Works Association, 104*, 73–81.

Blomquist, W., Schlager, E., & Heikkila, T. (2004). *Common waters, diverging streams: Linking institutions and water management in Arizona, California, and Colorado*. Washington, DC: Resources for the Future Press.

Carver, W. B., Cole, D., & Wingate, C. A. (2011). *Tapping the Tennessee River at Georgia's northwest corner: A solution to North Georgia's water supply crisis*. In: Proceedings, Georgia water resources conference, Atlanta.

Cohen, R., Nelson, B., & Wolf, G. (2004). *Energy down the drain: The hidden costs of California's water supply*. Oakland: Natural Resource Defense Council and Pacific Institute.

Commission Nationale du Débat Public. (2012). *Compte rendu du débat public Aquadomitia 15 septembre – 29 décembre 2011* (p. 182). Paris: CNDP.

Cox, W. E. (1999). *Interbasin water transfer*. In: Proceedings of the international workshop. Paris: IHP/UNESCO.

Davis, J. (2004). Corruption in public service delivery: Experience from South Asia's water and sanitation sector. *World Development, 32*, 53–71.

Dyrnes, G. V., & Vatn, A. (2005). Who owns the water? A study of a water conflict in the Valley of Ixtlahuaca, Mexico. *Water Policy, 7*, 295–312.

Environment Agency of England and Whales. (2011). *Large-scale water transfers: Position statement*. https://www.gov.uk/government/uploads/system/uploads/attachment_data/file/297318/geho0811btvr-e-e.pdf

Garrido, A., Rey, D., & Calatrava, J. (2013). Water trading in Spain. In L. De Stefano (Ed.), *Water, agriculture and the environment in Spain: Can we square the circle?* (pp. 205–216). Madrid: Botín Foundation.

Ghaffour, N., Missimer, T. M., & Amy, G. L. (2013). Technical review and evaluation of the economics of water desalination: Current and future challenges for better water supply sustainability. *Desalination, 309*, 197–207.

Ghassemi, F., & White, I. (2007). *Inter-basin water transfer: Case studies from Australia, United States, Canada, China and India*. Cambridge: Cambridge University Press.

Gleick, P. (2003). Global freshwater resources: Soft-path solutions for the 21st century. *Science, 302*, 1524–1528.

Gleick, P. H., Haasz, D., Henges-Jeck, C., Srinivasan, V., Wolff, G., Kao Cushing, K., & Mann, A. (2003). *Waste not, want not: The potential for urban water conservation in California*. Oakland: Pacific Institute.

Guillerme, A. (1983). Les Temps de l'eau, la cité l'eau et les techniques. Seyssel: Champ Vallon, coll. Milieux. 267 p.

Gumbo, B., & van der Zaag, P. (2002). Water losses and the political constraints to demand management: The case of the City of Mutare, Zimbabwe. *Physics and Chemistry of the Earth, 27*, 805–813.

Gupta, J., & van der Zaag, P. (2008). Interbasin water transfers and integrated water resources management: Where engineering, science and politics interlock. *Physics and Chemistry of the Earth, 33*, 28–40.

Hallegate, S. (2009). Strategies to adapt to an uncertain climate change. *Global Environmental Change, 19*, 240–247.

Hansen, K., Howitt, R., & Williams, J. (2008). Valuing risk: Options in California water markets. *American Journal of Agricultural Economics, 90*, 1336–1342.

Kenney, S. D., Goemans, C., Klein, R., Lowrey, J., & Reidy, K. (2008). Residential water demand management: Lessons learnt from Aurora, Colorado. *Journal of the American Water Resources Association, 44*, 192–207.

Lapuente, E. (2012). Full cost in desalination. A case study of the Segura River Basin. *Desalination, 300*, 40–45.

Luisa, S., Tiago, S., & João, P. (2012). In search of (hidden) Portuguese urban water conflicts: The Lisbon water story (1856–2006). In B. Barraqué (Ed.), *Urban water conflicts* (Urban water series – UNESCO-IHP, pp. 71–91). Paris/Boca Raton/London: UNESCO/CRC Press/Taylor & Francis (Balkema Book).

Maddaus, W., & Maddaus, M. (2004). *Evaluating water conservation cost-effectiveness with an end use model*. Proceedings of the AWWA Water Sources Conference, Austin.

Miller, W. G. (2006). Integrated concepts in water reuse: Managing global water needs. *Desalination, 187*, 65–75.

Molle, F., & Berkoff, J. (2009). Cities vs. agriculture: A review of intersectoral water re-allocation. *Natural Resources Forum, 33*, 6–18.

National Academy of Sciences. (2012). *Water reuse: Potential for expanding the nation's water supply through reuse of municipal wastewater*. Washington, DC: NAS.

Pincetl, S. (2012). Urban water conflicts in the Western US. In B. Barraqué (Ed.), *Urban water conflicts* (pp. 237–246). Paris: Taylor & Francis.

Pittock, J., Meng, J., & Chapagain, A. K. (2009). *Interbasin water transfers and water scarcity in a changing world: A solution or pipedream?* Frankfurt am Main: WWF Germany.

Pouget, L., Escaler, I., Giuiu, R., Mc Ennis, S., & Versini, P.-A. (2012). Global change adaptation in water resources management: The water change project. *Science of the Total Environment, 440*, 186–193.

Rinaudo, J., Maton, L., & Caballero, Y. (2010). Cost-effectiveness of a water scarcity management plan: Considering long term socio-economic and climatic changes. *Options Méditerranéennes, 95*, 183–190.

Rinaudo, J.-D., Neverre, N., & Montginoul, M. (2012). Simulating the impact of pricing policies on urban water demand: A Southern France case study. *Water Resources Management, 26*, 2057–2068.

Saurí, D., Hug, M., Gorostiza, S. (2014). Des ressources conventionnelles aux ressources non conventionnelles: l'approvisionnement moderne en eau de la ville de Barcelone. *Flux, 4/2014*(97–98), 101–109.

Shirley-Smith, C., Cheeseman, C., & Butler, D. (2008). Sustainability of water management in Zaragoza city. *Water and Environment Journal, 22*, 287–296.

Swyngedouw, E. (2007). Technonatural revolutions: The scalar politics of Franco's hydro-social dream for Spain, 1939–1975. *Transactions of the Institute of British Geographers, 32*(1), 9–28.

Tomkins, C., & Weber, T. (2010). Option contracting in the California water market. *Journal of Regulatory Economics, 37*, 107–141.

Tortajada, C. (2006). Water management in Singapore. *International Journal of Water Resources Development, 22*, 227–240.

Turner, A., & White, S. (2003). *ACT water strategy: Preliminary demand management and least cost planning assessment* (Report prepared for ACTEW, Institute for Sustainable Futures). Sydney: University of Technology Sydney. 45 p.

Vedachalam, S., & Riha, S. J. (2012). Desalination in northeastern U.S.: Lessons from four case studies. *Desalination, 297*, 104–110.

Waterwise. (2009). *Preston water efficiency initiative*. London: Waterwise.

Zekri, S., Ahmed, M., Chaieb, R., & Ghaffour, N. (2013). Managed aquifer recharge using quaternary-treated wastewater: An economic perspective. *International Journal of Water Resources Development, 30*, 246–261.

Zhou, Y., & Tol, R. S. J. (2005). Evaluating the costs of desalination and water transport. *Water Resources Research, 41*, W03003.

Chapter 9
Three Engineering Paradigms in the Historical Development of Water Services: More, Better and Cheaper Water to European Cities

Bernard Barraqué

1 Introduction

Paris is now a major metropolis, but unfortunately the Seine has become too small for it! This is not an isolated case: many large European cities experience a shortage of water for drinking, and also stormwater problems due to sealed surfaces. Some cities have been led to stop relying on technology only (pipes and water/sewage works), and have reconsidered the traditional separation between water management and land use planning. Indeed, over the twentieth century in Europe, the development of water transport and treatment technologies has allowed authorities to separate water services from water resources, as well as from urban policies and budgets. Increasingly, however, increased interference of large metropolises with nature has forced planners to reflect on 'urban water': the protection of nearby water resources, will reduce the cost of transporting and treating drinking water. Similarly, sewage treatment plants reduce the time and space needed to purify waste water, but stormwater perturbates the processes, or overflows directly into rivers; so it is better to treat stormwater before it enters the sewers. However, this implies a serious revision of urban design and planning.

Reticulated water systems expanded slowly on both sides of the Atlantic until WWI; afterwards, connections to drinking water systems were generalised in developed countries, but not in developing ones. Similarly, in the nineteenth century sewerage was basic, often limited to drainage, with most houses on cesspools or privies. But, over the course of the twentieth century, the richest western European States, the USA, and Australia managed to universalise these services, whereas in developing countries this did not quite happen. The difference was due not just to the wealth of industrialised countries, nor to technical developments or differences in

B. Barraqué (✉)
Centre International de Recherches sur l'Environnement et le Développement, Paris, France
e-mail: bernard.barraque@agroparistech.fr

Q. Grafton et al. (eds.), *Understanding and Managing Urban Water in Transition*, Global Issues in Water Policy 15,
DOI 10.1007/978-94-017-9801-3_9

demography or financial systems; it was also due to differences in engineering cultures and in how the job was allocated between local, regional, and national bodies (Barraqué et al. 2008). In this chapter we can only sketch how these various factors combined. Over the last century and a half, we can identify three socio-technical systems, and we treat these under the perspectives of civil engineering (Sect. 2), sanitary engineering (Sect. 3), and environmental engineering (Sect. 4). Indeed, the first two approaches were developed in Europe partly as answers to sustainability crises, and they enlarged the freedom of cities to access water resources. However, there are now signs of a new crisis which might make water services unsustainable: a downwards spiral could occur, due in part to the low acceptability of the 'full cost recovery' doctrine and subsequent prices increases. It is increasingly admitted that land-use based policies might settle what technology cannot do any more, and this is why the separation we mentioned between water services and water resources has become increasingly blurred. After making comparisons, we draw conclusions on the role of capacity building at the local level, but also on the need for multi-level and integrated governance of both water resources and water services.

2 The Paradigm of Water Quantity

In the nineteenth century, even before the Ebert, Koch, and Pasteur discoveries were popularised, hygienists and engineers promoted the idea of abundant running water to 'wash the city clean'. The mastering of pipes, taps, and welding led to the development of water systems and a multiplication of public fountains (Goubert 1986). Some engineers and industry leaders, notably in England, thought they could provide water services right into buildings by drawing water from nearby rivers or wells and using either hydropowered pumps, or, if they could afford it, newly invented steam engines. In some countries, pumping the alluvial aquifer was found more convenient. Other countries decided to get safer water from natural sources far from the cities. Some small cities developed gravity systems using water from upstream catchments. A few large cities, with the financial and institutional support of governments, built long-distance aqueducts. Even before 1850, New York and Madrid, for example, used such aqueducts to satisfy their thirst for water. Not long after, French engineer Belgrand supplemented the pumping stations that provided Paris with river water, with water from several distant sources (around 100 km), all at a higher elevation than the city. In 1859 Glasgow obtained water from a highlands lake 55 km away (Maver 2000).

Obtaining a water right or a concession on distant water sources required the legal intervention of national governments, who had sovereignty over water resources and their allocation (and this was the origin of the 'statutory water companies' in England: they were private, but had a protected concession and a monopoly for distribution). This led to a *de facto* government involvement in water supply; however, the high costs limited the number of aqueducts. Most cities went on taking water from local wells and nearby rivers (Guillerme 1988), and in Britain this choice was supported by the development of slow filtration. Begun as an experiment in Chelsea earlier in the nineteenth century, filtration was imposed on all water companies in London after the cholera outbreak of 1848.

Conversely, Paris chose long distance aqueducts: prefect Haussmann and engineer Belgrand thought filtration was not fully safe, and they decided to 'go like the Romans'. Thanks to clever hydraulic works, including siphons, Paris could get distant and clean spring water at low energy cost (Cebron de Lisle 1991). However, central government involvement remained limited to the capital until WWII. After the great crisis of the 1930s, States' intervention in the economy became more acceptable. In Mediterranean Europe, central government involvement was shaped through large hydraulic projects, in particular during authoritarian periods before or after World War II (Italy, Spain, and Portugal). Indirectly, this would lead to subsidised water supplies, keeping water cheap and affordable.

At first, distribution systems were developed mostly to serving fire hydrants, street-cleaning outlets, and public fountains, the latter supporting the traditional idea that people should have free access to water resources for their domestic needs. Paying for water was limited to paying the water vendor, and the cesspool emptying; and the notion of free water was supported by almost absent operating costs (apart from cases where there were energy costs related to pumping). Most of the expense came from installation of the infrastructure, which was paid for by public money or by the elites.

2.1 *The Failure of Private Concessions in Europe*

However, late in the nineteenth century, private companies sprung up whose purpose was, for a payment, to bring piped water to private residences, including condominiums. In many cities of the time, authorities considered this a luxury, and they did not want to get involved. In France, they granted concessions to these companies, who often were required to deliver water at no cost to fountains and to the public, with their profit coming from subscribers to 'private' water (Pezon 2000). Companies were left to themselves to produce and distribute water, having to solve several technological problems (leak control, metering, etc.). These initial ventures usually lacked both enough capital and political support to be able to take the risk and universalise the service. The initial model of the concession in turn engendered mistrust between companies and the populace (e.g. Glasgow, Newcastle, Berlin, and several cities in France). At the end of nineteenth century conflict grew when municipalities became convinced that water supply was not a luxury, but a fundamental public health issue. If operators took water from nearby rivers, it was of poor quality, but there were few options for getting it from a distance because of the financial and legal obstacles. Concession contracts were rigid and the 'price cap rule' crystallised growing conflicts between municipalities willing to expand the service to new and eventually poorer areas, and companies which were afraid of losing money. In most cases, it was the municipalities which set up complete water supply systems, after reclaiming operations in favour of public corporations. It was thanks to new treatment technologies that they managed to find local solutions. This new public model, developed in England after the well known example of Birmingham, flourished in France at the expense of initial concessions, and it was the rule in Germany. In Portugal, the government took over the water supply of Lisbon, and a similar thing happened in Turkey with Istanbul's supply.

Turning now to waste water, heavy investments were made on urban drainage, which led to the adoption of combined sewers (first in London). For public health reasons it was necessary for everybody to be connected. To achieve this usually required public procurement, with the role of private companies being limited to construction phases. Once access to water (i.e. the quantity problem) was broadly achieved, the concepts of high water quality and hygiene immediately became more important.

2.2 State Involvement and Large Hydraulic Schemes in Other Continents

Our hypothesis is that this civil engineering paradigm remained more dominant in the New World, and was extended to the Third World after World War II. The main factors were the co-occurrence of abundant water resources, international financing institutions offering cheap money, and various forms of support for national governments' intervention in multipurpose infrastructure provision (keynesian or socialist). The American federal government's involvement in large, multipurpose hydraulics during the first half of the twentieth century showed the way (Tennessee Valley Authority, Mississippi, Colorado–California), while many Anglo-Saxon experts also became convinced that local governments were ineffective in providing public services (e.g. in Britain, see Saunders 1983). In Brazil, for instance, an authoritarian government created a public water company in each state (Companhia Estadual de Saneamento Básico (CESB)), pushing local authorities to sign concession contracts with them. Initially these CESBs managed to extend the areas served by water supply, but for the sake of profitability, they increasingly chose to privilege dams and water transfers for global economic development rather than extend the distribution pipes in cities. Sewerage was clearly lagging behind, and sewage treatment remained almost nonexistent until the 1990s.

3 The Paradigm of Water Treatment

At the end of nineteenth century, industrialisation and urbanisation in north-west Europe led to a quicker crisis of the 'civil engineering era' than in the new world. As cities grew, water needs increased, and taking more water from further away generated conflicts with communities deprived of their local resources. This was the case for the long-distance aqueducts taking water to Paris, those transporting water from the Durance to Marseille, and those proposed to take water from Wales to London. In addition, because of their quest for more autonomy and decentralisation, cities eventually became reluctant to depend on legal and financial support from central governments. In Paris, a 30-year debate started in 1890 about building a 450 km aqueduct from Lake Geneva which, it was claimed, would solve all the city's water quantity problems forever (Duvillard and Badois 1900). Opponents of the scheme pointed to the geopolitical risks and the heavy costs, which made it impractical compared to smaller and closer projects like the Avre river 100 km away).

Then, following a trend set in England, local public services emerged in an atmosphere of decentralisation and the rise of what could be termed 'municipalism' (rather than the more dramatic French wording 'municipal socialism' and the derisory 'water and gas socialism' of the UK). Indeed, the welfare state and the involvement of central governments in the economy were anticipated by at least two generations of municipal welfare policy in England, followed by other cities in north-west Europe. With the emergence of a middle class and a qualified working class came the savings banks, institutions which could provide loans at cheap rates for local welfare purposes. Municipal bonds were attractive to the public, and governments would also subsidise projects. Together, direct management and cheap money allowed water supplies and sewerage systems to spread. Since operating costs were still modest in Britain at that early stage, the idea prevailed that water services should be covered by local taxes, i.e. by rates proportional to the rental value of houses and of industrial premises. This not only provided funds for the initial installation, but provided cheaper water to the poorer people, eventually played a significant role in the acceptance of domestic supply as the norm.

In Britain, however, water was not overabundant, and many cities had to take it from rivers. Some form of purification was therefore needed, and filtration was invented. In this way, thanks to sanitary engineering innovations, the overall problem of water quantity could be fixed by improving water quality. In Germany, groundwater was usually preferred when it was available, but in case not, cities would take water from the alluvial aquifers through the river banks. This natural filtration would eventually support local water management.

3.1 *Drinking Water Plants and Sewage Treatment Plants in Europe*[1]

Once it became known in the 1880s that water contamination by bacteria was a major cause of disease, even distant and pure water would eventually need to be treated. So sanitary engineering supplemented civil engineering and overcame shortages in the quantity of water available. Because medical cures for waterborne

[1] The history of drinking water treatment technologies largely remains to be written. For Paris, see Gaillard-Butruille (n.d.), for sewerage works see Boutin (1986).

diseases were not readily to hand in the mid nineteenth century, the issue was transferred to the sanitary engineers who supported water filtration and later added, around World War I, disinfection or chemical treatment (chlorine, ozone, ultraviolet, activated carbon). At that stage, taking water from just upstream of cities had economies of scale, and was a good return on investment. The Lake Geneva to Paris project was discarded after World War I for geostrategic reasons, but it had already become superfluous after the inauguration, in 1902, of a slow filtration plant at Ivry, operating with Seine water just upstream from Paris. Filtration won the day following a typhoid epidemic in 1899, which was attributed to the contamination of the Loing springs, reputedly the furthest and supposedly cleanest water source for the city.

However, water treatment created a serious rise in operating costs, which were better understood and accepted by both the public and city councils than investment costs. Hence the idea arose that water services could be at least partly covered by bills. This was an important change for the public, and charges were initially limited to cover only operation and maintenance. Slowly however, delivery of water directly to homes changed its status from a luxury good to a commodity, and made water billing normal. In Paris, for instance, metering and billing by volume slowly replaced former subscriptions based on permanent flows feeding cisterns (Chatzis 2006). In Europe, only Britain and the Republic of Ireland retained, until the end of the twentieth century, the ancient charging system based on rateable housing value.

Using billing to cover an ever-larger fraction of costs increased the self-financing capacity of water services and improved their sustainability. In the 1960s, many continental countries decided to include sewerage charges in drinking water bills so as to facilitate cost recovery, despite the compulsory character of sewer connections (the cost of which were then normally covered by local taxes). Sewage treatment, and even collection, thereby became a commercial service. Eventually, increased self-financing capacity reinforced the legitimacy of local authorities as providers, or at least as organisers of the services. But in order to achieve economies of scale, it was often necessary to join forces with neighbouring communes. This is how Dutch communes (700 in 1950) have progressively consolidated their water supplies into only ten drinking water companies, which they fully own together with the 12 provinces. Central governments usually allowed and indeed supported the creation of joint municipal boards in order to bring the institutional, technical, and management scales closer together. Back in the early twentieth century, the German Empire created a state water company to mutualise the costs of providing water to a vast urban area around Stuttgart. Concerned municipalities recovered full control of the company after WW2, but kept it going. With its myriad of small communes, France has become the foremost country for joint boards and delegations to private companies (Pezon and Canneva 2009), companies which have mastered water and waste water treatment technologies, but also the metering and billing part of this business.

This innovation in water and wastewater treatment has in turn supported the development of new territories and of new relationships between politics and expertise, between elected representatives and engineers. In several countries, citizens pay a water bill to a private company which is owned by a joint board of municipalities, sometimes involving a regional level institution (Barraqué 1995; Barraqué and Isnard 2014). The counterpart to this is that the relationship between water services, urban planning, and environmental management is usually weak: elected officials and urban planners always assume that water engineers 'will follow', and water engineers anticipate population growth and per capita increases with additional supply side solutions. However, at the end of the twentieth century such an approach led to new shortcomings.

3.2 *Technological Shortcomings*

In developed countries, the supply-side approach, combining the solutions of civil and sanitary engineering, ends up in a crisis, because it becomes too costly. Until recently, 'infrastructure' was hidden from the public action agenda: it offered more freedom to urbanites, as it saved time and space, and largely served plots of built or buildable land (i.e. landowners), thus increasing land rents which provided political legitimacy. City aldermen and elected representatives were competent on issues like valorisation (or devalorisation) of urban land, but they did not know much about water systems. At the same time, early sanitary engineers were convinced that public health urgently required water systems, in particular waste water collection and drainage, to evacuate the growing flows of contaminated water (Barles 1995). As it happened, they preferred to locate them under the streets, i.e. in public space. The choice of the *tout-à-l'égout* in Paris is typical (Dupuy and Knaebel 1982). Eventually, people became ignorant of the importance of these systems: 'out of sight, out of mind' and 'nimbyism' characterise the public's attitude, and they are at odds with a conservation attitude (Melosi 1981). Operators had no interaction with the public and with demand side problems: they just had to match the demand with more or less invisible infrastructure and that was it. But the impact on water costs and prices eventually brought infrastructure back on the public agenda.

3.3 *Water Prices Bound to Increase*

Charging for waste water collection and treatment in water bills (as was decided in Germany and France in the 1960s) led to water prices more than doubling. Additionally, in the same period, water supply itself became a mature business, i.e. it had to face the issue of renewing ageing infrastructure without receiving any more subsidies. We hypothesise that European municipalism had to adapt in various ways, but chiefly through turning to legal private status[2]; in this way it could

[2] Turning to private status does not mean privatisation: in many countries, municipalities or water districts are allowed to set up private companies which they fully own. In France this only became

depreciate assets and make provision for renewals, which is difficult under public accounting. But allowing for depreciation and provisioning meant another rise in water bills. The result today is that an increasing number of large water users (industry, services) either quit the public service system, invest in eco-efficient technologies, or fix their leaks. This explains the recent stagnation, and even decrease, in volumes sold (Barbier 2000; Barraqué et al. 2011). In some countries, domestic consumers too have reduced their demand, through changing fixtures, altering garden design, using rain tanks, or accessing alternative sources of water for non-drinking uses.[3] The problem is that, ultimately, this reduction in demand worsens the already fragile financial balance of collective services. Unit prices go up, explaining why in the most advanced countries waste water collection and treatment represent more than half the total water bill.

In addition, water suppliers have difficulty in complying with drinking water standards (DWSs) at reasonable cost. In developed countries, drinking water criteria set out in legislation often side with the traditional "no-risk" strategy regardless of economic considerations. Ironically, the multiplication of criteria is slowly making the situation too complex: chlorination by-products give cancer (Okun 1996), for example. There are many other examples, especially the emerging issue of micropollutants. The media now report that a growing proportion of people receive non-complying water, even though, over the long term, treatment quality is improving. To lower the risk of being unable to achieve the standards, water suppliers, and local, national, and European authorities, are turning to a new policy: water resources protection. This strategy is part of what is called environmental engineering.

4 The Third Paradigm of Sustainable Water

In the third paradigm, technology alone is insufficient to solve all urban water problems, and both demand management and territorial policies are needed. Under this scenario, the sustainability of the European model – strong local legitimacy and high quality standards, and allowing for high cost recovery in the face of heavy investment – could soon be under challenge. Once everybody is connected, how is it possible to maintain, in the long run, good service quality at an affordable cost, if full cost recovery[4] is required and investments are no longer subsidised? This is indeed the challenge with the Water Framework Directive of the European Union (2000/60/EC of 23 October 2000): water users should pay for the water services

possible in 2010, and it is one of the reasons why cities often keep a private operator under various delegation formulas.

[3] See Chap. 7 by Rinaudo et al., this volume, on the development of domestic boreholes in France and Australia.

[4] Full cost recovery includes operation and maintenance costs, plus a fair share of depreciation of invested capital, plus environmental costs (environmental impacts which are not internalised), and even users' costs including the impacts of resource scarcity and opportunity cost on water value.

they are rendered, and should support the cost of keeping the aquatic environment clean. If this were implemented, the resulting 'full cost pricing' would be much more expensive than present day prices.

4.1 *Environmental Engineering as Part of the Third Paradigm*

Deriving from American sanitary engineering, environmental engineering aims to protect not only populations from negative environmental factors, but also global and local environments from potentially dangerous human activities. Of course, understanding natural processes is an important issue, but programs also focus on urban issues and technical systems (Barraqué 1993). The common characteristic of the two first paradigms is a focus on supply-side solutions, while environmental approaches have to consider the 'demand side'. However, this does not imply finding market-type solutions (i.e. solutions involving pricing), because in the long run infrastructure represents fixed and heavy costs that constrain supply and demand. Sometimes causes interact with effects in feedback loops, provoking exponential outcomes. For example, an increase in water price is supposed to foster conservation, but if water consumption decreases, utilities receive less income. Since they are mandated to balance costs and revenues, they are obliged to increase prices; this in turn may lead to some water users to 'exit' the system. These are what are called 'network effects'.

With our modern 'systemic' eyes, we can see how the previous 'municipalist' model needs to be corrected with new, land-use based solutions. Sewer systems are designed so big in order to accommodate large volumes of stormwater, because planning regulations seldom include limitations on surface sealing. Additionally stormwater detention tanks and on-site infiltration areas reduce stormwater flows to treatment plants, thereby reducing the size of supporting infrastructure and making the plants' operation more efficient (because the sewage is now more concentrated). Similarly with drinking water, where its provision to urban areas is now reaching a diseconomy of scale, 'demand side' management offers solutions to the problem (e.g., the rise of non-conventional water uses such as rainwater harvesting for garden use, and land-use control at catchment level) – both help reduce the cost of water quantity/quality issues. Increasingly, in Germany, The Netherlands, and to a lesser degree in Denmark and France, utilities or their organising authorities sign cooperative agreements with farmers (Brouwer et al. 2003) so as to obtain raw water of better quality (through a reduction in the use of fertilisers and pesticides and compensation for corresponding losses in revenue). This policy has been criticised by some economists and ecologists, who argue that the polluters, and not the victims, should be the payers. Other economists (and lawyers) simply acknowledge that what they term 'payment for environmental services' is an efficient policy, and probably the most acceptable, at least during the 'social learning' phase (Salzman 2005).

Turning now to the internal governance of utilities, this leads us to redefine the very notion of 'operator'. If supply and demand constantly interact, this blurs the traditional separation between supply-side and engineers on the one hand, and demand-side and elected representatives on the other. A direct link between operators and the public is required, so as to escape from coarse, linear, and overoptimistic projections of demand. This is all the more important that drinking water demand has started to decline in central city areas for the first time since the inception of public services (Barraqué et al. 2011). Water services operators need to understand what citizens do with water, if only to better anticipate the potential evolution of demand. Real time information technology can support this new water business: 'going beyond the (smart) meter', so as to give advice to customers and at least help them fix their leaks. Beyond the refining of demand analysis, one can motivate users of the system to use water more efficiently and alleviate negative network effects. This implies a significant change in governance: the new grid-based systems operator needs to consider a complex mix of people and institutions, and the networks effects can only be mastered when interactions are taken into account. When a growing number of institutions jointly provide water services and manage water resources, information sharing becomes crucial to success.

4.2 The Case of the Paris Region

As with any large metropolis, Paris offers a very complex institutional situation. Communes are responsible for water services, but they cannot do it alone anymore. For historical reasons, Paris and its 2.2 million people has its own water services, and relies for half its needs upon distant springs, with the other half coming from abstraction points in rivers. Suburban communes are almost all part of large joint boards (the largest one, SEDIF, serves 4 million inhabitants in around 150 communes), an arrangement which rationalises their services: the joint boards benefit from economies of scale, with large plants to treat surface water; and the joining of several local authorities gives their councillors a better balance of power with private operators.

Since water supply in the suburbs of Paris relies heavily on rivers, and due to the dramatic increase in population, three upstream reservoirs totalling 700 million m^3 capacity had to be built after 1958 to sustain supplies during the low flows of summer[5]: Seine, Marne and Aube reservoirs. The first one (called Lac d'Orient) was built with central government funding taking advantage of a major flood in the

[5] Initially, four small reservoirs were constructed: Crescent (in 1931), Chaumençon (1934), Champaubert-aux-Bois (1938), and Pannecière-Chaumard (1949), but they were not sufficient to mitigate Seine floods and droughts. The largest, Pannecière-Chaumard, stored only 80 million m^3; so three additional larger reservoirs were constructed: Lac d'Orient (1966), Lac du Der (1974), Reservoir Aube (1990).

winter of 1952.[6] The next two were funded by Agence de l'eau Seine Normandie using levies paid by water abstractors. The new burden of responsibility of water users to fund dam construction caused a project for an additional reservoir to be discarded.[7] This turned out to be a good choice, since drinking water demand is now on a slight but steady decline. Clearly, the major issue remains drinking water quality, and since part of the treatment plants are now within the urbanised area, the big metropolis has an interest in better protecting surface water resources (from upstream cities, agriculture, and industry) and groundwater (from agriculture). Paris city has developed some cooperative agreements with farmers in the feeding areas of its springs. It also recently reclaimed its drinking water management fully into the public domain, to have better control on the whole drinking water chain; and other cities in the suburbs are considering this too. Indeed, the most important outcome of this renewed public–private debate is the realisation that there are now more than 15 drinking water plants serving the metropolis, on top of Belgrand's aqueducts. It is clear that a rationalisation is possible, eventually through the creation of some sort of regional water production body. This illustrates the need to re-consider technological solutions in line and simultaneously with the adoption of new territorial governance schemes.

The situation concerning sewage collection and treatment is even more complex. The initial scheme at the end of nineteenth century was to collect all waste water, and some rainfall runoff, and convey it downstream (in combined sewers at the time), where it was spread on sewage farms. Large interceptors were built and operated by central government staff at the *département* (county) level. But with the growth of the city, there was not enough space, and sewerage works with biological treatment were built in Achères just before World War II. Stormwater would overflow directly in the river, under the misapprehension that rainwater overflow was clean. This was a typical 'second paradigm' approach.

However, with the rise of environmental concerns, it was found that Achères, the second largest sewage works in the world, had a severe environmental impact on the Seine down to the estuary because there was not enough on-site space to fully remove or stabilise pollutants. In addition, stormwater was found to be heavily polluted: the post-war choice of separate sewers forced engineers to admit this fact. After several years of intense debate between the many stakeholders involved, it was finally decided to break the 'linear' scheme of engineer Belgrand: taking in water upstream and discharging it all downstream. The construction of modern and innovative sewage works upstream of Paris (in Valenton), and the enlargement of some smaller sewage works on Seine tributaries, will reduce waste water arriving in

[6]Which means it was decided on the basis of floods and funded with government grants, while it was needed mainly for scarcity reasons. The next ones were rightly funded by the Agence de l'eau, on the basis of water needs, not floods.

[7]The additional reservoir was planned by mayor J. Chirac and his experts, but was abandoned because of cost (in addition, the water supply companies said they had to purify the water anyway and the probability of having a major drought was very low). Furthermore, Paris water demand went down by 16 % between 1990 and 1998 (Cambon-Grau 2000). Following a brief pause, demand has again started to decline; today consumption is almost 30 % less than in 1991.

Achères by 30 %, which will allow better treatment. Increasingly, stormwater is being collected and stored in reservoirs or artificial ponds before reaching sewers. At times of peak flow, water is sent for treatment either to the sewage works or to a couple of special treatment plants. In the outer suburbs, municipalities now grant building permits only under 'no additional discharge' rules. All this illustrates a transition to the third paradigm.

The resulting institutional arrangements are quite complex: suburban communes are in charge of street sewers and (eventually) on-site storage, while the three *départements* of the inner ring around the city of Paris are in charge of larger interceptors and of stormwater control policy (in particular, separate stormwater sewers). Together with Paris city, they have formed SIAAP (Syndicat Interdépartemental d'Assainissement de l'Agglomération Parisienne), an inter-county board running the largest interceptors and all the sewage treatment plants. In the outer ring, there are some inter-communal joint boards which operate at the level of catchments of small tributaries, and which combine sewage collection and treatment with protection of the aquatic environment. Projects also benefit from the financial support of the Agence de l'eau, on the condition that they reduce environmental impacts. This implies an unprecedented need for multilevel governance, and it is a very sensitive issue in the region and in the country. The *agences de l'eau* get their income from water and sewer bills, but since 'people don't buy the rain', the *agences* cannot fund stormwater control projects unless it can be proved that quantity management has a positive impact on water quality. At the same time, communes do not receive much encouragement to reduce runoff from their territory by adopting constraints on public or private spaces.

It is now clear that the new water services policy needs to turn towards demand management, and we argue that this means developing citizen participation. Indeed, several institutions like the SIAAP, the city of Paris, and some suburban *départemental* councils have recently set up consultative bodies to share information with citizens and NGOs. This opens up water operations and investment budgets to public supervision. However high urban densities mean there is little opportunity for citizens to act directly to improve the water–land relationship; there are some 'eco-neighbourhood' projects but they will probably only stimulate 'eco-imagination', since they will remain more limited in number and scope than in Berlin.

5 Conclusion

In Europe, urban water services used to be able to run independently of how water resources were allocated, thanks to technological innovations in water works and sewage treatment plants. Today it has become preferable to combine technology with land-use based solutions to make water services more sustainable, rather than just rely upon technological solutions. The EU Water Framework Directive proposes to derive water policy first from the recovery and protection of the aquatic environment. For water users and public water services this implies returning to a situation where there is active participation in water resources management. A new awareness of the limits of urban technology also means extending the services to new areas, but with alternative technologies, e.g. with decentralised sewerage systems (in most rural areas, and in low density suburbs, it is economically impractical to connect everybody to a sewer). Advances in soil biology mean that septic tanks, if properly operated, can work better than small treatment plants. Even more striking today is the decision of some ecologically-minded cities to stop extending their water systems and to set up decentralised drinking water services on the outskirts. What we need to invent is a service in between the costly traditional centralised networked system and the fully self-reliant rural scheme.[8] Even for water supply, there are a lot of flexible, alternative (non-conventional) technologies that could be safely used if operated under an appropriate institutional set up (e.g. the new water services policy in the Stockholm archipelago). The biggest technological challenge is to better define the limits of centralised and decentralised water service technologies. The first paradigm has in particular to be questioned, since it is mobilising water but frequently fails to deliver it to domestic users. In many emerging countries like Brazil, as in some European countries, historical centralisation and government support of large hydraulic projects has made it more difficult to separate water services provision and water resources mobilization, resulting today in a crisis due to water resources contamination and even competition between electricity and water services (São Paulo and Rio de Janeiro). The irony is that people could now say, "water, water everywhere, and not a drop to drink!".

In the early twentieth century, water services were under the responsibility of local authorities. In Western Europe, they were part of local welfare through what is termed municipalism (including other local public services). Even though upper levels of government would intervene in support of local policies, there was little top-down centralisation; instead, there was frequent bottom-up concentration (e.g. joint boards). Conversely, in Brazil for instance, water policy was long dominated by the influential hydro-power sector. In addition, agricultural conversion, starting in the 1960s, with no land reform, led to an intense rural to urban migration of very poor people, while local authorities were weak and unprepared. During the military-dominated period starting in 1964, water services were centralised at the level of the states: a program named PLANASA met with initial success in extending water services, but failed to modernise services up to European standards. In Brazil, water use by the elites and the middle class is now higher per capita than in Europe, but the quality of services is lower and is kept low by a lack of self-funding capacity. Because the poor cannot afford to pay higher prices, the price of water services is low, and so the service remains below par. It seems that the first paradigm of civil engineering – to mobilise water resources for economic development – remained dominant, and was not sufficiently supplanted by the sanitary engineering paradigm.

[8] France has created the option of SPANC (service public de l'assainissement non collectif) in areas that, under the EU Urban Wastewater Directive, are specifically designed with no sewers.

Some European countries are still in an intermediate position between the two first paradigms. Portugal and Spain had authoritarian governments until 1974/75, and they ended up with water services which lagged behind, with public funding and United States aid focussed on large hydraulic projects developed for the sake of hydroelectricity and irrigated agriculture (Pato 2008). Joining the EU allowed for some decentralisation, and more recently they developed cooperation between central, regional, and local governments, allowing for an improvement of urban water services. In Brazil there is a debate on decentralisation, but it reveals more confrontation than co-operation between levels of government.

Our analysis leads us to think that the centralisation/decentralisation issue is more important than the public/private debate. Indeed, a striking institutional evolution in continental Europe is voluntary concentration of services at a supra-local level of government. In Great Britain, regionalisation and abolition of local authorities' control took place 15 years before privatisation (Saunders 1983), but both remain very controversial (Bakker 2003). Few people know that in The Netherlands only ten water supply companies remain, and less than 24 water boards (in charge of large sewers and sewage works). In Belgium, a strong concentration process has led to some sort of water services regionalisation. In Italy, communal water services have been merged into new integrated boards (water and sewer) organised at the level of ATOs (optimal territories), which usually correspond to the provinces. In Portugal, the government encourages communes to join large mixed boards where the national water company holds half the shares and communes have the other half. In Germany, the tradition of integrated municipal enterprise (serving water, gas, electricity, and public transport, plus other services here and there) is still strong, but shares of capital are increasingly exchanged with public–private companies operating at regional or national level (typically electricity). Many of these examples illustrate that the real challenge is to develop intergovernmental cooperation.

The resulting picture is often blurred: in developed countries there is good tap water, but people have been brought up to ignore just about everything about their services, so they almost never make the distinction between water resources and water from the tap. Conversely, in developing countries part of the population has to rely on untreated water. Unfortunately, in the global debate about privatisation, there is a tendency to conflate water resources and water services. In both developed and developing countries, people advocate, for the sake of fighting globalisation and related privatisation, that water should be public and free for the poor. This will not help establish sustainable financial management!

In addition, in large metropolitan areas around the world, water resources, water services, and other services like electricity have to be integrated, and they require intergovernmental solutions and public participation. One can point here again to the examples of Sao Paulo and Rio de Janeiro.

The expression "three engineering paradigms for the water industry" should not be understood as the next paradigm sequentially replacing the previous one: each brings cumulative innovations, which add more degrees of freedom to the services, but also new technology/territory articulations. Some projected dams and water transfers are still necessary and sustainable. And cities will go on needing potable water and sewage works. But new land use based solutions, requesting new and interdisciplinary approaches, will help. A comparative analysis helps develop a clear vision of what is at stake in water services provision today. There is a need for what in the social sciences are called 'hybrid forums', where stakeholders are confronted with scientific or technical issues and can build alternative or innovative 'advocacy coalitions' which can lead to more sustainable water policies (Sabatier 1993). We therefore advocate a broad interdisciplinarity in environmental engineering. Perhaps the most important thing is to give engineers a socio-economic and institutional culture for their actions, something that can help them see the consequences of a new motto: integrated and participatory water management.

References

Bakker, K. (2003). *An uncooperative commodity: Privatizing water in England and Wales*. Oxford: Oxford University Press. 260pp.

Barbier, J. M. (2000). Dossier: Evolution des consommations d'eau. *Techniques Sciences Méthodes: Génie Urbain Génie Rural, 2*, 14–59.

Barles, S. (1995). *La ville délétère: Médecins et ingénieurs dans l'espace urbain, XVIIIe–XIXe siècle*. Seyssel: Champ Vallon. 373pp.

Barraqué, B. (Ed.). (1993). *La ville et le génie de l'environnement*. Paris: Presses de l'ENPC. 249pp.

Barraqué, B. (Ed.). (1995). *Les politiques de l'eau en Europe*. Paris: La Découverte, coll. Recherches. 304pp.

Barraqué, B., Formiga Johnsson, R. M., & Nogueira de Paiva Britto, A. L. (2008). The development of water services and their interaction with water resources in European and Brazilian cities. *Hydrology and Earth System Sciences, 12*, 1153–1164. www.hydrol-earth-syst-sci.net/12/1153/2008/.

Barraqué, B., Isnard, L., Montginoul, M., Rinaudo, J. D., & Souriau, J. (2011). Baisse des consommations d'eau potable et développement durable. *Responsabilité & Environnement (série des Annales des Mines), 63*, 102–108.

Barraqué, B., Isnard, L., & Souriau, J. (2014). How water services manage territories and technologies: History and current trends in developed countries. Chapter 2, this volume.

Boutin, P. (1986). Eléments pour une histoire des procédés de traitement des eaux résiduaires. *Tribune du Cebedeau*, 511–517.

Brouwer, F., Heinz, I., & Zabel, T. (Eds.). (2003). *Governance of water-related conflicts in agriculture: new directions in agri-environmental and water policies in the EU*. South Holland: Kluwer. 236pp.

Cambon-Grau, S. (2000). Baisse des consommations d'eau à Paris: enquête auprès de 51 gros consommateurs. *Techniques Sciences Méthodes: Génie Urbain Génie Rural, 2*, 37–46.

Chatzis, K. (2006). Brève histoire des compteurs d'eau à Paris. In, *Terrains et Travaux (ENS-Cachan) no. 2*, pp. 159–178.

Cebron de Lisle, P. (1991). *L'eau à Paris au XIXe siècle*. PhD thesis, University Paris I, 691pp.

Dupuy, G., & Knaebel, G. (1982). *Assainir la ville hier et aujourd'hui*. Paris: Dunod. 92pp.

Duvillard, P., & Badois, E. (1900). *L'adduction des eaux françaises du Lac Léman à Paris et dans la banlieue (projet P. Duvillard, résumé des études, projet définitif)*. Paris: Librairie Polytechnique Ch Béranger. 191pp + 7 planches hors texte.

Gaillard-Butruille, S. (no date). *Evolution des traitements de potabilisation jusqu'en 1940, l'exemple de Paris*. Paris: SAGEP (Société Anonyme de Gestion des Eaux de Paris).

Goubert, J. P. (1986). *La conquête de l'eau: L'avènement de la santé à l'âge industriel*. Paris: Hachette, Pluriel. 302pp.
Guillerme, A. (1988). The genesis of water supply, distribution and sewerage systems in France. In J. Tarr & G. Dupuy (Eds.), *Technology and the rise of the networked city in Europe and America* (pp. 1800–1850). Philadelphia: Temple University Press.
Maver, I. (2000). *Glasgow: Town and city histories*. Edinburgh: Edinburgh University Press. 256pp.
Melosi, M. (1981). *Garbage in the cities, refuse, reform and the environment*. Austin: University of Texas Press. 268pp.
Okun, D. A. (1996). From cholera to cancer to cryptosporidiosis. *Journal of Environmental Engineering, 122*, 453–458.
Pato, J. H. (2008). *O valor da água como bem público*. PhD thesis, Instituto de Ciências Sociais, Universidade de Lisboa, 353pp + anexos.
Pezon, C. (2000). *Le service d'eau potable en France de 1850 à 1995*. Paris: Presses du CEREM. 450pp.
Pezon, C., & Canneva, G. (2009). Petites communes et opérateurs privés: génèse du modèle français de gestion des services d'eau potable. *Espace et Sociétés, 139*, 21–38.
Sabatier, P., & Jenkins-Smith, H. (1993). *Policy change and learning, an advocacy coalitions approach*. Boulder: Westview Press. 290pp.
Salzman, J. (2005). Creating markets for ecosystem services: Notes from the field. *New York University Law Review, 80*, 870–961.
Saunders, P. (1983). *The 'regional state', a review of literature and agenda for research*. Working paper, Urban and Regional Studies. Brignhton: University of Sussex. 82pp.

Chapter 10
Investing in the Water Infrastructure of Tomorrow

David R. Marlow, David J. Beale, Stephen Cook, and Ashok Sharma

1 Introduction

Water is a key factor in urban sustainability (Schaffer and Vollmer 2010). In most modern cities, water services are currently delivered via networks of buried pipes that connect customers to treatment works and, ultimately, to sources of water and sinks for wastewater. These systems have predominantly been developed under a centralised model wherein the construction and maintenance of infrastructure has been undertaken on behalf of communities with the intention of delivering cheap and reliable water services (Brown et al. 2009). This infrastructure represents a significant capital stock and future generations will inherit the lasting effect of our continuing investment decisions (Marlow et al. 2010a). The urban water sector is, however, facing increasing pressures associated with climate change, a volatile global economy, increasing energy prices, heightened environmental awareness, and complex regulatory and social circumstances (Marlow et al. 2010b). Local factors vary considerably, but the reality for many water service providers (WSPs) is that they must operate within constrained budgets, while being expected to deliver quality service at a low price. In such a business environment, it is important that WSPs develop strategies that address growing uncertainty with respect to operational, environmental, social, and economic constraints (Blackmore and Plant 2008; Pearson et al. 2010). Adapting business thinking to cope with these emerging realities implies a transition to a new operating paradigm – that of sustainability (Marlow and Humphries 2009).

D.R. Marlow (✉)
WISER Analysis, Melbourne, VIC, Australia
e-mail: david.marlow@wiseranalysis.com

D.J. Beale • S. Cook • A. Sharma
Commonwealth Science and Industrial Research Organisation,
Highett, P.O. Box 56, Melbourne, VIC 3190, Australia
e-mail: david.beale@csiro.au; stephen.cook@csiro.au; ashok.sharma@csiro.au

Q. Grafton et al. (eds.), *Understanding and Managing Urban Water in Transition*, Global Issues in Water Policy 15,
DOI 10.1007/978-94-017-9801-3_10

The term sustainability is somewhat overused and vague. At a fundamental level, it merely implies the ability to continue to do something into the future. A focus on a long term view is thus always implicit within any use of 'sustainability', but after that, the word is used in a variety of ways. For example, Klostermann and Cramer (2006) suggested that for most respondents 'sustainable' is now a label for all issues and activities previously labelled 'environmental'. Within the water sector, the term 'triple bottom line' (TBL) is becoming widely used as a synonym for 'sustainability' (e.g. Kenway et al. 2006). Advocates of the TBL concept argue that businesses can no longer focus on financial performance alone, and that business continuity requires appropriate attention to be given to social and environmental output measures (e.g. Elkington 1998). In a broader context, 'sustainability' often implies 'sustainable development', a widely quoted definition of which was given in the Brundtland Commission report (WCED 1987).

Unfortunately, political expedience and business pressures often mean that the economy is given more focus than these broader interpretations of sustainability. To provide counter arguments to this tendency, simple conceptual models have been proposed that emphasise that the natural environment, by providing essential ecosystem services, supports all economic and other human activity (Porritt 2005). Urban infrastructure can be conceptualised both as a key part of the economy and as a way of linking society with broader economic activity, and it is thus useful to extend these illustrative models to highlight the role that infrastructure plays in broader sustainability issues, as shown in Fig. 10.1. Each of the levels shown in the figure is important from an anthropogenic perspective, but the health of the biosphere is still ultimately an overriding and limiting factor.

From the perspective of this chapter, the reason for making the modification to the simple conceptual model shown in Fig. 10.1 is that while urban water infrastructure is crucial to the functioning of modern society, from the perspective of much of the community it provides services passively and automatically. As a result, in an analogous way to the biosphere, its role can also be undervalued and this in turn can lead to long-term underinvestment. Certainly, studies undertaken in the USA have

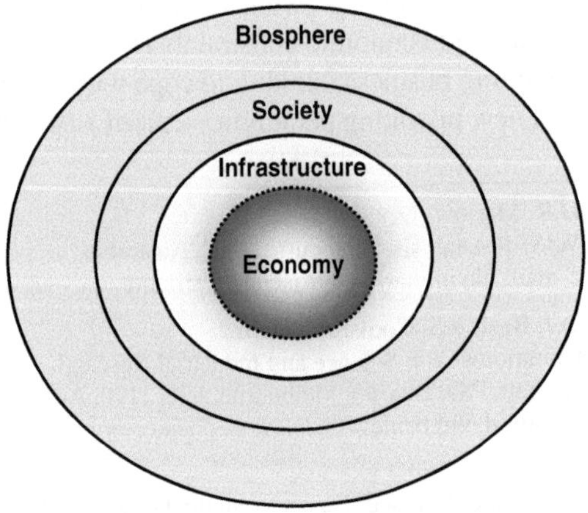

Fig. 10.1 Infrastructure as a key element of broader sustainability issues

highlighted that there is widespread deterioration of urban water infrastructure (Baird 2010; Copeland 2010), and the rating given in ASCE (2009) was D–, reflecting a 'poor' rating, and only one grade higher than 'inadequate/failing'. For many cities, the decay of urban infrastructure coincides with pressures from continued urban growth (Vlachos and Braga 2001). It is estimated that by 2050, the world will experience a doubling of the urban population to 6.2 billion, the implication being that there will need to be a doubling of urban capacity (i.e., housing, infrastructure, facilities) in the next 40 years. The World Bank has estimated global infrastructure investment needs will be US$35 trillion over the next 20 years, and the World Economic Forum's Global Risks Report 2010 indicated that failure to adequately invest in, upgrade, and/or secure infrastructure networks is one of the biggest threats facing society (WEF 2012).

While the need for much of this investment is associated with future demands, problems with existing assets have come about because of poor asset management (Burn et al. 2007). A key aspect of asset management is to ensure sufficient investment is made to maintain and enhance system capacity, both through the provision of new assets and the maintenance and renewal of existing ones (whether built or natural components of urban water systems). Australia, as an early adopter of asset management principles, has seen an improvement in urban water infrastructure since 1999, when Engineers Australia scored these assets as C– (C being adequate), to the position reached in 2005 when this infrastructure was scored as B–, indicating that it was generally in good condition with minor changes needed to meet future demands. This rating was maintained in 2010. This improvement was achieved through greater investment in renewing pipe networks and improving treatment, even though the renewal of assets was still not keeping up with the rate of deterioration (Engineers Australia 1999/2005/2010).

While investment in infrastructure and improvements in management strategies directed through asset management can increase the carrying capacity of an urban water system (Brush 1975), the challenges facing the sector are broader than this. In particular, there is significant uncertainty concerning the impacts of climate change on the environments used for water supply and wastewater disposal, such that there is a significant risk of systemic failure should conditions deteriorate (Speers 2009). As such, the capacity of the urban water sector to respond to future uncertainties has been called into question (Vlachos and Braga 2001; Newman 2001; Ashley et al. 2003; Pahl-Wostl 2007; Farrelly and Brown 2011; Brown et al. 2011). Alternative service provision paradigms have therefore been advocated as a means of meeting emerging challenges.

With these observations in mind, this chapter considers aspects of investment decision making couched in the context of sustainability, asset management, and integrated urban water management (IUWM) concepts. While it must be acknowledged that asset management is sometimes considered to be solely concerned with existing assets, acquisition of new assets is conceptualised here as the initial stage of the asset management cycle. This can be justified by noting that a holistic 'life cycle management' approach is required for effective infrastructure provision over the long term (WERF 2009). The remainder of this chapter is presented in the following manner. First, a brief overview of asset management and related investments is given, followed by a discussion of the *hybridisation of infrastructure* that is occurring through incremental investments in alternative technologies, a process which adds to the backbone of legacy infrastructure. Two case studies for investments made in Melbourne, Australia, are then presented; they illustrate the main components of decision support for investments in asset renewals and asset acquisitions, which together contribute to the process of system hybridisation.

2 Sustainability-Based Asset Management

Formalised asset management is an emerging discipline which aims to provide sustained services at a cost and level of risk that is acceptable to customers and other stakeholders, including regulators and investors. As noted in Sect. 1 and shown in Fig. 10.2 (WERF 2007), asset management is conceptualised here as a whole of life concept and thus involves a range of activities undertaken at different spatial and temporal scales (WERF 2009).

As suggested by Fig. 10.2, asset management is concerned with assuring that infrastructure capacity is maintained in the light of specified targets, although setting of these targets is not part of the asset management cycle per se. Instead, they are set in relation to the requirements of all relevant stakeholders and/or regulations. While such requirements can be narrowly scoped, they can also reflect the broader facets of sustainability discussed previously. This requires WSPs to adopt a TBL view of their business operations, giving more emphasis to environmental and community issues, and thereby contributing to societal aspirations for sustainable development

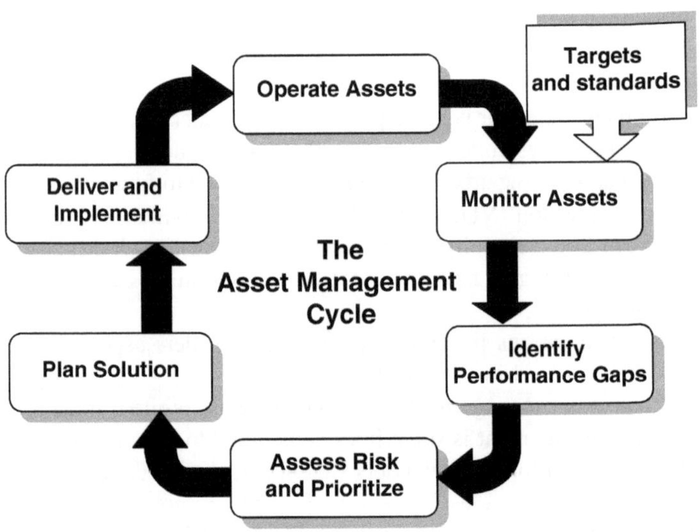

Fig. 10.2 The asset management cycle

(Marlow et al. 2010a, b). The decision to operate in this way is, however, effectively outside the scope of asset management. Nevertheless, if such sustainability objectives are adopted, many related activities will be directed through asset management. Certainly, asset management frameworks have, over time, transitioned so as to shift the focus away from a purely 'asset-centric' view to one that considers the generation of societal value through the provision of water services (Burn et al. 2010). A definition that encapsulates the main features of this sustainability-based asset management is: "*...the combination of management, financial, economic, engineering and other practices applied to assets with the objective of maximizing the value derived from an asset stock over the whole life cycle, within the context of delivering appropriate levels of service to customers, communities and the environment and at an acceptable level of risk*" (Marlow and Burn 2008).

2.1 *Investing in Infrastructure*

Making timely investments to maintain and augment system capacity is a key goal of asset management. Investment is the act of incurring a cost in the expectation of receiving future benefits. Creation and maintenance of water infrastructure incurs costs and, in return, benefits are provided in the form of services intended to have a positive effect on the economy, society, and/or the environment. Investments are often characterised as being either operational ('opex') or capital expenditure ('capex'). Accounting conventions allow expenditure to be charged to capex only if it creates an asset that will exist for more than one year (Kapelan et al. 2011). Effective asset management requires a focus on both kinds of expenditure. For example, expenditure on planned maintenance can ensure that the operational life of assets is extended, whereas a lack of such maintenance can lead to premature failure and require early asset replacement (WERF 2007).

In general, there are numerous and often competing investments that WSPs can make, so analysis is required to determine where investment *must* be made (e.g. to meet regulatory requirements or provide service to new developments) and, given sufficient budgets, which discretionary investments are worthwhile. From a strategic asset management perspective, this process can be considered to be a 'gap analysis' undertaken with respect to the capabilities of existing infrastructure and assessed in the light of service targets, other performance requirements, and planned developments (WERF 2009). A shortfall in capabilities can be related to service and/or cost, and can be real (associated with historical problems or the need to provide infrastructure) or probabilistic (e.g. the risk of service failure is deemed unacceptable). Once potential schemes for investment have been identified, the WSP must then identify feasible options and assess their relative worth.

Capital investments (i.e. involving capex) can be generally categorised as either *asset renewal* or *asset acquisition*. Asset renewal relates to the reinvestment in assets at the end of their life, which can involve either replacement or renovation (renovation implying that a capital investment is made to refurbish the existing asset in some way). The end of an asset's life can be reached in three distinct ways (WERF 2009), as follows.

1. *End of physical life* – an asset's physical state is below some 'acceptability threshold', which may be defined in terms of an unacceptable risk of failure or some measure of structural condition, including dereliction.
2. *End of service/capacity life* – an asset no longer provides required service due to changes in an asset's operational context or because service standards change (e.g. a change in discharge consents).
3. *End of economic life* – an asset ceases to be the lowest cost alternative to satisfy a specified level of performance or service level.

Asset acquisition can relate to greenfield (i.e., where there is no previous development), infill (subdivision of existing developments or construction of higher density dwellings), or brownfield sites (where a previous development, generally industrial or commercial, has been demolished and new developments, usually residential, are being constructed), as well as involving retrofitting of new or modified technologies to existing assets. The WSP may be directly involved in the acquisition, or it may be undertaken by developers or private entities, potentially in the light of technical and other requirements specified by the WSP. The case studies at the end of this chapter provide insights into the type of analysis undertaken to support both kinds of investment decision.

3 Transitions in Infrastructure Systems

As noted in Sect. 1, many urban water systems have been assessed as being in a poor state due to insufficient investment in asset renewals and system augmentation. In addition, the capacity of the sector to respond to the needs of the future has also been questioned. Together, such factors have led various commentators to argue for a transition in both the approach to infrastructure provision and the type of solutions used (Newman 2001; Vlachos and Braga 2001; Ashley et al. 2003; Pahl-Wostl 2007; Niemczynowicz 1999; Urich et al. 2011; Brown et al. 2009; Brown et al. 2011; Farrelly and Brown 2011; Binney 2012). Calls for change have been underpinned by the emergence of new urban water management paradigms, such as IUWM (Coombes and Kuczera 2003; Mitchell 2006; Maheepala et al. 2010; Burn et al. 2012). IUWM seeks to address broader sustainability, liveability, and productivity objectives through a holistic approach to the urban water cycle. In practice, this includes a diversification of infrastructure to incorporate alternative technologies including desalination, rainwater and stormwater harvesting (for both potable and non-potable use), indirect potable reuse, sewer mining, decentralised treatment of sewage, greywater systems, and various classes of water recycling (Newman 2001; Brown et al. 2008). As IUWM solutions are generally implemented and managed within the context of legacy infrastructure, there are many technical, institutional, regulatory, and social-acceptance challenges to overcome, such as those indicated in Table 10.1 (Pinkham 2004; Tjandraatmadja et al. 2009).

Table 10.1 Some challenges associated with investment in IUWM solutions

Issue	Details
Lack of appropriate governance frameworks	IUWM approaches have developed rapidly and in many cases there is a lag in guidelines and regulations
Lack of experiences and skills	IUWM presents a challenge for many practitioners as it requires a departure from conventional design and management approaches
Impact on WSP revenue base	IUWM solutions may reduce the traditional revenue base of WSPs. This is particularly problematic if traditional water systems are still designed as 'systems of last resort' such that there are no reductions in system sizing, etc. Properties connected to centralised infrastructure also pay service charges (a significant component of overall charges) even if they have reduced demand on the system, which reduces incentives for uptake at the household level
Operation and maintenance	IUWM solutions will often require management at the development or household scale. This presents a challenge to ensure adherence to appropriate operational and maintenance procedures
Lack of understanding	Limited information is available on the life-cycle performance of IUWM solutions
Public health risks	Alternative water sources may impose health and safety risks to the public, and related community concerns are a barrier for adoption

One constraint in modifying the infrastructure used to supply water services is that investment planning cycles are relatively short (for example, 5 years or less depending on the governance regime and/or political cycle), which presents a challenge when attempting to adapt to longer term pressures (de Graff and van der Brugge 2010). Another is that widespread adoption of a specific technological solution leads to 'lock-in' effects. Such effects occur because of the long life of infrastructure and the interdependence this creates between management and governance regimes. Lock-in effects are thus both institutional and technological (Foxon et al. 2002). Arthur (1994) identified four aspects of competitive advantage that generate lock-in effects: (1) economies of scale (unit production costs decline as these as spread over increasing production volume), (2) learning effects (which improve products or reduce their cost), (3) adaptive expectations (users and producers become increasingly confident about quality, performance, and longevity of the current technology), and (4) network economies (due to agents adopting the same technologies as others). At its worst, lock-in effects create a barrier to the adoption of more sustainable technologies (Foxon et al., 2002). Advocates for change argue that since legacy infrastructure is long-lived and expensive it locks out innovative alternatives. Brown et al. (2011) have in fact referred to this as 'entrapment'.

3.1 *Hybridisation Through Incremental Investments*

While the lock-in effect of existing infrastructure means that a broad transition in urban water systems is difficult to achieve, at a local scheme level, innovations do occur. From the perspective of the broader urban water system, these innovations are however technical 'add-ons'. In terms of infrastructure, the process of change can thus be considered as one of system hybridisation. This concept has some similarities with the satellite and decentralised strategies for water resource management described by Gikas and Tchobanoglous (2009), where decentralised solutions are integral and complementary to the existing centralised system.

In theory, opportunities for hybridisation arise whenever an investment is made, whether it is related to asset acquisition or renewal. In practice, however, the space for innovation often opens up when legacy approaches are shown to be sub-optimal (Balslev Nielsen and Elle 2000). For example, novel solutions are considered where existing infrastructure has insufficient capacity or would need to be extended, and the cost of doing this would be prohibitive (Pinkham 2004). More generally, system hybridisation is driven by the dynamic interplay between technology and societal values and expectations, underpinned by the lock-in effect of legacy solutions, the development of technological innovations, and a range of endogenous and exogenous pressures. Critical elements of this process can be described using two coupled conceptual models (Marlow et al. 2013). The first describes the flow of investment into the asset stock, while the second describes factors that influence the selection of options.

3.2 A Conceptual Model of Investment Flows

Figure 10.3 shows how accumulation of capital occurs in terms of flow of investment and other key influences. As illustrated, the flow of investment accumulates over time as a capital stock of constructed and natural infrastructure assets. These assets contribute to the delivery of urban water services. Service delivery is sustained through a combination of both business and asset capabilities. Hence, shortfalls in infrastructure can be managed to a certain extent through improved management and maintenance practices, but fundamentally the infrastructure places bounds on the level of service that can be delivered. The capital stock of constructed and natural assets is reduced over time due to deterioration, which is counterbalanced by capital maintenance. The main flow out of the asset stock can be conceived as a flow of services, which in turn contributes to a broader notional capital stock that delivers TBL outcomes such as community wellbeing, environmental health, and economic production in other sectors.

There are two main feedback loops influencing the flow of investment and thus the maintenance and growth of capital value of the asset stock and its ability to provide service. Firstly, from an asset management perspective, infrastructure systems are managed in the light of explicit (or sometimes implicit) service targets. Hence, a 'hard' metric-driven feedback loop exists wherein the gap between service delivered and that required is taken as a measure of system capacity. This informs the need for future investment, but there is also a requirement to take into account the inherent capacity of the system, including operational responses, the effect of endogenous factors (asset depreciation and backlog in investment), and the effect of

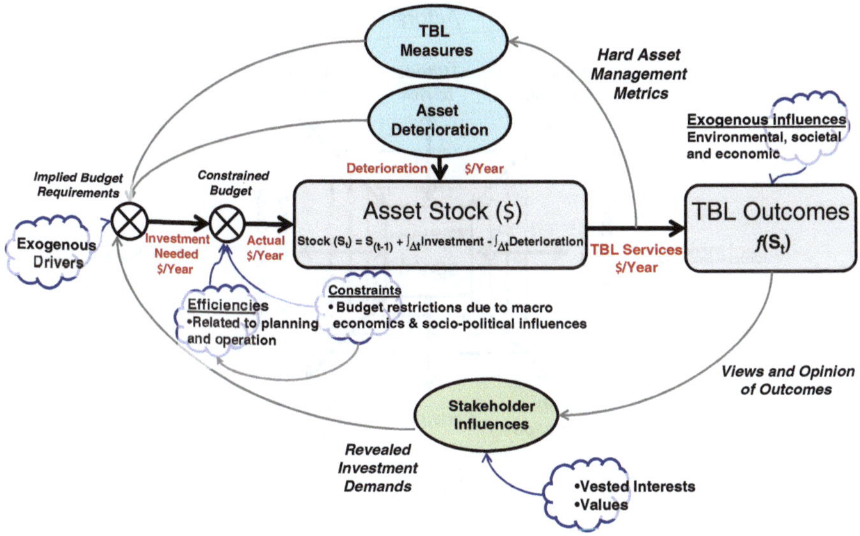

Fig. 10.3 A systems perspective of water sector investments

exogenous factors (climate change, population growth, etc.). Consideration of these factors requires both diagnostic and prognostic capacity (i.e. to understand the current capacity to deliver service and predict future capacity). Secondly, the system is subject to a 'soft' feedback loop from the various stakeholder groups (customers, regulators, suppliers, politicians, academics, etc.). This reflects the perception of broader outcomes delivered by the sector. Perceptions are influenced by complex issues related to vested interests and values, and vary according to the responsibilities and interests of each stakeholder group. For example, individuals and institutions with responsibility for, or interest in, waterway health and sustainability will have different perspectives to, say, an economic regulator driven by the requirement to maintain efficiency from the perspective of customer bills. The net result of these influences is an implied demand for investment, but this is generally constrained in the light of budget restrictions, which drive efficiencies that in turn can reduce the investment needed. Groups or individuals that consider the outcomes delivered by the water sector to be misaligned with their expectations will engage in advocacy for a change in both the way investments are justified and the type of infrastructure funded.

3.3 *Option Identification and Assessment*

The other perspective that drives the hybridisation process is how options are identified and considered. A key aspect in any investment decision is to identify feasible solutions. In this context, 'feasible solutions' imply infrastructure (or other) options

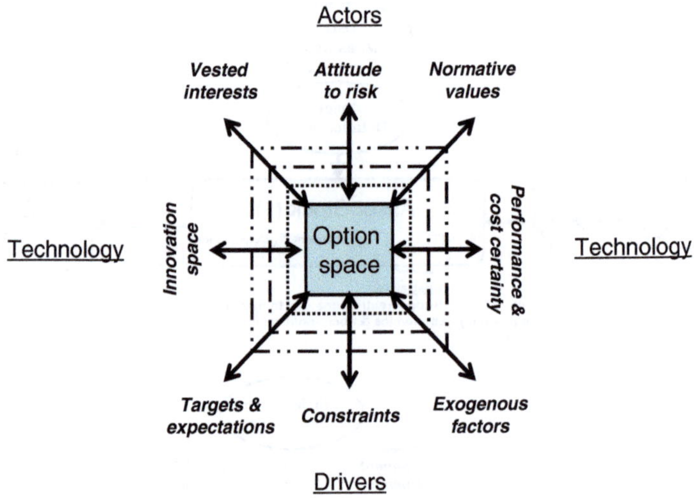

Fig. 10.4 Factors influencing the option space considered

that are able to provide service over an economic life. The 'option space' is that sub-set of solutions that is seriously considered for implementation. Depending on the predominant factors influencing the decision, the option space can be constrained to traditional solutions or embrace the full set of feasible solutions, including a range of innovative options associated with IUWM concepts. Uncertainties with respect to performance, cost, utility preferences, community acceptance, and related outcomes imply that there may need to be a stronger justification for innovative solutions compared to traditional ones. The various factors that influence the option space are shown in Fig. 10.4. Factors are grouped in terms of 'actors', 'technology', and 'drivers'. The individual factors are dynamic and interact with one another to create the option space considered for a particular investment or scheme. It is noteworthy that these same factors influence stakeholder perception of investment needs, as shown in Fig. 10.3.

The perspectives and attitudes of the actors involved are a key determinant in how the option space is defined. In this context, actors are taken to be the individuals or teams making the decision. The options considered will reflect vested interests, attitude to risk, and normative values within their organisation. However, other stakeholders with different values, interests, and attitudes to risk will assess, from their own perspective, both the options considered and the solution ultimately selected. The issue of risk perception is critical to understanding different perspectives on progress towards a more sustainable system. Decision maker(s) are responsible for the actual option selected, and thus are exposed to risks associated with making a poor choice. In contrast, while stakeholders may have an informed view of the technical and broader issues and a stake in the outcome, they are not making the decision and thus will have a different risk perception (the decision-making risks are actually borne by others).

Farrelly and Brown (2011) have stated that the continued use of large-scale technological solutions reflects a reticence to go beyond pilot trials, which are taken to be representative of both the fear of failure of actors and inherent conservatism, especially with respect to public health risks. An alternative view is that many innovative solutions remain unproven and have low community acceptance. From this perspective, any conservatism would then be considered as an appropriate assessment of the uncertainty associated with available alternatives, especially given that funds are not unlimited and that community preferences must be considered (Speers 2009). The fallback onto centralised solutions would then be considered a rational decision based on a pragmatic assessment of the need to invest in solutions that are acceptable to communities, and provide more certainty with respect to life-cycle performance and cost in comparison to alternative options.

Similar observations can be made about the changes in normative values that have, over time, shifted the outcomes the sector is expected to deliver. The transformation of values is, by definition, stronger within the niche group of individuals advocating a shift away from legacy solutions. The advocacy this engenders provides a driver for experimentation and infrastructure diversification, but this is counterbalanced by decision-makers' perceptions of certainty/uncertainty and exposure to risk. However, it should also be recognised that innovative solutions may be deemed higher risk simply because the institutional capacity to manage those risks has not yet been developed. Such capacity can only be built up if innovative schemes are implemented. This perception of risk is a real barrier to change, so the underlying challenge is to broaden the 'optioneering' to consider novel solutions and ensure these are given appropriate consideration. This implies undertaking a rational review of future costs (considering potential economies of scale once institutional capacity has been built), risks and benefits, and implementing innovative schemes where possible to allow a critical mass of core competencies to be built.

4 Case Studies

To this point, we have discussed the role that asset management plays in making investment decisions about both asset acquisitions and asset renewals, and highlighted how transitional processes – driven by concerns over the future, emerging paradigms like IUWM, and the need to manage legacy assets – are leading to a hybridisation of water systems. These issues are also providing drivers for a change in the way investment decisions are analysed (Pahl-Wostl 2007). To illustrate examples of the decision support process used in both asset renewal and acquisition, the remainder of this chapter presents two case studies. Both are real examples of analysis undertaken to support investment decisions in Melbourne, Australia. The first reflects the classic asset management challenge of determining the optimum (or at least a reasonable) time to invest in asset renewal. The second shows how IUWM solutions were compared to the base case of providing traditional infrastructure for a new development.

In both cases, the analysis undertaken extends the scope of issues taken into consideration beyond just the financial implications of the investment. However, while the renewal decision was considered from a TBL perspective, there was no attempt to consider an alternative means of service provision. This reflects the reality that the existing system often locks-in the type of solution considered. The second case study shows how new developments can provide more scope for considering innovative options. In the case presented, an innovative scheme was in fact selected because it aligned with broader sustainability objectives of the WSP and was shown to be economically competitive, at least when a societal view of costs was taken. When considered from the perspective of the infrastructure used to supply urban water services in Melbourne, the case studies thus represent on-the-ground examples of how the system hybridisation process discussed above occurs. More specifically, the decision to renew the water main had the effect of perpetuating the legacy approach to service provision, which remains the predominant model in the city, whereas the implementation of innovative IUWM solutions expanded the type of infrastructure used within that broader Melbourne system.

4.1 Case Study 1: Analysis of a Traditional Like-for-Like Renewal

This case study relates to the renewal of approximately 2.5 km of 250/300 mm diameter cast iron pipe (Marlow et al. 2010c; Marlow et al. 2011). The pipe in question ran along a busy street in Melbourne and passed through an iconic shopping area, which meant that a catastrophic failure of the asset had the potential to impose significant social disruption. Though the pipe was providing adequate service, these circumstances lead the WSP to commission an analysis to predict the remaining economic life of the pipe and so determine when renewal should be undertaken. This analysis involved the development of a physical probabilistic model that considered deterioration, structural capacity, and imposed loads (Moglia et al. 2008). This was then used to develop an economic model (Davis and Marlow 2008).

The hydraulic capacity of the existing pipe was considered sufficient, so the benefits associated with the service provision of both the existing and replacement pipes were the same and thus not considered in the analysis. However, pipe replacement was considered to accrue additional benefits associated with the avoidance of future pipe bursts. This concept is illustrated in Fig. 10.5, which shows a timeline from the current time, τ, to some future time, T_{MAX}, when the asset would be considered derelict and have to be replaced (i.e., end of physical life is reached at T_{MAX}). At any time in between, the asset could be replaced through a scheduled intervention (asset renewal).

As shown in Fig. 10.5, renewing the asset at any time later than the current time, τ, implies that the consequences associated with potential bursts (the star shapes in Fig. 10.5 which represent probabilistic rather than actual failures) would be incurred

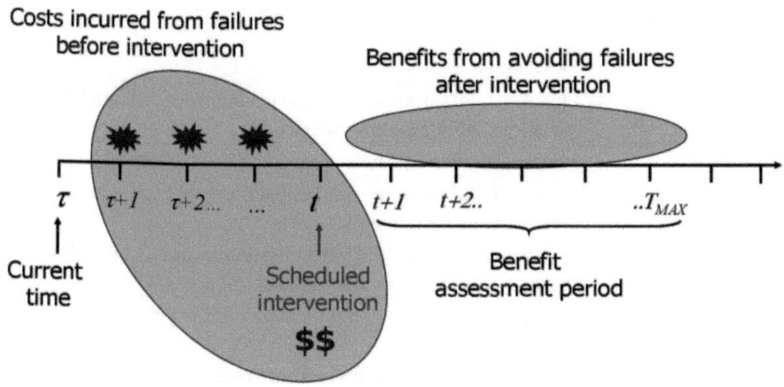

Fig. 10.5 Illustration of intervention costs and benefits

up until the time of replacement (time t). In contrast, replacing the asset earlier than T_{MAX} means that costs associated with potential bursts occurring after the replacement will be avoided. Within this framework, the optimum time to replace the pipe was taken as the time at which the Net Present Value (NPV) was a maximum. In application, future monetary flows were discounted (using a discount rate of 7 %) and the NPV calculated as:

$$NPV = PV_{Benefit} - PV_{Cost} \tag{10.1}$$

It is important to note that while the calculation algorithm is expressed in terms of NPV, the benefits are avoided failure costs. As such, the analysis is sensitive to how failure costs are taken into account.

In order to apply Eq. 10.1, the costs and benefits of renewal needed to be estimated. The total costs associated with renewal were considered to be the sum of two components: the costs of pipe failure up to intervention at time t and the cost of intervention (pipe renewal) at time t. The expected failure costs in each year (i.e. the failure cost weighted by the probability of failure) were estimated through the application of the physical probabilistic model and estimates of failure costs. As well as direct costs associated with the failure (repair, clean-up costs, etc.), pipe failure can result in a range of consequences, as categorised in Table 10.2. Some of these consequences impose costs on the WSP, but others (e.g. traffic disruption) are costs imposed on third parties (what economists refer to as externalities) and are not monetised (Marlow et al. 2011). Increasingly, WSPs are taking into account these broader failure consequences in their renewal decisions by placing a notional monetary value against them and treating them in the same way as tangible costs. In practice, this brings the economic time of the renewal forward.

In this case, disruption to transport and related social disruption were identified as potential failure consequences, and related externality costs were thus calculated using the figures shown in Table 10.3. With these assumptions, the externality costs per burst were estimated at $121,620 for failures occurring on a normal road section

Table 10.2 Examples of failure consequences

Category	Description
Service failures	Operational impacts such as service interruptions, quality or quantity variations, poor customer service, etc.
Customer impacts	Impacts of interruption and customer dissatisfaction
Environmental impacts	Unacceptable environmental impacts considering natural and constructed environments
Externalities	Traffic disruptions, third party discomfort, aesthetic and recreational impacts, health impacts, ecological impacts, etc.
Health and safety impacts	Death or injury to staff and members of the public
Public relations impacts	Public relations issues arising from asset failures

Table 10.3 Estimates used in the analysis of externalities

Cost component	Basis of calculation
Traffic disruption	Value of time: $20 per hour per vehicle
	Traffic frequency (one road): 1,200 vehicles/h
	Traffic frequency (intersection): 2,920 vehicles per hour
	Time of disruption: 6 h
	Time of traffic delay: 0.33 h
Public transport	Passengers: 190
	Cost of fare: $6.50
	Frequency: 5 per hour
	Time of disruption: 6 h
	Routes disrupted (one road): 2
	Routes disrupted (intersection): 3

or $226,782 for failures occurring at or near an intersection. In combination with estimates of failure probability, these figures justified immediate renewal of the pipe. For reference, the cost of pipe replacement was around $4,000 per metre, so the overall replacement cost was approximately $10 million.

While the values shown in Table 10.3 were those used by the WSP to justify immediate renewal, sensitivity analysis showed that, depending on the monetary value placed on externalities, the optimum time for replacement ranged from ‘replace now’ to ‘replace in 57 years’, the latter being the replacement time if only internal costs borne by the WSP (e.g. reactive maintenance costs) were considered (i.e., assuming externalities had no value). This highlights how important it is to place meaningful estimates on the societal value of externalities and other non-monetary consequences. While it may appear that in the case of such like-for-like renewals the issue is merely a matter of timing (the pipe will eventually be replaced), in reality the time value of money generates significant opportunity costs if renewals are premature.

4.2 *Case Study 2: Analysis of Asset Acquisition Using IUWM Concepts*

The second case study focuses on an investment decision for a large greenfield development on the peri-urban fringe of Melbourne (Sharma et al. 2010). The development was on land previously used for grazing, and intended to house approximately 86,000 people with a mix of land uses, including residential (62 %), commercial (10 %), industrial (27 %), and open space (1 %). Given the limited capacity of existing water infrastructure, as well as nitrogen discharge limitations, there was a need to determine whether IUWM options would be more cost-effective and sustainable. The WSP thus commissioned analysis of various servicing scenarios as part of the strategic planning processes (Sharma et al. 2009). Fig. 10.6 presents a summarised version of the assessment methodology used.

Information on the proposed development layout, allotment size, anticipated occupancy rate, and available water resources was available. To establish development-specific objectives, data on climate, topography, and water usage were also collated. The objectives were specified in consultation with stakeholders and shaped by location-specific factors including ecological conditions of the area, community concerns, proximity to and available capacity of centralised water and wastewater infrastructure, and the availability of alternative and fresh water sources. By identifying objectives and constraints, the feasible option space was reduced to four potential infrastructure solutions:

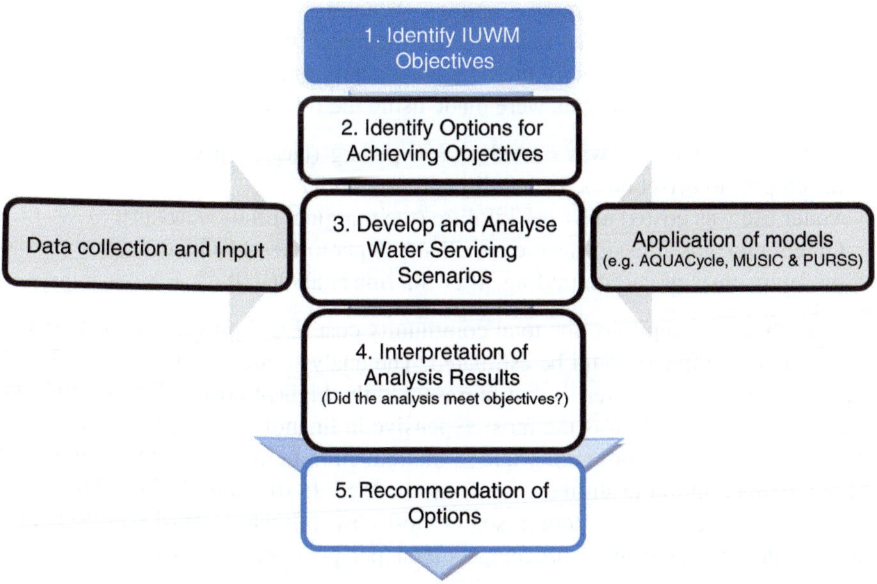

Fig. 10.6 Assessment methodology

Table 10.4 Summary of LCC for various IUWM options

Scenario	Water supply system (k$)	Wastewater system (k$)	Stormwater system (k$)	LCC (k$)[a]
A	250,140	275,610	121,270	647,030
B	342,260	258,580	121,270	722,120
C	331,360	269,740	121,270	722,390
D	408,430	575,000	62,720	1,046,900

[a]Analysis period of 50 years with a discount rate of 6.5 %

Scenario A conventional centralised supply and wastewater (base case)

Scenario B third pipe system with reclaimed water

Scenario C third pipe system using stormwater

Scenario D self-contained servicing configuration drawing upon greywater reuse, rainwater tanks, and on-site wastewater treatment.

The life-cycle cost (LCC) of service provision was calculated over an analysis period of 50 years (Table 10.4). The base system (A) was least expensive, whereas the self-contained scenario (D) was most expensive due to the provision of rainwater tanks and on-site wastewater systems. The reuse scenarios (B and C) with dual-pipe systems were in the middle range.

Four environmental indicators were also selected for use in the assessment of the options: (1) greenhouse gas emissions (GHG); (2) eutrophication potential (nutrient release to waterways); (3) the amount of imported potable water, and (4) solid waste going to land fill. The alternative options (Scenarios B, C, and D) significantly reduced both the nutrient load to the environment and import of water. To integrate the environmental impacts with the LCCs, monetary valuations of greenhouse gas emissions, water, and nutrients were made using the following figures:

- Nitrate to waterways was costed at $60 per kg (based on charges applied to developers to offset costs for treating nitrate runoff)
- Water use was costed at $1 per kilolitre (the marginal retail water price)
- Greenhouse gas emissions were costed at $50 per tonne of CO_2 equivalent (based on future costs of carbon trading, after Sharma et al. 2010).

With these assumptions, the total community cost (LCC plus monetised cost of environmental impact) could be estimated. The analysis indicated Scenario A was the cheapest in financial terms, but represented the highest cost to the community. In contrast, Scenario D was the most expensive in financial terms, but represented the lowest cost to the community due to the reduction in environmental impact. At the same time, option D shifted costs from the WSP to the householder. This illustrates a challenge in that a 'total resource' test may indicate IUWM solutions provide greatest value from a broader societal perspective, but current models of charging for water services can transfer costs to decentralised entities (households, body corporate, etc.), which is an impediment to their uptake (Pinkham 2004; Tjandraatmadja et al. 2009).

5 Conclusions

Decisions relating to urban water infrastructure investments have direct and indirect impacts on water resources and communities, including imposition of opportunity costs and externalities associated with use of resources and asset failures. Through these mechanisms, poorly conceived investments can impose significant welfare costs on society and lead to an economically inefficient use of resources. As such, it is important for WSPs to target and time investment so as to minimise negative impacts and, as far as is practical, maximise economic value delivered. With this challenge in mind, this chapter has considered investment decisions made in the light of asset management concepts, considering the influence of increasingly important concepts like sustainability and IUWM.

Various factors mean that, from an infrastructure perspective, investments in innovative solutions are leading to an incremental hybridisation of water systems. To illustrate this process, two coupled conceptual models specific to the urban water sector have been presented. The factors that influence this hybridisation vary according to individual views, vested interests, and attitudes/exposure to decision-making risks. There are two relevant questions that face the sector in the future. Firstly, whether a step change is needed in the light of climate change and other exogenous drivers, and secondly, whether such a step change can be afforded in view of other demands on limited resources and budgets.

At the asset or scheme level, investment decisions need to be supported through appropriate analysis. Two case studies have been presented that correspond to real investment decisions made in Melbourne, Australia. Together these illustrate how WSP decisions both perpetuate traditional solutions and lead to the implementation of innovative solutions that are added to the back-bone of the legacy system.

Importantly, both case studies involved analysis which considered factors that extended beyond the traditional financial scope of asset investments. The first demonstrates that the value placed on social and other impacts can significantly influence the timing of like-for-like asset renewals. In the second, the assessment framework included consideration of both economic and environmental implications of alternative options, which allowed an innovative solution to be justified. Overall, the case studies illustrate that the relative worth of options considered in both asset renewals and asset acquisitions depends strongly on how broader environmental and social factors are considered, including the valuation placed on externalities. Such factors are increasingly influencing infrastructure investment decisions in the transition to more sustainable urban water systems.

References

Arthur, B. (1994). *Increasing returns and path dependence in the economy*. Ann Arbor: University of Michigan Press.

ASCE. (2009). *Report card for America's infrastructure*. American Society of Civil Engineers. http://www.infrastructurereportcard.org/index

Ashley, R. M., Blackwood, D. J., Butler, D., Davies, J. A., Jowitt, P., & Smith, H. (2003). Sustainable decision making for the UK water industry. *Engineering Sustainability, 156*(ES1), 41–49.

Baird, G. M. (2010). How to eat an elephant: The infrastructure investment gap. *Journal of the American Water Works Association, 102*(6), 26–33.

Balslev Nielsen, S., & Elle, M. (2000). Assessing the potential for change in urban infrastructure systems. *Environmental Impact Assessment Review, 20*, 403–412.

Binney, P. D. (2012). A framework for developing sustainable water utilities in the coming decades. In C. Howe & C. Mitchell (Eds.), *Water sensitive cities* (pp. 82–94). London: IWA Publishing.

Blackmore, J. M., & Plant, R. A. J. (2008). Risk and resilience to enhance sustainability with application to urban water systems. *Journal of Water Resources Planning and Management, 134*, 224–233.

Brown, R. R., Keath, N., & Wong, T. (2008). *Transitioning to the water sensitive city: Historical, current and future transition states*. In Proceedings of the 11th international conference on urban drainage, Edinburgh

Brown, R. R., Keath, N., & Wong, T. H. F. (2009). Urban water management in cities: Historical, current and future regimes. *Water Science and Technology, 59*(4), 847–855.

Brown, R., Ashley, R., & Farrelly, M. (2011). Political and professional agency entrapment: An agenda for urban water research. *Water Resources Management, 25*(15), 4037–4050.

Brush, S. B. (1975). The concept of carrying capacity for systems of shifting cultivation. *American Anthropologist, 77*, 799–811.

Burn, S., Marlow, D., Moglia, M., & Buckland, P. (2007). Asset management for water infrastructure. *Water Asset Management International, 3*(2), 12–18.

Burn, D., Marlow, D., & Tran, D. (2010). Modelling asset lifetimes and their role in asset management. *Journal of Water Supply: Research and Technology – AQUA, 59*(6–7), 362–377.

Burn, S., Maheepala, S., & Sharma, A. (2012). Utilising integrated urban water management to assess the viability of decentralised water solutions. *Water Science and Technology, 66*(1), 113–121.

Coombes, P. J., & Kuczera, G. (2003*). Analysis of performance of rainwater tanks in Australian capital cities*. In: 28th international hydrology and water resources symposium: New South Wales, Australia.

Copeland, C. (2010). *Water infrastructure projects designated in EPA appropriations: Trends and policy implications*. Washington, USA: US Congressional Research Service report RL32201.

Davis, P., & Marlow, D. (2008). Quantifying economic lifetime for asset management of large diameter pipelines. *Journal AWWA, 100*(7), 110.

De Graaf, R., & van der Brugge, R. (2010). Transforming water infrastructure by linking water management and urban renewal in Rotterdam. *Technological Forecasting and Social Change, 77*, 1282–1291.

Elkington, J. (1998). *Cannibals with forks: The triple bottom line of 21st century business*. Gabriola Island: New Society. ISBN 0-86571-392-8.

Engineers Australia. (1999/2005/2010). *Australian infrastructure report card*. Canberra: The Institution of Engineers Australia. http://www.engineersaustralia.org.au/irc/

Farrelly, M., & Brown, R. (2011). Rethinking urban water management: Experimentation as a way forward? *Global Environmental Change, 721*, 721–732.

Foxon, T. J., McIlkenny, G., Gilmour, D., Oltean-Dumbrava, C., Souter, N., Ashley, R., Butler, D., Pearson, P., Jowitt, P., & Moir, J. (2002). Sustainability criteria for decision support in the UK water industry. *Journal of Environmental Planning and Management, 45*(2), 285–301.

Gikas, P., & Tchobanoglous, G. (2009). The role of satellite and decentralized strategies in water resources management. *Journal of Environmental Management, 90*(1), 144–152. doi:10.1016/j.jenvman.2007.08.016.

Kapelan, Z., Banyard, J. K., Randall-Smith, M., & Savic, D. A. (2011). Asset planning and management. In D. A. Savic & J. K. Banyard (Eds.), *Water distribution systems* (pp. 227–261). London: ICE Publishing. ISBN 978-0-7277-4112-7.

Kenway, S., Howe, C., & Maheepala, S. (2006). *Triple bottom line reporting of sustainable water utility performance*. AwwaRF report 91179. Denver: AwwaRF

Klostermann, J. E. M., & Cramer, J. (2006). The contextual meaning of sustainable development: The case of the Dutch drinking water sector. *Sustainable Development, 14*, 268–276.

Maheepala, S., Blackmore, J., Diaper, C., Moglia, M., Sharma, A., & Kenway, S. (2010). *Integrated urban water management planning manual*. Denver: Water Research Foundation.

Marlow, D., & Burn, S. (2008). Effective use of condition assessment within asset management. *Journal AWWA, 100*(1), 54–63.

Marlow, D., & Humphries, B. (2009). Sustainability within the Australian water industry: An operational definition. *Australian Water Association Journal, February*, 46–50.

Marlow, D., Beale, D., & Burn, S. (2010a). A pathway to a more sustainable water sector: Sustainability-based asset management. *Water Science and Technology, 61*(5), 1245–1255.

Marlow, D., Beale, D., & Burn, S. (2010b). Linking asset management with sustainability: Views from the Australian sector. *Journal AWWA, 102*(1), 56–67.

Marlow, D., De Silva, D., Beale, D., & Marney, D. (2010c). Effective management of large diameter pressure pipes. *Journal AWWA, 102*(4), 104–112.

Marlow, D., Person, L., Whitten, S., Hatton MacDonald, D., & Burn, S. (2011). A framework for considering externalities in urban water asset management. *Water Science and Technology, 64*(11), 2199–2206.

Marlow, D., Moglia, M., Cook, S., & Beale, D. (2013). *Towards sustainable urban water management: A critical reassessment, water research*, accepted for publication in Sept 2013 special edition: Urban Water Management to Increase Sustainability of Cities.

Mitchell, V. G. (2006). Applying integrated urban water management concepts: A review of Australian experience. *Environmental Management, 37*(5), 598–605.

Moglia, M., Davis, P., & Burn, S. (2008). Strong exploration of a cast iron pipe failure model. *Reliability Engineering and System Safety, 93*, 885–896.

Newman, P. (2001). Sustainable urban water systems in rich and poor cities – Steps towards a new approach. *Water Science and Technology, 43*(4), 93–99.

Niemczynowicz, J. (1999). Urban hydrology and water management: Present and future challenges. *Urban Water Journal, 1*, 1–14.

Pahl-Wostl, C. (2007). Transitions towards adaptive management of water facing climate and global change. *Water Resources Management, 21*(1), 49–62. doi:10.1007/s11269-006-9040-4.

Pearson, L. J., Coggan, A., Proctor, W., & Smith, T. F. (2010). A sustainable decision support framework for urban water management. *Water Resources Management, 24*(2), 363–376.

Pinkham, R. D. (2004). *Valuing decentralized wastewater technologies: A catalog of benefits, costs, and economic analysis techniques*. Prepared by Rocky Mountain Institute for the U.S. Environmental Protection Agency, Nov 2004.

Porritt, J. (2005). *Capitalism: As if the world matters*. London: Earthscan Publications.

Schaffer, D., & Vollmer, D. (2010). *Pathways to urban sustainability: Research and development on urban systems. Summary of a workshop*. Washington, DC: National Academies Press.

Sharma, A. K., Grant, A. L., Grant, T., Pamminger, F., & Opray, L. (2009). Environmental and economic assessment of urban water services for greenfield development. *Environmental Engineering Science, 5*, 921–934.

Sharma, A. K., Tjandraatmadja, G., Grant, A. L., Grant, T., & Pamminger, F. (2010). Sustainable sewerage servicing options for peri-urban areas with failing septic systems. *Water Science and Technology, 62*, 570–585.

Speers, A. (2009). Urban water futures. In P. W. Newton (Ed.), *Transitions: Pathways toward more sustainable urban development in Australia*. Melbourne: CSIRO Publishing.

Tjandraatmadja, G., Cook, S., Sharma, A., Diaper, C., Grant, A., Toifl, M., Barron, O., Burn, S., & Gregory, A. (2009). *ICON water sensitive urban developments*. Melbourne: CSIRO Publishing. http://archive.nwc.gov.au/library/topic/urban/icon-water-sensitive-urban-developments-dec-2009. Accessed Nov 2012.

Urich, C., Bach, P. M., Hellbach, C., Sitzenfrei, R., Kleidorfer, M., McCarthy, D. T., Deletic, A., & Rauch, W. (2011). *Dynamics of cities and water infrastructure in the DAnCE4Water model*. In 12th international conference on urban drainage, Porto Alegre.

Vlachos, P. E., & Braga, P. B. (2001). The challenge of urban water management. In C. Maksimovic & J. A. Tejada-Guibert (Eds.), *Frontiers in urban water management, deadlock or hope* (pp. 1–34). London: IWA Publishing.

WCED: World Commission on Environment and Development. (1987). *Our common future (the Brundtland report)*. Oxford: Oxford University Press.

WEF. (2012). *Global risk report 2010: A global risk network report*. Geneva: World Economic Forum.

WERF. (2007). *Condition assessment strategies and protocols for water and wastewater utility assets*. Report prepared by D. R. Marlow et al. for Water Environment Research Foundation, Alexandria, Project Ref. 03-CTS-20CO.

WERF. (2009). *A state of the art review: Remaining asset life*. Report prepared by D. R. Marlow et al. for Water Environment Research Foundation, Alexandria, Project Ref 06-SAM-1 CO.

Part II
Water Demand and Water Economics

Chapter 11
Long-Term Water Demand Forecasting

Jean-Daniel Rinaudo

1 Introduction

In the years following World War II, the priority of many governments and water utilities around the world was to develop water supply and increase the percentage of households connected to the mains. Predicting the intermediate- and long-term evolution of the demand for water was thus not a major concern for managers of drinking water utilities. Water was viewed as an inexpensive commodity, and developing excess capacity was considered a much better option than risking a shortage. This tactic worked well – as long as economic and population growth continued and water resources were readily available. This perception gradually changed, however, as the marginal cost of developing new resources progressively rose throughout the 1970s and 1980s, notably in arid regions of developed countries (the western US states, in particular). In other regions, an unanticipated decline in per capita water use led to costly oversized water supply systems. Due to the mounting cost of prediction errors, improving the accuracy of future water demand forecasting became crucial for optimising the expansion of water supply systems. Indeed, building oversized facilities based on an 'upper bound' demand estimate would lead to significant extra cost burdens that would have to be passed on to customers through tariff increases (Beecher and Chesnutt 2012). Conversely, underestimating future water demand would result in shortages, in this way imposing costs in the form of losses in garden landscaping, convenience of water use, and extra constraints on residential and economic development.

Awareness grew that better water demand forecasts meant gaining a better understanding of the factors creating that demand. Early research in the field (Howe and Linaweaver 1967) was quickly followed by the development and diffusion of operational tools and software packages such as IWR-MAIN, released in the 1980s

J.-D. Rinaudo (✉)
Water Department, BRGM, French Geological Survey, Montpellier, France
e-mail: jd.rinaudo@brgm.fr

Q. Grafton et al. (eds.), *Understanding and Managing Urban Water in Transition*, Global Issues in Water Policy 15,
DOI 10.1007/978-94-017-9801-3_11

by the US Army Corps of Engineers (Boland 1997; Bauman et al. 1998). Uninterruptedly for three decades, an abundant flow of scientific literature ensued. This literature can be divided into two main streams. The first comprised contributions from economists who mainly studied the effect of price level and tariff structures on water demand using econometric methods (for a review, see Espey et al. 1997; Arbués et al. 2003; Dalhuisen et al. 2003; Worthington and Hoffman 2008, among others). The second stream consisted of multidisciplinary contributions from civil engineers and modelers (for a review, see Donkor et al. 2012). Both have led to the development of a variety of innovative forecasting tools, based on a variety of modeling techniques (statistical, econometric, neural networks, agent-based models, etc.) and intended to support short-, intermediate-, and long-term forecasting.

Within practitioner communities these tools have become generalised at a pace that has varied from country to country. Overall, a gradual shift from 'water requirement' to 'water demand' models has occurred. Requirement models assume that the amount of water needed per consumer is absolute and constant over time, unaffected by socioeconomic changes. Demand models consider that water use can be altered by pricing and other water conservation policies such as information and awareness-raising campaigns, water device retrofit incentives, etc. This on-going transition, which began in the south-western states of the USA in the 1980s and appeared in the UK at the start of the millennium, has not been adopted in all countries, particularly those where water is still perceived as an abundant resource (for a French example, see Rinaudo 2013).

A number of books and scientific articles review research in the field (Donkor et al. 2012) or provide technical guidelines on how to implement forecasting methodologies (Billings and Jones 2008). This literature, however, does not describe how forecasting methodologies have been implemented *in practice* by water utilities around the world. This chapter presents an attempt at (partially) filling this gap. Based on detailed analyses of three case studies in the USA, the UK, and Australia, it illustrates the diversity of practices, discusses the challenges faced by water utilities, and identifies issues that should be addressed by future research. It is structured as follows: in the next section, we give a brief overview of existing long-term water demand forecasting methodologies. Subsequently, we discuss some of the challenges faced by water utilities when developing water demand forecasts. We then present three case studies illustrating how methods are deployed in operational (as opposed to research) contexts. The chapter concludes by identifying research perspectives to improve forecasting methodologies.

2 An Overview of Forecasting Methodologies

2.1 What Is Long-Term Water Demand Forecasting About?

This chapter uses the common interpretation of water demand as the observed amount of water consumed by residential, public, commercial, and industrial customers connected to a public water distribution system. Water demand strongly depends on prevailing economic conditions, particularly on water rates and tariffs, population income, and economic activity. Water losses in distribution systems (due to leakage) are not included in demand (Merrett 2004).

Water demand forecasting can be conducted for varying horizons. Short-term forecasting aims at anticipating water demand over the coming hours, days, or weeks, so as to optimise the operation of water systems (reservoirs, desalination plants) while factoring in changes in weather and consumer behaviors. Short-term demand forecasting can help estimate revenues from water sales and plan short-term expenditures. Intermediate-term forecasting (1–10 years) focuses on the variability of water consumption by a fixed or slowly increasing customer base. It considers changes driven by weather cycles, changes in the composition or characteristics of the customer base, or economic cycles. Long-term forecasting, the focus of this chapter, consider horizons of 20–30 years. This is the timeframe taken into account when building long-lifespan water supply infrastructures such as desalination plants, storages, or large-capacity inter-basin transfers. In long-term planning, many factors of change are liable to modify both the customer base and per unit water consumption. Uncertainty is a key issue in long-term water demand forecasting.

Water utilities are not the only players for which such forecasts are a concern. Demand forecasts can be conducted at state or national levels to assess whether, and to what extent, water is likely to become a limiting factor for economic development in the future (see the example of Western Australia below). The result of a forecasting study may determine investment decisions such as the construction of a large regional inter-basin transfer. The level of detail does not require the development of a water forecast for each provider in the state. It may also support decisions concerning water allocation between sectors or regions liable to compete for scarce water resources in the future. Such is the case in Nevada, for instance, where the State Engineer considers the future water demand forecast of a region before authorising export of water to other regions. Here, forecasting serves to protect the interests of rural counties, securing water resources needed for their future development (HRBWA 2007).

2.2 A Typology of Water Demand Forecasting Methodologies

A variety of models have been developed and implemented by water utilities and their consultants to forecast the future evolution of water demand (Bauman et al. 1998; Billings and Jones 2008). These models can be classified into five main types (Table 11.1).

2.2.1 Temporal Extrapolation Models

This modeling approach is based on the assumption that the future evolution of demand can be deduced from past tendencies. Several mathematical models can be used, including moving average, exponential smoothing, or Bow–Jenkins models

Table 11.1 Main characteristics of water demand forecasting methodologies

Method	Principle	Applications	Data requirements	Shortcomings
Temporal extrapolation models	Projection of past observed tendencies	Development of a “business as usual scenario” assuming a continuation of prevailing socio-economic conditions	Time series of water consumption	Limited predicted capability – does not account for changes in socioeconomic context
Unit water demand analysis	Estimation based on “unit water demand coefficient” multiplied by the number of users in each category	Development of sectoral demand forecast accounting for expected future population growth, change in economic activity per branch. Demand can easily be represented spatially (link with GIS)	Unit water consumption coefficients (per type of users) Estimated future number of users per category	Does not account for possible future changes in unit water consumption due to evolving water tariffs, household income, etc.
Multivariate statistical models	Estimates per capita consumption as a function of explanatory variables such as water rates, household income, level of economic activity (employment/turnover), housing characteristics), weather conditions, etc.	Allows forecasting future demand considering changes in (i) population and economic activity and (ii) changes in socio- economic variables (water rates, households’ characteristics and income, etc.)	Time series for water consumption and all explanatory variables Estimated future number of users per category	Does not account for changes in plumbing code or campaigns to promote water conservation
Micro-component modeling	Simulation of end-use by domestic customers	Demand forecast considering future changes in household appliances and indoor/outdoor water use practices Ex-ante evaluation of the efficiency of water conservation policies	Widescale households survey to assess customer appliance ownership, frequency of use, and volumes used	Mainly adapted to residential water demand. Often used in combination with a multivariate statistical model
Land use based models	Demand assessed on the scale of uniform spatial entities using unit ratio	Spatially accurate water demand forecast, integrated with urban planning	Long range urban planning scheme Unit consumption ratio per category of urban development	Does not account for changes in economic conditions (prices, income) nor evolution of technologies/ plumbing code

(Billings and Jones 2008; Donkor et al. 2012). The projection of the tendencies may be applied globally at the scale of a single drinking-water utility or of a region, or be refined by reasoning according to types of consumers (domestic users, services sector, industry). Sophisticated geostatistical methods that simultaneously consider time and space variability have also been used to map future water demand (Lee et al. 2010). The advantage of the extrapolation approach is that the only data required are time series of the variable being forecasted. However, its predictive capability is quite limited because it is unable to take into account changes in the socioeconomic context (tariffs, employment, population and urban patterns) and the occurrence of discontinuities (e.g., changes in technology, plumbing codes, or water conservation policies).

2.2.2 Models Based on 'Unit Water Demand'

This method typically consists in tying future needs strictly to the number of users. It relies on the use of 'unit water demand' coefficients determined per inhabitant, per customer, per employee, or per unit of industrial output. Demand is estimated by multiplying these coefficients by the number of users the water utility is liable to serve in the future. Applications of the method can be differentiated according to the level of customer disaggregation. The first level of disaggregation generally consists in a breakdown into domestic, commercial, industrial and public-sector uses (sectoral forecasting). Domestic demand may further be decomposed according to housing type, estimating separately multiple dwellings and single-family homes and houses with or without meters. Likewise, the demand of industrial and commercial users may be broken down according to activity sector (see the California and UK examples later). One can consider the consumption coefficients as variable with time, extrapolating their future direction from past tendencies. This approach is useful where little or no data are available. It may also suffice when a rough estimate is required for preliminary planning purposes. One of its advantages is transparency, and so it is easily understood by stakeholders. For all these reasons, this method is probably the most commonly used.

2.2.3 Multivariate Statistical Models

This method recognises that change in demand stems from many factors, including water rates and tariffs, household income, climate, economic activity, water conservation programs, etc. The method consists in estimating the statistical relationship between per capita consumption (the dependent variable) and a set of explanatory variables. The main explanatory variables are the cost of water, household income, the level of economic activity (employment or turnover), housing characteristics (proportion of single-family versus collective dwellings, urban density), and possibly weather conditions and the like. The model is generally built using panel data, i.e., a sample of municipalities for which data over a 5–10 year interval is available.

Subsequently, the model can be used for prediction purposes to calculate the demand that would be obtained under a hypothetical evolution of the explanatory variables, supposing that the model coefficients (estimated on the basis of a past time window) hold true over the future time window considered. The development of this model type is reflected by an abundant scientific literature (for a review, see Espey et al. 1997; Arbués et al. 2003; Dalhuisen et al. 2003). The main weakness of statistical models for long-range forecasting is their out-of-sample predictive capacity (Fullerton and Molina 2010).

2.2.4 Micro-Component Modeling

This method, also termed 'end-use modeling', assesses total consumption by simulating in detail the different ways that consumers use drinking water (Froukh 2001). Applied mainly to domestic use, the approach estimates the volumes of water associated with each of the main water use devices: showers, bathtubs, lavatories, sanitary facilities, household appliances (washing machines and dishwashers), kitchen taps, and outdoor devices (hoses and sprinklers, swimming pools). In this model, each use is the product of (i) device ownership percentage, (ii) frequency of use, and (iii) volume per use.[1] In turn, these factors are recognised as capable of being affected by economic prosperity, type of housing, occupancy, climate, and technical developments in water-using devices. This method's main advantage is that it enables the long-term effect of technological evolution to be simulated: appliance performance, decreased volume of toilet flush, etc. These models are thus more prospective, allowing the effects of water conservation policy incentives to be estimated. The method is widely used by the UK water Industry (see the Thames Water example below) and in the USA (see, for instance, Levin et al. 2006, but examples also come from other countries such as South Africa (Jacobs and Haarhoff 2004).

2.2.5 Estimation Based on Projections for Urbanisation and Land Use

This method consists in basing the estimated future drinking water demand on urban planning documents and ordinances. The demand forecasting model is integrated into a geographic information system (GIS). Drinking water demand is assessed on a uniform scale of spatial entities (quarters or housing developments for single-family homes, economic activity areas) using unitary consumption ratios for each type of entity. This method can only be implemented if a relatively detailed urban planning scheme is available, one which is regularly updated and takes into account the target timeframe of the projection exercise.

[1] Some sophisticated models also account for device leakage (taps dripping or losses due to poor fittings, pipe connections, etc.).

2.2.6 Composite Models

In practice, many of the models developed or applied by consultants and/or water utilities are hybrid tools combining several of the methods described above. Note that water planning agencies (such as the UK Environment Agency or the California Bay–Delta Authority) recommend adopting such composite approaches (Davis 2003; Environment Agency et al. 2012). This is also the case for water demand forecasting software packages such as IWR-MAIN, which has been intensively used in the USA (Wurbs 1994; Bauman et al. 1998). IWR-MAIN (standing for Institute for Water Resources – Municipal And Industrial Needs) includes a variety of forecasting models, including extrapolation models, statistical models, unit water demand models, and end-use models. This software has been used by more than 40 large American cities and state organisations (such as the California Water Department), and elsewhere around the world (Mohamed and Al-Mualla 2010). A number of other hybrid tools have been developed and tested as part of research projects, such as the demand forecasting and management system described in Froukh (2001), but none, to our knowledge, are routinely called on by the water industry.

3 Key Issues and Challenges

Despite significant progress achieved during the last three decades, a number of challenges still have to be addressed by water demand forecasters. Three of these are discussed in the following paragraphs. The first one lies in the need to better integrate water demand forecasting with urban development planning, recognising that uncertainties with new users' water use are intrinsically different than those of existing users. The second challenge consists in improving our understanding of the potential impact of climate change on future demands. The third entails quantifying uncertainties attached to water demand forecasts and developing new procedures to help water managers take robust decisions based on this information.

3.1 Integrating Land Use Planning and Water Demand Forecasting

Several statistical studies have pointed out that residential water use is strongly influenced by urban development characteristics, housing density in particular. Per capita water consumption tends to be much higher in urban areas where single-family units (with gardens) represent a large share of the total housing stock. Also, outdoor water use tends to increase with lot size, since larger properties have larger irrigated gardens. A study conducted in Barcelona, Spain, showed that per person water use varied from 120 L per day in high-density housing to more than 200 L/d in low-density housing (Domene and Saurí 2006). In the UK, a survey in Yorkshire estimated water use at 370, 280, and 170 L/p/d respectively for detached, semidetached, and terrace houses and flats (Clarke et al. 1997). In California, a study conducted by the Public Policy Institute of California showed that single-family homes use about twice as much landscaping water as multifamily units (Hanak and Davis 2006). This phenomenon is particularly significant under a dry and hot climate, where outdoor uses related to swimming pools and garden irrigation can represent as much as 50–70 % of total water use, as reported in some states in the western USA (Hanak and Davis 2006; Wentz and Gober 2007). Conversely, urban densification may result in decreasing per capita consumption. This tendency is notably reported in Seattle where large single family lots are being converted into new condominiums or smaller town-houses with little or no yard space (Polebitski et al. 2011). A similar trend is also observed in large Asian cities where traditional houses are progressively replaced by condominiums (Bradley 2004).

The accuracy of forecasting studies could therefore be enhanced by taking into account the type of urban development to be expected in the future. Analysts should make explicit assumptions about patterns of future dwellings (single- or multiple-family units, flats), average lot size, and characteristics (type of vegetation, percentage of houses equipped with swimming pools, etc.). This can easily be accommodated by existing multivariate regression models, single-coefficient models, or micro-component models (Jacobs and Haarhoff 2004) if adequate data are introduced. An example is provided by Patterson and Wentz (2008), who assessed future residential water demand for Phoenix, Arizona, using four urban development scenarios. Scenarios differed according to the statistical distribution of lot size and their spatial distribution. The authors showed that a reduction in average lot size can lead to a 7 % reduction in total water consumption compared to a baseline scenario, assuming a continuation of existing land use patterns (Patterson and Wentz 2008). A more sophisticated approach in which a water demand model was coupled to an urban simulation model was presented by Polebitski et al. (2011). Model coupling allowed integration into the water demand forecast of assumptions about the position of urban growth boundaries, changes in transportation networks, and new land use policies. The authors showed that certain changes in building characteristics (denser suburban areas) and spatial features of demographic growth lead to demand reductions of about 4 % (over a 20-year planning period). Urban development modeling can also be fully integrated into the water demand model. An example can be found in Galán et al. (2009), who use an agent-based model to simulate urban development and water consumption over a 10-year period in the city of Valladolid, Spain.

3.2 *Accounting for Climate Change*

While a substantial body of literature deals with how climate change impacts water *supply*, fewer papers have looked at the possible consequences of global warming on long-term urban water *demand*. Climate change is likely to affect both indoor and outdoor water demands. Indoors, rising temperatures will lead to more frequent showering and more recourse to cooling. Outdoors, higher temperatures, evapotranspiration, and declining rainfall will increase irrigation water needs in gardens and evaporation from swimming pools.

Several methodological approaches have been used to assess this impact on residential water demand. The first entails developing a statistical model that includes weather variables in the explanatory variables. The model can subsequently be used in simulation to predict evolution in water demand while factoring in changes in weather variables. A study conducted in the UK (Goodchild 2003) concludes that, by 2020, climate change will increase residential water demand by 2 %. A similar study conducted in Seattle, WA, showed that climate change could result in a 7 % increase in water demand by 2030 and up to 15 % by the end of the century (Polebitski et al. 2011).[2] A caveat of the statistical approach is that it assumes people will react to weather variations in the future as they do now, without considering possible changes in water use practices (e.g., allowing lawns to brown in summer or changing landscaping).

A second methodological approach consists in calculating turf irrigation requirements and swimming pool evaporation using climatic and agronomic models. Relying on process-oriented models, the approach is likely to be more robust than the statistical one. It was implemented in southern France by Desprats et al. (2013). Using high-resolution satellite images, the authors quantified the area of irrigated lawns and the presence of swimming pools in a sample of 45,000 detached houses. An agro-climatic model was then used to assess lawn irrigation requirements and swimming pool evaporation under present and future climate conditions. The results showed that residential water demand of single-family homes would increase by 8–10 %. All other things equal, this represents a 4–5 % increase in total urban demand (Desprats et al. 2013).

A third approach, implemented in the UK, consists in using a micro-component (or end-uses) model (Downing et al. 2003). Some uses are assumed to be insensitive to climate change. Others, such as shower use, garden watering, and swimming pool refilling, are affected, and the model parameters (use frequency, ownership rate, unit use) are modified accordingly. These changes are based on simple assumptions linking frequency of use with climate parameters such as accumulated degree days (cumulative time being that during which the temperature exceeds a given threshold). The UK study concluded that the impact of climate change will remain modest, ranging between 1 % and 1.8 % in 2020 and between 2.7 % and 3.7 % in 2050.

A problem common to the three aforementioned approaches is the uncertainty attaching to climate change scenarios. Different GCMs tend to predict very different changes in temperature, rainfall, and evapotranspiration. The effects of model uncertainties per se are accentuated when different IPCC emission scenarios are adopted. The general conclusion must be that considering climate change in demand forecasting can lead to very different conclusions depending on the chosen climate scenario and the GCM. This was clearly illustrated by Boland (1997) in a case study

[2] Note that this increase in demand would occur solely in summer months, putting additional stress on water resources and aquatic ecosystems during low-flow periods.

focusing on Washington, DC, where the impact of climate change on water demand is estimated to range from −4 % to +11 %. The most common response to this problem consists in adopting an ensemble approach and considering multi-model average climatic scenarios (Goodchild 2003; Polebitski et al. 2011; Desprats et al. 2013).

An analysis of recent urban water management plans selected from various countries shows that climate change effects on water demand is progressively being taken into consideration in long-term planning. In the UK, most water companies have considered climate change when developing water demand forecasts, albeit in a somewhat simplistic manner. Thames Water, for instance, applied a percentage increase over the 'normal year' forecasts in line with the findings of a national study on "Climate Change and the Demand for Water" (Downing et al. 2003). According to Charlton and Arnell (2011), who reviewed the Water Resources Plans of 21 companies operating in England, most water companies assume an increase in demand ranging between 2 % and 5 % (as of 2030). They note that this is minor compared to other drivers of change in demand and the effect of climate change on supply.

In the previous paragraphs, we focused on direct impacts of climate change on water demand. But climate change may also have significant indirect impacts by affecting water demand through structural economic effects. This is nicely illustrated by the water demand forecast study conducted by the State of Western Australia, described in Sect. 6. Using a macro-economic model, this study assessed the possible impact of global warming on the level of economic activity in the main economic sectors. The indirect impact on water demand was then estimated using a simple 'use coefficient' approach (Thomas 2008). Results showed that this indirect impact far exceeds the direct impact on water use.

3.3 *Dealing with Forecast Uncertainty*

Many water consumption prediction models have been developed and used in a deterministic context despite the presence of uncertainty in assumed model structures and parameters. It is now widely recognised that water use forecasts, regardless of the timeframe or the forecast method employed, are likely to be always highly inaccurate (Osborn et al. 1986; Fullerton and Molina 2010). It is thus crucially important to give consideration to model uncertainties. Two alternative approaches can be called on to do so: the use of contrasted scenarios and the probability approach based on Monte Carlo simulations.

3.3.1 The Scenario Approach

Using a limited number of contrasted scenarios is one way to account for the uncertainty attached to future evolution of water demand. Scenarios consist of a narrative description of ways society might develop and use water in the future. Scenarios are expected to help water utilities assess the performance of alternative strategies under different plausible future conditions. In the UK, this approach was initiated by the Environment Agency in 2001 (Westcott 2004) and further developed since (Environment Agency 2009). The approach was based on more comprehensive scenarios developed for the Environment Agency and Defra to explore pressures on the UK environment and possible changes in them by 2030. Scenarios depict four plausible futures that differ in two main dimensions: first, the type of society (conservationist through to consumerist) and, secondly, the type of governance (growth-focused through to sustainability-focused). Water experts were asked to describe how key factors influencing water demand would be affected by these global scenarios. They defined consistent quantitative assumptions related to water demand drivers which were subsequently used to assess future demand for all resource zones in England and Wales (Environment Agency 2009). Results obtained are illustrated in Fig. 11.1. The result is an envelope within which future water demand is likely to sit. Water companies are then expected to consider these scenarios to identify strategies that perform well under these different plausible representations of the future and to understand the risks inherent in the different alternatives. This approach, however,

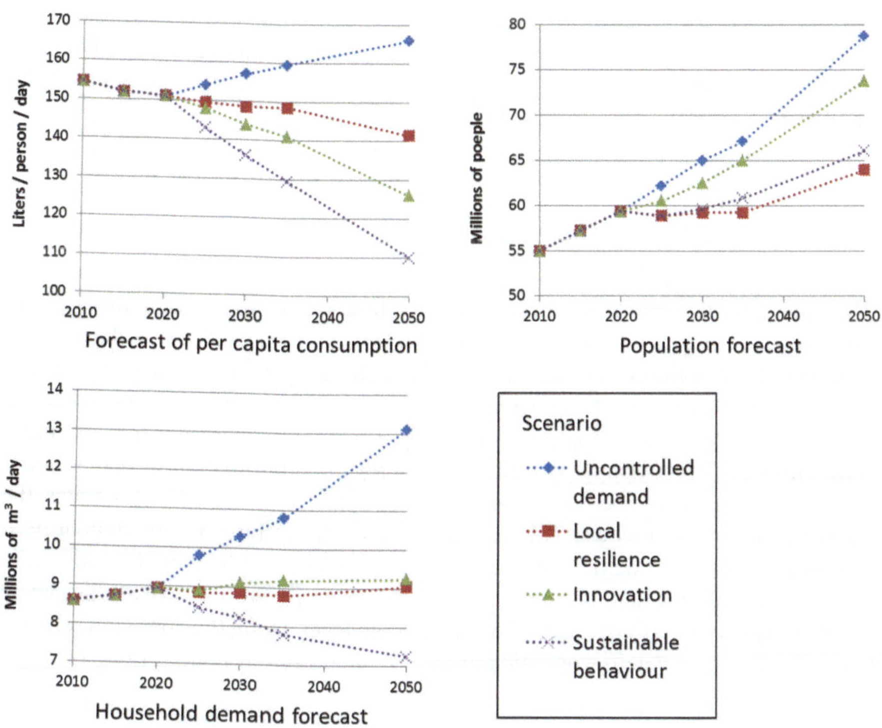

Fig. 11.1 Population, per capita consumption, and residential water demand forecasts according to four scenarios (Source: Adapted from Environment Agency (2009): pp. 21–24.)

has been criticised on the grounds of the selection bias for assumption that all forecasters suffer from, based on their own knowledge, beliefs, and ideology.

3.3.2 The Probabilistic Approach

A number of authors criticise forecasts that use only a small number of scenarios, suggesting that uncertainty should really be assessed by considering hundreds of alternative representations of the future. They argue that in situations of great uncertainty, decision-makers need to seek robust rather than optimal (i.e., lowest cost) strategies. Robustness is defined as the ability of a strategy to perform well in a large number of plausible futures (Lampert et al. 2003).

The probabilistic approach consists in running models repeatedly using uncertain input variables randomly chosen from a defined probability distribution. It comprises four steps: (i) establish the probability distribution of key factors determining future demand; (ii) sample their values based on randomised techniques; (iii) calculate water demand for a large number of samples; and (iv) compute a statistical distribution of future water demand. The main sources of uncertainty considered are population forecast, economic forecast (employment), water use parameters (end use or total use coefficients) and climate. This approach is now widely implemented by the water industry, using software packages such as @RISK (Palisade Corporation).[3] Applications of this method to the cases of Thames Water (UK) and Tampa Bay Water District (USA) can be found in Thames Water (2010) and Hazen and Sawyer (2004).

3.4 *Presentation of Case Studies*

Three case studies are presented to illustrate how the various methodologies are implemented in practice and how the challenges described above have been addressed by practitioners (Table 11.2). We deliberately focus on advanced situations. The first example is taken from Southern California, where demand forecasting is conducted at two geographic levels by a regional water importer and retail water utilities. This example describes an interesting combination of econometric and end-use models. The second example, chosen from the UK, illustrates the potential of end-use models. This case study also shows how a very standardised forecasting procedure defined at a national level can be implemented by all water companies. The third and last example was drawn from Western Australia, where water demand forecasting is carried out at the state level. Here, the approach selected strongly relies on a macroeconomic model.

[3] http://www.palisade.com/

Table 11.2 Main characteristics of case studies

Location	Forecasting scale	Forecasting method
Southern California, USA	Regional level (*Metropolitan water district of Southern California*)	Statistical multivariate model to forecast per capita consumption (PCC). Combined with end-use model to assess the effects of water conservation measures and impact of changing plumbing code (regional level)
	Public utility level (*Eastern municipal water district, riverside county, 700,000 inhabitants*)	Demand forecast mainly based on a spatialized database (GIS) referencing all potential future residential/commercial constructions/ developments
Thames Water, UK	Water resource zone (sub-regional scale, 13.5 million inhabitants)	Residential demand forecast using a micro-component modeling used to forecast future PCC values, considering changes in plumbing code, increase percentage of metered houses, and water efficiency measures (including water conservation oriented rates).
		Non-residential demand forecast using a statistical multivariate model
Western Australia	State level	Multiple water use coefficient model (60 types of users, 19 regions) combined with population and economic activity forecast models. Allows forecasting the impact of macro-economic changes on total water demand

4 Multi-level Forecasting in Southern California

4.1 The Context in Southern California

In California, the Urban Water Management Act mandates that each drinking water service shall prepare an urban water management plan, the purpose of which is to ensure a long-term balance between water demand and the resources available and to provide for emergency management measures in the event of exceptional drought conditions. These plans, covering a 30-year time span, must be updated every 5 years and submitted to the California Department of Water Resources. They must contain one demand forecast per category of user, including a description of water conservation measures planned by the utility.

In the Los Angeles metropolitan area, this water demand projection is calculated at several geographical levels. At the metropolitan (regional) level, a global demand forecast is established by the water wholesaler, the Metropolitan Water District of Southern California (MWD). MWD imports water from distant sources (Colorado, Northern California, Owens Valley) to supply 26 inter-municipal services (retailers) in the Los Angeles region. MWD's projections are based on a statistical model that seeks to assess the overall needs of the 18 million customers supplied, as well as needs for imported water after the resources available locally have been used up. At a more local scale, each of the 26 retailers does its own projections using methods based on a survey of future development projects. These forecasting tools are described below in greater detail.

4.2 Demand Forecast by the Regional Importer

To project the long-term evolution of drinking water demand, MWD calls on a sophisticated forecasting model developed from the IWR-MAIN program, which allows projected population and economic growth to be translated into drinking water demand, while simultaneously integrating the effect of programs to promote water conservation measures. The tool relies on a combination of two types of models (MWD 2010): (i) an econometric (statistical) model that simulates the evolution of consumption ratios; and (ii) a model of end uses that simulates the effect of conservation programs.

The statistical model decomposes the forecast according to types of use (household, commercial, industrial, and public), geographical sector (over 50 sectors at MWD), and season. As to household needs, these are assessed separately in terms of housing type (single-family home, large and small collective dwellings, trailers, rural properties). The model allows the evolution over time of the unitary ratios (m^3 per capita, m^3 per job) to be simulated in line with the hypotheses regarding the evolution of household size and revenue, the service rates (price level and structure), the characteristics of new housing (single-family or collective, density), and climate (precipitation). The coefficients of this statistical model were determined by statistical processing (meta-analysis) of the results from 60 case studies carried out in the US. Industrial and commercial needs are assessed by decomposing the demand corresponding to the main branches of economic activities, for which a unitary consumption coefficient (m^3 per job) is used. This complex statistical model is first used to construct a baseline scenario of total demand considering demographic and economic hypotheses.

A micro-component model is then used to estimate the decrease in demand that can be obtained via water conservation programs. The estimated conserved water is then subtracted from the baseline scenario. The micro-component model decomposes demand into elementary uses such as toilet flushes, washing machines, lavatories and showers, watering gardens, car and floor wash, etc. Hypotheses are made concerning household equipment, how such equipment is used, and leaks (faucets, toilets, garden watering systems). These hypotheses can be adjusted to simulate the effects, at regional scale, of the proactive water-saving policies engaged by MWD, consistent with the policy defined at state level (California Water Conservation Council). An example of these is the distribution of water efficiency kits (shower heads, aerators, low-volume toilet flushers), rebates for the replacement of low-efficiency equipment (US$100 for washing machines, for instance), conducting consumption audits on private or commercial users with a view to reducing outdoor use, etc. The model also allows the trend effect of the evolution of factory standards (plumbing code) for the equipment being simulated (for example, prohibiting the sale of toilet flushers with a capacity exceeding 6 l).

4.3 Demand Projection at the Scale of Water Services

At a more local level, water retail utilities (municipal water districts) have developed forecasting methods that are more detailed in terms of both their spatial and temporal resolution. The main objective is to make a global assessment of resource needs so as to plan investments in new resources (a plant for desalinisation of brackish groundwater, for example) that can be substituted for water purchased from MWD. Forecasting also aims to determine the spatial distribution of future needs in order to plan for reinforcement or development of the distribution network and storage infrastructure. As the projections are developed in largely the same way by a majority of services, we are presenting here the approach implemented by one of these services, the Eastern Municipal Water District (EMWD).

Located in Riverside County, some 120 km east of Los Angeles, this service supplies water and sewage treatment to approximately 700,000 inhabitants. Because of the saturation of urban zones nearer Los Angeles, the area has been experiencing vigorous population growth for a number of decades. Growth often exceeded 10 % per year between 1980 and 1990, before leveling off at 3 % per year between 1990 and 2010. It is expected to remain at this level up to 2025. The development of low-density single-family housing generates a strong demand for water due to outdoor uses (gardens and swimming pools). The extremely rapid growth in demand resulting from this necessitates a strong anticipatory ability; otherwise investments made could become poorly suited even before they are amortised. In the worst case, a failure to meet supply needs would entail considerable cost to the local economy.

To assess future needs, EMWD relies on existing prospective studies, which it supplements with its own analyses. To predict the long-term demographic trend at the scale of its territory, EMWD adopts the population growth predictions prepared by the Southern California Association of Governments in the framework of the Transportation Plan. An additional study is then assigned to a specialised consultant, who determines the characteristics and spatial distribution of housing liable to be built to accommodate incoming population to 2030. Changes in the urban environment (types of housing erected) is considered an essential factor in determining future water demand. The study consists of an in-depth analysis of the dynamics of the real estate market, which integrates macro-economic factors (employment, revenue, credit) as well as more local ones (distance to centers of employment and the attractiveness of the territory, including crime levels, the quality of schools, real estate prices, compared to competing territories).

To complete these projections, EMWD is developing a spatial database (GIS) that allows the potential for building new housing to be estimated from urban planning documents. This base, the Database of Proposed Projects, also makes it possible to identify and follow up on the advancement of all the residential or economic development projects in its territory, from the design phase through to when the meters are installed. In 2005, this database described 150,000 housing units, both single-family and collective, as well as some 10,000 acres of commercial, industrial, or public land (parks, establishments serving the public). This approach makes

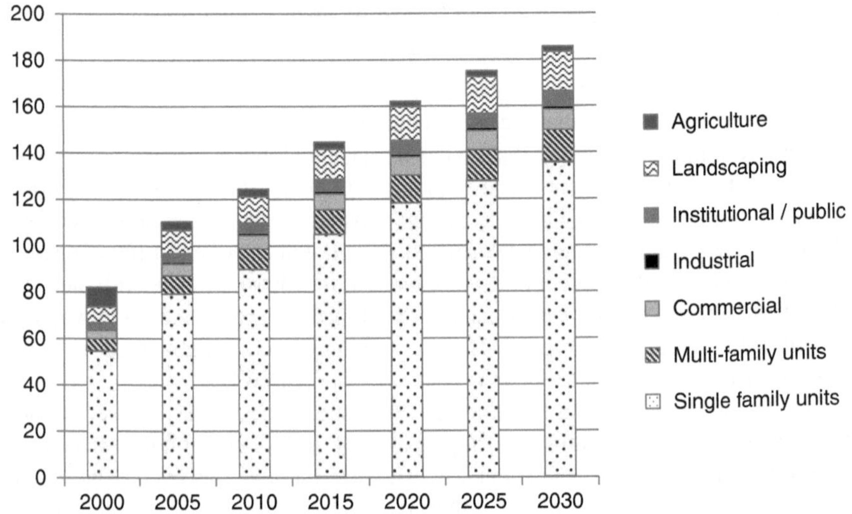

Fig. 11.2 Projection of the evolution in water demand according to user category (millions of m^3 per year) (Source: Adapted from the Eastern Municipal Water District Urban Water Management Plan, 2005.)

it possible to anticipate in detail the future demand for water some 2–5 years hence. It supplements the other approaches with a more distant timeframe. The approach is updated every 5 years in conjunction with planning conducted at a larger geographic scale by the regional importer, the Metropolitan Water District.

EMWD then takes the population and urban development projections and uses them to calculate future water demand, using per capita consumption coefficients that are adjusted according to the type of housing development (density, lot area, price, average household revenue, etc.), bearing in mind a steady rise in income of its population and a decrease in the number of individuals per household. These ratios are estimated on the basis of a detailed analysis of billing databases. Ultimately, demand is estimated for seven categories of users: domestic users in single-family, collective housing, commercial customers, industrial customers, public users, institutional users, or green areas and farmland (see Fig. 11.2). EMWD also estimates the water conservation that may be achieved in the future, either passively (improved performance of the water appliances sold) or actively (via programs of specific measures intended to modify practices and behaviors – in particular, the establishment of tariff incentives by increasing tiers). EMWD also regulates new developments requiring water-efficient landscaping. Their Water Use Efficiency Regulations[4] are similar to other areas that design efficiency into new developments. Future water demand will, by design, differ from the historical one. EMWD's methodology allows this refinement to be included in the forecasts by having the demand from future customers adjusted by municipal or county codes.

[4] See EMWD Water Use Efficiency Regulation at: http://www.emwd.org/index.aspx?page=91

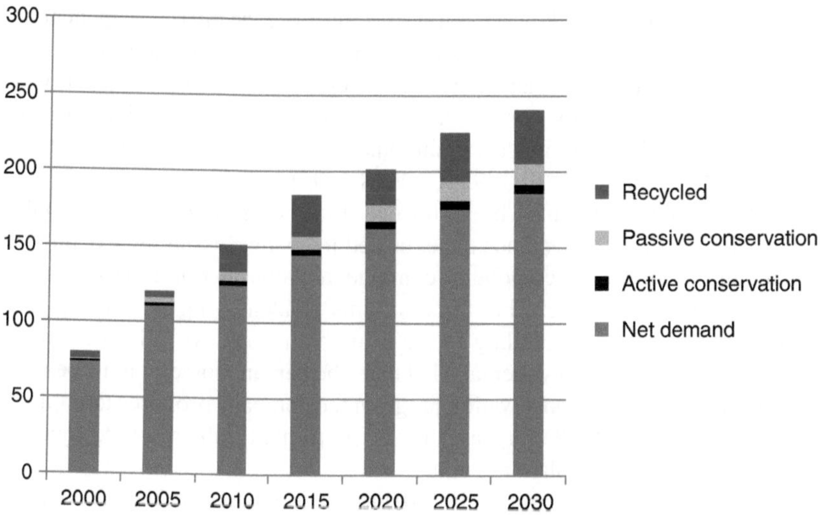

Fig. 11.3 Estimate of expected water savings by the year 2030 (millions of m^3 per year) (Source: Adapted from the Eastern Municipal Water District Urban Water Management Plan, 2005.)

EMWD also issues a projection of sales of recycled waste water that it has been developing for specific uses since 2000 as a substitute for drinking water (Fig. 11.3). Globally, the projection method developed by EMWD takes a very large number of factors explicitly into account: demographic and economic growth; the evolution of housing types; the effect of rate changes; the establishment of programs encouraging water conservation; a downward trend in consumption resulting from the evolution in performance of the materials sold; and the development of substitution resources.

5 The United Kingdom: The Micro-Component Approach

5.1 Regulatory Frameworks and National Guidelines for Demand Forecasting

In England and Wales, long-run water demand forecasting is mandatory for all water service providers. According to the Water Industry Act of 2003, companies are under the obligation to develop a water resources management plan (WRMP) showing how they propose to balance supply and demand for a 25-year period. Plans must be revised every 5 years.

The regulators (Ofwat and the Environment Agency) publish and update detailed guidelines specifying the methodology for preparing the plan (Environment Agency et al. 2012). Water companies are required to forecast three main components of demand: household use (metered or not), industrial and commercial use, and leakage.

As to household demand, water companies must use the micro-component modeling approach. This approach makes it possible to estimate how advances in technology, changes in society, and the role of regulation will influence growth or decline in water use over the coming 25 years. This methodology derives from work by Herrington (1995), and has been implemented by the water industry since the mid-1990s (NRA and UKWIR 1995; UKWIR 1997).

Two types of forecast must be performed. First, a baseline forecast should be established to show how demands are expected to change, assuming existing management and water efficiency policies continue, and considering trends in technology and behavioral change. This forecast aims at identifying potential planning problems (e.g., demand exceeding future available resources). Where a company predicts a deficit in its supply–demand balance, the demand forecast must be revised by incorporating a water conservation program the company proposes to implement over the 25-year period. The forecast shall be performed at the resource zone level, which is the fundamental planning unit.[5]

These WRMPs were submitted in 2010 to the regulators by the 23 companies providing water services to domestic, commercial, industrial, and agricultural consumers in England and Wales. The demand forecasting methodology developed by one of these companies, Thames Water, is presented below by way of illustration, based on an analysis of the latest version of their plan (Thames Water 2010).

5.2 *Thames Water Forecasting Methodology: Overview*

Thames Water is the UK's largest water and wastewater services company. It serves 13.5 million customers in London and the Thames Valley, supplying an average of 2.6 million cubic meters of drinking water per day. Household consumption accounts for approximately 50 % of total water use, non-household consumption 20 %, and unbilled and operational use 2 %, while leakage is estimated at 28 % of total water use.

The forecasting methodology developed and implemented by Thames Water is based on a modular modeling platform as illustrated in Fig. 11.4. The first module assesses future population and number of households in the water resource zone; the second module estimates present and future residential per capita consumption (baseline scenario); the third module assesses the impact on per capita consumption (PCC) and total demand of policy options (metering program, innovative tariffs); a separate model is developed to forecast non-domestic demand. The modules are integrated at the water resource zone level and run throughout the 25 years of the planning timeframe.

[5] A resource zone is defined as one where water taken from anywhere within the zone can be supplied to any other location in the zone. There are 68 resource zones in England and Wales. Resource zones are therefore relatively large units compared to what is found in other European countries where water services are still frequently operated at a local (municipal) level (e.g. France).

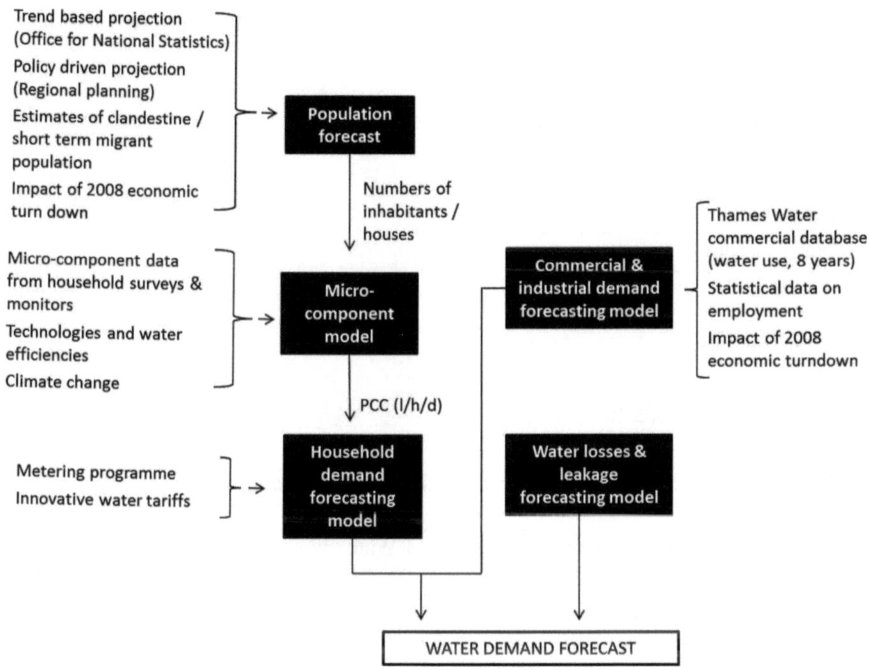

Fig. 11.4 Overview of the demand forecasting methodology implemented by Thames Water (Source: own elaboration)

5.2.1 Population Forecast

The approach developed by Thames Water to forecast future population is of interest insofar as it combines several methods. The first method reconciles the sub-national trend based population projections developed by the Office for National Statistics (ONS) with policy-driven housing projection defined in the regional plans (regional spatial strategies).[6] This work, entrusted to a consulting company, led to the definition of a "most likely" population forecast, a forecast subsequently adjusted by including estimated clandestine and short-term migrant populations (estimated to exceed 280,000 persons). Finally, Thames Water commissioned a study to assess the impact of the 2008 economic crisis on employment and population growth. It was assumed that the effects of the crisis would be temporary and that by 2021 recovery would be full, achieving the levels of population and household numbers projected in the most likely scenario. Overall, Thames Water predicts a rise from 8.5 million inhabitants in 2007/08 to 10.2 million by 2034/35, this increase taking place essentially in the London Water Resource Zone (1.3 million).

[6] Regional spatial strategies (RSSs) set out how many homes are needed to meet the future needs of the population in the region. They are policy driven and include planned development initiatives by local authorities.

Table 11.3 Current water use per micro-component, for measured and non measured households all water resource zones of Thames Water

Use per micro-component (in % of total household consumption)	Measured households	Unmeasured households
Toilet flushing	22	23
Bath use	12	12
Shower	19	10
Clothes washing	8	7
Dish washing	6	5
Garden use	10	8
Miscellaneous use	21	20
Wastage	2	9

Source: Thames Water (2010)

5.2.2 Per Capita Consumption Modeling and Forecasting

In line with industry best practice, Thames Water has assessed present and future household PCC at the micro-component level, examining the ownership, frequency of use, and volume per use of a range of water-using appliances. Water use per person is influenced by several factors, the main ones being: household occupancy; water consumption of appliances, fixtures and fittings in the property; householders' water use behavior; garden use; and whether the property is metered. The PCC model is calibrated using household data collected through wide-scale surveys carried out every few years (2003, 2007).[7] Additional information procured by monitoring water use in 78 households is also used, as well as results of studies conducted in other regions of England and Wales. Current water use is depicted in Table 11.3 below.

This model is then run to forecast values of PCC increase over a 30-year period, assuming changes in future usage based on research and survey data. The main assumptions of the baseline scenario are: replacement of older-model toilets with more efficient ones every 15 years; increased equipment with power showers; reduction of wastage which is higher for unmetered households than metered ones (9 liters/person/day against 2.6 L/p/d). Thames Water also considered that domestic consumption would decrease to 125 L/p/d by 2015 in all new properties, following the introduction of new building regulations for fixtures and fittings. The impact of decreasing household size is also considered.

Over the planning period out to 2035, Thames Water assumes an increase in water consumption for showering (+24 %) due to more people using showers as opposed to baths (−8 %), as well as the increasing popularity of power showers. Other micro-components are decreasing, such as dish washing (−1 %), clothes washing (−3 %), and toilets (−10 %) due to improved technology integrated by the manufacturers of these devices. Toilet flush volumes also decrease over the period

[7] In 2007, some 60,000 questionnaires were sent out, with 9650 returns.

as water fittings regulations and the increased availability of lower flush volume and dual flush toilets take effect. Overall, short-term reductions resulting from natural replacement of inefficient appliances with newer, more efficient ones are expected to be counterbalanced in the intermediate to long term by an increased ownership of power showers. Climate change is also expected to increase outdoor uses by 4–6 % by 2035 (garden watering). PCC is forecast for the 25 years of the planning period; in 2035, it is expected to exceed today's value (157 L/p/d) by 6 L/p/d.

5.2.3 Final Household Demand Forecast

The PCC model outputs are imported into the final household demand model, where additional assumptions can be formulated as to water efficiency policies the company intends to introduce during the 25-year period. The installation of meters is expected to result in a 10 % decrease in PCC, and even up to 20 % if automated meter reading (AMR) equipment is installed, as these can help detect leaks and wastage in individual properties (using devices likes *LeakFrog*). Another option considered is the introduction of sophisticated tariffs such as increasing block tariffs (IBT) or seasonal/peak tariffs, which are assumed to produce an additional 5 % reduction in demand. Thames Water expects that such tariffs will only be able to be implemented after 2017, when the level of meter penetration has exceeded 50 %.

5.2.4 Non-household Demand Forecast

Non-household demand forecast is based on a simple econometric model that establishes a linear statistical relationship between water use and employment. Data used to estimate this model are those in the Thames Water commercial database (measured commercial water deliveries) and statistical information on employment for an 8-year period. Data for industries were pooled into two main groups: service and non-service. For both groups, the elasticity of demand for water with respect to employment is approximately +0.8. In other words, this means that a 10 % increase in employment will lead to an 8 % increase in water demand for water, all else remaining equal. A separate equation is then estimated for 35 main categories of economic activity (classification based on standard industrial classification codes), assuming a +0.8 employment elasticity and estimating a specific intercept value for each activity.

The econometric model was then used to simulate the impact on water demand of economic and employment scenarios, which were prepared by a consultant. Overall, Thames Water expects the downward trend in demand from non-service industries to continue (mainly food, drink, and tobacco industries), falling by almost 25 % between 2007/08 and 2025/26. This is compensated by an increase in service industries, resulting in a slight increase of industrial water demand.

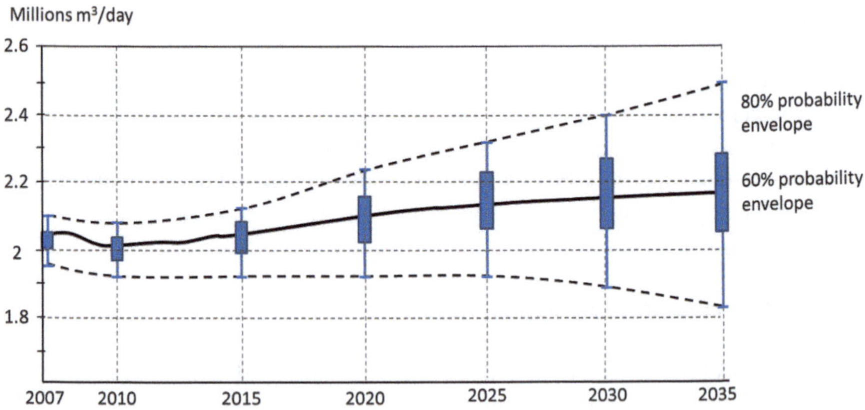

Fig. 11.5 Example of uncertainty profile of water demand for London (Adapted from Thames Water (2010), p. 96.)

5.2.5 Uncertainty Assessment

The uncertainty of the model presented above was assessed by Thames Water consultants using a Monte Carlo approach as described in Sect. 3. Figure 11.5 depicts the type of output obtained with this probabilistic approach. It provides an envelope within which future water demand is likely to fit.

6 State-Level Forecasting in Western Australia

Western Australia is the third example we have selected to illustrate current practices in urban water demand forecasting. Here, the forecast is based on a much more global model that estimates water demand for 60 different economic sectors, regardless of where the water they use comes from (self-supplied users or customers of public schemes). This global model was developed by the Department of Water and its results used by water companies such as the Water Corporation of Western Australia, which supplies drinking water to approximately 1.7 million people throughout the state of Western Australia (1.4 million in Perth).

6.1 *Overview of the Forecasting Approach*

Water demand forecasting is based on the development of a customised modeling tool which was applied to the entire State (Thomas 2008). This model relies on a multiple coefficient approach consisting of: (i) estimating water use coefficients for

60 types of users and (ii) assessing future evolution of these uses, in terms of population (residential demand) or economic model (added value and employment).

Concerning commercial and industrial uses, water use coefficients (liter/job/day or liter/$ of added value/day) are estimated for a base year using a licensing database and water companies' customer databases. These are assumed to remain constant throughout the planning period. The projection of added value and employment is based on a large-scale regionalised economic model[8] which produces three main outputs: the industry added value (sum of wages, salaries, and profits) measured in Australian dollars; industry employment; and total population. All model parameters and predictions are differentiated according to 19 geographical regions.

As to domestic water use, coefficients are adjusted to account for a declining trend in per capita water use. It is assumed that households will manage to reduce water use from a current 290 liter/person/day level (average from 2002 to 2008) to 275 L/p/d.[9] This represents a simple attempt to anticipate the effects of demand management intervention. However, the model does not allow the specific effect of each alternative water conservation measure to be simulated, as shown in the Thames Water example above.

Once total demand has been estimated for each of the 60 sectors and 19 regions, the model calculates the probable water demand for public schemes in each of the 19 regions. This is based on the assumption that the proportion of total water demand that is met through a scheme supply will remain the same as in 2008 (for each of the 60 sectors). The model does not therefore anticipate possible changes involving substitution of scheme water for self-extracted water or vice versa.

Compared to other modeling approaches presented earlier, the strength of this model resides in two aspects. The first is its ability to link future water demand to anticipated impacts of macro-economic trends on the demand for water.[10] This is a key advantage for forecasting global water demand at the regional level, which is the scale at which the Water Corporation of Western Australia operates. It would, however, be of little relevance for use at the level of small to medium water utilities.

The second strength of the model lies in its ability to assess the proportion of demand met by private sources. This information is often absent in water demand forecasting studies, possibly resulting in an overestimation of demand for scheme water. This is notably the case for residential water demand when the proportion of households equipped with private wells is significant. This applies to Western Australia (see chapter 7 in this book), where water drawn by households from private wells amounts to half that supplied by public schemes, but also to some

[8] A dynamic general equilibrium model was used in this case. Water demand forecasting reports, however, do not provide details concerning it.

[9] These demand assumptions reflect gains made in water use efficiency over the past 10 years: water use was about 500 L/p/d in 2000/01.

[10] Note however that the model cannot predict the development of totally new activities.

European regions where the number of private wells is rapidly increasing (Montginoul and Rinaudo 2011).

6.2 *Accounting for Uncertainty with Scenarios*

Projected growth rates for added value, employment, and populations are estimated for several contrasting scenarios to account for uncertainty over future economic development. Four economic and population growth scenarios for future water use have been developed. Note that these scenarios are quite optimistic, given that they were designed prior to the 2008 economic crisis.

The medium growth scenario assumes that the current rate of development of the resource-based industry continues until around 2014, after which the growth rate declines to historical levels; water use per unit of output is assumed to remain constant. The high-growth scenario assumes that the resource boom continues longer, resulting in a high economic growth rate sustained through to 2030; as in the first scenario, water use per unit of output is assumed to remain constant. The low-growth scenario considers a decline in growth rate with stabilisation close to historical levels in 2020; water use per unit of output also remains constant. Finally, a climate change scenario assumes changes in the production pattern at the regional level (decline of agriculture and related industries, increase in forestry, mounting investment in defensive expenditures, particularly in sectors affected by sea-level rise; water use per unit of output increases due to declining rainfall). The range of results obtained is depicted in Fig. 11.6.

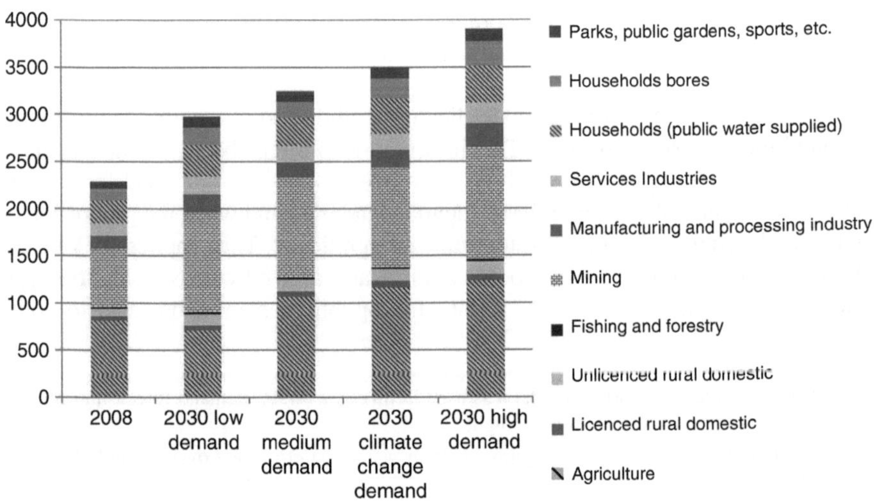

Fig. 11.6 Total water demand forecast in Western Australia under different macroeconomic and climate scenarios (Source: Adapted from Thomas (2008))

7 Discussion and Conclusion

7.1 An Outlook for the Water Industry

The scientific interest for water demand forecasting methodologies was spurred by water supply problems encountered by states in south-western USA during the 1970s and 1980s. Its development has benefited from contributions by economists, engineers, and system modelers, producing a wide range of tools, many of which have been tested and adopted by practitioners. This chapter has illustrated, via three case studies, how forecasting tools can be implemented in practice. The three examples show that there is no single 'off the shelf' forecasting tool that can be applied universally. Different modeling tools can be used depending on the regulatory context, the water scarcity level, the geographic scale at which they are deployed, and the technical background of water utilities and their consultants. For instance, the San Francisco Public Utility Commission used an end-use model, while the Metropolitan Water District of southern California used a customised application of IWR-MAIN, despite the fact that they are in a relatively comparable situation as public water wholesalers. Similarly, the same attention has not been devoted to the impact of climate change in demand forecasts prepared in California, the UK, and Western Australia, this issue being approached using very different methodological perspectives. Uncertainty is also dealt with via very different approaches in our three examples: while not formally assessed in regional water demand forecasts in California, it is analysed through Monte Carlo simulations by UK water companies and through a discrete number of contrasted scenarios in Australia. And while land use development planning and water demand forecasting are well integrated in the Eastern Municipal Water District, this issue is given little consideration in the UK and Australia.

Considering climate change and prospects for increased water scarcity in some parts of the world, one might expect that increasing numbers of utilities will have to invest in forecasting activities in the near future. In other areas characterised by declining per capita water use, improved forecasting capacity is needed to avoid costly investment decision errors (Beecher and Chesnutt 2012). It is likely that composite tools combining econometric models with micro-component models will become a reference in the water industry. Indeed, their strength lies in their ability to simultaneously account for economic changes and water conservation policies, and to allow for an assessment of uncertainty related to population forecast and future climate.

Such tools will not be developed in a uniform manner. Water utilities will only invest in forecasting modeling if the related benefits are clearly perceived. This will mainly happen in areas where the customer base is evolving rapidly, where investment options considered involve large sums of money, and where decisions cannot easily be revised in time. Also, the development will be facilitated where clear methodological guidelines are produced by regulators, as in the UK for instance.

This will pave the way for the development of standardised software packages and the emergence of broad-based expertise within consulting companies.

The development of forecasting practices will be facilitated in large utilities, or where public agencies can carry out forecasting studies on their behalf (see the case of Nevada, for instance). Indeed, developing, calibrating, and regularly updating water demand forecasting models requires significant financial and technical capacities that many utilities around the world are not able to mobilise. For instance, the forecasting study conducted by the San Francisco Public Utility Commission (supplying a population of 2.4 million) mobilised over 100 persons working on the project for 2 years (Levin et al. 2006).

Similarly, certain types of models require detailed data that are difficult or costly to acquire by any single utility. For instance, end-use models require conducting large-scale household surveys (or monitors) to estimate appliance ownership rates, use frequency, and average use per load. Such models are likely to be more easily adopted if governments or the water industry produce generic information that can be used by water utilities. Institutions like the Research Foundation of the American Water Works Association and the UK Water Industry Research have played a key role in that respect, which can explain the widespread use of complex models in these two countries.

7.2 Research Perspectives

Most of the models described in this chapter also suffer from the same serious weakness, namely that their structures and parameters rely on statistical evidence derived from past observations. In the context of rapid technological, climatic, economic, and societal changes, some authors argue that the predictive power of these methods may be rather limited (Galán et al. 2009). They insist on the need to develop models that capture the underlying causal relations determining the evolution of water demand under changing structural conditions. This implies explicitly modeling how users (households in particular) take decisions concerning water use practices, the purchase of appliances, and investment in alternative water supply sources such as rainwater harvesting systems, in-house grey-water recycling, or the drilling of private wells.

Research is presently on-going in that field, using different approaches. Rosenberg et al. (2007) have modeled household water use by explicitly describing multiple sources that can be used at different prices and with different water qualities suited to specific uses. Their model then looks for the least cost combination of actions that allows a household's water needs to be satisfied in a stochastic environment. They show how this can help quantify demands for indoor and outdoor uses and how customers may respond to water conservation incentives embedded in a tariff structure. Micro-economic models have also been used to simulate a household's water supply decision, assuming that users are trying to minimise the cost of water supply through an optimal combination of water sources. This approach was implemented in France, for instance, to simulate how households decide to drill wells and use cheap untreated groundwater as a substitute for tap water (Montginoul and Rinaudo 2011).

Social modelers adopt a much broader perspective by explicitly taking into account social phenomena that affect water demand, and the interaction between them. As compared to microeconomic approaches, they attempt to describe intra-population dynamics. This is mostly done through the development of agent-based models (ABMs) that aim to simulate the functioning of a society based on a detailed representation of individual agents' decisions and the interactions between them. Water demand ABMs can simulate households' decisions in terms of technology change (diffusion of innovation), compliance with regulations (irrigation bans), and volume use per load or per activity (showers). One of the main features of ABMs is their assumption that households are community-oriented agents, meaning that their decisions and actions are strongly influenced by the community around them, their neighborhood, and their social environment (Athanasiadis et al. 2005). For instance, households may agree to voluntarily reduce water use during droughts, but they may quickly shift back to their initial practices if they realise that most of the community is making no effort. Another key feature consists in assuming heterogeneity of agent characteristics and behavioral motivations, resulting in a greater diversity of responses to regulatory and economic policy signals and incentives. For instance, Athanasiadis et al. (2005) consider three main types of consumers: (i) households sensitive to water conservation objectives, who are directly impacted by information campaigns and actively promote the diffusion of innovative practices in their social networks (opinion leaders); (ii) households indifferent to public awareness campaigns and insensitive to social issues, who will have a negative attitude towards conservation; and (iii) households who act as opinion followers and who will engage in conservation as a result of their interactions with opinion leaders. ABMs can be combined with other models. In the DAWN model developed by Athanasiadis et al. (2005), for instance, an econometric model is nested within an ABM; the econometric model is used by agents to calculate their baseline water consumption before agents take decisions related to water conservation activities based on social interactions with their neighbors. This modeling approach has only been used in a limited number of research studies in the UK (Bartélémy 2008), Greece (Athanasiadis et al. 2005), and Spain (López-Paredes et al. 2005; Galán et al. 2009). To our knowledge, it has not been used as a planning tool by water utilities, but rather as tools to "aid in advancing our knowledge about the complex dynamic of the whole water management system" (Galán et al. 2009). Integrating such models with existing tools and promoting operational applications and deployment by water utilities represents a real challenge which, we believe, should be taken up by researchers.

Acknowledgments This chapter was prepared as part of the Eau&3E project funded by the French Research Agency (ANR), grant VD-08-321989. I warmly thank Antony Pack, Chairman of EMWD, and Tim Brick, chairman of Metropolitan Water District, as well as their staff for sharing their invaluable experience during a visit to California in September 2010. I also express my appreciation for comments made by Bernard Barraqué and Tom Chesnutt on an earlier version of this paper. The usual disclaimers apply.

References

Arbués, F., Garcia-Valinas, M. A., & Martinez-Espiñeira, R. (2003). Estimation of residential water demand: a state-of-the-art review. *Journal of Socio-Economics, 32*, 81–102.

Athanasiadis, I. N., Mentes, A. K., Mitkas, P. A., & Mylopoulos, Y. A. (2005). A hybrid agent-based model for estimating residential water demand. *Simulation, 81*, 175–187.

Bartélémy, O. (2008). Scenarios for water demand forecast. In A. Lopez Paredes & C. Hernandez Iglesias (Eds.), *Agent based modelling in natural resource management*. Valladolid, Spain: INSISOC.

Bauman, D. D., Boland, J. J., & Haneman, W. M. (1998). *Urban water demand management and planning*. New York: McGraw-Hill.

Beecher, J. A., & Chesnutt, T. W. (2012). *Declining sales and utility revenues: A framework for understanding and adapting*. Alliance for Water Efficiency, white paper for National Water Rates Summit, Racine, Wisconsin.

Billings, R. B., & Jones, C. V. (2008). *Forecasting urban water demand*. Denver: American Water Works Association.

Boland, J. (1997). Assessing urban water use and the role of water conservation measures under climate uncertainty. *Climatic Change, 37*, 157–176.

Bradley, R. (2004). Forecasting domestic water use in rapidly urbanizing areas in Asia. *Journal of Environmental Engineering, 130*, 465–471.

Charlton, M. B., & Arnell, N. W. (2011). Adapting to climate change impacts on water resources in England: An assessment of draft water resources management plans. *Global Environmental Change, 21*, 238–248.

Clarke, G. P., Kashti, A., McDonald, A., & Williamson, P. (1997). Estimating small area demand for water: A new methodology. *Journal of the Chartered Institute of Water and Environmental Management, 11*, 186–192.

Dalhuisen, J. M., Florax, R. J. G. M., De Groot, H. L. F., & Nikamp, P. (2003). Price and income elasticities of residential water demand: A meta analysis. *Land Economics, 79*(2), 292–308.

Davis, W. Y. (2003). *Water demand forecast methodology for California water planning areas: Work plan and model review*. Sacramento: California Bay–Delta Authority.

Desprats, J., Rinaudo, J.-D., & Montginoul, M. (2013). *Controlling residential water demand through urban planning: lessons learnt from two French case studies*. Proceedings, 3rd international IWA conference on water economics, statistics and finance, Marbella, Spain. London: International Water Association.

Domene, E., & Saurí, D. (2006). Urbanisation and water consumption: Influencing factors in the metropolitan region of Barcelona. *Urban Studies, 43*, 1605–1623.

Donkor, E., Mazzuchi, T. A., Soyer, R., & Roberson, J. A. (2012). Urban water demand forecasting: A review of methods and models. *Journal of Water Resources Planning and Management*. doi:10.1061/(ASCE)WR.1943-5452.0000314.

Downing, T. E., Butterfield, R. E., Edmonds, B., Knox, J. W., Moss, S., Piper, B. S., & Weatherhead, E. K. (2003). *Climate change and the demand for water*. Oxford, UK: Stockholm Environment Institute.

Environment Agency. (2009). *Water for people and the environment: Water resources strategy for England and Wales*. Bristol: Environment Agency.

Environment Agency, Ofwat, Defra & Welsh Government. (2012). *Water resources planning guideline: The technical methods and instructions*, 187 pp. Online at http://cdn.environment-agency.gov.uk/geho0612bwpe-e-e.pdf. Accessed 11 Nov 2013.

Espey, M., Espey, J., & Shaw, W. D. (1997). Price elasticity of residential demand for water: a meta-analysis. *Water Resources Research, 33*, 1369–1374.

Froukh, M. L. (2001). Decision-support system for domestic water demand forecasting and management. *Water Resources Management, 15*, 363–382.

Fullerton, T. M., & Molina, A. L. (2010). Municipal water consumption forecast accuracy. *Water Resources Research, 46*, W06515.

Galán, J. M., López-Paredes, A., & del Olmo, R. (2009). An agent-based model for domestic water management in Valladolid metropolitan area. *Water Resources Research, 45*, W05401.

Goodchild, C. W. (2003). Modelling the impact of climate change on domestic water demand. *Water and Environment Journal, 17*, 8–12.

Hanak, E., & Davis, M. (2006). Lawns and water demand in California. *California Economic Policy, 2*(2), 1–22.

Hazen, & Sawyer. (2004). *The Tampa Bay water long-term demand forecasting model*. Tampa: Tampa Bay Water.

Herrington, P. (1995). *Climate change and the demand for water*. London: Institute of Hydrology and the Department of the Environment.

Howe, C. W., & Linaweaver, F. P. (1967). The impact of price on residential water demand and its relationship to system design and price structure. *Water Resources Research, 3*, 13–32.

HRBWA. (2007). *Forecasting water demand in the Humboldt basin: Capabilities and constraints*. Carson City: Humboldt River Basin Water Authority.

Jacobs, H. E., & Haarhoff, J. (2004). Application of a residential end-use model for estimating cold and hot water demand, wastewater flow and salinity. *Water SA, 30*, 305–316.

Lampert, R. J., Popper, S. W., & Bankes, S. C. (2003). *Shaping the next one hundred years: New methods for quantitative long term policy analysis*. Santa Monica: RAND Corporation.

Lee, S.-J., Wentz, E., & Gober, P. (2010). Space–time forecasting using soft geostatistics: A case study in forecasting municipal water demand for phoenix, Arizona. *Stochastic Environmental Research and Risk Assessment, 24*, 283–295.

Levin, E., Maddaus, W. O., Sandkulla, N. M., & Pohl, H. (2006). Forecasting wholesale demand and conservation savings. *Journal of the American Water Works Association, 98*, 102–111.

López-Paredes, A., Saurí, D., & Galán, J. M. (2005). Urban water management with artificial societies of agents: The FIRMABAR simulator. *Simulation, 81*, 189–199.

Merrett, S. (2004). The demand for water: Four interpretations. *Water International, 29*, 27–29.

Mohamed, M., & Al-Mualla, A. (2010). Water demand forecasting in Umm Al-Quwain (UAE) using the IWR-MAIN specify forecasting model. *Water Resources Management, 24*, 4093–4120.

Montginoul, M., & Rinaudo, J.-D. (2011). Controlling households' drilling fever in France: an economic modeling approach. *Ecological Economics, 71*, 140–150.

MWD. (2010). *The regional urban water management plan*. Los Angeles: Metropolitan Water District of Southern California. 407pp.

NRA., & UKWIR. (1995). *Demand forecasting methodology: Main report*. 95/WR/01/1. ISBN 1 84057 120 9. London: UKWIR.

Osborn, C. T., Schefter, J. E., & Shabman, L. (1986). The accuracy of water use forecast: Evaluation and implications. *Journal of the American Water Resources Association, 22*, 101–109.

Patterson, J., & Wentz, A. E. (2008). *Forecasting single-family residential water consumption for phoenix and paradise valley, Arizona*. Phoenix: Arizona State University, geoDa Center.

Polebitski, A., Palmer, R., & Waddell, P. (2011). Evaluating water demands under climate change and transitions in the urban environment. *Journal of Water Resources Planning and Management, 137*, 249–257.

Rinaudo, J.-D. (2013). Prévoir la demande en eau potable: une comparaison des méthodes utilisées en France et en Californie. *Eau, Agriculture et Territoires, 10*, 78–85.

Rosenberg, D. E., Tarawneh, T., Abdel-Khaleq, R., & Lund, J. R. (2007). Modeling integrated water user decisions in intermittent supply systems. *Water Resources Research, 43*, W07425.

Thames Water. (2010). *Planning for the future: Revised draft water resources management plan*. Thames Water Utilities Limited. Accessed at: http://www.thameswater.co.uk

Thomas, J. F. (2008). *Water futures for Western Australia 2008–2030* (State report, Vol. 1). Perth: Resource Economics Unit, Department of Water. 67pp.

UKWIR. (1997). *Forecasting water demand components: Best practice manual.* Report 97/WR/07/1. ISBN 1 84057 040 7. London: UKWIR

Wentz, E., & Gober, P. (2007). Determinants of small-area water for the city of phoenix, Arizona. *Water Resources Management, 21*, 1849–1863.

Westcott, R. (2004). A scenario approach to demand forecasting. *Water Science & Technology, 4*, 45–55.

Worthington, A. C., & Hoffman, M. (2008). An empirical survey of residential water demand modelling. *Journal of Economic Surveys, 22*, 842–871.

Wurbs, R. A. (1994). *Computer models for water resources planning and management.* Alexandria: US Army Corps of Engineers.

Chapter 12
The Use of Non-pricing Instruments to Manage Residential Water Demand: What Have We Learned?

Maria A. Garcia-Valiñas, Roberto Martínez-Espiñeira, and Hang To

1 Introduction

Water plays a crucial role in the development and growth of communities. Rising populations and global warming – which potentially mean bigger demand for water and less supply – make water management an increasingly complex policy issue. A range of solutions have been proposed to manage imbalances between water availability and water demand. The traditional approach to address water shortages was to exploit new supply sources. However, there is increased interest in measures of demand-side management (DSM), which aim to promote residential water conservation. Equally, researchers (e.g. Renwick and Archibald 1998; Howarth and Butler 2004) and environmental organizations (e.g. UK Environment Agency 1997; USEPA 2002) have recommended an 'integrated water resource management' approach in which DSM policies to reduce consumption are combined with supply-side policies.

The literature on water demand management deals extensively with market-based policies aimed at moderating consumption, in particular water pricing

M.A. Garcia-Valiñas (✉)
Oviedo Efficiency Group, Department of Economics, University of Oviedo,
Avda. Del Cristo, s/n, 33006 Oviedo, Spain
e-mail: mariangv@uniovi.es

R. Martínez-Espiñeira
Department of Economics, Memorial University of Newfoundland,
A1A 5S7 St John's, NL, Canada
e-mail: rmartinezesp@mun.ca

H. To
Crawford School of Public Policy, The Australian National University,
Building 132, Lennox Crossing Acton, Canberra, ACT 0200, Australia
e-mail: hang.to@anu.edu.au

Q. Grafton et al. (eds.), *Understanding and Managing Urban Water in Transition*, Global Issues in Water Policy 15,
DOI 10.1007/978-94-017-9801-3_12

(Arbués et al. 2003; Dalhuisen et al. 2003; Worthington and Hoffman 2008). However, non-pricing policies have also been considered as an alternative, or complementary, tool to control residential water consumption (Ferrara 2007; Worthington and Hoffman 2008). The installation of water-efficient devices (water-efficient washing machines, low-volume/dual-flush toilets, low-flow shower heads, etc.) has been shown to potentially be an effective way of reducing water consumption. Educational campaigns oriented to change water use habits (turning off the tap when brushing one's teeth, or waiting until the washing machine or dishwasher is full before use, etc.) can also achieve considerable water savings.

In Sect. 2 we assess, based on a survey of the empirical literature, the impact on residential water consumption of the main DSM non-pricing tools. In Sect. 3 we look at the advantages and disadvantages of different policies and analyse the relationships between pricing and non-pricing policies, followed by a conclusion in Sect. 4.

2 Non-pricing Policies and Their Impact on Residential Water Demand

DSM policies can be classified into two main categories: regulatory policies and market-based policies. The former employ instruments such as restrictions on water use, quotas, education, persuasive messages, and moral suasion. Market-based policies include water pricing and monetary incentives such as subsidies aimed at encouraging households to make water-saving choices. These two categories correspond to the classical view (e.g. Weitzman 1974; Stavins 1996), which differentiates between quantity-based and price-based policy instruments.

Most research in the field of water demand management has focused exclusively on the effect of price, climate, and socio-economic characteristics when modelling residential demand. We know much less about the effectiveness of public information campaigns, education campaigns, voluntary retrofit schemes, water-efficiency labelling schemes, and other non-price conservation programs in encouraging households to adopt water-saving devices and undertake water conservation.

One key issue when trying to evaluate the effectiveness of non-pricing water conservation policies is that, in practice, demand-side management campaigns often involve the simultaneous implementation of a number of tools, both of the pricing and non-pricing type (Kenney et al. 2008). In order to be able to identify the saving potential of a given tool, a high level of information disaggregation is required (Inman and Jeffery 2006). It is also useful to have a control sample available in order to understand relative water savings (Turner et al. 2005), because absolute savings in a given season will not account for other factors such as climate or extra restrictions which can significantly change household demand from year to year.

We classify non-pricing tools according to four main categories: technological, educational, informational, and regulatory (Inman and Jeffery 2006). Next we describe these DSM tools and the results from the empirical literature on their impacts on residential water demand.

2.1 Technological Tools

The so-called 'engineering approach' (Campbell et al. 2004) involves retrofit programs of water-efficient devices based on the distribution, rebates, subsidies, or installation of water-efficient devices such as water-efficient washing machines, low-volume or dual-flush toilets, water flow restrictor taps, low-flow shower heads, etc. A lot of indoor residential water consumption happens in the bathroom, with showering and bathing accounting for a large proportion of total indoor domestic use. Kitchen activities also require large volumes of water. Efficient washing machines and dishwashers are well-known water-saving technologies, as well as systems that directly waste water from the hand basin or shower/bathtub to the toilet. Some shower systems give an option to recycle the water that allows people to have longer showers with minimal waste. The main saving technologies for outdoor uses focus on water-efficient irrigation techniques.

In practice, the water savings achieved from installing water-saving devices vary for several reasons. Not only is it important to consider the difference in water use before and after retrofitting an appliance, one must also take into account that water users react in different ways to the installation of a water-saving device.

Several studies have evaluated the effectiveness of policies aimed at promoting the installation of water-saving devices. For example, Grafton et al. (2011) used data for 10 OECD countries to build a model of residential water demand and found that the presence of a low-volume or dual-flush toilet can reduce household water consumption by 9–25 %. Renwick and Archibald (1998) studied the effectiveness of both pricing and conservation programs in Southern California during the 1987–1995 drought and found that increasing the number of low-flow toilets and showerheads by one unit decreased residential water consumption by 10 % and 8 % respectively. This study also showed that pricing and conservation policies can have different effects on low income households compared to high income households. Renwick and Archibald (1998) identified the main advantage of retrofitting domestic plumbing fixtures as securing significant long-term savings while requiring no change in household behaviour.

In another study based in California, Corral et al. (1999) examined the effect of pricing and non-price conservation programs on water use and conservation behaviour in three water districts in the San Francisco Bay Area; the study focused on a 10-year period that included drought years. The authors considered a variety of conservation programs, including billing information, conservation education, use restrictions, landscaping programs, and plumbing retrofit programs; the work showed that pricing can effectively reduce water use but that its effect is mitigated when non-price conservation measures are accounted for.

Michelsen et al. (1999) investigated the effectiveness of the major non-price conservation programs implemented during an 11-year period in seven cities in the south-western United States. The technological tools considered in this study included widespread distribution and selected installation of retrofit devices, residential audits, and water-efficient appliance rebates. They found significant reductions in water use ranging between 1.1 % and 4.0 % per program. However, due to lack of information, they failed to distinguish the effectiveness of specific types of programs, since utilities do not always maintain detailed and consistent information regarding the implementation of non-price tools.

Among the most popular policies they considered were subsidies to encourage adoption of more water-efficient technologies. Renwick and Green (2000) reported that the effect of retrofit programs was to reduce household water demand by 9 %. This is a substantial reduction compared with the effect of prices. For example, Renwick and Green (2000) found that a 10 % increase in water price leads to a 4 % reduction in water consumption. Similar results were found by Renwick and Archibald (1998), with a reduction of about 2 %.

Keeting and Styles (2004) found that a program to replace and install efficient toilets led to a reduction of 16 % and 5 % in water consumption in old houses and new constructions, respectively. Turner et al. (2005) reported that the Sydney Water Company's *Every Drop Counts* retrofit program achieved savings of approximately 8 %. Similarly, Kenney et al. (2008) showed that participation in an indoor rebate program reduced household demand by approximately 10 %. However, in a number of cases in the UK and USA, retrofit programs involving the installation of low-flush toilets were shown to be ineffective.

In Arizona, Campbell et al. (2004) estimated a reduction in water use of about 3.5 % due to a regulation imposing the installation of low-flow fixtures and devices. However, they found increases of about 3.8–4.6 % in water use after the installation of free retrofit device kits. Another policy based on having similar devices installed during in-house visits involving person-to-person communication led to significant water savings of 2.4–6.4 %. The authors suggest that the result of these policies may depend on whether the equipment is distributed for free or the resident actually makes investment to install it. Results in Campbell et al. (2004) raise the issue of a possible rebound effect, in which water use increases after the installation of water-efficient equipment. That is, after installing water-saving technologies, households adapt their water use practices and behaviours in such a way that the overall effect is an increase in water use. However, Bennear et al. (2011), using data from North Carolina regarding high-efficiency toilet installation, found no evidence of a rebound effect. However, they did show that water savings attributable to the rebate program were less than one-half the actual savings associated with the installation.

Renwick and Archibald (1998) showed that significant variations in outdoor water use can be attributed to lawn reticulation systems. Households with automatic sprinkler systems consume more water than households that use manually operated irrigation systems (Syme et al. 2004). This inverts the standard thinking that automatic reticulation systems promote outdoor water conservation.

More recently, Lee et al. (2011) studied the impact of water conservation incentives (including rebates and unit exchange programs for showerheads, toilets, and clothes washers) for residential users in Miami–Dade County, Florida, for the first 4 years after implementation. Although reductions in water demand of 6–14 % occurred during the first 2 years, savings during the third and fourth years were at a lower rate.

2.2 *Educational Tools*

Public information programs often convey messages to consumers about the importance of water conservation and of methods to achieve it, believing that the deeper the knowledge about environmental problems and the way to solve them, the higher the probability that consumers will choose to protect the environment (Kollmuss and Agyeman 2002).

Public information and education programs can be designed with an informative and/or moral-suasion orientation. Both aim to change households' routines and habits. However, the former only provides information on how to reduce water consumption and/or conserve or reuse water, while the latter stresses the importance of conserving water. In general, economists are rather cautious regarding the effects of moral suasion, although it is commonly used policy tool.

Engaging the public in water conservation is an important objective of DSM policies. Nieswiadomy (1992) estimated the effects of public information campaigns on water conservation for four regions in the USA, and found that the public information campaigns only had a significant impact on reducing water demand in the west of the country. UKWIR (1998) showed that the water savings in the UK resulting from public information campaigns was about 7.6 %. Renwick and Green (2000) reported that public education campaigns reduced water consumption by about 8 %. Martínez-Espiñeira and García-Valiñas (2013) found that educational campaigns appear to have a positive effect on both the decisions to adopt water efficient investments and habits.

However, not all studies show that public information campaigns are so effective (Howarth and Butler 2004). Syme et al. (2000) found that the effectiveness of water-saving media campaigns in drought situations, as indicated by regression-based analyses, to be very limited; however, possibly more accurate qualitative reviews and quasi-experimental evaluations have led to greater savings estimates (roughly 10–25 % savings in overall consumption in the short term). One key difference between technological tools and education tools involves the differences between short-run and long-run effects. For example, Fielding et al. (2013) describe an experimental study involving 221 households designed to test the long-term impact of three different interventions on household water consumption in South East Queensland. Each household was allocated to either a control group or one of three interventions groups: water saving information alone, information plus a descriptive norm manipulation, and information plus tailored end-user feedback. Smart water metering technology was then used to observe changes in behaviour and to test the effectiveness of the different demand management interventions. All intervened groups reduced consumption (11.3 l per person per day on average) during the interventions and for some months afterwards, leading to significant water savings. However, in all cases the effect of the intervention eventually dissipated and consumption returned to pre-intervention levels after about a year.

2.3 Informational Tools: Labelling Schemes

There is a very wide range in the water efficiency of water-using appliances, so poor access to water-efficiency information will mean that a user's ability to save water is severely limited. Therefore, a labelling scheme that provides information on water efficiency and/or water use of appliances (such as showerheads, washing machines, dishwashers, and toilets) can be regarded as a non-pricing policy. Often labelling schemes specify a rating for level of efficiency and the water-efficiency level required for each rating; they sometimes include a label design that encourages and enables buyers to compare the water efficiency of different appliance models. Water labelling is expected to raise public awareness about water issues and to change purchasing choices towards the targeted water-efficient products, thus reducing water consumption at the consumer level. Labelling schemes may be mandatory or, more commonly, voluntary.

Australia is an international leader in the labelling of water-using products and its scheme has inspired the design of labelling schemes in other countries (Ministry of Consumer Affairs 2007). Experiences from voluntary labelling schemes, such as the Irish Water Conservation Label and the Australian 5-A Scheme, show that few suppliers and retailers participate, and so voluntary schemes achieve only limited success (European Commission 2009). Given that mandatory water-labelling schemes have only been introduced fairly recently, little information on their impact on water savings is available, and most of the data on water-savings from labelling schemes are based on projections rather than actual savings.

In Australia, the mandatory water-efficiency labelling scheme began in 2005. It requires that toilets, washing machines, dishwashers, urinals, taps, and showers display a 'star rating' of their water efficiency at the point of sale. The mandatory scheme replaced the voluntary labelling scheme introduced in 1988. George Wilkenfeld and Associates (2004) reported that this voluntary labelling scheme was not effective in achieving water savings, since only a small proportion of the available models were labelled. On the other hand, Chong et al. (2008) reported a significant amount of water savings resulting from the mandatory scheme. They estimated that, compared to the voluntary labelling scheme, the total water savings resulting from its mandatory counterpart amounted to 1,457 ML (1 ML = 10^3 cubic meters) in 2006. They projected that, for the period 2005–2021, the total water savings achieved through the application of this scheme throughout Australia would be approximately 800 GL (1 GL = 10^6 cubic metres), with 36 % of savings derived from shower use and 34 % from washing machines. Toilets and urinals were expected to contribute about 23 % of water savings and approximately 6 % of savings from taps and dishwashers combined.

New Zealand's water-efficiency labelling scheme, similar to Australia's, came into effect in April 2011 and became mandatory in April 2013. Actual savings from this scheme are not available. However, a cost–benefit analysis concluded that New Zealand's labelling scheme could save only a small amount of water, due to the absence of volumetric water pricing in many parts of the country (Covec 2004).

In the United States, the voluntary 'WaterSense' label was introduced in 2006 by the Environmental Protection Agency to promote the efficient use of water. The labels are available for taps, showerheads, toilets, and urinals in the form of an endorsement or mark of approval for water-efficient products. It has been estimated that WaterSense-labelled faucets and faucet accessories can save about 30 % or more of water use without sacrificing performance (Kaps and Wolf 2011).

Within Europe, in 2009 the 'Ecodesign' directive established water requirements for energy-using products such as washing machines and dishwashers and there are plans to extend it to energy-related products including taps and showerheads. Kaps and Wolf (2011) estimate the potential water savings across Europe from Ecodesign at about 20 % of water use from taps and showerheads.

2.4 *Legislative Tools*

Under this category, we identify permanent ordinances and regulations which force households to adopt a particular behaviour or investment. Some of those tools make engineering solutions compulsory. For example, some building ordinances require the installation of water-saving equipment or individual metering in new or refurbished buildings.

Additionally, temporary ordinances and regulations restrict certain types and/or times of water use or ration the level of residential use to a specified amount. Such impositions might ban residential water consumption for some particular (usually outdoor) uses or involve some kind of hourly supply cuts.

Water restrictions are perhaps the most intrusive tool among non-price conservation measures (Michelsen et al. 1999), but also one of the most effective (Martínez-Espiñeira and Nauges 2004; Roibás et al. 2007). Restrictions such as periodic bans on watering gardens and washing down pavements or driveways in peak water demand periods were found to reduce water consumption by 29 % in California (Renwick and Archibald 1998) and by 25–35 % in Greece (Kanakoudis 2002).

Kenney et al. (2004) studied the effects of the restrictions on outdoor water use – mainly on limiting the frequency of lawn watering – during the 2002 drought in Colorado. Mandatory restrictions were shown to be an effective tool for coping with drought in this case, leading to savings measured in expected use per capita of 18–56 %, compared to just 4–12 % savings during periods of voluntary restrictions.

Further evidence on this differential effect between mandatory and voluntary restrictions is provided by both Coleman (2009) and Halich and Stephenson (2009). The former found not only that voluntary restrictions in Salt Lake City (Utah) achieved limited reductions in residential water use, but also that water-use restrictions implemented without making a concerted effort to enforce them have negligible, or moderate, effects on water use. Similarly, Halich and Stephenson (2009) showed that, during the 2002 drought in Virginia, USA, residential water-use in 21 municipalities declined in proportion to the level of information and enforcement effort.

Some countries apply minimum efficiency standards for residential and commercial water-using appliances and fixtures. In the United States, minimum efficiency standards apply to nearly all toilets, showerheads, faucets, and urinals manufactured after 1994. By 2026, water consumption is projected to drop from about 550 L/day to 250 L/day for the average 2.63 person household. Australia is considering establishing minimum water-efficiency standards, and it is expected that a mandatory water-efficiency standard for shower heads and washing machines would be about 10 times more effective than the rebate option in terms of projected volume of water saved (Chong et al. 2008).

3 Pricing Policies Versus Non-pricing Policies: Advantages and Disadvantages

As Stavins (1996) pointed out, policymakers might choose to set an environmental goal and then select a policy instrument to achieve it. He also warned that "in the presence of simultaneous uncertainty in both marginal benefits and marginal costs and some statistical dependence between them, benefit uncertainty expressed through the covariance term can make a difference for identifying the efficient policy instrument. A positive correlation tends to promote the quantity instrument, and a negative correlation favours the price instrument" (p. 229).

In general, economic instruments are seen as more flexible (e.g. Tietenberg 1990), since they leave it up to the user to choose the best way to face the price; consequently they provide marginal incentives for efficiency in both the short run and the long run. Price-based management tools also collect revenue, whereas many non-price management tools only involve spending taxpayers' money, so the resulting deadweight loss in taxation must be counted as part of their cost.

Most of the regression-based studies that compare pricing and non-pricing instruments include a separate variable for each kind of instrument. However, this approach sometimes ignores that there could be important synergies among them. Some authors (Syme et al. 2000; Van Vugt 2001) suggest that educational campaigns oriented towards improving household's knowledge of their water consumption and prices might affect their motivation to respond to pricing schedule changes.

However, explaining the relationships between pricing and non-pricing instruments is a complex task. According to Inman and Jeffrey (2006), "this is indicative of the difficulty in distinguishing between people's different motivation for displaying conservation behaviour" (p. 127).

Most non-price instruments are more effective at reducing water consumption, and can achieve results with more certainty, than their price-based counterparts, at least in the short run (Renwick and Archibald 1998; Renwick and Green 2000). The substantial water price increases necessary to obtain significant reductions in demand, and the associated increases in revenue (since water demand is usually price-inelastic), are often problematic for water providers – because there is public and political opposition to water rate increases and because regulatory constraints often prevent non-profit organizations from earning excess revenue (Michelsen et al. 1999; Kenney et al. 2008). In extreme cases, price-based measures to manage water demand are politically unacceptable because some people can pay for overusing the resource while others cannot, whereas the resource should be shared equally regardless of income. The effectiveness of water tariffs is poor if the tariffs are not well understood, or if habits are strong and the barriers to adopting water-using appliances are high.

Pricing policies have often been contrasted with water restrictions, that is, quantity-based mechanisms to ration demand. Droughts lead to the need for rationing and, although there are various ways to implement it (Winpenny 1994), public authorities frequently resort to water use restrictions and hourly supply cuts. Several studies have focused on the effects of policies aimed at moderating water consumption (e.g. Moncur 1987; Woo 1992, 1994; Renwick and Archibald 1998; Renwick and Green 2000; García-Valiñas 2006; Roibás et al. 2007; Grafton and Ward 2008).

Most studies identify *some* rationing regulations as a very effective tool for reducing water consumption. For example, Martínez-Espiñeira and Nauges (2004) found that a 9 % increase in price would achieve the same reduction in water use as an additional hour of daily supply restriction. However, some rationing regulations carry undesirable consequences in terms of efficiency and equity. Roibás et al. (2007), for instance, concluded that "the management of water shortages by implementing supply interruptions is a policy which has clearly regressive effects" (p. 239). Additionally, the majority of studies that have compared pricing with supply interruption have shown that the welfare loss associated to a supply cut is larger than the loss linked to an equivalent price increase.

Compared with non-price tools, the theoretical advantage of price-based demand management tools is a saving in the economic cost of achieving a certain water use reduction. However, very few empirical studies have compared the estimated costs of achieving a given reduction in consumption through different demand management tools. Timmins (2003) showed that a mandatory low-flow appliance regulation in 13 cities in California would be less cost-effective than a modest water tax (or water price increase) in reducing groundwater aquifer lift height in the long run. Similarly, Mansur and Olmstead (2011) estimated that, compared to residential outdoor watering restrictions, the welfare gains from a price-based approach added up to approximately US$96 per household during a lawn-watering season, which represented about 29 % of average annual household expenditures on water in their sample of urban areas in the United States and Canada. Brennan et al. (2007) calculated that the economic costs of a 2-day-a-week sprinkling restriction in Perth, Western Australia: allowing hand watering only cost just under AU$100 per household per season, while the cost of a complete outdoor watering ban ranged from AU$347 to AU$870 per household per season. Grafton and Ward (2008) calculated that mandatory water restrictions in Sydney, Australia, resulted in economic losses of AU$235 million in 1 year (about AU$150 per household, which was about half the average Sydney household's water bill in the year studied). Similar results suggesting the advantages of price-based conservation tools over restrictions were reported by Byrnes et al. (2010).

The cost savings resulting from the use of price-based tools compared to restriction tools can be explained by two factors (Olmstead and Stavins 2009): (1) households facing higher prices rather than quantity restrictions can decide which water usage to reduce; and (2) responses to price-based tools can vary between households, so substitution of scarce water from those households who value it less to those who value it more results in a more efficient allocation of the resource. Olmstead and Stavins (2009) conclude that "using price increases to reduce demand, allowing consumers to adjust their end uses of water, is more cost effective than implementing non-price demand management programs" (p. 3).

Halich and Stephenson (2009) and Renwick and Green (2000) show how the effectiveness of many non-price measures of water demand management depends on the level of their enforcement. It can also be argued that many non-price measures require intrusive levels of monitoring in order to be effective. Thus, apart from being less efficient, non-price demand management measures can also be intrusive and unpopular, which can make them politically unacceptable as long-term measures to deal with demand imbalances.

4 Conclusions

Our study has reviewed non-pricing policy instruments in DSM and their impacts on residential water demand. In general, those instruments are considered more effective in reducing water consumption (at least in the short run) and can achieve results with more certainty than their price-based counterparts. However, pricing is seen as a more flexible and efficient tool, allowing for revenue collection.

We considered different kind of non-pricing tools – technological, educational, informational, and regulatory. Setting out the advantages and disadvantages of each instrument offers a clear guide for steering public policies and seeing new possibilities for saving water. Although voluntary schemes appear to achieve limited success, mandatory schemes have the potential to bring about much larger water savings.

References

Arbués, F., García-Valiñas, M. A., & Martínez-Espiñeira, R. (2003). Estimation of residential water demand: A state-of-the-art review. *The Journal of Socio-Economics, 32*, 81–102.

Bennear, L. S., Lee, J. M., & Taylor, L. O. (2011). *Participation incentives, rebound effects and the cost-effectiveness of rebates for water-efficient appliances.* Paper EE11-10, Duke environmental economics working paper series, Duke University, Durham, US.

Brennan, D., Tapsuwan, S., & Ingram, G. (2007). The welfare costs of urban outdoor water restrictions. *Australian Journal of Agricultural and Resource Economics, 51*(3), 243–261.

Byrnes, J., Crase, L., Dollery, B., & Villano, R. (2010). The relative economic efficiency of urban water utilities in regional New South Wales and Victoria. *Resource and Energy Economics, 32*(3), 439–455.

Campbell, H., Johnson, R., & Larson, E. (2004). Prices, devices, people, or rules: The relative effectiveness of policy instruments in water conservation. *Review of Policy Research, 21*(5), 637–662.

Chong, J., Kazaglis, A., & Giurco, D. (2008). *Cost effectiveness analysis of WELS: The water efficiency labelling and standards scheme.* Prepared for the Department of the Environment and Heritage, Australia.

Coleman, E. A. (2009). A comparison of demand-side water management strategies using disaggregate data. *Public Works Management & Policy, 13*, 215–223.

Corral, L., Fisher, A. C., & Hatch, N. (1999). *Price and non-price influences on water conservation: An econometric model of aggregate demand under nonlinear budget constraint.* Working paper no. 881. Berkeley: Department of Agricultural and Resource Economics and Policy, University of California.

Covec. (2004). *Water efficiency labelling for New Zealand: Cost benefit analysis.* Prepared for Ministry for the Environment, New Zealand Government.

Dalhuisen, J. M., Florax, R. J. G. M., De Groot, H. L. F., & Nijkamp, P. (2003). Price and income elasticities of residential water demand: A meta-analysis. *Land Economics, 79*(2), 292–308.

European Commission. (2009). *Study on water efficiency standards.* http://ec.europa.eu/environment/water/quantity/pdf/Water%20efficiency%20standards_Study2009.pdf. Retrieved 21 Feb 2013.

Ferrara, I. (2007). Residential water use: A literature review. In *Household behaviour and environment: Reviewing the evidence.* Paris: Organization for Economic Cooperation and Development (OECD).

Fielding, K. S., Spinks, A., Russell, S., McCrea, R., Stewart, R., & Gardner, J. (2013). An experimental test of voluntary strategies to promote urban water demand management. *Journal of Environmental Management, 114*, 343–351.

García-Valiñas, M. A. (2006). Analysing rationing policies: Drought and its effects on urban users' welfare. *Applied Economics, 38*(8), 955–965.

George Wilkenfeld and Associates. (2004). *Regulation impact statement: Proposed national system of mandatory water efficiency labelling for selected products.* Prepared for the Department of the Environment and Heritage, Australia.

Grafton, R., & Ward, M. (2008). Prices versus rationing: Marshallian surplus and mandatory water restrictions. *The Economic Record, 84*(Suppl 1), S57–S65.

Grafton, R. Q., Ward, M., To, H., & Kompas, T. (2011). Determinants of residential water consumption: Evidence and analysis from a 10-country household survey. *Water Resources Research, 47*, W08537. doi:10.1029/2010WR009685.

Halich, G., & Stephenson, K. (2009). Effectiveness of residential water-use restrictions under varying levels of municipal effort. *Land Economics, 85*(4), 614–626.

Howarth, D., & Butler, S. (2004). Communicating water conservation: How can the public be engaged? *Water Supply, 4*(3), 33–34.

Inman, D., & Jeffery, P. (2006). A review of residential water conservation tool performance and influences on implementation effectiveness. *Urban Water Journal, 3*(3), 127–143.

Kanakoudis, V. K. (2002). Urban water use conservation measures. *Journal of Water Supply Research and Technology, 51*(3), 153.

Kaps, R., & Wolf, O. (2011). *Development of European Ecolabel and green public procurement criteria for sanitary tapware: Taps and showerheads*. Report prepared for first AHWG: Meeting for the development of ecological criteria for sanitary tapware (taps and showerheads), European Commission.

Keeting, T., & Styles, M. (2004). *Performance assessment of low flush volume toilets: Final report for Southern Water and the Environment Agency*. http://www.greenbuildingstore.co.uk/es4report.pdf

Kenney, D. S., Klein, R. A., & Clark, M. P. (2004). Use and effectiveness of municipal water restrictions during drought in Colorado. *Journal of the American Water Resources Association, 40*(1), 77–87.

Kenney, D. S., Goemans, C., Klein, R., Lowrey, J., & Reidy, K. (2008). Residential water demand management: Lessons from Aurora, Colorado. *Journal of the American Water Resources Association, 44*, 192–207.

Kollmuss, A., & Agyeman, J. (2002). Mind the gap: Why do people act environmentally and what are the barriers to pro-environmental behaviour? *Environmental Education Research, 8*(3), 239–260.

Lee, M., Tansel, B., & Balbin, M. (2011). Influence of residential water use efficiency measures on household water demand: A four year longitudinal study. *Resources, Conservation and Recycling, 56*, 1–6.

Mansur, E. T., & Olmstead, S. M. (2011). The value of scarce water: Measuring the inefficiency of municipal regulations. *Journal of Urban Economics, 71*(3), 332–346.

Martínez-Espiñeira, R., & García-Valiñas, M. A. (2013). Adopting versus adapting: Adoption of water-saving technology versus water conservation habits in Spain. *International Journal of Water Resources Development, 29*(3), 400–414.

Martínez-Espiñeira, R., & Nauges, C. (2004). Is really all domestic water consumption sensitive to price control? *Applied Economics, 6*(15), 1697–1703.

Michelsen, A., McGuckin, J., & Stumpf, D. (1999). Nonprice water conservation programs as a demand management tool. *Journal of the American Water Resources Association, 35*(3), 593–602.

Ministry of Consumer Affairs. (2007). *Proposed implementation of mandatory water efficiency labelling*. Discussion document for consultation under the Fair Trading Act 1986. Wellington: New Zealand Government.

Moncur, J. (1987). Urban water pricing and drought management. *Water Resources Research, 23*(3), 393–398.

Nieswiadomy, M. L. (1992). Estimating urban residential water demand: Effects of price structure, conservation and public education. *Water Resources Research, 28*(3), 609–615.

Olmstead, S. M., & Stavins, R. N. (2009). Comparing price and nonprice approaches to urban water conservation. *Water Resources Research, 45*, W04301.

Renwick, M. E., & Archibald, S. O. (1998). Demand-side management policies for residential water use: Who bears the conservation burden? *Land Economics, 74*(3), 343–359.

Renwick, M. E., & Green, R. D. (2000). Do residential demand side management policies measure up? An analysis of eight California water agencies. *Journal of Environmental Economics and Management, 40*(1), 37–55.

Roibás, D., García-Valiñas, M. A., & Wall, A. (2007). Measuring welfare losses from interruption and pricing as responses to water shortages: An application to the case of Seville. *Environmental and Resource Economics, 38*(2), 231–243.

Stavins, R. N. (1996). Correlated uncertainty and policy instrument choice. *Journal of Environmental Economics and Management, 30*(2), 218–232.

Syme, G. J., Nancarrow, B. E., & Seligman, C. (2000). The evaluation of information campaigns to promote voluntary household water conservation. *Evaluation Review, 24*(6), 539–578.

Syme, G. J., Shao, Q., Po, M., & Campbell, E. (2004). Predicting and understanding home garden water use. *Landscape and Urban Planning, 68*(1), 121–128.

Tietenberg, T. H. (1990). Economic instruments for environmental regulation. *Oxford Review of Economic Policy, 6*(1), 17–33.

Timmins, C. (2003). Demand-side technology standards under inefficient pricing regimes: Are they effective water conservation tools in the long-run? *Environmental and Resource Economics, 26*(1), 107–124.

Turner, A., White, S., Beatty, K., & Gregory, A. (2005). Results of the largest residential demand management program in Australia. *Water Science and Technology: Water Supply, 5*, 249–256.

UK Environment Agency. (1997). *Saving water: Taking action. The responses to the consultation report on water conservation and demand management*. Prepared by the National Water Demand Management Centre.

UKWIR (1998). *Evaluating the impact of restrictions on customer demand*. Report 98/WR/06/2, UK Water Industry Research Limited, London.

USEPA. (2002). *Cases in water conservation: How efficiency programs help water utilities save water and avoid costs*. United States Environmental Protection Agency Office of Water, 4204M, EPA832-B-02-003. http://www.epa.gov/WaterSense/docs/utilityconservation_508.pdf. Accessed 13 Nov 2013.

Van Vugt, M. (2001). Community identification moderating the impact of financial incentives in a natural social dilemma: Water conservation. *Personality and Social Psychology Bulletin, 27*, 1440–1449.

Weitzman, M. L. (1974). Prices vs. quantities. *The Review of Economic Studies, 41*(4), 477–491.

Winpenny, J. (1994). *Managing water as an economic resource*. London: Routledge.

Woo, C. K. (1992). Drought management, service interruption and water pricing: Evidence from Hong Kong. *Water Resources Research, 28*(10), 2591–2595.

Woo, C. K. (1994). Managing water supply shortage: Interruption versus pricing. *Journal of Public Economics, 54*(1), 145–160.

Worthington, A. C., & Hoffman, M. (2008). A state of the art review of residential water demand modelling. *Journal of Economic Surveys, 22*(5), 842–871.

Chapter 13
The Social Determinants of Water Consumption in Australian Cities

Patrick Troy

1 Introduction

Most cities in Australia periodically face acute shortages of potable water. In most of them the supply of potable water comes from sequestering supplies from ecosystems away from but close to that in which the urban development occurs. The response usually is to try to husband the water sequestered and stored in dams by managing the demand to 'get' through the dry periods.

This approach to the management of an essential supply worked well enough during the latter part of the nineteenth and first three quarters of the twentieth century. A combination of factors including the consequences of climate change that appears latterly to have reduced rainfall 'run off', and thus the water captured in city catchment areas in some regions. The increase in the population of most cities together with increasing per capita consumption in them is now pressing state governments that have the constitutional responsibility to manage water resources to find ways of responding to demand.

Australia is generally acknowledged as the driest continent after Antarctica and as the national poem suggests it is a nation that is known for its 'droughts and flooding rains' (Mackellar 1918) meaning that Australian cities experiences extreme variations in the supply of potable water yet their citizens have one of the highest levels of per capita consumption. The cities now, however, face a challenge: How do they respond to this dilemma of high levels of consumption with a high degree of uncertainty in supply?

P. Troy (✉)
Fenner School of Environment and Society, The Australian National University,
0200 Canberra, Australia
e-mail: patrick.troy@anu.edu.au

Q. Grafton et al. (eds.), *Understanding and Managing Urban Water in Transition*, Global Issues in Water Policy 15,
DOI 10.1007/978-94-017-9801-3_13

This is an obvious research question for those exploring present Australian cities and how they might secure supplies of potable water to best meet their demands.

In attempting an answer we must first understand a little of the history of the supply of urban water services in Australia.

2 A Little History of Urban Water Services

In mid nineteenth century urban settlements in Australia faced critical water shortages. All the easily 'exploitable' surface water resources progressively failed to meet the demand for potable water as the urban populations grew. They also increasingly faced significant difficulties in managing waste disposal. Outbreaks of disease occurred due to polluted water supplies and due to the accumulation of wastes, including domestic waste, the breakdown of which provided 'breeding grounds' for vectors of disease. We do not have the space here to document all of the failures in provision of water and sanitation services for all urban settlements but offer the evidence presented in Lloyd et al. (1992) on Newcastle and that found in Dingle and Rasmussen (1991) on Melbourne as typical illustrations of the pressures in and experiences of other urban centres in the same period. The Sydney City Council for example received many submissions from contemporary medical advisors on the water related sanitation and health issues from mid-nineteenth century following the path breaking work of Edwin Chadwick (1842) in England whose report drew attention to the need to provide a secure supply of potable water and a separate system for the management of wastes became the fundamental argument for the development of water services in Australian towns and cities. For a fuller account of the Chadwickian solution, see Dingle (2008).

3 Water Supplies

The nascent water supply agencies in the nineteenth century quickly developed the capacity to capture and store secure supplies of potable water from ecosystem beyond the boundaries of urban development. They also developed the capacity to deliver that water to the burgeoning urban settlements as they grew into cities. The adequacy and reliability of supplies led to increased consumption. The adoption by urban management authorities of the 'elegant' solution of using the cheaply available water as the medium to transport troublesome human and domestic wastes to treatment and or disposal sites in turn led to increased flows of water borne wastes.

Water supply agencies quickly developed the capacity to collect waste flows by developing sewerage systems that delivered the wastes to central 'treatment' plants. In the process the sanitary conditions in central areas were reflected in significant improvement in the health of the population. Chadwick (1842) had recognized the opportunity this represented and saw the systems as a way of returning the treated sewage flows to the near countryside for use in agriculture. Apart from minor experiments in Adelaide and Sydney, the only Australian city that attempted to follow Chadwick's schema was Melbourne which delivered the sewage to a 'farm' at Werribee, to the west of the city, that produced beef for official consumption. The system ultimately failed to cope with the massive increase in sewage flows following the population growth of Melbourne necessitating the construction of an additional sewerage outfall to the east.

The consequence of this improvement in secure supplies of potable water and the associated development of the sewerage systems was a massive increase in water consumption. As the cities grew the sewage flows increased. The very success of the highly centralized engineering services to supply water and to manage the resulting waste flows seemed secure even as the engineering solutions created ever greater challenges to the environment (challenges also arose from the management of storm water runoff that, paradoxically, in most cities remained the responsibility of local government authorities).

Two features of this 'perfect system' in which agencies supplied potable water then provided waste management services drove demand for 'urban water'. The first was that all properties within a specified distance of the water mains were required to pay for the service whether they were connected to it or not. Associated with this was the move by the water supply agencies in some states to secure their income base by having rainwater tanks declared illegal on health grounds - arguing that rainwater tanks collected detritus off the roofs of buildings that led to unhealthy level of lead in the water supplies (from the paint and lead flashing) and also contained bird and rodent faeces that caused health problems. It was also argued that the design of the runoff systems did not prevent rodents and birds from drowning and decomposing in the tanks which also increased health risks. The consequence of these measures was that water consumption increased dramatically and the cisterns and rainwater tanks on which households had previously relied were often removed (increasing the storm water runoff from developed areas). The water supply agencies enjoyed high standing because they had been able to deliver the secure reliable supplies of water at a low cost and the urban populations experienced greatly improved health outcomes.

4 Water Consumption: Domestic

The second feature that drove demand was that in response to the easy availability of water a new array of facilities and services blossomed. Flush toilets revolutionized the domestic management of human body wastes - we note that the Romans used continually running water in their toilet systems a little earlier. The development of hoses in mid-nineteenth Century led to increased water use in gardens. Bathing became more easily available because new systems were developed to heat and deliver water to bathrooms, the easy availability of showers replaced bathing which had been a more arduous way of improving personal hygiene.

The development of clothes washers (and later, dishwashers) also increased the consumption of water.

The easy supply of water also meant households could plant and maintain, through the drier seasons, gardens for food production and pleasure. Pat Mullins (1981a, b) claimed that the separate house and garden was a significant factor in the health and physical development of Australians who enjoyed a better diet than their forbears. Andrea Gaynor (2006) observed that home gardens (made possible by reliable supplies of water) were a significant source of food for households.

Note 1
It should be noted that changes in the use of lead in 'flashing' or in paints ceased 50 years ago and lead piping in buildings ceased a century ago thus eliminating those sources of pollution. The elimination of lead additives in petrol also eliminated air borne lead as a health threat in rainwater tanks. New ways of connecting roof drainage to rainwater tanks make it virtually impossible for rodents or birds to gain access to tanks so that source of risk has also been eliminated. The problem of faeces being washed into tanks from rainfall has been greatly reduced by the use of 'first flush' bypass systems that ensure that the first 10 min of rainfall 'wash the roofs' of detritus and is not captured for tank storage. All these innovations remove the 'case' for making rainwater tanks 'illegal'.

In some towns and cities the supply was 'generous enough' to allow households to invest in swimming pools which also increased consumption. In the 1960s and 1970s backyard pools proliferated and cheap above ground pools allowed battlers to ape the affluent.

For a long period the engineering solution seemed to be without limit. Little attention was paid to the stresses experienced in the ecosystems from which the water was harvested or into which waste flows were discharged. As the limit of the capacity of near city ecosystems to provide secure supplies of potable water was being reached the engineering based water authorities sought new engineering solutions to meet the demand. The Earth Summit of 1992 – also known as the Rio Convention of 1992 – The United Nations Framework Convention on Climate Change (No 28, Treaty Series 1995), began the process of reconsideration of traditional engineering responses to demand for water, however, little regard was given to the larger environmental consequences that increased supplies of potable water sought from the ecosystems that were being exploited would bring.

Supplies were also sought from additional ecosystems - generally further afield – again with little regard for the stresses such an approach created. As these resources were being exploited the response of water agencies in the late twentieth –early twenty-first Centuries was also to try to increase the efficiency of appliances.

Low-flow shower heads and dual flush toilets were developed and households were encouraged to install them. Lower water consumption 'front loading' clothes washers were encouraged. Households were encouraged to develop more 'water wise' gardening practices to reduce consumption. All of these measures helped slow the rate of growth in demand without requiring any fundamental reconsideration of the demand for potable water.

Water agencies, following their 'engineering traditions', sought to increase supply by inventing new ways of increasing the delivery of potable water. The development of 'new technology' such as desalination plants to 'manufacture' potable water from sea water e.g. Plants in, Perth, Western Australia in 2006, Gold Coast, New South Wales in 2009, the Kurnell (Sydney) New South Wales in 2010, Wonthaggi (Melbourne) Victoria in 2012, Port Stanvac (Adelaide) in 2011. Reclaimed water obtained by treating and recycling sewage to produce potable water in Brisbane, Queensland and in Sydney, New South Wales were the avenues of choice often with little regard for larger environmental concerns and, more dangerously, with little exploration of the long term epidemiological challenges these new technologies posed. Little regard seems to have been held for the consequences of such 'new' supplies for the increasing complexity of the waste water flows that not only included the human wastes that carried the drugs people consumed and an increasing array of nanoparticles that are stable and not removed in normal sewage treatment plants but contained effluent from increasingly complex industrial processes. In some areas (Western Australia and South Australia) water authorities, without the benefit of long term epidemiological studies, have sought to inject simply treated waste water flows into ground water resources as part of a 'managed aquifer recharge' in the belief that this would provide a felicitous increase in potable water supplies because of the 'natural' processes that occurred in the aquifers in which the water was found would result in an increased supply of potable water. Limited research has been undertaken to explore the efficacy of this approach nor was the precautionary principle observed.

In none of the 'new approaches' were ethical considerations of effectively forcing people to drink 'manufactured' water taken into account. No one, least of all the water authorities, seemed to challenge, or at least question, the efficacy of the engineering approach developed out of the nineteenth century response to the demand for water. Fewer paid attention to the consequences of the simple 'collect, deliver, then discharge the consequential waste water flows to the environment' or single pass through system that had been so dramatically effective in improving public health.

Throughout the period of this form of exploitation of potable water sources the greater level of concern seemed to be over narrow, even short term, 'economic' considerations. Such considerations loomed larger as neoliberal preoccupations with a search to find ways of reducing public sector funding and control of basic urban infrastructure grew in the last quarter of the twentieth century and early twenty-first century accompanied by resurgence in some quarters of support for the corporatization or privatization of such infrastructure. The resurgence was driven in part by governments that were unable to maintain the pace of infrastructure provision in line with the rate of population growth and with increasing expectation on the part of the increasing populations that they wanted the services and were not prepared to wait for them until they could be 'afforded'. The change in community expectations coincided with a prevailing attitude in some state governments e.g. New South Wales in the late 1970s that the reserves that water service authorities had prudently accumulated in the form of 'sinking funds' to replace ageing or obsolete infrastructure could be 'raided' to be used for general purposes. This had the effect of forcing states into the quasi-privatization of water services supply. The quasi-privatization also led to the departure from the traditional view that water services should be provided as a 'community' service, albeit on the basis on which all costs were recovered, to one in which they were seen as an avenue for general revenue. The corporatization of water supplies followed in 1994 (Water Board (Corporatisation) Act 1994) and was a takeover of investment in the water services of New South Wales enabling the state to charge water authorities for use of water extracted from the various ecosystems as environmental services and for the discharge of sewage (usually partly treated) to receiving water bodies (generally the near oceans in the case of the major cities).

5 Water Consumption: Industrial

The development of engineering services and manufacturing establishment (including manufacture of food products) frequently required a reliable supply of water as an essential element of the processes employed (in the case of food products the water was required to be at potable quality standards). The apparently endless supplies of cheaply available potable water encouraged industry entrepreneurs to take water from the public systems. The development of the piped sewerage systems also meant that such establishments could conveniently discharge their wastes to the sewer. In many cases the requirement that discharges to the sewer had to be of acceptable standard to avoid compromising the sewerage treatment was observed in the breach. Discharges were made to local water courses with significant deleterious effect on their ecosystems. Enforcement of the sewage quality standards reduced these effects and often led to increased efficiencies in production processes but led to increased flows in the sewerage systems.

6 Environmental Concerns

But there were two additional aspects that seemingly drew little attention but which raised serious 'meta' environmental issues.

The first was that the sewage flows from modern societies was vastly more complex than that of the mid to late nineteenth century. In the last hundred years new industries have developed that have been able to enjoy the security of a supply of high quality water. They have been able conveniently to discharge vast volumes of waste water flows into the urban sewerage systems and basically 'forget' about the problems of managing their waste flow. In some cases they have had to pay a higher price for the service but the stresses the flows cause to the environment by the increasing complexity of the sewage 'mix' have never been fully taken into account.

Additional complications arose from new drugs for the treatment of a variety of illnesses and for infection control. A range of endocrine altering drugs for medical and social reasons came into use as did an increasing array of nanoparticles, some of which occur naturally, in drugs, cosmetics, paints, food preparation and preservation and in water treatment to improve its quality. Many of the drugs and nanoparticles are chemically stable and end up in the sewer line and may only inefficiently be 'taken out' during sewage treatment processes – even by the more sophisticated membrane technology processes.

One of the benefits of the use of nanoparticles is that they can be designed to deliver drugs to specific organs in the human body. The 'downside' of their use is that nanoparticles may have serious but uncharted impacts on the environments in which they are used and may have serious impacts on the health of those who are inadvertently 'medicated' by consuming water containing them (there is a rich and growing research literature on nanoparticles in drinking water e.g.: Heidarpour et al. (2011), Kaegi (2009), Qu et al. (2013), Trouiller et al. (2009)) that draws attention to the need for more research into the efficacy of nanoparticles on human health as well as their effect on the environment.

The operation of many of the processes to 'manufacture' water including those relying on membrane technology themselves create serious challenges to the environment through their use of energy as well as the production of 'by-wash' flows that are toxic. In some cases the standards applied to the quality of the water 'product' from the processes have been changed and the public assured that the water meets acceptable health standards. E.g. following the adoption of desalination as a method to boost drinking water supplies the level of boron permitted in potable water in Australia was increased in 2011 from the 1.0 mg of boron per litre established under the European standard or 2.4 mg of boron per litre standard set by the World Health Organisation to 4.0 mg of boron per litre under Australian Drinking Water Guidelines (NHMRC, NRMMC 2011). Little additional Australian epidemiological research seems to have been undertaken to warrant the reduction in standards leaving the initiative open to conjecture that the reduction in standard was set for the convenience of the desalination plant operators (Ockhams Razor 2012).

Major cities now 'produce' large volumes of sewage, mostly treated to rudimentary standards, that is discharged to the oceans increasing local acidity that further reduces their capacity to take CO_2 out of the atmosphere as well as creating local environments that are inhospitable for marine life. The discharge from 'inland' towns of treated sewage to local riverine systems is not generally treated to high standards. In most cases the simple filtering processes do not remove the more complex pollutants leaving them in the receiving water flows some of which may be drawn on further downstream for use in and around dwellings.

7 An Alternative Approach?

In exploring the nature of the demand for water services it is clear that there are alternative approaches to meeting the demand. But the prior question is: How can the demand for potable water be reduced while retaining the level of comfort and security modern communities have come to expect?

8 Domestic Demand

The variation in urban domestic consumption between Australian States is high and raises questions about the way the estimates are made for each 'use'. Nonetheless, it is clear that a small proportion of the potable water delivered to households for domestic consumption – namely that used for drinking, food preparation, cooking and cleaning of utensils and crockery must be of the highest quality. The water used in health facilities should also be of the highest quality. In total, approximately 10 % of the domestic consumption needs to be of the highest quality.

(In most cases the water used in health facilities carries waste streams indicating that it cannot easily be used for a 'second' purpose although modern technology might be able to use such flows efficiently for purposes such as selected industrial processes, gardening and park irrigation).

Most of the water used domestically for bathing, showering, toilets and clothes washing can be of lower quality with little risk to health. Little of the water used in gardens needs to be of potable water standard and there is even greater certainty that the water used to flush toilets can be of less than potable quality.

The point has been reached in the major cities where all the readily available supplies of potable water have been exploited and governments have begun to 'reach out' for supplies further afield. In so doing cities have sought to extract water previously used in agricultural production or to support eco-system services on the grounds that its use in them has a higher economic value. The proposals to divert waters from the regions in Victoria for consumption in Melbourne in the *Our Water Our Future* plan was one such exercise (Victoria. Department of Sustainability and Environment 2004) see also Grafton (2010).

Cities have also chosen to pursue production of 'manufactured' water by investing in desalination plants and processes to recover potable water from sewage flows. But the composition of the demand for domestic consumption suggests that one way of responding to the water 'crisis' is to re-examine the behavioral aspects of water consumption in and around people's homes.

Conventional wisdom, and indeed much current water management policy, is founded on the notion that those who live in houses use more water than those who live in medium density housing (flats, apartments, etc.) or other higher density forms of accommodation. Much of the justification supporting higher density housing policy is based on the notion that those who live in houses consume more water. Earlier research on domestic water consumption indicated that this might not be true

Table 13.1 Estimated average annual per-capita water consumption by location of use in 2010–2011 (kl)

	NSW	VIC	QLD	SA	WA	TAS	ACT	AUST
Total	72	55	68	70	132	135	69	75

ABS 4610.0 Water Account 2010–2011

(Troy et al. 2005). A Canberra study based on a large sample of dwellings strongly suggests that, in that city, this might also not be true. The fact that there is no statistically significant difference between the per capita water consumption of those who live in traditional houses compared with those living in higher density housing may be seen as counter-intuitive (Talent et al. 2013) but begins to delineate a possible strategy to reduce consumption.

It is unlikely that household members in differing forms of accommodation in the different Australian cities actually ingest less water although the crude estimate available from ABS (2004) figures (Table 13.1) indicate substantial differences between States. It is presumed that they use similar amounts for food preparation and cooking although those in higher density accommodation are more likely to 'eat out' so use less in the kitchen. Nonetheless, the volume of water ingested or used in food preparation and cleaning of equipment and utensils or used for 'health' services is estimated to be no more than, say 15 l per person per day or 6 kl per year. There may be some ways in which cooking and cleaning utensils could be carried out more efficiently leading to reduction in consumption but the possibility of significant reduction water consumption in this area of use of say 30 % is unlikely.

An aspect of domestic consumption that relates to behaviour is the way modern plumbing affects consumption. The first relates to use of hot water. Whether the hot water is delivered from a storage reservoir or from an 'instant service' there is usually wastage until the 'dead' line is filled with hot water. The wastage occurs each time the hot water supply is turned on and once it is turned off the water cools. Another version of this process occurs each time a 'mixer' tap is used when the 'default' position of the tap is at the mid-point meaning that hot water is drawn into the line to be 'mixed' with the cold water supply even when the user only wants cold water. We have no estimate of the wastage from either of these practices other than to note that it is determined by the length of the 'dead line' and may be significant if the supply of hot water is some distance from the point of use.

Individual members of households in different cities are also likely to use similar volumes of water for the disposal of human body wastes. So far it has been possible to gain some economies in sewerages services largely by reduction in the volume of water used to flush the toilet bowl. With current technology and prevailing aesthetic considerations the possibility of further significant reductions in water consumption from this aspect of our behavior is limited, although encouraging households not to flush routinely for urine discharges is one possibility. Rough estimates are that using water to transport human body wastes results in consumption of about 60 l per person per day or 20 kl per year that could be reduced by 25 %. Adoption of modern composting toilets would offer greater opportunities for reduction in 'black water' flows. Adopting human body waste management systems that separated most of the urine from the fecal material would create opportunities to recover valuable elements of the urine flow and would eliminate most of the health and environmental risks arising from contemporary management of waste. It could also lead to composting of fecal material to produce a useful soil conditioner (Although it should be noted that there is and there is likely to be a continuing aversion to composting – an attitude that is likely to remain so long as some locations and activities may continue to 'enjoy' flush toilets). Another benefit from developing a more environmentally sensitive approach to management of human body wastes is that it would lead to significant reduction in waste flows into rivers, lakes and oceans that currently compromise the quality of such water bodies.

This leaves a significant proportion of the domestic consumption of water being used in bathing, laundry and garden uses. Changes in the domestic capture and re-use of water used for these activities is probably the major source of reduction in consumption.

While low flow shower heads may reduce water consumption there is a view that savings from such equipment may be offset by an increased propensity to increase shower time. It is therefore unlikely that a major reduction in water consumption for bathing can be achieved given the, focus on personal hygiene and the undoubted pleasures of bathing and showering. Nonetheless, waste flows from bathing and showering are low in pollutants suggesting that they can be reduced sufficiently to provide water of quality suitable for toilet flushing, laundry use and or garden watering.

We note, however, that the public education programs and water efficiency programs such as the use of low flow shower heads, dual flush toilets appear to have led to continuing but relatively small levels of reduction in water use in Canberra (Talent et al. 2013).

Assuming that households use potable water supplies only for kitchen and bathroom usage the potable water consumption could be reduced by as much as two thirds with the 'waste flows' from kitchen and bathroom being treated on site to produce water for toilet flushing and laundry usage the 'surplus' water from such consumption would be available for garden usage. Maguire (2008) provides a simple model of how such a system could be developed. One benefit of his proposal is that it could be retrofitted relatively easily in existing suburban developments and even more easily in new traditional housing developments. If no such water was used in domestic garden maintenance it would be discharged to the sewerage system for treatment prior to discharge to the environment or available for local community open space/garden usage or appropriate industrial usage. There is a range of dry composting toilets that could be appropriately installed in traditional density neighbourhoods which would also lead to significant reductions in domestic water consumption.

9 Reduction in Domestic Water Consumption

The traditional model of consumption recognized that detached dwellings had more garden space which, it was assumed, led to greater water consumption to maintain them. It was also based on the assumption that there were economies of scale in many household activities that led to lower water consumption in these dwelling types.

The crude average consumption of different dwelling types, to an uncritical eye, do suggest that flats use less water than houses. But once these averages are mediated by the households that live in them and the comparative consumption is adjusted to take account of the size of households in the various dwelling types the differences largely disappear and even suggests that those in high density dwellings consume slightly more water per capita than those who live in medium and low density dwellings. The obvious point is that households in detached dwellings have, on average, 2.4 people per house compared with 1.5 for those who live in high density dwellings (Randolph and Troy (2008), Talent et al. (2013)).

It would seem possible to reduce domestic water consumption by about two thirds if dwellings were required to have a simple way of capturing the relevant flows from baths, showers, laundries and hand basins and treating the stored sullied water to an appropriate standard to account for the differences in time of the demand for use in toilet flushing laundries and gardens.. A relatively small rainfall storage tank, plumbed into the dwelling, would give a high level of security over the supply of water for the different uses. It would have the incidental effect of reducing storm water runoff problems.

In most Australian cities the daily demand for water for domestic purposes is, in total, relatively constant while variable throughout the day across their area. Water supply services are, however, highly centralized. By developing on site processing and storage capacity for each area we could reduce the amount of water pumping, spread the demand across the day and across the city region, minimize storm-water runoff, produce locally treated water for local public parks and gardens and increase local resilience while reducing the stress on the ecosystems from which water is extracted and on those into which waste water currently is discharged.

In an earlier study Randolph and Troy (2008) argued that 'attitudinal' causes of environmentally significant behavior identified by Stern (2000) 'have the greatest predictive value for behaviours that are not constrained by context or personal capabilities'. There is also some evidence that changing attitudes to environmental concerns may result in reduced consumption. An extension of the study of water consumption of Canberra households indicated that social attitudes may have resulted in lower per capita consumption of water in households who highly valued environmental concerns. A similar saving in water consumption among households in conventional detached houses identified as 'Water Wise' was noted in a Sydney survey (IPART 2011). Note that this small household survey did not include households in flats but assumed house dwellers used more water 'but only because they use water outdoors' (IPART 2011, p. 3).

Detached dwellings are more likely to have gardens that consume high volumes of water, especially in summer. But many flat dwellers maintain a variety of pot plants that have high levels of water consumption. Many high density dwellings also have garden beds or landscaped elements that require season watering. Moreover, over recent years the housing policy pursued by the planning agencies has led to a significant reduction in the size of detached dwelling allotments which, paradoxically, has been accompanied by an increase in house size, thereby reducing significantly the size of any garden. One of the consequences of the reduction in allotment size is that dwelling are less likely to be surrounded by trees and large shrubs that have a local climate moderating effect. The more recent subdivisions have had greatly reduced street widths, and thus reduced 'nature strips' This in turn means that there is less opportunity to grow street trees that help moderate the local climate thus leading to housing that is more dependent on 'air conditioning' that often increases water consumption. that the more recent subdivisions have had greatly reduced street widths, and reduced 'nature strips' thus leading to development of 'heat islands' within the city (Stone (2012)) as a consequence increasing the propensity to use 'air-conditioning' including evaporative coolers which consume significant volumes of water.

Households in high density dwellings are more likely to be renting their dwelling and are less likely to have their own water consumption directly metered, creating a 'tragedy of the commons' environment. High density dwellings typically have only 1 meter for the block, so water consumption is apportioned evenly across dwellings. Occupants thus have no pressure on them to be more careful in their use of water because they cannot reduce their water bill by moderating their consumption (Dwellings could be provided with separate meters but this would be expensive because of the increased plumbing.).

Moreover, rented accommodation is less likely to have modern water efficient fittings and appliances or for them to be maintained as well as those in privately owned housing. Approximately two thirds of higher density dwellings are rented making this form of accommodation more likely to have higher consumption.

There is also evidence from other studies that those who live in high density dwellings (with fewer occupants per dwelling) tend not to wait until the clothes and dishwashers are full before using them. There is anecdotal evidence that those who live in high density dwellings are less likely to be required to moderate their behaviour because they are under less social pressure to 'share' than larger households in conventional houses and are likely to have longer showers which in turn tends to increase water consumption. This is more likely to be the case in new high density dwellings that have no storage heated hot water compared with detached dwellings, especially the older dwellings that have relatively small 'off peak' hot water storage, in which the household develops strategies for 'sharing' the hot water for showering (this is commonly referred as a problem families face in trying to get teenagers to 'behave' in a socially responsible way).

Low volumetric water prices offer little incentive to economise on consumption. The current imposition of high fixed charges relative to the metered consumption charge also provides little opportunity for households in conventional housing to economise on their use of water (Grafton et al. 2011). Households in higher density developments that usually have only 1 m per development have even less opportunity to reduce their water consumption charges. This leads to a 'tragedy of the commons' effect which may even increase the apparent consumption of households. The smaller size of households that live in flats also generally means there is less social pressure to 'share' hot water supplies (commonly provided in conventional houses by 'off peak' hot water tanks) that results in longer showers and therefore higher per capita consumption in shower consumption among flat dwellers. Water authorities could, and possibly should, be forced to change their approach by reducing or eliminating most of the fixed charges by including the 'fixed charges' onto the price of a litre of water which would make the rewards of economizing clearer to consumers. At the moment 'economisers' subsidise the profligate by paying more than their share of fixed costs. Such a reform would encourage more households to capture more of the rain falling on their roofs for their own domestic use.

10 Conclusion

The domestic consumption of water is undeniably influenced by the conditions under which people live. It is clear however that physical determinist assumptions related to city form are not confirmed by evidence of their actual consumption. Social attitudes to the significance of environmental concerns is reflected in household consumption leading to approximately 10 % reduction in water use per capita but this only applies to a small proportion of the population. The 'water cycle' that followed from the breakthroughs in water consumption proposed by Chadwick (1842) a century and a half ago was felicitous but now, given the increase in population, the increase in per capita consumption and the consequential increasing stress on the environment a new paradigm is urgently needed in the management of urban water services. This implies that it is important to explore new ways of achieving the comforts offered by urban life and in the way wastes are managed while reducing the consumption of water and thus reducing the stresses on the ecosystems that sustain urban living.

References

Australian Bureau of Statistics. (2004). *Water account Australia 2000–01*. Catalogue number 4610.0, Canberra.

Chadwick, E. (1842). *The sanitary conditions of the laboring population of Great Britain*. Edinburgh: Edinburgh University Press (with an Introduction by Flinn, M. W. (1965)).

Dingle, T. (2008). The life and times of the Chadwickian solution. In P. Troy (Ed.), *Troubled waters*. Canberra: ANU E Press.

Dingle, T., & Rasmussen, C. (1991). *Vital connections: Melbourne and its board of works 1891–1991 McPhee Gribble*. Melbourne: Penguin Books Australia.

Earth Summit. (1992). See also No 28, Treaty Series, (1995).

Gaynor, A. (2006). *The harvest of the suburbs: An environmental history of growing food in Australia cities*. Crawley: University of Western Australia Press.

Grafton, R. Q. (2010). How to increase the cost-effectiveness of water reform and environmental flows in the Murray-Darling Basin. *Agenda, 17*(2), 17–40.

Grafton, R. Q., Ward, M. B., To, H., & Kompas, T. (2011). Determinants of residential water consumption: Evidence and analysis from a 10-country household survey. *Water Resources Research, 47*, W08537.

Heidarpour, F., Ghani, W. W. A. K., Fakhru'l-Razi, A., Sobri, S., Heydarpour, V., Zargar, M., & Mozafari, M. R. (2011). Complete removal of pathogenic bacteria from drinking water using nano silver-coated cylindrical polypropylene filters. *Clean Technologies and Environmental Policy, 13*(3), 499–507.

IPART. (2011). *Determinants of residential energy and water consumption in Sydney and surrounds, regression analysis of the 2008 and 2010 IPART household survey data electricity, gas and water — research report*. Independent Pricing and Regulatory Tribunal, Sydney, N.S.W. ISBN 978-1-921929-61-8 RP-34.

Kaegi, R. (2009). Nanoparticles in drinking water. *Eawag News, 66*, 7–9.

Lloyd, C., Troy, P., & Schreiner, S. (1992). *For the public health: The hunter district water board 1892–1992*. Sydney: Longman Cheshire.

Mackellar, D. (1918). My country. In W. Murdoch (Ed.), *The Oxford book of Australasian verse*. London: H. Milford, Oxford University Press.

Maguire, S. (2008). *Water not down the drain*. Melbourne: Alternative Technology Association.

Mullins, P. (1981a). Theoretical perspectives on Australian urbanisation 1. Material components in the reproduction of Australian labour power. *Australian and New Zealand Journal of Sociology, 17*(1), 65–76.

Mullins, P. (1981b). Theoretical perspectives on Australian urbanisation: II. Social components in the reproduction of Australian labour power. *Australian and New Zealand Journal of Sociology, 17*(3), 35–43.

NHMRC, NRMMC. (2011). *Australian drinking water guidelines paper 6 national water quality management strategy*. National Health and Medical Research Council, National Resource Management Ministerial Council, Commonwealth of Australia, Canberra.

No. 28, Treaty Series. (1995). *United Nations framework convention on climate change*. New York, 9 May 1992.

Ockhams Razor. (2012). Australian broadcasting commission, Radio National 29 Jan 2012.

Qu, X., Alvarez, P. J. J., & Li, Q. (2013). Applications of nanotechnology in water and wastewater treatment. *Water Research, 47*(12), 3931–3946.

Randolph, B., & Troy, P. (2008). Attitudes to conservation and water consumption. *Environmental Science and Policy, 11*, 441–455.

Stern, P. C. (2000). Towards a coherent theory of environmentally significant behaviour. *Journal of Social Issues, 56*(3), 407–424.

Stone, B. (2012). *The city and the coming climate change in the places we live*. New York: Cambridge University Press.

Talent, M., Troy, P., & Dovers, S. (2013). *Canberra residential energy and water consumption*. Report to ACT Department of Climate Change Energy and Water. Fenner School of Environment and Society, Australian National University, Canberra.

Troy, P., Holloway, D., & Randolph, B. (2005). Water use and the built environment: Patterns of water consumption in Sydney. *City futures research centre research paper no. 1*. Sydney: University of New South Wales City Futures Research Centre.

Trouiller, B., Reliene, R., Westbrook, A., Solaimani, P., & Schiestl, R. H. (2009). Titanium dioxide nanoparticles induce DNA damage and genetic instability in vivo in mice. *Cancer Research, 69*(22), 8784–8789.

Victoria. Department of Sustainability and Environment. (2004). *Securing our water future together: Victorian government white paper*. Departmentt of Sustainability and Environment, Melbourne.

Water Board (Corporatisation) Act (1994) No. 88 (NSW).

Chapter 14
Non-household Water Demand: The Industrial and Commercial Sectors

Steven Renzetti

1 Introduction

Major businesses are increasingly turning their attention to water. Some of these firms view water as a source of risk, while many others see it as a potential business opportunity (Carbon Disclosure Project 2011). In addition, many governments and municipal water supply agencies are turning their attention to non-residential water uses as potential sources of water conservation and savings (Vickers 2001). In Canada, for example, an increase of 1 % in the volume of water internally recirculated by all manufacturing firms would release enough water to supply the inhabitants and businesses of a city of 500,000 people (Statistics Canada 2008). These trends suggest it would be valuable to consider what is known regarding the economics of non-residential water use.

Non-residential water use refers to water supplied to industrial and commercial firms and to institutions such as government offices, hospitals, and schools. The economic features of the water demands of these businesses and organizations have not received the same research attention as has residential water use. This is largely due to the challenges in acquiring industrial water use data and because there is a greater degree of diversity among commercial and industrial water users. For example, this category includes the water used to make coffee in a lawyer's office as well as the water used to cool petroleum distillates in a petrochemical plant. This range in the technologies associated with these different water uses poses a challenge to modellers and analysts seeking to understand the water use of the sector.

The purpose of this chapter is to examine what is known regarding the economic features of water use by industrial and commercial firms and to consider how this

---

S. Renzetti (✉)
Department of Economics, Brock University,
500 Glenridge Avenue, L2S 3A1 Saint Catharines, ON, Canada
e-mail: srenzetti@brocku.ca

Q. Grafton et al. (eds.), *Understanding and Managing Urban Water in Transition*, Global Issues in Water Policy 15,
DOI 10.1007/978-94-017-9801-3_14

information can be used to promote efficient and sustainable water use. The focus of the chapter is on water use by commercial and manufacturing firms, and we do not consider firms in primary industries such as mining and natural resource extraction and institutions such as government offices and schools.[1] The next section points out some of the more important characteristics of industrial and commercial water use. Of particular interest are the ways in which this sector's water use differs from residential water use. Section 3 considers the results of the application of economic modelling techniques to measuring industrial and commercial water use. Section 4 presents a number of case studies of efforts to promote water conservation in these sectors.

2 Features of Industrial and Commercial Water Use

Firms use water as an input in their production processes. The uses to which water is put vary widely within the industrial and commercial sectors:

- Cooling of intermediate outputs (e.g. petroleum distillates)
- Steam production
- Moving and cleaning raw materials (e.g. meat products)
- Producing electricity
- Inclusion in final output (e.g. canned peaches)
- Sanitation
- Lawn watering.

It is surprisingly difficult to get accurate data on industrial and commercial water use. This is in part due to the fact that some of this water use is supplied by municipal water systems and, thus, may be recorded as part of aggregate municipal water use. In the European Union, for example, non-residential and residential water use are aggregated into 'urban' water use (Lallana and Marcuello 2004). In the most recent report on water use from the United States, total estimated urban water use is divided into domestic and "All other uses and system losses" (Kenny et al. 2009, p. 17). In Canada, a survey conducted of municipal water systems indicates that, in 2006, the shares of municipally supplied water going to residential, commercial and institutional, and industrial were 57.0 %, 19.3 %, and 11.5 %, respectively (the remaining 12.8 % of output was considered as system losses) (Environment Canada 2010). Recent estimates from Australia indicate that 20–30 % or publicly supplied water goes to commercial and industrial customers (Water Services Association of Australia 2009). Similarly, Zhou and Tol (2005) estimate that industrial water use represents 22 % of overall water use in China.

A second complicating factor is that, for those industrial facilities which obtain their water directly from rivers, lakes, or aquifers, intake may be estimated rather

[1] Readers interested in the economic dimensions of water use in these industries should consult the National Roundtable on the Environment and the Economy (2011).

than directly measured. The United States Geological Survey derives its estimate of total self-supplied industrial water use by combining state-level estimates from licence data with estimated imputed water use figures (calculated by multiplying the number of employees by coefficients of estimated water use per employee). In fact, most national governments do not directly monitor and report water withdrawals by self-supplied industrial facilities. An exception is Canada where an Industrial Water Use Survey is carried out every 5 years in which manufacturing and mining facilities are sampled in order to collect plant-level observations of water use and expenditures (Statistics Canada 2008).

Despite these data challenges, there are a variety of observed features of commercial and industrial water use which are noteworthy. These features may play a role in determining firms' choices regarding water use and how firms respond to changes in prices and other relevant factors. A number of these features also demonstrate the differences between non-residential and residential water use.

The first feature concerns the sources for commercial and industrial water. Firms may obtain intake water from several potential sources. Most commercial water demand is supplied by municipal water systems (with the possible exception of purchases of bottled water in small quantities). Furthermore, most smaller industrial water users (e.g. small manufacturing plants) are also likely to be supplied by municipal water systems. Larger industrial facilities such as oil and gas refineries, steel plants, and chemical producers may obtain their water through a combination of self-supplied water and water supplied by a municipal water system. In general, the larger the industrial facility, the larger its water demands and the greater is the likelihood that it is self-supplied.

The second feature relates to the potential for in-plant recirculation. In a number of industrial production processes, firms may choose to recapture water and reuse it. This may be done for a variety of reasons including the desire to capture and remove something valuable in the water (e.g. heat, chemicals, raw materials), the desire to avoid having to increase water intakes, or the desire to avoid water discharges (which may be subject to environmental and other regulations). The decision to recirculate water must be preceded by the firm investing in the capital, technologies, and worker training to make recirculation feasible. Once that is done, the firm may optimize over the volume of water to recirculate just as it optimizes over the amount of water to bring into the plant. This potential for water recirculation is quite important for at least two reasons. First, it is one of the main ways in which industrial water use differs from most observed residential water use (although the possibility of households reusing 'grey water' is increasing). Second, it has the potential to reduce water intake to the extent that firms view recirculated water as a substitute for intake water. We will return to this point below.

The third feature concerns the relationship between firms' choices regarding water use and how they regard the employment of other inputs. It has long been appreciated in applied economics studies of production technologies that a firm's profit-maximizing use of one input is likely to be closely tied to not only its own price but the prices of other inputs. In this way, a firm's use of water may depend not only on its own price but the prices of capital, labour, materials, and other inputs. Furthermore, efforts by a firm to conserve a particular input – for example, energy – may have implications for its water use. For example, in an effort to reduce its energy expenditure, a firm may switch from air cooling to water cooling of intermediate inputs.

The fourth feature relates to the diversity of water use within the commercial and industrial sector. As indicated above, water use in the sector varies from the small amounts of water used in offices to larger volumes used in restaurants and finally to the very large volumes used in industrial facilities such as chemical plants. This diversity stands in contrast to residential water use where, for the most part, households use water for largely the same purposes. This diversity has several implications. First, the relationship between water and other inputs will vary across firms. Second, the risks associated with water supply disruptions or variations in raw water quality will also vary significantly across firms. It is important to note, however, that it is not just 'heavy' water users who are most vulnerable to water supply curtailments. Commercial establishments cannot operate without adequate water for fire protection and any commercial operation involving the handling and production of food-related products relies critically on water supplies.

The fifth feature concerns the role of regulations governing non-residential water use. In some cases, industrial water users may have their water use decisions influenced through one of two types of government regulations. The first concerns water intakes for self-supplied water users. In most jurisdictions, self-supplied water users must hold a government-issued licence that sets out the conditions of water use: location, volume, timing, etc. These licences will also sometimes indicate how a user's water allotment may be decreased in times of shortage. Common methods include using the seniority of the licence to determine the order in which users' withdrawals are curtailed (typically the most senior licence-holder is the last to be rationed), and reducing all water licence-holders by equal proportions. A second type of regulation that may influence water use concerns the quality of discharge waters. Water quality regulations may constrain the firm's decision-making by limiting the volume, mass, and/or concentration of contaminants contained in wastewater flows.

The final feature concerns firms' expenditures on water. In the cases where firms are supplied by municipal water systems, billing records can in principle provide analysts with the data needed to calculate expenditures. The heterogeneity of scale and water uses among non-residential customers, however, makes generalizations quite difficult. Another source of difficulty is that large-scale self-supplied industrial water users may face both external and internal costs associated with their water use. In some jurisdictions, self-supplied firms must pay administrative fees to government or obtain their water through market transactions. In addition to these external costs, self-supplied firms typically also experience internal costs associated with pumping, treatment (prior to and after use), and on-site storage. There is remarkably little known about these internal costs. In Canada, the Industrial Water Use Survey sheds some light on these internal costs and survey responses suggest that the cost share of water in the manufacturing sector rarely rises above 1 % (Dupont and Renzetti 2001). Reynaud (2003) found that the French manufacturing sector's cost share associated with water varied from 1.2 % to 1.9 %, according to the subsector.

3 Economic Determinants of Commercial and Industrial Water Use

Firms combine and transform inputs such as labour, capital, energy, materials, and water into goods and services for sale. In doing so, economists assume firms' decisions related to the quantities and types of inputs and outputs are guided by their goals (typically to maximize profits or to minimize the cost of a given level of production) and are constrained by external forces (market prices, consumer preferences, competitors' actions, government regulations). One of the outcomes of these decisions is the firm's demand for intake water – that is, the relationship between the firm's desired level of water use and all of the other factors which influence that demand. Specifically, it can be shown (Renzetti 2002b) that a firm's representative cost-minimizing demand for intake water is a function of the following factors:

- Price of water
- Prices of other inputs
- Level of output
- Production process
- Regulations (e.g. licences governing quantity of water that may be withdrawn or levels of contaminants in wastewater flows).

A change in any one of these factors is predicted to alter the desired level of water use. For example, one would expect that increases in the level of output and decreases in the price of intake water will both increase the demand for water. The effect of changes in other input prices on the demand for water will depend on whether those inputs are substitutes or complements to intake water in the firm's production process.

In order to measure the strength and nature of these relationships, economists apply statistical techniques to input and output data in order to estimate the water use–demand relationship. From that estimated relationship, a commonly computed parameter is the demand elasticity. For example, the price elasticity of intake water demand is calculated as the predicted proportional change in the desired quantity of water intake in response to a small change in the price of intake water. Thus, a price elasticity of –0.5 indicates that a 10 % increase in the price of water is predicted to reduce water intake by 5 %. Similarly, the cross-price elasticity of intake water demand is the change in the desired quantity of water intake that occurs in response to a small change in the price of another input. If a cross-price elasticity is estimated to be positive, then the two inputs are deemed to be substitutes, since an increase in the price of one (e.g. energy) leads to a greater use of the other (e.g. water).

Economists have carried out only a limited number of empirical studies that apply statistical methods to estimate the commercial and industrial water demands and the relative importance of each of the factors identified above. Surveys of this literature are available in Renzetti (2002a, b), de Gispert (2004), and Worthington (2010). As those surveys highlight, less attention has been paid to commercial and industrial water demands compared to residential and agricultural water demands.

This is probably due to the fact that these sectors do not usually constitute the dominant share of water use, to challenges of obtaining data, and to the heterogeneity of uses for water observed in commercial and industrial processes. The features and major empirical findings of this literature are summarized in Table 14.1. Based on that summary, this section will point to several important findings in the literature.

The factor which has received the most attention as a potential determinant of water use in studies of commercial and industrial water demands is the price of water. In an early study, Babin et al. (1982) estimated water demands for US manufacturing industries. Price elasticities of water demand varied by manufacturing subsector from 0.00 to −0.81. In a series of studies on plant-level observations of Canadian manufacturing firms carried out by the author, estimated price elasticities water intake ranged widely: −0.1534 to −0.5885 (Renzetti 1992) and −0.8098 (Dupont and Renzetti 2001). More recently, Zhou and Tol (2005) estimated a model

Table 14.1 Sample of commercial and industrial water demand studies

Author	Sector	Data	Method	Estimated elasticities
Babin et al. (1982)	U.S. manufacturing	1973 cross section of 2-digit SIC level observations	Systems of equations	Intake: 0.54 to −0.66
Schneider and Whitlatch (1991)	Commercial	Time series observations on aggregate commercial customer class water use from 16 Ohio cities	OLS	Intake (short-run): −0.234 Intake (long run): −0.918
Renzetti (1992)	Canadian manufacturing	1985 cross section of plant-level observations	Two-stage systems of equations	Intake: −0.153 to −0.588; Recirculation: −0.508 to −1.48
Lynn et al. (1993)	Commercial	1985 cross section of firm-level observations	OLS	Intake: −0.24 (motels) to −1.33 (department stores)
Dupont and Renzetti (2001)	Canadian manufacturing	3-year panel of 2-digit SIC level observations	Two-stage systems of equations	Intake: −0.79 to −0.81; Recirculation: −0.75 to −0.82
Reynaud (2003)	French manufacturing	Panel of plant-level observations 1994–1996	Seemingly unrelated regressions and feasible GLS	Intake (public supply): −0.10 to −0.79; Intake (self-supply): −0.90 to −2.21
Zhou and Tol (2005)	Chinese manufacturing	Province-level panel data 1997–2003	Feasible GLS	Intake: −0.22 to −0.35
Féres and Reynaud (2005)	Brazilian manufacturing	1999 cross section of plant-level observations	Seemingly unrelated regressions	Intake: −1.085

of industrial water demands using aggregate Chinese data and calculated an estimated price elasticity of −0.35. Thus, most industrial water demand studies find that water demands are inelastic, but elasticity values have tended to be higher than those found in residential water demand studies.

Given the relatively small number of studies and broad range of technologies and countries found in the literature, it is difficult to make generalizations regarding the factors influencing the estimated price elasticity of demand for intake water. Two studies (Renzetti 1993; Reynaud 2003) find that firms have different price elasticities for demand for publicly supplied networked versus self-supplied water. Furthermore, Féres and Reynaud (2005) suggest that recent industrial water demand studies from Brazil, China, and India all exhibit relatively high intake price elasticities. This may be a feature of the relatively newer technologies embodied in industrial water demands in developing economies or some other feature of the data (such as a preponderance of large firms in the samples).

It can be expected that commercial and industrial water use will be influenced by the level of a firm's production. As Worthington (2010) points out, comparisons across studies are difficult because a number of measures of output (value added, number of employees, etc.) have been used in various studies. Nonetheless, the limited number of studies available suggests that non-residential water demands may be characterized by output elasticities that are relatively large. Dupont and Renzetti (2001) report an output elasticity for the Canadian manufacturing sector's water demand as 0.7, while Reynaud (2003) reports values of 0.34 and 0.58 for French manufacturing's demand for publicly supplied and self-supplied water, respectively. This stands in contrast to the common finding in studies of residential water demands of income elasticities being positive but quite small (Renzetti 2002a).

Another potentially important factor in determining industrial and commercial water demands is the price of energy. In industrial applications, water is often used to either cool an intermediate input or heated to clean inputs or produce steam. In these applications, one could expect that water demands would be closely tied to energy prices. Unfortunately, relatively few studies of commercial and industrial water use have included energy prices as an explanatory variable in their models. In the recent cases where this has been done (Dupont and Renzetti 2001; Féres and Reynaud 2005), intake water and energy have been found to be substitutes.

A small number of research studies have considered commercial water use. The work done in this area suggests that this sector's water use is somewhat sensitive to water prices and the firm's level of output, although these factors may not play as important a role as found in residential and industrial water use. Schneider and Whitlatch (1991) estimate price and output elasticities for municipally supplied commercial water users and estimate the sector's price elasticity of water demand to be −0.918. Lynn et al. (1993) studied commercial firms in Miami, Florida, and estimated firms' responses to price changes on water use. Estimated price elasticities were the following: −1.33 (department stores), −0.76 (grocery stores), −0.12 to −0.24 (motels and hotels), and −0.174 (restaurants). While these estimates indicate that commercial firms will conserve on water use when confronted with price increases, the diversity of water uses in the commercial sector and water's typically small cost share make it difficult to give general conclusions regarding the nature of commercial firms' water use.

An alternative to the econometric approach to studying water use by commercial firms is to conduct detailed studies of the 'end uses' of water in different types of establishments (restaurants, hotels, offices, etc.) and then develop Best Management Practices that establish benchmark levels of water use for the different types of water use (kitchens, laundry, bathrooms, outdoor uses, etc.). This approach has proven popular with water supply agencies and governments who are interested in promoting water conservation in these sectors (Dziegielewski 2000). It also facilitates planning for future economic development and implied growth in water demand since the bases for forecasts (number of sinks per restaurant, number of bathrooms in a hotel) are usually known and, as a result, can form the basis for water demand forecasts.

While it is fairly clear that the available econometric studies of commercial and industrial water demands demonstrate that economic factors (such as the price of water and the level of output) play a role in determining water use rates, conclusions from this small set of studies should be drawn with caution. Many of the studies of manufacturing water use are based on samples in which large-scale, self-supplied industrial facilities predominate. In addition, the limited number of recent studies and the heterogeneity of water use among commercial establishments mean that further analysis is required.

Studies cited thus far have focused primarily on investigating firms' responsiveness to the price of water and the level of output, all within the context of a neoclassical model of production technologies. To a lesser degree, past research has also investigated the role played by the prices of other inputs. These empirical models are valuable in assisting analysts seeking, for example, to anticipate firms' responses to changes in input prices. The models, however, have largely neglected some features of industrial and commercial water use which are relevant to efforts to predict reactions to changes in a firm's operating environment, changes which relate to the factors identified above that differentiate firms' water demands from residential water demands. Four specific examples of these neglected features are the following: in-plant water recirculation, the potential for switching water sources, the role played by technological change, and the role played by water quality regulations.

The ability to recirculate water within a production process is an important characteristic of many manufacturing processes and it is a feature that distinguishes industrial water use from most other instances of water use. Furthermore, changes in recirculation practices by industrial water users can potentially have significant impacts on water availability for ecological and human needs. Despite this, there has been limited analysis of the factors influencing recirculation decisions – largely because of the difficulty of obtaining observations of the volumes of water recirculated and other relevant variables. A recent paper (Bruneau et al. 2010) describes an econometric model that considers two facets of firms' recirculation behaviour: first, the discrete decision of whether to recirculate, and second the decision of how much to recirculate. The model applies a two-stage estimation procedure to cross-sectional data from Environment Canada's 1996 Industrial Water Use Survey. Estimation results indicate that, in the first stage, relative water scarcity and production technologies influence the decision of whether to recirculate water. In the second stage, the prices of intake water and water recirculation, as well as the scale of operations, are found to influence the choice of the optimal quantity of water to recirculate.

Large industrial water users may also have more than one source of water supply and, as a result, may possess the capacity to choose how much water to draw from each source. For example, in response to a significant increase in the price of water, an industrial facility might switch from relying on publicly supplied water to being self-supplied, or switch between one public supply network and another. There is limited evidence for firms having this capacity (Renzetti 1993), but the potential for it to happen is important because a large industrial user leaving a water supply network could have significant implications for a water agency's revenues. In addition, firms with the capacity to switch water sources may be able to negotiate exemptions from rate increases. Indeed, recent work by the water industry regulator in England and Wales suggests that this possibility – of large industrial water users switching suppliers – is one of the factors to be exploited in efforts to promote increased innovation and competition among water suppliers (United Kingdom Office of Water 2012).

The applied economics literature has a long history of concern about the impact of technological innovation and input use. This line of research has, for example, documented the role played by technological change in reducing the manufacturing sector's use of energy relative to the value of output (Fatih and Yeddir-Tamsamani 2010). Unfortunately, while there is some anecdotal evidence of the impact of innovation on non-residential water demands (Vickers 2001; National Roundtable on the Environment and the Economy 2011), this issue has not yet been carefully studied. Solley et al. (1999, p. 64) contend that "the decrease [in water use] from 1980 to 1995 can be attributed, in part, to the following major factors… New technologies in the industrial sector that require less water, improved plant efficiencies, increased water recycling, higher energy prices, and changes in laws and regulations to reduce the discharge of pollutants resulted in decreased water use and less water being returned to the natural system after use." Unfortunately, the authors did not measure the relative importance of each of these factors on industrial water demands. As a result of this lack of empirical analysis, the role of technological change in industrial water use is a particularly important gap in the literature.

The final feature concerns non-residential facilities' water discharges. The amount of water leaving an industrial facility must equal the amount entering the facility minus the amount lost or consumed during production. This equality, however, neglects the possibility that the quality of water may be changed as a result of its use. Discharges from manufacturing and other industrial processes contain a wide variety of contaminants including organic materials, metals, and ammonia. All of these pose significant potential threats to human and ecosystem health (Culp et al. 2001). Recognizing these dangers, most jurisdictions have enacted regulations that limit the quantities and/or concentrations of contaminants in discharge waters.

There is some evidence that water quality regulations may have been an influential factor behind observed reductions in industrial water intake in the U.S. and Europe (Solley et al. 1999). Market conditions, such as rising sewage treatment costs, may have reinforced these effects. For example, Renzetti (1992) finds that, for the Canadian manufacturing sector as a whole, the own price elasticity of water discharge is –0.97. Thus, Canadian evidence suggests increasing the cost of discharges through regulations or fees may reduce both intake and discharges. Murty (2002) investigated the impacts of water pollution regulations on a sample of Indian manufacturing firms. The author found that regulations imposed costs upon the firms and that these costs varied significantly across firms and industrial sectors. Féres and Reynaud (2005), however, found that more stringent regulations regarding the quality of industrial discharge waters may increase water intake for Brazilian manufacturing firms. The authors suggest that a combination of increased intake prices and more stringent discharge requirements would induce firms to reduce both intakes and discharges.

4 Promoting Conservation

There is a growing interest on the part of governments and municipal water suppliers in promoting conservation among water users. Much of the attention has focused on residential water use and has resulted in some agencies adopting conservation-oriented pricing, which provides subsidies for the installation of water-conserving appliances and the adoption of water-saving practices. The city of Guelph, for example, is one of Canada's fastest growing cities and depends almost entirely on groundwater for its water supply. The city has set a goal of reducing overall water use by 20 % by 2025 and has established a consumption target of using less residential water per capita than all other comparable Canadian cities. In order to achieve these goals, Guelph has adopted a comprehensive water conservation plan which includes a 20 % increase in water and wastewater user rates, subsidies for residential water retrofits, and an education and awareness program (Brandes et al. 2010).

While perhaps not as widespread as the interest in promoting residential water conservation, there are also programs which seek to encourage greater efficiencies in commercial and industrial water use (Vickers 2001). In some cases, conservation programs are organized by municipal water suppliers. The water utilities in Seattle and neighbouring cities combined to promote water conservation among all customer classes (www.savingwater.org). The consortium employed a number of policies to encourage water use among commercial and industrial customers, including the following:

- Subsidies for retrofits of inefficient water-using appliances and equipment
- Assistance with water use audits
- Publication of Best Management Practice guidelines and case studies which set benchmarks for common water uses.

In Australia, many municipal water suppliers have introduced a range of programs and policies to promote commercial and industrial water conservation. In a review of those efforts (Water Services Association of Australia 2009), the national association of municipal water suppliers identified key features necessary for success:

- Leadership: managers and staff within a firm or institution require positive signals from senior management that water conservation initiatives will be supported
- Measurement: while most commercial and industrial water users are metered, it may be necessary to introduce submeters throughout a facility in order to obtain a more accurate measure of water use in specific functions and processes
- Staff involvement: all members of an institution or firm must be part of water conservation efforts, especially given the amount of technical knowledge held by operational staff
- Target setting: setting specific targets has been found to be more persuasive than adopting poorly defined and open-ended conservation or improved efficiency goals.
- Benchmarking: the identification of water use reductions that are achievable in comparable settings provides a strong motivating force to encourage innovation among staff and managers.
- Funding: small and medium-sized firms operating in competitive environments may have difficulty investing in water conserving processes and technologies. The experience from Australia indicates that uptake by these firms can be accelerated through relatively small cost-sharing subsidies.

In other cases, state and national governments have developed water conservation plans and manuals for firms seeking to reduce their water use. The United Kingdom's Environment Agency has developed a six-step template for firms to follow when developing their own water management plan (United Kingdom Environment Agency 2006). The main elements of that template are the following:

- Obtain management and staff support
- Establish water and sewage expenditures for the organisation
- Identify and measure water use
- Reassess water use to identify potential savings
- Evaluate potential water efficiency measures (e.g. in terms of payback period) and develop a plan to implement chosen measures
- Implement the plan, monitor results, and report.

California's Urban Water Conservation Council provides commercial, institutional, and industrial water users with detailed information on potential water savings arising from adoption of Best Management Practices in their sectors; it also supplies manuals that allow calculation of the costs and benefits of installing water-conserving capital and adopting water-saving processes (California Urban Water Conservation Council 2002). Furthermore, jurisdictions that have traditionally been viewed as water-abundant are encouraging commercial and industrial firms to explore water conservation. The State of North Carolina recently released a manual providing both general and sector-specific guidelines for firms to design and implement a water-efficiency program (North Carolina Department of Environment and Natural Resources 2009).

These programs attest to the widespread and growing interest in promoting water conservation in the commercial and industrial sectors. A significant challenge arises when attempting to assess these programs as their impacts are rarely carefully documented. Thus, while there is a growing body of literature that measures the impacts of water conservation programs on residential water demands (Millock and Nauges 2010), there is little comparable research for commercial and industrial water users. Nonetheless, in a comprehensive analysis of the potential for water conservation in California, Gleick et al. (2003, p. 10) conclude with respect to commercial and industrial water use that "Overall, we estimate that the range of potential savings is between 710,000 AF/year [acre-feet/year] and 1.3 MAF/year over current use. Our best estimate of practical savings in the CII sector is about 975,000 AF, or 39 % of total current annual water use."

5 Conclusions

In many jurisdictions, water use by industrial plants, retail businesses, hotels, and other firms constitutes an important share in the total demands placed on municipal water supply and wastewater treatment infrastructures. However, there has been a limited amount of analysis of the role water plays in industrial and commercial processes. The available evidence suggests that most facets of industrial and commercial water use are responsive to economic forces such as input prices and the level of output. Unfortunately, the degree to which the various facets of industrial and commercial water use are sensitive to, for example, price changes, has not been as well established as in the cases of other sectors' water use. Furthermore, the roles of factors such as the price of energy are not fully understood. In addition, the time series data sets needed to assess the important issue of technological change and its impacts of water demands have not been assembled. Given the history of real energy price increases over the last several decades and the attendant efforts to develop energy-conserving technologies, it would be particularly interesting to determine the implications of these efforts for water demands.

References

Babin, F., Willis, C., & Allen, P. (1982). Estimation of substitution possibilities between water and other production inputs. *American Journal of Agricultural Economics, 64*(1), 148–151.

Brandes, O., Renzetti, S., & Stinchcombe, K. (2010). *Worth the charge: A primer on conservation-oriented water pricing. POLIS project on ecological governance.* Victoria: University of Victoria.

Bruneau, J., Renzetti, S., & Villeneuve, M. (2010). Analysis of the factors influencing water recirculation decisions by Canadian manufacturing firms. *Canadian Journal of Agricultural Economics, 58*(4), 515–530.
California Urban Water Conservation Council. (2002). *Implementation of best management practices*. http://www.cuwcc.org/home.html. Accessed 18 May 2011.
Carbon Disclosure Project. (2011). *CDP water disclosure global report 2011*. https://www.cdproject.net/water. Accessed 8 Feb 2012.
Culp, J., Ciborowski, J., Dubé, M., Liber, K., Munnkittrick, K., & Parker, W. (2001). Industrial point source discharges. In Environment Canada (Ed.), *Threats to drinking water and aquatic ecosystem health in Canada* (National Water Research Institute scientific assessment report no. 1, pp. 41–51). Burlington: National Water Research Institute.
De Gispert, C. (2004). The economic analysis of industrial water demand: A review. *Environment and Planning C, 22*, 15–30.
Dupont, D., & Renzetti, S. (2001). Water's role in manufacturing. *Environmental and Resource Economics, 18*(4), 411–432.
Dziegielewski, B. (2000). *Commercial and institutional end uses of water*. Denver: AWWA Research Foundation.
Environment Canada. (2010). *2010 Municipal water use report*. http://www.ec.gc.ca/Publications. Accessed 24 Aug 2011.
Fatih, K., & Yeddir-Tamsamani, Y. (2010). Is technological change biased toward energy? A multi-sectoral analysis for the French economy. *Energy Policy, 38*(4), 1842–1850.
Féres, J., & Reynaud, A. (2005). Assessing the impact of environmental regulation on industrial water use: Evidence from Brazil. *Land Economics, 81*(3), 396–411.
Gleick, P., Haasz, D., Henges-Jeck, C., Srinivasan, V., Wolff, G., Cushing, K., & Mann, A. (2003). *Waste not, want not: The potential for urban water conservation in California*. The Pacific Institute, Seattle. http://www.pacinst.org/reports/urban_usage/waste_not_want_not_full_report.pdf. Accessed 16 Feb 2012.
Kenny, J., Barber, N., Hutson, S., Linsey, K., Lovelace, J., & Maupin, M. (2009). *Estimated use of water in the United States in 2005*. United States Geological Service Circular 1344. http://water.usgs.gov/watuse/. Accessed 24 Jan 2012.
Lallana, C., & Marcuello, C. (2004). Indicator fact sheet (WQ2): Water use by sectors. European Environment Agency, Copenhagen. http://www.eea.europa.eu/themes/water. Accessed 4 Jan 2012.
Lynn, G. D., Luppold, W. G., & Kiker, C. (1993). Water price responsiveness of commercial establishments. *Journal of the American Water Resources Association, 14*, 719–729.
Millock, K., & Nauges, C. (2010). Household adoption of water-efficient equipment: The role of socio-economic factors, environmental attitudes and policy. *Environmental and Resource Economics, 46*(2), 539–565.
Murty, M. N. (2002). Measuring the cost of environmentally sustainable industrial development in India: A distance function approach. *Environment and Development Economics, 7*, 467–486.
National Roundtable on the Environment and the Economy. (2011). Charting a course: Sustainable water use by Canada's natural resource sectors, Ottawa. http://nrtee-trnee.ca/charting-a-course-sustainable-water. Use-by-canadas-natural-resource-sectors. Accessed 12 Dec 2011.
North Carolina Department of Environment and Natural Resources. (2009). Water efficiency manual for commercial, industrial and institutional facilities, Raleigh. http://savewaternc.org/Documents/water-efficiency-business.pdf. Accessed 16 Jan 2012.
Renzetti, S. (1992). Estimating the structure of industrial water demands: The case of Canadian manufacturing. *Land Economics, 68*(4), 396–404.
Renzetti, S. (1993). Examining the differences between self and publicly supplied firms' water demands. *Land Economics, 69*(2), 181–188.
Renzetti, S. (2002a). *The economics of water demands*. Norwell: Kluwer Academic Press.
Renzetti, S. (2002b). Introduction. In S. Renzetti (Ed.), *The economics of industrial water use*. Cheltenham: Edward Elgar

Reynaud, A. (2003). An econometric estimation of industrial water demand in France. *Environmental and Resource Economics, 25*, 213–232.
Schneider, M., & Whitlatch, E. (1991). User-specific water demand elasticities. *Journal of Water Resources Planning and Management-ASCE, 117*(1), 52–73.
Solley, W., Pierce, R., & Perlman, H. (1999). Estimated use of water in the United States in 1995. United States Geological Survey Circular 1200. http://water.usgs.gov/watuse/pdf1995/html/. Accessed 16 July 2011.
Statistics Canada. (2008). Industrial water use 2005. Catalogue no. 16-401-X, Ottawa.
United Kingdom Environment Agency. (2006). Waterwise: Good for business and good for the environment. London: The National Water Demand Management Centre. http://www.environment-agency.gov.uk/research/library/publications/33993.aspx. Accessed 9 Jan 2012.
United Kingdom Office of Water. (2012). Water today, water tomorrow: Future price limits – a consultation on the framework, London. http://www.ofwat.gov.uk/consultations/pap_con-201111fpl_app11.pdf. Accessed 4 Jan 2012.
Vickers, A. (2001). *Handbook of water use and conservation*. Los Angeles: Californian Urban Water Conservation Council.
Water Services Association of Australia. (2009). Meeting Australia's water challenges: Case studies in commercial and industrial water savings. WSSA Occasional Paper #23. http://www.waterforlife.nsw.gov.au. Accessed 22 Jan 2012.
Worthington, A. C. (2010). Commercial and industrial water demand estimation: Theoretical and methodological guidelines for applied economics research. *Estudios de Economica Aplicada, 28*(2), 237–258.
Zhou, Y., & Tol, R. (2005). Water use in China's domestic, industrial and agricultural sectors: An empirical analysis. *Water Science and Technology, 5*(6), 85–93.

Chapter 15
Integrating Social Aspects into Urban Water Pricing: Australian and International Perspectives

Noel Wai Wah Chan

1 Introduction

The notion of water as an 'economic good' has been widely accepted among water resource managers (Savenije and van der Zaag 2002). Like other commodities, water can be scarce, competed for, placed on or excluded from a market, and its ownership transferred. It is an important input for agricultural and industrial production as well as for human consumption. With the increasing incidence of water scarcity and to fulfil competing demands, the price of water is becoming an important demand management tool (Griffin 2006).

Water is also an essential good for drinking, food preparation, hygiene, and sanitation. Cost–benefit studies also demonstrate that significant economic and social benefits flow from providing quality water and sanitation services (Hutton et al. 2007; Haller et al. 2007). However, the 2008 Global Water Tariff Survey demonstrated that global water prices increased by about 7 % in 2009 (OECD 2010, pp. 33–62) and the trend is likely to have continued. Taking a global perspective, the Organization for Economic Cooperation and Development (OECD) has suggested that higher water prices are necessary in order to increase awareness of water scarcity, to conserve precious water resources, to reduce pollution, and to provide financial stability for the water sector (OECD 2010).

Over recent years, urban water reform has tended to place increasing emphasis on economic efficiency, environmental conservation, and financial sustainability. The challenge then is to achieve economic objectives while keeping the cost of water equitable and affordable. Instead of relegating social concerns to secondary

N.W.W. Chan (✉)
Crawford School of Public Policy, The Australian National University,
Canberra, Aust Capital Terr, Australia
e-mail: noel.chan@anu.edu.au

Q. Grafton et al. (eds.), *Understanding and Managing Urban Water in Transition*, Global Issues in Water Policy 15,
DOI 10.1007/978-94-017-9801-3_15

consideration, this paper argues that it is possible to integrate social dimensions into urban water pricing policy. We use examples from various countries to show how social aspirations can be integrated into urban water pricing. It is argued that the social aspect of water pricing policy has four major considerations: principles, processes, outcomes, and existing social policies for the disadvantaged. It then provides a comparative analysis of water affordability for urban water and sewerage services across major Australian cities. Finally it followed with comparing the benefits and pricing discounts of government-offered water concessions in different jurisdictions.

2 Overview of Social Equity Aspects in Urban Water Policy

The right to water has been explicitly recognised in several human rights conventions (COHRE 2007) and is also implicitly recognised in the International Covenant on Economic, Social and Cultural Rights (ICESCR, General Comment 15). It is absolutely essential that individuals have access to a certain minimum of safe drinking water. Estimates of the minimum have been given as 20 litres per person per day (L/p/day) (Chenoweth 2008), 50 L/p/day (Gleick 1996), or 100 L/p/day (Howard and Batram 2003), depending on geographical location, abundance of water, gender, economic activity, and cultural setting. When urban water prices are set, the expectation is that individuals and households in the society can afford the minimum amount of safe drinking water for drinking, food preparation, and hygiene needs, regardless of their social status, income, or ability to pay. Smets (2000) suggests that the right to affordable water should be incorporated in all urban water pricing principles and objectives.

Setting a water price typically involves multiple objectives, including revenue sufficiency and stability, economic efficiency, resource conservation, administrative simplicity, and legality, as well as equity and fairness (Griffin 2006, p. 251; Boland 1993; Whittington 2003). These factors can be distilled down to four sustainability dimensions: financial, economic, environmental, and social sustainability (OECD 2010, p. 24; Martins et al. 2010), the last being crucial to the discussion here. If the social dimension of water pricing is not given due weight, water reforms are likely to lead to 'price shock', a label which encapsulates the twin issues of affordability and public acceptability. Figure 15.1 depicts a suggested framework by which social principles can be incorporated into urban water pricing. Social concerns enter through four separate layers: principles and objectives, processes, outcomes, and relevant social policies. If principles of social justice and equity can be successfully integrated into urban water pricing, it will make the public more accepting of proposed water pricing reforms.

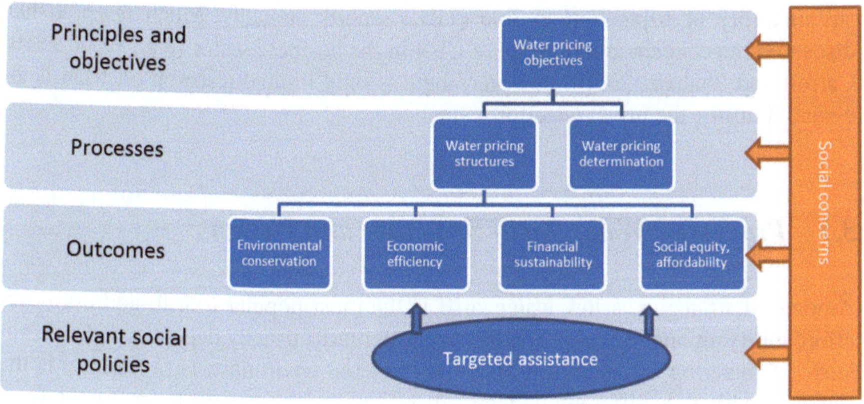

Fig. 15.1 Four levels by which social concerns can be integrated into urban water pricing.

3 Water Pricing Principles and Objectives

Social concerns and other sustainability objectives can sometimes contradict each other, although sometimes they can be complementary. For instance, it is sometimes argued that for equity reasons water should be provided free or at a very low price. However, the OECD (2001) suggests that low-cost water not only fails to achieve full cost recovery and meet conservation objectives, it also causes social inequality – because it provides insufficient capital for maintaining or expanding water infrastructure to remote communities. Today, free domestic water is only available in a few cities – such as Ashgabat (Turkmenistan), Dublin (Ireland), Cork (Ireland), Belfast (UK), and Tripoli (Libya) (GWI 2010). Bithas (2008) proposes that, in the long run, water should be priced at full cost so as to achieve the four dimensions of sustainability.

Nevertheless, social equity remains important and one way to give more weight to it is via the top level of Fig. 15.1. For example, social objectives can be built into a water company's vision through developing a corporate social responsibility (CSR) strategy. CSR involves four main strands: financial, economic, environmental, and social responsibility. It promotes stakeholder engagement, meaning that investors, board of directors, and employees at one end become engaged with customers or communities at the receiving end of the services. For instance, Manila Water have devised a CSR vision in which they commit to the idea of empowering people, building community capacity, and providing water and sewage service at an affordable rate (Manila Water 2012). In Cambridge, U.K., "protecting vulnerable customers" is one of Cambridge Water's commitments (Cambridge Water 2012). Water businesses can also develop a 'residential customer charter', such as in the case of Yarra Valley Water (Melbourne, Australia), in which their customer service commitments and the rights and obligations of their customers are clearly stated (YVW 2012). To reflect their CSR commitments, companies usually provide sustainability or triple bottom line (TBL) reports annually. Much research has discussed the concept and practice of CSR in the business sector (e.g., Wood 2010; Carroll and Shabana 2010); however, there is only limited research on what is an essential utility, the urban water provider.

3.1 Processes: Water Tariff Structure and Design

Among all strategies, a low water tariff is the most popular tool to address water affordability among the poor. The form of water tariff usually depends on the prevalence of metering infrastructure. For non-metered environments, a flat rate is the standard, although differential rates can be applied according to customer type, dwelling type, location, and property size. Even in developed countries, not all households have water meters installed, especially in areas with relatively abundant water resources. For example, the proportion of metered customers in England and Wales was just over 30 % in 2008 (Godley et al. 2008, p. iv). In Iceland, Ireland, Norway, and Scotland, meter penetration is very low (ABS Energy Research 2006). In Canada and New Zealand, only 55 % and 25 %, respectively, of single dwellings have meters installed. Although a fixed charge provides stable and predictable revenue for the water utility, the drawback is that customers have no incentive to conserve water or be aware of their water consumption behaviour.

Some countries have 95–100 % meter penetration, such as Japan, Korea, Hong Kong, Greece, Latvia, Estonia, Spain, and Turkey (ABS Energy Research 2006). In Australia, Austria, France, Finland, Germany, and the United States, almost all single dwellings have meters installed (although not individual apartments). When a metering system is in place, a two-part tariff is usually applied. It includes a fixed supply charge and a volumetric usage charge. The fixed supply charge provides revenue stability and allows sunk costs to be recovered. In addition, customers are also charged according to the amount of water consumed. The International Benchmarking Network for Water and Sanitation Utilities (IBNET) has provided a performance and tariff database which contains information from 2,000 utilities from 85 countries (IBNET 2012). Below we will discuss some of the popular tariff structures, keeping in mind that some of them provide better social equity than others.

3.1.1 Decreasing Block Tariff (DBT)

Typically, water and sanitation services involve a significant sunk investment. Because of economies of scale, we expect the average cost of water to decrease as consumption increases. The DBT reflects this, and decreases the volumetric rate with successive consumption blocks. DBTs provide stable revenue to water utilities but also offer a lower per capita water charge for large users. Currently, DBT is applied to industrial water charges in OECD countries such as Belgium, Canada,

France, and the United Kingdom (OECD 2010, pp. 56–57). It is seldom applied to residential water charges due to water conservation concerns.

3.1.2 Two-Part Tariff (TWT)

A single volumetric rate is considered economically efficient, easy to administer, and understandable by customers. Most of the major cities in China (e.g., Tiangjin, Beijing, Shanghai, Guangzhou), Russia (e.g., Novosibirsk, Krasnoyarsk, Omsk), Canada (Ottawa, Vancouver), Mongolia (Ulan Bator), the United Kingdom (e.g., Newcastle, London, Birmingham, Manchester), and some cities in the United States (e.g., Washington DC, New York, Memphis) use a single volumetric rate (IBNET 2012). Sibly (2006a, b) suggested that applying a single volumetric rate is consistent with the economic equity principle in which everyone pays the same marginal price regardless of consumption.

3.1.3 Increasing Block Tariff (IBT)

IBT is a so-called 'social tariff', and is the most popular tariff structure adopted by governments or water utilities seeking to assist those who find it difficult to meet basic needs. The water rate for IBT is low for the first block, and then increases in successive blocks as the amount of water consumed increases. In most cases, the size of the first block represents a basic, essential amount of water and this rate is usually very low and subsidised. The subsequent block rates are usually much higher and provide stronger conservation incentives. However, there has been increasing criticism over IBT recently in regard to economic inefficiency, administrative complexity, and unfairness to large households (Crase et al. 2007; Sibly 2006a), in these terms questioning its social equity. Most countries in Asia apply IBT (e.g., Hong Kong, Taiwan, Korea, Philippines, Japan, Indonesia, Vietnam), as do some in Europe (e.g., Italy, Spain, Turkey, Portugal), and some cities in the United States (e.g., Denver, Atlanta, Columbus, Las Vegas) (IBNET 2012).

3.1.4 Scarcity Pricing

Scarcity pricing, or dynamic efficient water pricing, was proposed by Grafton and Kompas (2007) and Grafton and Ward (2008) in their analysis of water pricing, water rationing, and investment decisions in response to the drought in Sydney from 2006 to 2008. They suggested that pricing water depending on both the availability of water and household demand (i.e., the rate depends on the dynamic, day-by-day dam level) could sensitively reflect the marginal cost of water and would be an efficient way to balance water supply and demand. To achieve social equity, they suggested providing rebates to targeted households.

3.1.5 Adapted IBT

To address fairness and equity issues, some cities have adjusted their IBT water rates according to household characteristics (see Table 15.1). For instance, in Spain (e.g., Barcelona, Madrid, Seville) the size of the first block of the IBT increases when household size is larger than four. In Los Angeles, there is a two-tiered IBT system for residential water rates. The first tier is adjusted based on location (temperature zones), lot size, household size, season, and consumption, and the second tier is much higher and universal for all. In Malta, there is also a two-tiered IBT water rate which is calculated on a per person basis, and the price is different depending on whether the consumer is registered as a Maltese resident or not. In Belgium, the first block (15 m^3/person/year) is free and any additional use is charged at full price. In Murcia (Spain) and several Greek cities (Athens, Thessaloniki, Larissa), volumetric water rates are adjusted according to the number of children.

Adapted IBT systems can address equity and fairness concerns across different types of households. However, there is a sacrifice in economic efficiency and environmental sustainability because it takes away financial incentives for water conservation, particularly in times of water scarcity. If the adjusted IBT has a very large first block which is either free or very low in price, utilities find it difficult to recover the costs of operation and maintenance, and generate sufficient revenue. Solving these problems requires increasing the fixed supply charge, increasing the price of the second tier or subsequent blocks, or applying cross-subsidisation between residential and industrial customers. In addition, adapted IBTs are complex and increase administration costs because they require an accurate social reporting system to adjust for different household characteristics. Inevitably, there is revenue loss from misreporting of household size. Despite its limitations, adapted IBT has been successfully adopted by a range of countries. It provides an alternative way of achieving social equity and other objectives by modifying the more popular and publicly acceptable IBT structure.

3.2 *Outcomes: Water Affordability Analysis*

Assessing the social outcomes of various water pricing policies can provide useful inputs to designing water tariffs and determining rates. Key social indicators that can be built into the price structure are (1) the ratio of connected properties, (2) the number of households unable to pay their bills on time, (3) the number of disconnections, and (4) water affordability. The ratio of connected properties is an important indicator of 'water access poverty', which is a number usually reported in a water utility's annual report. The reason it reflects poverty is that, among very poor households, saving enough money to pay for a water connection is a significant challenge (Hutton 2012). The indicator is therefore a measure of the proportion of households who can afford water and sanitation services.

Table 15.1 Example of adapted IBTs

City/country	Year	Description
Adjusted by household size (H)		
Spain		
Barcelona[a]	2002	$H>4$: for each extra person, 1st and 2nd blocks each increase by 4.5 m^3/quarter on basic 18 m^3/quarter
Madrid[a]	2002	$H>4$: if consumption less than 30 m^3/quarter, 1st block extended from 15 m^3/qtr to actual consumption
		$H>5$: if consumption less than 40 m^3/quarter, 2nd block extended from 30 m^3/qtr to actual consumption
Seville[a]	2002	$H>4$: for each extra person, 1st block increased by 4 m^3/month on basic 16 m^3/month
USA		
Los Angeles[b]	2012	2-block system: 1st block is based on location/temperature zone, season, lot size, household size, and consumption
		$H \geq 6$: household can apply for first-tier allowance increase
Malta[c]	2010	Tariffs are calculated per person and whether the person is registered on the premises (NoP>0) or not (NoP=0)
		Domestic (NoP =0): 0–33 m^3, € 2.30/m^3; >33 m^3, € 5.41/m^3
		Domestic (NoP>0): 0–33 m^3, € 1.47/m^3; >33 m^3, € 5.41/m^3. It has cheaper rates and residents are usually Maltese passport holders
Belgium		
Flanders[a]	1997	First 15 m^3/person/year is free, all other use is charged at full price
Adjusted by number of children (C)		
Spain		
Murcia[a]	2002	In 5-block systems, adjustment of 1st block size for
		$C<3$: 0–20 m^3/bi-monthly
		$C=3$: 0–45 m^3/bi-monthly
		$C=4$: 0–54 m^3/bi-monthly
		$C=5$: 0–63 m^3/bi-monthly
		$C=6$: 0–72 m^3/bi-monthly
		$C>6$: all consumption blocks equal to the price of 1st block
Greece		
Athens[a]	1993	5 block system: adjustment of 1st block
		$C=0,1,2$: 0–15 m^3/qtr
		$C=3$: 0–45 m^3/qtr
		Each extra child, extra 9 m^3/qtr
Thessaloniki[a]	n.a.	$C=0,1,2$: normal tariff
		$C>2$: 1st block increases by 50 % (?)
Larissa[a]	n.a.	$C=3,4$: 0–50 m^3/qtr, only half is charged
		$C=5,6,7$: 0–80 m^3/qtr, only half is charged
		$C>7$: 0–100 m^3/qtr, only half is charged

[a]OECD (2003, pp. 85–86)
[b]www.ladwp.com
[c]http://www.wsc.com.mt

A related water affordability indicator is the ratio of expenditure on water and sanitation services to household income or expenditure. There are some international guidelines about the limits to water affordability. According to the World Bank's guidelines for funding water projects, water should not cost more than 3–5 % of disposal income or household expenditure (OECD 2010). The UNDP uses 3 %, the OECD 4 %, while the Asian Development Bank and the African Development Bank use a 5 % figure (Smets 2009; Fankhauser and Tepic 2007). Some countries also set affordability limits in their national laws. For both water and sanitation, the limits are: Lithuania (2 %), Northern Ireland and France (3 %), Venezuela (4 %), Chile and Kenya (5 %), and Mongolia (6 %). For water only, the figures are United States (2 %), Argentina, Venezuela, and United Kingdom (3 %), and Indonesia and Mongolia (4 %) (Smets 2009). Although the threshold varies by country, comparing water affordability across time and across regions can give us a quantitative indication of how water pricing affects social outcomes.

There are a number of international studies comparing water affordability (e.g., OECD 2003, 2010; Fankhauser and Tepic 2007; Smets 1999, 2000, 2009). The OECD (2003) survey showed that in most OECD countries water and sewerage (W&S) charges ranged from 0.5 % to 2.4 % of household income or expenditure. In a later study in 2008 (OECD 2010), water affordability was found to range from 0.2 % to 1.2 %. In the case of transitional European economies, Fankhauser and Tepic (2007) found that water expenditure accounted for less than 3 % of income in most countries, except in Hungary (4.1 %), Romania (3.1 %), and Russia (3.5 %). However, it should be remembered that if the water price were required to generate enough revenue for full cost recovery, the estimated water affordability index would be considerably higher, even over 10 % in certain countries.

So far, we have been talking in terms of 'macro-affordability', a broad analysis which uses a statistical mean to compare water affordability at national and international levels. The drawback of such an aggregate indicator is that it does not reflect the burden of water charges across different income groups, geographical locations, or household types.

Micro-affordability analysis seeks to break down the macro-affordability indicator into various factors of interest such as income group, region, and family type, or even over time (OECD 2003, p. 37). The OECD (2003) survey found that, as income goes down, the proportion of expenditure spent on water increases. So, for the highest income households, expenditure on water is usually less than 1 % of total expenditure, but for the lowest quintile it accounts for more than 3 %, as shown from surveys in Mexico and England and Wales. The updated 2008 survey (OECD 2010) also found that for the poorest of Polish households, W&S expenditure accounted for 7.9 % of household expenditure.

In measuring the level of water poverty, Smets (2009) found that 1 % of French households spent more than 4.8 % of their income on W&S bills, and 2 % of UK households spent more than 8 % of their income on W&S. The affordability issue was even more serious among the median income and lowest income households in Poland, Rumania, Armenia, El Salvador, Argentina, Jamaica, and Africa (Smets 2009). In general, the main factors contributing to water affordability problems are poverty, water scarcity, and inadequate water infrastructure (OECD 2003, pp. 32–33).

The dilemma is that poor water infrastructure requires significant investment, and improvements cause a rapid increase in water prices, so poor people end up facing a significant financial burden.

3.3 *Relevant Social Policies*

Instead of having a price structure that applies to all water consumers, it is possible for governments to address water affordability and social equity through approaches outside the tariff scheme, such as targeted social measures involving direct aid. Governments or water utilities identify certain target group(s), and provide them with one or more forms of assistance such as free water allowances, output-based water concessions or subsidies, income support payments, or debt relief for their water bills. Measures such as these have been used in a variety of ways in different countries to assist water affordability.

3.3.1 Types of Targeted Measures

Free Water Allowance

Instead of providing a universal free water quota, such as happens in South Africa (Muller 2008), an alternative is a free water allowance that is only allocated to identified households which have a low income or specific need. For instance, in Uruguay free water and sanitation services are provided to the elderly who consume less than 10 m^3 per month; in Niger, indigent households with especially low incomes receive 6 m^3 of water per month for free (Smets 2009).

Price Reduction with Conditions

Targeted households can be provided with price reductions under certain conditions. For example, in Porto Alegre, Brazil, low income water users who consume less than 10 m^3 per month receive a 60 % reduction in their water bill (Smets 2009). In Gabon (Libreville), subscribers who consume less than 15 m^3 per month can apply for a 50 % discount on their water tariff (Smets 2009). These measures balance both social equity and conservation objectives.

Cross-Subsidisation and Price Differentiation

Utility subsidies can be provided to poor households to relieve financial hardship using a system of cross-subsidisation, such as in the case of Colombia (Gómez-Lobo and Meléndez 2007, p. 17). The poorest Colombian households pay for water at below-average cost, while higher income households, businesses, and industries pay a surcharge as a contribution to finance the subsidy (if there is deficit the government pays). The amount of subsidy that a household receives depends on the dwelling type and the amount of water consumed. Recently, the UK government introduced a policy that water companies need to provide social tariffs to 'target' households (DEFRA 2012). To generate sufficient revenue, companies are allowed to charge differential tariffs among their different customer classes in order to subsidise the scheme.

Debt Aid

Households experiencing financial hardships in paying water and sewerage bills can sometimes seek assistance from the government or social organisations. In the Walloon region of Belgium, a social fund for water has been established through a tax on water consumption of € 1.25 cents/m^3 to assist households having payment difficulties. Assistance is delivered via municipal social assistance centres (Smets 2009). Similarly, in France a housing solidarity fund is co-financed by the government and the private water utilities to assist households who have trouble paying their water bills (Smets 2009). In Australia, households encountering financial hardship can approach community organisations, such as St Vincent de Paul, to obtain water vouchers. Water utilities in some countries also assist households in financial difficulty by working out a payment plan without disconnecting or restricting their water use, such as happens in Australia and in some parts of the United States (e.g., San Antonio).

Output-Based Subsidies

Governments can implement an output-based consumption subsidy to make water more affordable for the poor. For example, Chilean water consumers in the lower income quintile who spend more than 5 % of their income on water and sanitation are entitled to a discount on their water bills (Gómez-Lobo 2001). To be eligible for the subsidy, households have to apply to the municipality on a scoring system, called CAS, which produces a score for each household based on a personal interview at their dwelling. The process identifies household members, living conditions, health conditions, occupations, income and assets, and other socio-economic variables. It is the main target instrument in Chile for distributing means-tested subsidies, pensioner payments, and other benefits. Eligible households can apply for their water bills to be partially or totally paid for by the government, with a cap of 15 m^3 of water a month. In 1998, 13 % of households received an average of US$10 per month at a total cost of US$33.6 million. The subsidies represent almost 8 % of the income of the lowest income group (Gómez-Lobo 2001).

Water Concessions

Provision of water concessions can be integrated with other utility allowances and the existing social security system. For instance, in Australia most of the water concession schemes are funded at both Commonwealth and State levels, although administered by the latter and delivered by utility companies. In 2010/11, the total expenditure on water and sewerage concessions in the State of Victoria was $135.3 million, and 6.7 million (35 %) of Victorian households benefited from the water concession (Victorian DHS 2012, p. 32). The following section discusses State water concession policies in Australia.

In summary, social measures can be devised to give households relief from financial hardship or assist in paying for essential water and sanitation services. This improves affordability and reduces social inequity. Unlike tariff-based solutions, targeted measures can be means-tested and directed to households or individuals with specific needs. There is no direct distortion of the water price, and consumers remain responsive to water price signals, leading to less economic inefficiency. Collectively, a disadvantaged group are helped to meet their financial, social, and other objectives.

4 Australian Urban Water Pricing and Concession Policy

This section provides an example of water affordability analysis and the impacts of concession policies applied in a range of major Australian cities. The cities under comparison are Sydney (New South Wales), Melbourne (Victoria), Canberra (Australian Capital Territory), Brisbane (Queensland), Adelaide (South Australia), Perth (Western Australia), and Darwin (Northern Territory). The cities are located in various climatic zones, they are under different jurisdictions, administrations, and institutions, and their regulatory frameworks for urban water pricing and concession policies differ. Table 15.2 summarises the average maximum temperature, average rainfall, average household water consumption, median household income, and average water and sewerage bills in 2011/12. The following analysis and comparison was based on the median household income and average residential water consumption of each city for the period 1995/96 to 2011/12.

4.1 Urban Water Regulatory and Pricing Framework

In Australia, all retail water utility companies in the major cities are state-owned companies. Except in Melbourne, all the cities have only one main retail water operator. In the case of Melbourne, the city went through a phase of devolving its water business in a reform of the urban water sector in 1990s (Godden 2008). Now Melbourne Water manages the water resources and bulk water supply infrastructure,

Table 15.2 Summary of water utility and residential consumption (2011/12)

City	Average max. temperature (°C)[a]	Average rainfall (mm)[a]	Average water use (kL/property)[b]	Average water & sewerage bill (AU$)[b]	Median HH gross weekly income (AU$)[c]	Median equivalised household disposable weekly income (AU$)[c]
Sydney	22.3	1,276.5	193	1,090	1,726	851
Melbourne[d]	20.1	654.4	144	910	1,568	816
Canberra	19.7	630.0	180	1,073	2,124	1,065
Brisbane	25.3	1,194.0	139	1,013	1,534	858
Adelaide	22.1	563.0	179	1,148	1,308	765
Perth	24.5	745.3	250	1,128	1,695	882
Darwin	32.1	1,847.1	471	1,417	1,969	962

Data sources
[a]ABS Year Book Australia 2012, Cat. 1301.0
[b]NWC (2013) National Performance Report 2011/12: Urban water utilities
[c]ABS Household Income and Income Distribution, Australia, 2011/12, Cat. 6523.0
[d]Melbourne water consumption and billing data those reported by Yarra Valley Water

while three retail water companies, Yarra Valley Water, City West Water, and South East Water, purchase water from Melbourne Water and resell it to residential and business customers. Because the three retailers service customers in distinct geographical locations, there is no direct competition between the companies. However the Productivity Commission (2011, p. 333) recognised that the horizontal segregation of water businesses could encourage benchmarking and comparative best practices, ultimately benefitting customers. Queensland, by way of contrast, went through urban water sector reform by horizontal integration in 2008, leading to a South East Queensland (SEQ) Water Grid. Under the SEQ Water Grid arrangement, the water retail business for the Brisbane region was transferred from Brisbane City Council to a single company, Queensland Urban Utilities.

To explain the differences in the water business models, it needs to be recognised that each Australian State or Territory has established its own independent economic regulator to manage the licensing of the water utilities (Table 15.3). However, as of 2009, the independent regulators have set water and sewerage prices only in Sydney (NSW), Melbourne (Victoria), and Canberra (ACT). In these cities, water utilities submit proposed tariffs for the next 3–5 years, with supporting evidence of forecast water consumption, revenue, expenditure, and costs for infrastructure investment or upgrade. The independent regulators assess the proposed pricing and announce their final price determination which takes into consideration cost recovery, financial sustainability, environmental conservation, and affordability impacts on water customers. As suggested by the Productivity Commission (2011), the urban water sector is able to contribute to the social objective of universal and affordable access to water and wastewater services "by ensuring that service delivery costs are no higher than necessary" (p. 53).

Table 15.3 Independent economic regulators who determine residential water pricing in Australia (as of 2009)

State (city)	Urban water retailers	Economic regulator	Water price setting authority
NSW (Sydney)	Sydney Water	Independent Pricing and Regulatory Tribunal (IPART)	IPART
Victoria (Melbourne)	Yarra Valley Water, City West Water, South East Water	Essential Services Commission (ESC)	ESC
ACT (Canberra)	ActewAGL	Independent Competition and Regulatory Commission (ICRC)	ICRC
QLD (Brisbane)	Brisbane Water/ Queensland Urban Utilities	Queensland Competition Authority	Brisbane City Council
NT (Darwin)	Power and Water Authority	Utilities Commission	Treasurer (Regulatory Minister)
SA (Adelaide)	SA Water	Essential Services Commission of South Australia (ESCOSA)	SA Cabinet
WA (Perth)	Water Corporation – Perth	Economic Regulation Authority (ERA)	Western Australia Cabinet

Source: National Water Commission – Australian Water Governance 2009

4.2 *Urban Water Tariffs in Major Cities*

In terms of water tariff structure, two-part tariffs (TPT) and increasing block tariffs (IBTs) were introduced between 1995 and 2012 in all major cities. Table 15.4 shows that all cities have experienced a change in water tariff structure over the last 15 years. Both Sydney and Melbourne water retailers changed from TPT in the 1990s to IBT in 2005/06 in response to drought conditions. Melbourne retailers have kept the IBT structure since then and Sydney Water changed back to TPT after 2009. Brisbane has changed from TPT to IBT since 2008/09. Adelaide, Perth, and Canberra applied IBT over all that period, while Darwin applied TPT throughout this time. In regard to sewerage charge, all cities applied a fixed charge, except Melbourne where retailers applied TPT. In this way, Melbourne households received a price signal for both water use and sewerage disposal. Due to the differences in water tariffs across cities, we will compute the affordability ratio and value of concessions in accordance with the pricing and concession rules.

Table 15.4 Water and wastewater tariff structures over time

		Australian capital city						
		Sydney	Melbourne	Brisbane	Adelaide	Perth	Darwin	Canberra
1995/96	**Water**	TPT	TPT	TPT	IBT	IBT	TPT	IBT
					(3 steps)	(2 steps)		(2 steps)
	Wastewater	FC	TPT	FC	FC	FC	FC	FC
2005/06	**Water**	IBT	IBT	TPT	IBT	IBT	TPT	IBT
		(2 steps)	(3 steps)		(2 steps)	(5 steps)		(3 steps)
	Wastewater	FC	TPT	FC	FC	FC	FC	FC
2011/12	**Water**	TPT	IBT	IBT	TPT	IBT	TPT	IBT
			(3 steps)	(3 steps)		(6 steps)		(2 steps)
	Wastewater	FC	TPT	FC	FC	FC	FC	FC

Source: *WSAA facts* (1995–2005); NWC National Water Performance report (2006/06–2011/12)

4.3 *Trends in Water Consumption and Expenditure*

As shown in Fig. 15.2, the level of average residential water consumption has, over the period 1995/96 to 2012/13, increased then decreased across all major Australian cities. The reduction corresponds to the introduction of water restrictions, community water conservation and education programs, and increases in water prices (Halich and Stephenson 2009; Renwick and Green 2000). Since Darwin has the highest average maximum temperature and rainfall, the average water consumption there has always been highest. Brisbane and Melbourne have the lowest average water consumption, with 139 kL and 144 kL respectively in 2011/12. Responding to drought condition during the period 1997–2009, almost all major cities implemented water restrictions. Sydney imposed water restrictions in 2003 and upgraded them to level 3 in 2008, which remained in force until June 2009. Similarly, Melbourne implemented Stage 1 restrictions in August 2006 and gradually ramped them up to Stage 3a in 2007. Under water restrictions, household residents were only permitted to water their garden and lawns with hand buckets on alternate days; while washing cars or driveways or refilling swimming pools was not permitted. With increased rainfall, water restrictions were lifted in 2012 and have been replaced with other mandatory water conservation measures in all cities. Average water consumption has therefore bounced back slightly in 2012/13, but it is still much lower than in 1996/97. This demonstrates that water efficiency and education programs implemented during the drought period have successfully reduced people's water consumption patterns.

However, reduction in average water consumption does not necessarily translate to lower water bills (Fig. 15.2). After accounting for a rise in the consumer price index (CPI), the real average water and wastewater expenditure in all Australian cities have slightly decreased and then increased since 2007/08. For the latest 2011/12 figures, an average household in Darwin is now charged the most ($1,417), while Melbourne households pay the least ($793). In general, however, over the

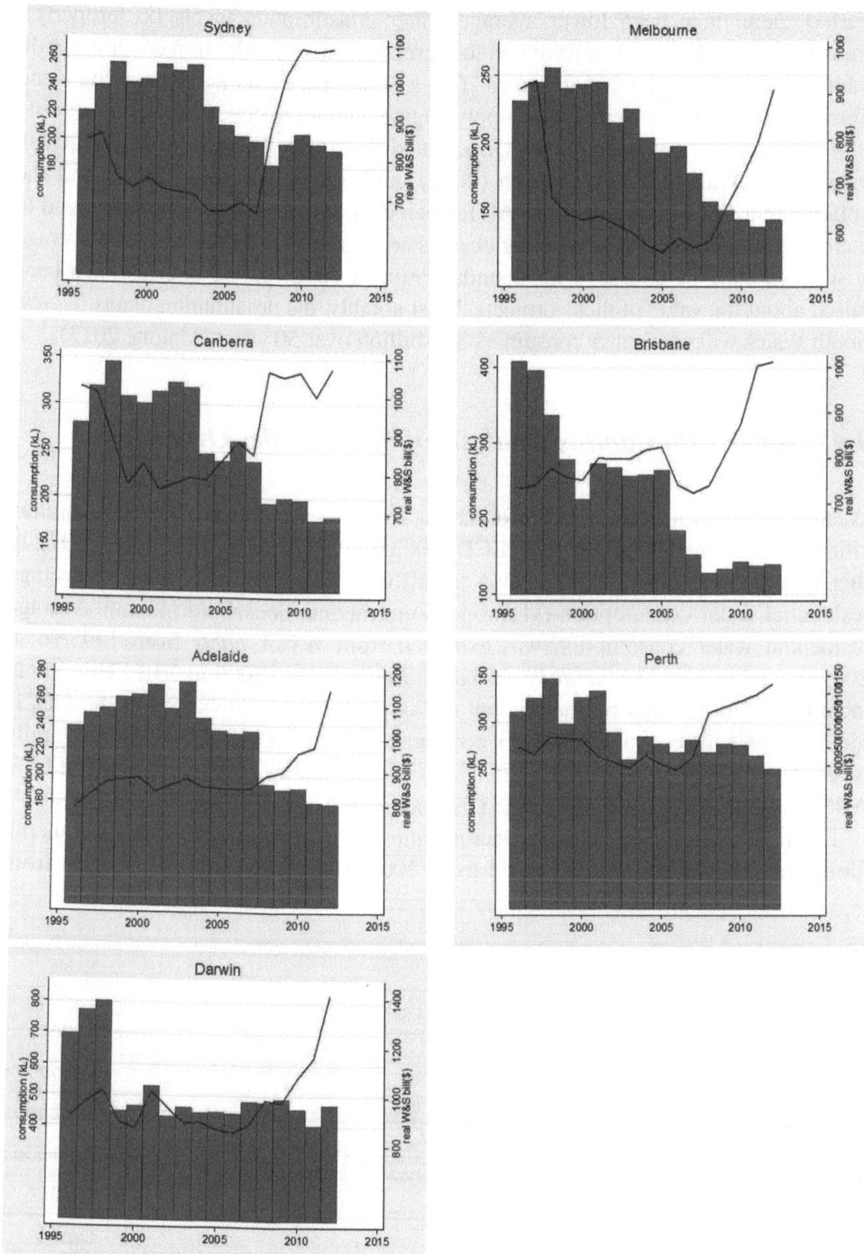

Fig. 15.2 Average water consumption (bars, left axis) and average real water and sewerage expenditure (line, right axis) of major Australian capital cities from 1996 to 2013

period there have been lower average water consumption levels (kL/property), meaning that water utilities have had to increase tariffs in order to recover costs. In addition, the prolonged drought from 2003 to 2007 has encouraged many State and Territory governments to invest in supply augmentation projects, such as the building of desalination plants in southeast Queensland (in 2009), Western Australia (2006), New South Wales (2010), Victoria (2012), and South Australia (2012); and the enlargement of Cotter Dam in the ACT. Such capital investments inevitably lead to increased water tariffs because water charges are determined by the weighted average cost of capital. Of course, with abundant rain in 2012, public queries have been raised about the value of these projects. Most notably, the desalination plants in New South Wales will cost water consumers $10 billion over 50 years (Malone 2013).

4.4 *Water Affordability Analysis of Major Cities Over Time*

We now set out to compare water affordability over time for the seven major Australian cities using the method applied in OECD (2003) and Fankhauser and Tepic (2007). In this analysis, water affordability (also called 'water burden') is the ratio of average residential water consumption (kL/property) to median household income. Average residential water consumption was extracted from *WSAA Facts* from 1995/96 to 2005/06 and in *NWC National Performance Reports* from 2006 to 2013. Household income is also adjusted by the amount of tax paid and is equivalised to the OECD modified scale to account for different economic resources required by different family types. The weekly median equivalent disposable income in each city was taken from ABS Household Income and Income Distribution in various years.

Figure 15.3 shows water burdens across cities from 2000/01 to 2011/12. In general, the average water burden decreased from 2000/01 but then increased again from

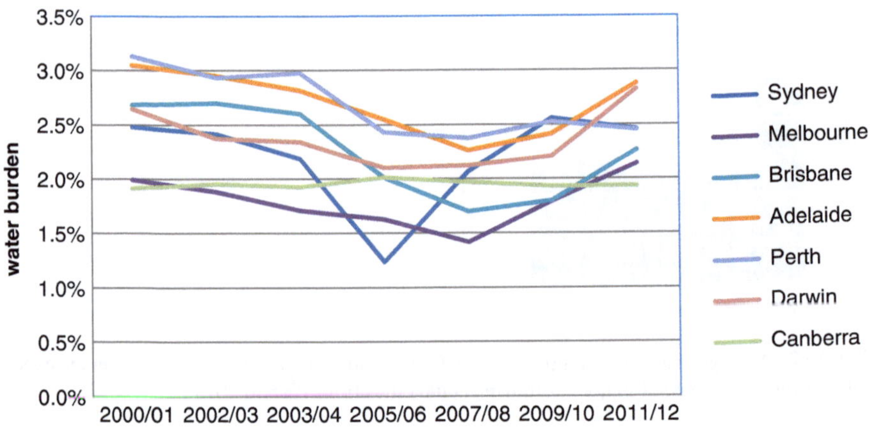

Fig. 15.3 Average water burden (% of average income) of major Australian cities from 2000/01 to 2011/12

2006/07 to 2011/12. The increase in water burden largely corresponds with the rapid rise of water tariffs in most cities during the drought period. In 2000/01, the average water burden in Perth and Adelaide was above 3 %, it then declined to below 2.5 % in 2007/08 and is now 2.8 % in 2011/12. Melbourne residents have enjoyed the lowest average water burden in most years except 2011/12. Canberra residents, who have the nation's highest average income, had the lowest average water burden in 2011/12. After all, all cities except Darwin, have a lower average water burden in 2011/12 compared to 10 years ago.

4.5 *Water Affordability Across Income Quintiles*

It is possible to compare water affordability of major Australian cities across income groups. This micro-affordability analysis is based on average water consumption and equivalent median household disposable income for each income quintile reported by ABS (2013).

Figure 15.4 shows the distribution of average water burden across household income quintiles based on an average water consumption in 2009/10. As expected, the average water burden increases as household income decreases across all capital cities. For the lowest income quintile households (Q1), water burdens were over 4 % for all cities, the highest being 6.6 % in Sydney and 6.2 % in Perth. The second

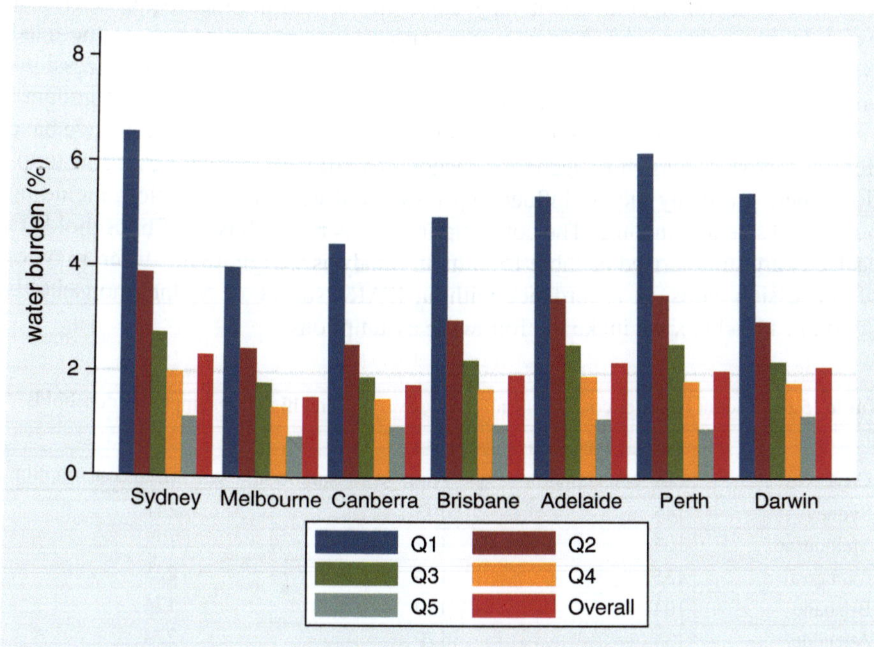

Fig. 15.4 Water burden by quintiles in major Australian cities in 2009/10

quintile households (Q2) also had water burdens over 3 % in Sydney, Brisbane Adelaide, Perth, and Darwin. Melbourne households experienced the lowest water burden in all income quintiles, with Canberra households having the second lowest.

The water burdens for the lowest quintile ranged between 4 and 6.6 %, while the figure for the highest quintile households ranged from 0.8 % to 1.2 %. In terms of equity, water burden for the lowest quintile was 2–2.5 times heavier than for those in the top quintile. However, if we consider the 3 % water affordability benchmark, as applied in Fitch and Price (2002) and Fankhauser and Tepic (2007), this result suggests that households in the bottom 40 % of income distribution (Q1 and Q2) should receive extra assistance to reduce their water burden.

4.6 *Water Concessions in Australian Cities*

In Australia, all States and Territory governments provide water concessions to help eligible households afford water (see Appendix). The concession scheme is usually funded and administered by government and delivered via water retailers as part of their customer service obligations. The eligibility and entitlements vary from state to state but are mostly tied to the possession of concession cards issued by the Commonwealth government. In the case of Victoria, almost 25 % of the Victorians held at least one concession card and almost 35 % of the households received water concession in 2006/07 (Vic DHS 2007).

Water concession entitlement may come in vary forms (see Appendix). For example, in Canberra and Sydney, water concessions are applied only to the fixed water supply charge only, whereas in Victoria and the Northern Territory concessions are applied as a discount on the total water and sewerage charges (to a capped amount).

In order to compare the values and impacts of State water concessions, we have done an analysis using three scenarios each with a different level of water consumption. There are many factors influencing residential water consumption, including household size and income. The consumption levels for each type of households in each city are summarised in Table 15.5. In this analysis we ignore the different types of concession cards. In accordance with an IPART survey of Sydney households (2010, p. 116–117), we make the following assumptions.

Table 15.5 Amount of water consumption for low, average, and high consumption households

	Annual water consumption (kL)		
City	Low consumption	Average consumption	High consumption
Sydney	145	193	241
Melbourne	107	142	178
Canberra	135	180	225
Brisbane	104	139	174
Adelaide	134	179	224
Perth	188	250	313
Darwin	377	471	589

- Low water consumption households use 25 % less than the average water consumption in their city. It represents consumption for a 1–2 person household.
- Average consumption level represents the water use for a 3–4 person household.
- High consumption households use 25 % more than the average. It represents consumption for large family household.

Based on the water concession rules in each city in 2010/11 (see Appendix), Fig. 15.5 shows, for different water consumption levels, the water and sewerage (W&S) bills before and after concessions. The red areas represent the amount of water concession received by eligible households.

4.6.1 Scenario 1: Low Water Consumption Households

Among households with low consumption, who use 25 % less than the average consumption, Darwin households had the highest W&S bills in 2010/11, both before ($1,271) and after ($680) water concessions were applied. Melbourne households had the lowest bills before concessions were applied ($720), whereas after water rebates were applied, eligible Perth households had the lowest bills ($397). The value of water concessions ranged from $231 to $591 across cities. Eligible households in Adelaide and Melbourne had the lowest levels of water rebates, $231 and $270 respectively. Darwin and Sydney households had the largest water rebate, $591 and $576 respectively. Overall, water concessions entitled eligible consumers with a 28–58 % discount from the original bill. The largest figure (58 %) applied to eligible Sydney households, while eligible households in Darwin, Brisbane, Perth, and Canberra had almost a 50 % reduction. However Adelaide households who had low water consumption received the smallest discount (28 %).

4.6.2 Scenario 2: Average Water Consumption Households

Similarly, for average water consumption levels, Darwin households had the highest W&S bills in 2011/12, both before ($1,423) and after ($765) water concessions were applied. Melbourne households had the lowest bill before rebates ($821), while Perth households had the lowest total bill after rebates ($493). The value of water concessions ranged from $262 to $658. As for Scenario 1, Darwin and Sydney received the largest amount of water rebates, while Adelaide and Perth households received the smallest. Percentage discounts from water concessions ranged from 28 % to 53 %. Eligible Sydney households had the largest percentage reduction (53 %) while Adelaide households received the smallest (28 %).

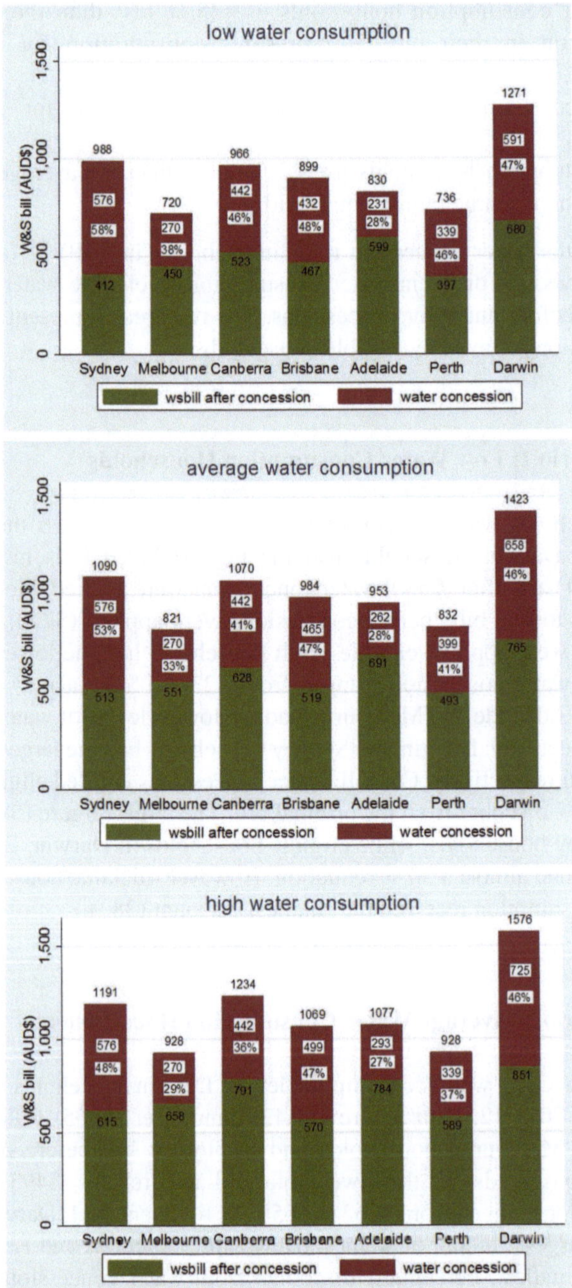

Fig. 15.5 Water and sewerage (W&S) bills before and after water rebates (2010/11) for households with different levels of water consumption (low, average, and high)

4.6.3 Scenario 3: High Water Consumption Households

For households with high levels of water consumption (25 % more than the average), the W&S bills in Darwin, Sydney, and Canberra were all above $1,000 before concessions were applied. After water rebates, the bills were reduced by half. Both Melbourne and Perth households still had the lowest total bill before concessions were applied ($928), while Brisbane households had the lowest bills after water rebates ($570). The value of water concessions ranged between $270 and $725 across all cities. Melbourne consumers received the smallest amount of rebate while Darwin consumers received the largest amount for high levels of water consumption. Percentage discounts from water concessions ranged between 27 % and 48 %. At this consumption level, Sydney, Brisbane, and Darwin had the biggest discount (46–48 %) while Adelaide and Melbourne had the smallest (27–29 %).

By and large, water concession policies in all major cities appear to be effective in reducing water bills for eligible households by significant amounts. In Sydney, Melbourne, Canberra, and Perth, the value of water concessions (i.e., the amount of rebate) remained about the same for different levels of water consumption. This is because water rebates only apply to fixed supply charges (in the cases of Sydney and Canberra) or to a percentage reduction and with a low cap amount (in the cases of Melbourne and Perth). In other cities, the amount of water rebates increase with increasing water consumption. From the perspective of economic efficiency, Sibly (2006a) and Whittington (2003) recommend that rebates should target the fixed supply charge without distorting the marginal price of water.

Nevertheless, in terms of percentages, the discounts generally become smaller with increasing levels of water consumption, regardless of city, in accordance to the relevant water concession rules. Eligible Darwin, Brisbane, and Adelaide households have slight reduction in rebate discount (%) even water consumption increases. In other cities, high consumption households receive much less in rebate discount than low consumption households. It aligns to the objective of water conservation that the concession policy does not encourage excessive water use. However in terms of 'horizontal' equity, larger households, who have higher levels of water consumption, receive relatively less concession benefits (i.e., percentage reduction). A review by the Productivity Commission (2011, p. 203) concluded that current water concessions were to a certain extent inefficient and inequitable. When designing a concession policy, striking a balance between different objectives therefore poses a challenge, both to water pricing regulators and social policy makers.

5 Conclusions

With rapid rises in urban water prices around the world, addressing the social equity and affordability issues which surround the cost of water is increasingly important. This chapter provides Australian and international perspectives on how social concerns about urban water costs can be tackled through the use of water pricing principles and policies.

Designing and implementing residential water pricing is complex. There are multiple objectives involving financial sustainability, economic efficiency, environmental conservation, social equity, and affordability. Some of these objectives conflict with each other, although others are complementary. Different pricing designs will achieve different equity balances. Among tariff-based solutions, the increasing block tariff (IBT) is the most popular method and has been used in both developed and developing countries (IBNET 2012). However it is debatable whether it achieves the most equitable outcome. Adapted IBT have been proposed in order to achieve a better balance.

To assess the effect of pricing policy, water affordability analysis can assist in identifying those water consumers who would encounter affordability problems if real prices were to increase. To better identify target households and address water poverty issues, the use of metrics such as household size and income level could be helpful. In all cases, the aim and outcome of such targeted solutions needs to be equitable and efficient. Targeted policies are more efficient than most other solutions, but they do require an advanced social security system such as in Australia and Chile.

Successful social policy needs to achieve five 'E's: equity, effectiveness, employment, efficiency, and economy (Herscovitch and Stanton 2008). When comparing the Australian and Chilean systems of assisting households pay for water, the output-based system in Chile is more comprehensive, fairer, and targeted to those in need. The value of water subsidies provided to eligible Chilean households is based on their socio-economic characteristics and consumption levels. In Australia, by contrast, eligibility to receive water concession largely relies on possessing Commonwealth and State concession cards. However, possession of those cards does not fully reflect the true mix of household characteristics, economic status, and consumption level.

This paper has provided a brief analysis of the Australian State water concession policies and their impacts on eligible households. As more data becomes available, further analysis would be useful. Before any change to water pricing is made, a review of government assistance should be done as to address equity and affordability. In this way, we can ensure that disadvantaged groups are not financially hurt by price increases.

Appendix: State-by-State Water Concession Policies in Australian Cities, 2011/12

Cities	Eligibility	Concession	Source
Canberra (ACT)	Centrelink Pensioner Concession Card holder	68 % discount in water and sewerage supply charge	www.actewagl.com.au
	Veterans' Affairs Gold Card holder	68 % discount in water and sewerage supply charge	
	Health Care Card holder	Rebate on water charges only	

(continued)

Cities	Eligibility	Concession	Source
Sydney (NSW)	Owner-occupiers with Pensioner Concession Card, Dept of Veterans' Affairs Gold Card, Veterans' Affairs Blue Card – Pensioner Concession, or receiving DVA intermediate rate pension	Water: 100 % discount on the standard quarterly service charge to maximum of $36.22. Reduction of 33 % on water use charges to a maximum of 100 kL a year (for resident pensioners who have a water service only)	www.sydneywater.com.au
		Sewerage: 83 % discount on the standard quarterly service charge	
Melbourne (Victoria)	Centrelink Pensioner Concession Card, Centrelink Health Care Card, DVA Concession Card, DVA Gold Card	50 % discount on water and sewerage charges up to max of $270.20 per year	www.yvw.com.au
		Water only: 50 % discount on water charges up to max of $138.50 per year	
Adelaide (South Australia)	Owner-occupier or tenants with Pensioner Concession Card; Seniors Card; DVA Gold Card; full-time student; Centrelink benefit or allowance receiver; low income earner	25 % discount on water charges over a year subject to minimum and maximum amounts	www.dcsi.sa.gov.au
		Water concession:	
		Owner occupier: min $155, max $265	
		Tenant: min $90, max $200	
		Sewerage concession: max $110 per year	
Brisbane (Queensland)	Owner-occupier or life tenant with Pensioner Concession Card or DVA Gold Card	Subsidy up to a max of $120 off the cost of water charges per year from Queensland Council.	www.communities.qld.gov.au
		Brisbane City Council provided Pension remission up to 40 % discount of net charges in total bill to max $476 per year.	
Perth (Western Australia)	Pensioner Concession Card, state concession card	Rebate of up to 50 % of annual service charges and 50 % of water usage charge up to 150 kL per year.	www.watercorporation.com.au
	WA Seniors Card	Rebate of up to 25 % (capped) of annual service charges	
	Both WA Seniors and Commonwealth Seniors Health Card	Rebate of up to 50 % on annual service charges, or may be eligible to defer those charges	

(continued)

Cities	Eligibility	Concession	Source
Darwin (Northern Territory)	Centrelink Pensioner Card; DVA Gold Card; DVA Concession Card; Centrelink carer allowance receiver; non-pensioner aged war service veteran; low income superannuants; senior citizens	Daily water concession: water fixed charge = $0.407 per day; water usage charge = $0.725 per kL, sewerage fixed charge = $0.754 per day.	www.health.nt.gov.au

Source: Adapted from Productivity Commission (2011, Table 8.4) and updated information from the government websites listed

References

ABS Energy Research. (2006). *World water meter report* (5th ed.). London: ABS Energy Research.

Australian Bureau of Statistics (ABS). (2012). *Year book Australia 2012*. Cat. 1301.0

Australian Bureau of Statistics (ABS). (2013). *Household income and income distribution, Australia, 2011–12*. Cat. 6523.0

Bithas, K. (2008). The sustainable residential water use: sustainability, efficiency and social equity. The European experience. *Ecological Economics, 68*, 221–229.

Boland, J. J. (1993). Pricing urban water: Principles and compromises. *Water Resources Update, 92*, 7–10.

Cambridge Water. (2012). *Corporate social responsibility report*. http://www.cambridge-water.co.uk/customers/csr

Carroll, A. B., & Shabana, K. M. (2010). The business case for corporate social responsibilities: A review of concepts, research and practice. *International Journal of Management Reviews, 12*(1), 85–105.

Chenoweth, J. (2008). Minimum water requirement for social and economic development. *Desalination, 229*, 245–256.

Centre on Housing Rights and Evictions (COHRE). (2007). *Manual on the right to water and sanitation*. Geneva: COHRE, AAAS, SDC, and UN-HABITAT.

Crase, L., O'Keefe, S., & Burston, J. (2007). Inclining block tariffs for urban water. *Agenda, 14*, 69–80.

Department for Environment Food and Rural Affairs (DEFRA). (2012). *Company social tariffs: guidance to water and sewerage undertakers and the water services regulation authority under section 44 of the flood and water management Act 2010*. London: Social Tariffs Consultation, DEFRA.

Fankhauser, S., & Tepic, S. (2007). Can poor consumers pay for energy and water? An affordability analysis for transition countries. *Energy Policy, 35*, 1038–1049.

Fitch, M., & Price, H. (2002). *Water poverty in England and Wales*. London: Centre for Utility Consumer Law and Chartered Institute of Environmental Health.

Gleick, P. H. (1996). Basic water requirements for human activities: Meeting basic needs. *Water International, 21*(2), 83–92.

Global Water Intelligence (GWI). (2010). World water prices rise by 6.7 %. *Global Water Intelligence, 9*(9), 34–36.

Godden, L. (2008). Property in urban water: Private rights and public governance. Chapter 8. In P. Troy (Ed.), *Troubled Waters: Confronting the water crisis in Australia's cities*. Canberra: ANU E Press.

Godley, A., Ashton, V., Brown J., & Saddique, S. (2008). *The costs and benefits of moving to full water metering*. Science Report SC070016/SR1 (WP2). Bristol: UK Environment Agency.

Gómez-Lobo, A. (2001). Making water affordable: output-based consumption subsidies in Chile. In P. J. Brook & S. Smith (Eds.), *Contracting for public services: output-based aid and its applications*. Washington, DC: World Bank and International Finance Corporation.

Gómez-Lobo, A., & Meléndez, M. (2007). *Social policies and private sector participation in water supply: the case of Colombia*. Working papers wp252, University of Chile, Department of Economics.

Grafton, R. Q., & Kompas, T. (2007). Pricing Sydney water. *Australian Journal of Agricultural and Resource Economics, 51*, 227–241.

Grafton, R. Q., & Ward, M. B. (2008). Prices versus rationing: Marshallian surplus and mandatory water restrictions. *Economic Record, 84*, S57–S65.

Grafton, R. Q., Chu, L., Kompas, T., & Ward, M. (2014). Volumetric water pricing, social surplus and supply augmentation. *Water Resources and Economics, 6*, 74–87.

Griffin, R. C. (2006). *Water resource economics: The analysis of scarcity, policies, and projects*. Cambridge, MA\London: MIT Press.

Halich, G., & Stephenson, K. (2009). Effectiveness of residential water-use restrictions under varying levels of municipal effort. *Land Economics, 85*(4), 614–626.

Haller, L., Hutton, G., & Bartram, J. (2007). Estimating the costs and health benefits of water and sanitation improvements at global level. *Journal of Water and Health, 5*, 467–480.

Herscovitch, A., & Stanton, D. (2008). History of social security in Australia. *Family Matters, 80*, 51–60.

Howard, G., & Bartram, J. (2003). *Domestic water quantity, service level and health*. Geneva: World Health Organization.

Hutton, G. (2012). *Monitoring 'affordability' of water and sanitation services after 2015: Review of global indicator options*. Paper submitted to the United Nations Office of the High Commission for Human Rights.

Hutton, G., Haller, L., & Bartram, J. (2007). Global cost–benefit analysis of water supply and sanitation intervention. *Journal of Water and Health, 5*, 481–502.

IBNET. (2012). *The 2011 IBNET water tariff database*. http://www.ib-net.org/en/tariffs_map.php

IPART. (2010. December). *Residential energy and water use in Sydney, the blue mountains and Ilawarra*. Results from the 2010 household survey. (Electricity, gas and water – research report)

Manila Water (2012). *Corporate social responsibility*. http://www.manilawater.com/Pages/CorporateSustainability.aspx. Retrieved 15 Feb 2013

Martins, R., Pimentel, A., Quintal, C., Barata, E., & Cruz, L. (2010). *Water pricing and social equity in Portuguese municipalities*. ISEE 2010 conference on advancing sustainability in a time of crisis. Oldenburg/Bremen, Germany

Malone, U. (2013). New South Wales desalination plant deal to cost consumers $10 billion over 50 years. *ABC News*, 27 September 2013. Available at http://www.abc.net.au/news/2013-09-27/nsw-desalination-plant-deal-costing-customers-10-billion/4985168

Muller, M. (2008). Free basic water – a sustainable instrument for a sustainable future in South Africa. *Environment and Urbanization, 20*(1), 67–87.

National Water Commission (NWC). (2009). *Governance at a glance: Water pricing and economic regulation. Australian water governance 2009*. Canberra: NWC.

National Water Commission (NWC). (2006–2013). *National performance report. Major urban water utilities*. Canberra: NWC

OECD. (2001). *The price of water: Trends in OECD countries*. Paris: OECD.

OECD. (2003). *Social issues in the provision and pricing of water services*. Paris: OECD.

OECD. (2010). *Pricing water resources and water and sanitation services*. Paris: OECD.

Productivity Commission. (2011). *Australia's urban water sector*. Inquiry report no. 55. Vol 1, 31 August 2011. Melbourne: Productivity Commission.

Renwick, M. E., & Green, R. D. (2000). Do residential water demand side management policies measure up? An analysis of eight California water agencies. *Journal of Environmental Economics and Management, 40*, 37–55.

Savenije, H., & van der Zaag, P. (2002). Water as an economic good and demand management: Paradigms with pitfalls. *Water International, 27*(1), 98–104.
Sibly, H. (2006a). Efficient urban water pricing. *Australian Economic Review, 39*, 227–237.
Sibly, H. (2006b). Urban water pricing. *Agenda, 13*, 17–30.
Smets, H. (1999). *Implementing the right to drinking water in OECD countries*. OECD seminar social and environment interface proceedings, Paris, 22–24 Sept 1999, pp. 107–148.
Smets, H. (2000). The right to water as a human right. *Environmental Policy and Law, 30*(5), 248–250.
Smets, H. (2009). *Is drinking water affordable for all?* Paper presented in 5th world water forum, Istanbul. 16–22 March 2009.
Victorian Government Department of Human Services (Vic DHS). (2012). *State concessions and hardship programs 2010–11*. Melbourne: DHS.
Victorian Government Department of Human Services. (2007). *State concessions and hardship programs 2006–07*. Melbourne: DHS.
Water Services Association of Australia (WSAA). (1995–2005). *WSAA facts*. Melbourne: WSAA.
Whittington, D. (2003). Municipal water pricing and tariff design: A reform agenda for South Asia. *Water Policy, 5*, 61–76.
Wood, D. J. (2010). Measuring corporate social performance: A review. *International Journal of Management Reviews, 12*(1), 50–84.
Yarra Valley Water (YVW). (2012). *YVW residential customer charter*. Melbourne: YVW.

Chapter 16
Does Residential Water Use Depend on Water Quality? Some Answers from a French Case Study

Arnaud Reynaud and Marian A. Garcia-Valiñas

1 Introduction

Since water is considered a basic human right and essential for life and health (United Nations 2002), water quality problems have emerged as a significant environmental issue. Beach closures, destroyed habitat, unsafe drinking water, fish kills, and many other severe environmental and human health problems result from water pollution. Several countries have made advances to clean up the aquatic environment by controlling point source pollution entering waterways from single, identifiable sources, such as a pipe or a ditch. However, nonpoint source pollution remains an important source of water quality problems. This kind of pollution occurs when rainfall, snowmelt, or irrigation runs over land or through the ground, picks up pollutants, and deposits them into rivers, lakes, or coastal waters or introduces them into ground water. It also includes adverse changes to the vegetation, shape, and flow of streams and other aquatic systems.

Pollutants found in water bodies come from households, industry, or agricultural activities. Some common pollutants are pesticides, pathogens (bacteria and viruses), salts, oil, grease, toxic chemicals, and heavy metals. Other pollutants (typically coming from nonpoint source pollution) are sediment and nutrients. There are also some emerging issues like the presence of pharmaceuticals in tap water. Reports of

---

A. Reynaud (✉)
Toulouse School of Economics (LERNA), Université Toulouse 1 Capitole,
21 Allée de Brienne, 31000 Toulouse, France
e-mail: areynaud@toulouse.inra.fr

M.A. Garcia-Valiñas
Oviedo Efficiency Group, Department of Economics, University of Oviedo,
Avda. Del Cristo, s/n, 33006 Oviedo, Spain
e-mail: mariangv@uniovi.es

Q. Grafton et al. (eds.), *Understanding and Managing Urban Water in Transition*, Global Issues in Water Policy 15,
DOI 10.1007/978-94-017-9801-3_16

trace concentrations of pharmaceuticals in the water cycle have raised concerns among various stakeholders –such as regulators, governments, water suppliers, and the public – over the potential human health risks from exposure to very low levels of pharmaceuticals in drinking water (World Health Organization 2011). This issue is difficult to address since some pollutants are not easy to measure, and their consequences for human health are not totally clear.[1]

With increasing levels of pollution in water bodies and high media coverage of tap water contamination episodes,[2] it is not surprising that households have become more suspicious about tap water. In 2000, Ortalda and Hatchuel (2000) surveyed a representative sample of the France population. They found that 42 % of the sample did not drink tap water. Among these people, the main reason for this choice (44.7 %) was the bad taste of tap water; contamination by toxic substances and associated health issues was the second main reason (23.6 %). According to a Sofres/CIEAU survey in 2009, 22 % of the French population have a negative view of tap water quality, especially because of its bad taste, and 20 % consider that controls of tap water quality are not sufficient.

As a result, a few scholars have tried to understand how residential water users are influenced by the quality of raw and tap water, and this chapter adds to this literature. More specifically, we investigate the possible link between residential water consumption and the quality of both raw water and tap water. As will be discussed later, there are various difficulties that must be overcome. First, consumption of water for drinking and cooking purposes (potentially affected by tap water quality) represents only a small part of total household water use, making variations in tap water consumption induced by low water quality difficult to identify.[3] Second, perceived water quality might be as important as observed water quality when it comes to influencing household behaviour. Third, data on the quality of raw water and drinking water are difficult to get.

The chapter is organised as follows. First, we survey the relevant literature on residential water use in relation to water quality. In the next section, we discuss the methodology and develop a model which assumes that water quality can affect both residential water demand and prices. Finally, we present an empirical analysis of a sample of municipalities located in the Southwest of France (the Adour–Garonne basin), and conclude with a summary of the main findings.

[1] As Shaw (2005, p. 67) pointed out: "an excellent current example of an uncertain effect is from exposure to methyl tertiary-butyl ether (MTBE), which is a fuel additive used to oxygenate fuels to decrease air pollution, but which is being found in groundwater supplies … there is still debate about whether MTBE causes cancer or not".

[2] For instance, in January 2013 the French non-governmental association 'France Liberté' and the consumer association '60 millions de consommateurs' published a map of municipalities allowed to deliver tap water which does not comply with water quality standards.

[3] In France, drinking water represents about 1 % of households' tap water consumption. This figure rises to 7 % if we include water used for preparing food.

2 Water Quality and Its Impact on Residential Water Consumption: A Brief Review

In developed countries, most studies have focused on modelling the impact of several economic (and non-strictly economic) instruments on residential water demand. Despite the importance of quality issues, only a few studies have analysed the impact of water quality on households' water consumption.

In the United States, Piper (2003) reported that water quality had a strong impact both on residential water demand and on the cost of water. García-Valiñas (2006) and Roibás et al. (2007) analysed the effects of drought and restrictions on residential water demand in Seville, Spain. In their model, they used a dummy variable which was put equal to 1 when chemical parameters of the water did not comply with the minimum requirements defined by water quality regulations. They found that water quality deterioration led to appreciable reductions in water consumption. In a similar way, Monteiro (2010) included a quality variable into a residential water demand model, where the variable was defined as the percentage of water analyses failing to comply with mandatory quality parameters. He found that, for Portugal, there was a negative relationship between the quality variable and residential water demand.

A few studies have focused on the relationship between tap water consumption and raw water quality (in rivers, lakes, or aquifers). For instance, in constructing water demand models, Reynaud et al. (2005) used average biochemical oxygen demand (BOD) as the regressor, whereas Reynaud (2010) took the proportion of rivers classified as bad quality for this role. Both papers found a significant negative impact of raw water quality on residential water demand, suggesting that households react to the quality of water in their environment.

The perceived quality and the perception of health risk associated with residential water use have both been shown to be important factors.[4] As with some others, Ford and Ziegler (1981) found a negative relationship between quality perceived by households and residential water demand. Garn (1998) showed that water quality perceptions have an influence on residential water demand. However, as discussed by Nauges and Whittington (2010), including perception of quality in a water demand equation can create some endogeneity and collinearity issues.

A substantial number of papers have focused on how bad tap water quality causes people to substitute bottled water for tap water. Households have been shown to undertake preventative actions to protect themselves against the risk of drinking contaminated water (Alberini et al. 1996; Abrahams et al. 2000; Bontemps and Nauges 2009; Zivin et al. 2011; Johnstone and Serret 2012).[5] In France, Bontemps

[4] For instance, according to the Sofres/CIEAU survey in 2009, 62 % of the French population believe that low environmental water quality will result in low tap water quality.

[5] Households can also boil water, at some cost in terms of energy and time (Shaw 2005). As Jalan et al. (2009) pointed out, households can adopt different measures to avoid bad water quality

and Nauges (2009) found that raw water quality had a significant impact on the choice between bottled and tap water: for households living in a municipality where raw water quality was low, the probability of drinking tap water decreased by 9 %. Similarly, based on a sample of 10,000 households from ten OECD countries, Johnstone and Serret (2012) reported that negative perceptions of tap water quality affected the decision to purchase bottled water. In the United States, Zivin et al. (2011) identified a significant increase in bottled water sales (by 17–22 %) as a result of several tap water quality violations. They estimated that US consumers were willing to pay nearly US$60 million to avoid such violations.

Finally, a household's willingness to pay for water quality improvements has also been investigated.[6] Using the contingent valuation method, Yongsung et al. (2005) have estimated a Minnesota household's willingness to pay in order to improve their drinking water quality. They showed that, on average, individuals were willing to pay US$5.25 per month to reduce the level of iron and US$4.33 per month to reduce the level of sulfate in their water. Using the method of virtual prices, Roibás et al. (2007) found that households located in the city of Seville, Spain, were willing to support an increase of around 40 % in the price of water to improve the quality and reliability of their water supply during drought periods. Again using contingent valuation, Genius et al. (2008) have estimated an average willingness to pay of €10.64 per household per year (around 17 % of the average residential water bill) to avoid water supply shortages and low drinking water quality in the municipality of Rethymno, Greece, during peak water demand periods. Also in Greece, Polyzou et al. (2011) found that households were willing to pay €5 per month for an improvement in drinking water quality.

3 French Case Study

3.1 Regulation of Tap Water Quality in France

In France, tap water quality standards are defined in the 'Code de la Santé Publique'. Water at the tap must comply with regulatory quality requirements set by a ministerial decree covering about sixty parameters. Water treatment aims to eliminate any pathogenic microorganisms and undesirable substances such as chemical pollutants and excessive mineral salts; it is also designed to maintain the microbiological and

impacts, ranging from inexpensive or low-grade technologies (straining with a cloth, using chlorine, and safe storage vessels) to relatively sophisticated and expensive technologies (e.g., electrically powered filters that use ultraviolet to remove pathogens). In developing countries, there is a broad variety of studies of such defensive techniques (among others, Aini et al. 2007; Bukenya 2008; Jalan et al. 2009).

[6]Um et al. (2002), Yongsung et al. (2005), Whitehead (2006), Genius et al. (2008), and Polyzou et al. (2011) are some examples of recent studies on this topic.

physicochemical quality standards of the water in the distribution system up to the consumer's tap.

Water quality is monitored by companies producing and distributing water and by ongoing health inspections from regional health agencies. According to the Code de la Santé Publique, mayors are required to post the results of all quality analyses in the city hall within two working days. Also, water service providers must, at least once a year, send their customers a report on water quality in the area covered by the service. In addition to key indicators such as microbiological water quality and concentration of nitrates and pesticides, this report must include sanitary recommendations.

Non-compliance with quality norms must be disclosed to consumers. At the European level, Directive 98/83/EC states that it is necessary to provide consumers with adequate information on the quality of the water supply; it also determines a set of actions to be taken by member states in case of non-compliance with quality standards. At the national level, the Code de la Santé Publique requires disclosure of information to users. In case of quality problems involving risk to human health, information disclosure must be immediate. Information must also be passed to consumers if a water service is allowed by the local authorities (Préfet) to distribute, under a derogation regime, water violating quality limits.

3.2 *Area of Interest*

Our area of interest is the Adour–Garonne river basin. The Adour–Garonne river basin occupies one-fifth of French territory (115,000 km^2) and around 11 % of the French population (seven million inhabitants in 2005). Population density in Adour–Garonne is relatively low (60 inhabitants per km^2 compared to approximately 100 for France as a whole). The Adour–Garonne river basin includes both rural and urban areas. Some 35 % of the population is concentrated in two French 'départements' (Gironde and Haute-Garonne) corresponding to the two main urban areas in the Adour–Garonne river basin (Bordeaux and Toulouse).[7]

3.3 *Data Sources*

In this section we describe the main sources of data. We restrict our analysis to municipalities located under the jurisdiction of the Adour–Garonne Agency. Our sample includes 711 municipalities observed in 2004. A description of our variables is provided in the Appendix.

---

[7] France is divided into 101 administrative units called 'départements'.

3.3.1 Water Consumption and Water Utility Characteristics

Data on water price and tap water consumption come from the survey undertaken at the *municipality level* in 2004 by the French water agencies along with IFEN (the French Institute for the Environment) and SCEES (the Statistical Unit of the French Ministry of Agriculture).[8] From this survey, we compute the average household water consumption in each municipality by dividing the total residential water use by the number of residential connections. For the water price, we consider the average residential water price corresponding to a consumption equal to 120 m^3 per year. This database also provides some technical information on water services in each municipality (network length, type of water treatment implemented, customer density, etc.).

3.3.2 Data on Water Quality at the Tap

All information about quality of water at the tap comes from the SISE–Eaux database (Environmental Health Information System – Water). This database is managed by the French Ministry of Health, and it is the main source of data used by the French Institute for Public Health Surveillance (InVS) and the French Food Safety Agency (AFSSA) for assessment of drinking water quality. This database contains more than 70 million observations from the testing of all parameters used to monitor water quality. In the SISE–Eaux database, the unit of observation is called 'Unit of distribution' (UDI). A UDI is a geographical area where the water network is operated by a single firm or administrative unit and where water quality can be considered homogeneous. The size of a UDI is highly variable and may correspond to a suburb, a village, a town, or even a group of towns. In 2005, there were 4,415 UDIs in the Adour–Garonne river basin. For each UDI, the SISE–Eaux database indicates the municipality where the main measurement point is located, which allowed us to merge databases using the municipality identifier.

We first consider the *bacteriological quality* of water at the tap. Bacteriological quality is measured through three parameters: concentration of coliforms (*n*/100 mL), concentration of enterococcus bacteria (*n*/100 mL), and concentration of *Escheria coli* or *E. coli* (*n*/100 mL).[9] Over the period 2003–2009, an improvement of the bacteriological quality has been observed in the Adour–Garonne river basin. The average concentration of coliforms, enterococcus bacteria, or *E. coli* has been reduced from 3.16 *n*/100 mL in 2003 to 0.99 *n*/100 mL in 2009. The percentage

[8] About 5,000 French local communities have been surveyed about the technical, financial, and organisational characteristics of their water and sanitation services. All municipalities with a population greater than 10,000 inhabitants have been surveyed, while the smaller municipalities have been randomly selected.

[9] General coliforms indicate that the water has come in contact with plant or animal life. Fecal coliforms, particularly *E. coli*, indicate that there are mammal or bird feces in the water. Enterococcus bacteria also indicate that there are feces from warm-blooded animals in the water. Enterococcus is a type of fecal streptococci.

of UDI with a frequency of bacteriological contamination lower than 0.05 has increased from 73.6 % in 2003 to 78 % in 2009. Over the same period, the number of UDIs with chronic bacteriological contamination has decreased from 13 % to 5 %.

Next, we consider the contamination of tap water by nitrates. The presence of nitrates (or nitrites) in infants under 6 months old creates a condition known as methaemoglobinaemia or 'blue baby syndrome', a potentially fatal illness in which blood lacks the ability to carry sufficient oxygen to individual body cells. This extremely unusual illness only occurs at very high nitrate concentrations. In the long term and in case of a lifetime exposure at high levels, nitrates and nitrites have the potential to cause diuresis, increased starchy deposits, and haemorrhaging of the spleen. The current regulatory standard of 50 mg/L nitrate is derived from the European Union's drinking water directive 98/83/EC on the quality of water intended for human consumption. That standard is intended to ensure that drinking water will not cause methaemoglobinaemia. In the Adour–Garonne river basin, the average concentration of nitrates has been slightly reduced from 17.1 mg/L in 2003 to 16.3 mg/L in 2009 and it remains below the EU standard of 50 mg/L. The percentage of UDIs for which the average concentration of nitrates exceeds the EU standard has decreased from 1.5 % in 2003 to 1 % in 2009.

3.3.3 Data on Groundwater Quality

Concerning information related to groundwater quality, we have used the ADES database which is the French national groundwater database developed by the Ministry of Ecology and Sustainable Development, the Ministry of Employment, Solidarity and Health, and the water agencies. A number of parameters (nitrates, nitrites, pesticides, etc.) are collected through a network of measurement points. We have selected all groundwater quality measures realised in the Adour–Garonne river basin from 2000 to 2004 (594,587 measures realised from 4,353 different points). The ADES database indicates in which municipality each measurement point is located. This information has been used to build two variables measuring groundwater quality at the municipality level. Variable *BadGwPesticides* is a dummy variable equal to 1 if the average concentration in pesticides has exceeded the limit defined by the Code de la Santé Publique over the period 2000–2004, and zero otherwise. Variable *BadGwNitrates* is a dummy variable equal to 1 if the average concentration in nitrates has exceeded the limit defined in the Code de la Santé Publique (50 mg/L), and zero otherwise.

3.3.4 Data on Surface Water Quality

For supplying water in the Adour–Garonne basin, a lot of municipalities rely on surface water. Information on the quality of surface water comes from the 'système d'information sur l'eau' (SIE) of the Adour–Garonne Water Agency. This database contains all quality measurements made by monitoring stations in the basin since 1971. We have selected all surface quality measures reported in the Adour–Garonne river basin and computed an average value over the period 2000–2004. The SIE indicates in which municipality each measurement point is located and measures of surface water quality have been built for each municipality.

3.3.5 Other Data

Water data have been complemented by data coming from various sources. First, we have used data from the French National Statistical Institute (INSEE), including data on household taxable income and data on demographic characteristics of the population in each municipality (housing characteristics and household characteristics). Since residential water consumption has been shown to be highly dependent on climate conditions, we consider some climatic variables defined at the municipal level (monthly and annual mean temperatures, maximum temperature, monthly and annual mean rainfall, etc.). Those variables have been provided by Météo-France (the French meteorological institute).

3.4 Model and Estimation Strategy

The proposed empirical specification is based on three ideas: first, that both raw and tap water quality may have an impact on tap water consumption; second, that the price of water at the tap depends on both the quality of raw water and the quality of water at the tap; and, third, that the quality of water at the tap is chosen by each municipality (taking into account the constraints set by the current regulations).

As discussed in the previous section, tap water consumption is influenced by tap water quality (hardness, colour, taste, concentration of pollutants). However, households may not have perfect information about the quality of tap water. This will clearly be the case if information on tap water quality is not available or too costly to acquire. In such a case, their tap water consumption may be influenced by their perception of tap water quality (Ford and Ziegler 1981; Garn 1998). Here, we assume that the perception of tap water quality is related to the quality of raw water. The idea is that if a household lives in a municipality where the quality of water in rivers and aquifers is publically known to be very good, it is likely that his/her priors for the quality of tap water will be very high. On the contrary, a household living in a place where surface water and groundwater have been polluted for a long time by nitrates, pesticides, or bacteriological elements may be much more suspicious of the quality of water at the tap, even if the quality measures of water at the tap reported by public authorities are good. In this kind of setting the quality of raw water (surface water and groundwater) may be interpreted by households as an indicator of tap water quality.[10] As a result, both the quality of tap water and raw water might be a determinant of tap water consumption.

In the second equation of our model, we assume that water price is determined by characteristics of the water service (size, characteristics of water consumers, seasonal variability of water consumption, etc.). We also take into account that the tap water price depends both on the quality of raw water and on the quality of tap water. Depending on the quality of raw water sources, different chemical treatments might be required to obtain a given level of quality of water at the tap. Thus, it is expected that the gap between water quality at the tap and quality of raw water may have an impact on water prices.

Finally, we take into account that water supply managers may decide on a particular level of water quality at the tap (within the range of levels allowed by current quality regulations). Water managers might take into account household preferences for high quality at the tap when they decide upon some investments that will have an impact on the quality of tap water. For instance, one may expect that a municipality where households are wealthier might be inclined to provide a high tap water quality since the willingness to pay for this high quality might be high.

In sum, our model includes three equations: one for tap water consumption, one for tap water price, and one for tap water quality. We estimate the following system of simultaneous equations:

$$\begin{aligned} x &= \Phi\left(p, pol^{tap}, pol^{raw}, I, z_1\right) \\ p &= \Gamma\left(x, pol^{tap}, pol^{raw}, I, z_2\right) \\ pol^{tap} &= \Delta\left(x, pol^{raw}, I, z_3\right) \end{aligned} \tag{16.1}$$

where x denotes residential water consumption per household, p is the tap water price paid by residential users, pol^{tap}, pol^{raw} represent two vectors of variables related respectively to the quality of water at the tap and the quality of raw water, and I is the average household income. In the system of Eq. 16.1, z_1, z_2, and z_3 are vectors which include some other variables (such as climatic factors or socio-demographic characteristics of households[11]) assumed to have an impact on the endogenous variables. To control for the endogeneity of tap water consumption, tap water price, and quality of water at the tap, we estimate the system of simultaneous equations using three-stage least squares (3SLS). Compared to 2SLS with instrumental variables, it allows calculation of efficiency gains (or cross-equation tests) with a consistent estimator of equations in case of endogenous regressors.

---

[10] Pollution of water is highly discussed in French newspapers. The title of an article published on 20 March 2012 in *Le Monde* (a major national French newspaper) was "Agriculture is responsible for water pollution in France in two cases out of three". One may expect that a French household reacts to this type of information, even if the objective tap water quality is not directly affected by pollution.

[11] See, among others, Arbués et al. (2003), Worthington and Hoffman (2008), and Schleich and Hillenbrand (2009) for surveys of the main determinants of residential water demand.

3.5 Descriptive Statistics

In Table 16.1 we report some descriptive statistics for the main variables of interest. A more formal definition of these variables is provided in Table 16.5 (Appendix).

The average water consumption per household in the Adour–Garonne river basin was 131.77 m^3 per year in 2004, a figure not too far from the French average for the

Table 16.1 Descriptive statistics for selected variables

Variable	Mean	Std dev	Min	Max
Water consumption and price				
WatCons – water consumption (m^3/hh/year)	131.77	44.90	20.00	296.88
Price – average water price (euros/m^3)	2.52	0.96	0.24	5.58
IBR – dummy for increasing block rate	0.09	0.29	0.00	1.00
Household characteristics				
Income – average income (euros/hh/year)	14527.14	3357.03	7929.30	43310.78
Density – population density (inhabitants/km^2)	238.73	544.62	1.45	5019.61
Principal – share of main residences	0.79	0.16	0.17	0.99
House – share of households living in a house	0.86	0.16	0.16	1.00
TouristicCapacity – number of hotel and camping rooms per inhabitant	0.05	0.17	0.00	2.16
Quality of water at the tap, in groundwater, and in surface water				
TapBactPol – level of bacteriological contamination at the tap	0.09	0.14	0.00	0.67
BadSurfOrganic – dummy if surface water contamination by organic matter	0.16	0.36	0.00	1.00
BadSurfNitrates – dummy if surface water contamination by nitrates	0.15	0.36	0.00	1.00
BadGwPesticides – dummy if groundwater contamination by pesticides	0.64	0.48	0.00	1.00
BadGwNitrates – dummy if groundwater contamination by nitrates	0.22	0.42	0.00	1.00
Main characteristics of water services				
Customers – number of domestic connections to the water service	1819.47	3998.30	20.00	74787.00
Groundwater – dummy if groundwater used	0.62	0.49	0.00	1.00
NetworkDensity – number of connections per km of network	22.78	20.81	1.00	222.71
ProtectAll – dummy if all water withdrawal points are protected	0.21	0.41	0.00	1.00
TreatNo – dummy if no treatment is applied to raw water	0.003	0.05	0.00	1.00
Climate data				
Temperature – average maximum temperature (in degrees Celsius)	25.26	1.47	18.07	27.82

Table 16.2 Residential water consumption and type of water contamination

	Average residential water consumption		*p*-value for two-sample *t*-test
	Contamination		Null hypothesis of no statistical difference
Type of contamination	Yes	No	
Bacteriological contamination at the tap	123.84	137.44	0.075
Surface water contamination by organic matter	126.82	136.24	0.097
Surface water contamination by nitrates	119.67	137.44	0.000
Groundwater contamination by pesticides	146.76	125.37	0.000
Groundwater contamination by nitrates	157.96	128.95	0.000

Note: The last column reports the *p*-value of a mean test between the average residential water consumption (in m^3/year/hh)

same year (136 m^3) according to Garcia-Valiñas et al. (2010). The average water price was 2.52 euros per m^3 compared to 2.99 euros per m^3 on average in France. Around 9 % of municipalities have set increasing block rates.

Regarding quality variables, the fraction of municipalities with surface contamination by organic materials and nitrates is around 15 %. Groundwater quality problems are more frequent, since 64 % and 22 % of municipalities report cases of pesticide and nitrate pollution, respectively, probably a consequence of farming activities. Concerning tap water quality, the average proportion of analyses which do not comply with bacteriological regulation is 9 %.

Table 16.2 provides some preliminary evidence of a link between water contamination (at the tap and in the environment) and residential water consumption. In this table, we relate the average residential water consumption per household to whether the tap water has been contaminated by bacteriological pollution. We also assess the relationship between the quality of the raw water (surface and ground water) and water consumption. Average water consumption per household is 137.44 m^3 per year when there is no bacteriological pollution whereas it is only 123.84 m^3 per year when bacteriological pollution regularly occurs (we reject the null hypothesis of no statistical difference at the 5 % level of significance). The lower consumption might be related to restrictions of water use implemented by public authorities, or to modifications of households' consumption habits. A significant impact related to the poor quality of surface water is apparent. When surface water has been contaminated by organic matter or by nitrates, the residential water consumption per household is lower (*t*-tests reveal significantly different mean consumptions). This might be interpreted as an averting behaviour by households, in which case households are substituting tap water with bottled water for certain uses. However, we find as well that, in the case of groundwater, contamination also significantly impacts residential water consumption, but in this case, surprisingly, residential water consumption appears to be higher when groundwater is affected by pollution. Since the characteristics of households in areas with and without groundwater contaminations might be different, one should however be careful with interpreting this last result as a

Table 16.3 Number of quality measures and bacteriological contamination status

UDI size	Average annual number of quality measures		p-value for two-sample t-test
	Bacteriological contamination		Null hypothesis of no statistical difference
	No	Yes	
0–49 inhabitants	2.47	2.34	0.16
50–499 inhabitants	2.78	2.81	0.75
500–1,999 inhabitants	2.61	4.46	0.02
2,000–4,999 inhabitants	4.90	4.76	0.90
>5,000 inhabitants	7.21	5.00	0.45

Note: The last column reports the *p*-value of a mean test between average number of measures per UDI

causality relationship.[12] A more detailed analysis of the impact of surface and groundwater pollution on tap water use will be provided in the next section.

Using tap water contamination in a residential water demand function is subject to caution since a number of tap quality measures may be endogenous. Hence, one may expect tap water quality to be more frequently tested in municipalities where pollution has been detected. To address this issue, we have done some mean comparison tests of the number of measures done by an UDI, looking to see whether the UDI is in a state of chronic bacteriological contamination (Table 16.3).

It should be mentioned that the number of measures to be conducted at the tap by each UDI is defined in France by an administrative document called 'Arrêté du 11 janvier 2007'. This document determines both the frequency of measures and the type of analyses. The frequency is defined by the size of the UDI. For instance, the smallest UDI (serving between 0 and 49 inhabitants) must perform at least one measure in the distribution network and between two and four at the tap. At the other extreme, the largest UDI (serving more than 625,000 inhabitants) must perform 144 measures in the distribution network and around 800 at the tap. So the frequency of measures is strongly driven by the size of the UDI. This is why we report, in Table 16.3, the results of the *t*-tests by size of UDI. With the exception of UDIs serving between 2,000 and 4,999 inhabitants, we do not reject the null hypothesis of no statistical difference in the number of measures by UDI. This means we do not have a problem of endogeneity of the number of tap sample tests.

3.6 Empirical Results

Table 16.4 shows the main results of the 3SLS model estimates. Various specifications of the system of simultaneous Eq. 16.1 have been considered. In what follows, we report only the estimates with the best quality of fit.

[12] Another explanation could be a price effect. Hence, it is well known that the tap water price tends to be lower in municipalities supplied by groundwater (compared to surface water).

Table 16.4 Model estimates by 3SLS

	Coef	Std error	z	$p>\|z\|$
Tap water demand equation (endogenous variable is logarithm of *WatCons*)				
Price (in logs)	−0.325***	0.072	−4.526	0.00
IBR	−0.083**	0.034	−2.481	0.01
TapBactPol	−0.708***	0.212	−3.344	0.00
BadSurfOrganic	−0.012	0.031	−0.386	0.70
BadSurfNitrates	−0.033	0.037	−0.883	0.38
BadGwPesticides	−0.072***	0.027	−2.648	0.01
BadGwNitrates	−0.009	0.031	−0.285	0.78
Groundwater	0.022	0.023	0.952	0.34
Income (in logs)	0.030	0.07	0.434	0.66
Density (in logs)	−0.024	0.015	−1.598	0.11
Pop0014s (in logs)	−0.050	0.053	−0.930	0.35
Pop0003s (in logs)	0.003	0.027	0.110	0.91
Single (in logs)	−0.136**	0.055	−2.464	0.01
TouristicCapacity (in logs)	0.038***	0.009	4.293	0.00
Principal (in logs)	0.558***	0.070	7.990	0.00
House (in logs)	−0.469***	0.067	−7.016	0.00
OldHouse (in logs)	−0.044*	0.023	−1.903	0.06
RecentHouse (in logs)	0.005	0.023	0.202	0.84
Temperature (in logs)	0.601***	0.229	2.628	0.01
Constant	3.033***	0.980	3.097	0.00
Tap water price equation (endogenous variable is logarithm of *price*)				
WatCons (in logs)	−0.647***	0.118	−5.496	0.00
Income (in logs)	0.050	0.083	0.607	0.54
Customers (in logs)	0.157***	0.016	10.065	0.00
NetworkDensity (in logs)	−0.059***	0.020	−2.915	0.00
TouristicCapacity (in logs)	0.057***	0.010	5.679	0.00
Principal (in logs)	0.435***	0.093	4.675	0.00
ProtectAll	0.022	0.030	0.727	0.47
Groundwater	−0.014	0.053	−0.263	0.79
TapBactPol	−2.192***	0.502	−4.363	0.00
BadSurfOrganic	0.019	0.042	0.454	0.65
BadSurfNitrates	0.023	0.052	0.436	0.66
*BadGwPesticides*Groundwater*	−0.072*	0.044	−1.648	0.10
*BadGwNitrates*Groundwater*	−0.036	0.057	−0.631	0.53
TreatBasic	0.070	0.081	0.858	0.39
TreatStandard	−0.019	0.087	−0.223	0.82
TreatAdvanced	−0.008	0.092	−0.091	0.93
TreatMixed	0.001	0.074	0.008	0.99
Constant	3.257***	0.996	3.270	0.00

(continued)

Table 16.4 (continued)

	Coef	Std error	z	$p>\|z\|$
Bacteriological pollution at the tap equation (endogenous variable is *TapBactPol)*				
WatCons (in logs)	–0.050	0.032	–1.537	0.12
Income (in logs)	–0.002	0.023	–0.071	0.94
Temperature (in logs)	–0.301***	0.094	–3.202	0.00
TreatBasic	–0.121***	0.017	–7.132	0.00
TreatStandard	–0.123***	0.018	–6.735	0.00
TreatAdvanced	–0.128***	0.019	–6.721	0.00
TreatMixed	–0.045*	0.024	–1.867	0.06
ProtectAll	0.005	0.011	0.470	0.64
Groundwater	–0.018	0.018	–1.010	0.31
BadSurfOrganic	–0.003	0.014	–0.186	0.85
BadSurfNitrates	0.045***	0.015	2.980	0.00
*BadGwPesticides*Groundwater*	0.016	0.015	1.091	0.28
*BadGwNitrates*Groundwater*	–0.015	0.020	–0.760	0.45
Constant	1.425***	0.321	4.435	0.00

***, **, * for significance at 1 %, 5 %, and 10 % level, respectively

The global adjustment of the model to the data is correct. The pseudo R^2 obtained with the 3SLS estimates for the tap water demand equation, the tap water price equation, and the bacteriological pollution at the tap equation are 0.28, 0.43, and 0.12, respectively.

3.6.1 Tap Water Demand Equation

Price-elasticity of residential water demand is estimated at –0.32 for our sample, which is in the range of what has been estimated earlier using data from France (Nauges and Thomas 2000; Nauges and Reynaud 2001). This estimate also broadly agrees with several previous studies of urban water demand. Residential water use is usually found to be a normal good, and the bulk of water demand studies worldwide also find estimates of price-elasticities below 1 in absolute value (Arbués et al. 2003; Worthington and Hoffman 2008).

Additionally, we find that water consumption is significantly lower (–0.92 m^3 per year per household) in municipalities which have implemented increasing block rates (IBRs). This result is in line with previous studies having reported lower residential water consumption in the case of IBR (Reynaud et al. 2005).

Regarding quality issues, we find that a higher level of bacteriological pollution at the tap (*TapBactPol*) significantly decreases tap water consumption. It is not possible, however, to identify whether this impact of bacteriological contamination results from restrictions in water use implemented by public authorities or from long-term modification of water consumption habits. When bacteriological pollution increases by 10 % (which means in our case that the frequency of water analyses revealing a bacteriological contamination at the tap increases by 10 %), consumption will decrease on average by 0.6 m^3 per year per household. This amount represents a little less than 0.5 % of annual tap water consumption. Such a modification of household behaviour is a priori plausible since in France drinking water represents approximately 1 % of total water consumption. Our results indicate that residential water use is affected by objective tap water quality. Omitting some variables characterising tap water quality into the water demand function might lead to a biased estimate of the demand function.[13] However, the relationship between tap water quality and tap water consumption appears less strong than what has been reported previously. For example, García-Valiñas (2006) and Roibás et al. (2007) estimated that a severe reduction in tap water quality leads to a reduction in residential water consumption of 6–12 %.

In addition, we have introduced some variables describing the quality of raw water (surface water and groundwater) in order to determine if residential water demand is affected by a household's perception of raw water quality. We found only a reaction of tap water consumption to groundwater quality. More precisely, the variable *BadGwPesticides* is significant at 1 % with a negative sign. Pesticide contamination of groundwater reduces tap water consumption by 0.93 m^3 per year. If we suppose substitution with bottled water, this represents a substantial additional cost. It should be mentioned that the quality of surface water (in terms of organic matter and nitrates) was found not to be a significant determinant of tap water consumption, although the signs of the estimated coefficients for surface water contamination by organic matter or by nitrates are negative, as expected according to the empirical results presented in Table 16.2.

We have also identified other determinants of tap water consumption. For example, the tourism capacity in the municipality (*TouristicCapacity*) has a positive significant impact on water consumption. This result is in line with previous studies (Martínez-Espiñeira 2002; Essex et al. 2004; Garcia-Valiñas et al. 2010). One possible explanation could be that, due to specific types of water use (swimming pools, air conditioning, etc.) seasonal residents use more water on average than those residing there year round (Gössling 2006). Additionally, the share of permanent housing (*Principal*) has a positive significant impact on average residential water consumption.

Households who are living in a house (instead of an apartment) have significantly lower tap water consumption. We find that if the proportion of households living in a house increases by 10 %, one may expect an average reduction of 6 % in tap water consumption. Access to private wells for households living in a house may explain this unexpected result. In this context, Montginoul and Rinaudo (2011) found that water price increases have led French households to drill their own garden bores.

---

[13] We have estimated the same model without including the tap bacteriological contamination variable in the two first equations, and we find some significant changes in the coefficient estimates. For example, variable *BadSurfNitrates* becomes significant in the first equation, with a negative sign.

Lastly, residential water demand is related to climate. When the average monthly maximum temperature increases by 10 %, tap water consumption increases by 6 %. Such an impact of climate is in line with the previous literature (Nauges and Reynaud 2001; Martínez-Espiñeira 2002; Garcia-Valiñas et al. 2010).

3.6.2 Tap Water Price Equation

First, the tap water price used by a municipality is related to the average water consumption. The higher the average tap water consumption, the lower the average tap water price. This reflects the fact that most of the costs of a water utility are fixed. Hence, the constraint of balancing the budget leads municipalities to set higher water prices if the tap water consumption per household is lower (the level of fixed charges is constrained by French law).

Second, the water price is driven by some technological characteristics of the water utility. We find that the customer density (*NetworkDensity*) is significant, with a negative sign, which indicates the presence of economies of customer density. When the customer density increases by 10 %, the average water price reduces by 0.6 %. High customer density can be associated with low average variable costs (energy costs in the distribution network), but also low average fixed costs (network length). This may explain the negative impact of customer density on average water price, although a high customer density has been shown to generate extra costs such as congestion costs.[14]

The tourism capacity of a municipality is associated with a higher average water price. That capacity is a key driver of the peak water demand, which is a strong determinant of the water utility cost function. These results are in line with those obtained by García-Valiñas et al. (2010), who found a positive impact of tourism on the ratio of total water bill to municipal average income (where the water bill corresponds to what an average individual household would be expected to pay for the 'lifeline'). That fact strengthens the hypothesis that water services are less affordable in tourism-based municipalities.

Additionally, as expected, both quality of tap water and raw water are shown to have a significant impact on the average tap water price. First, we find that a high level of bacteriological pollution is associated with a low average unit price. This reflects the fact that water utilities wishing to achieve a higher tap water quality must spend more financial resources, in particular in the treatment of raw water. It then seems that there is a trade-off between high tap water quality, which requires high production costs, and low tap water quality allowing low water prices. The optimal quality of tap water (within the range of quality regulation defined by public authorities) must then be determined by each municipality based on a cost–benefit analysis. Second, we find that a bad quality of groundwater (in terms of pesticide

[14] Garcia-Valiñas et al. (2009), based on a sample of medium to large Spanish municipalities, found a positive impact of population density on residential water bills.

concentration) has a negative impact on the average price, which is significant at 10 %.

3.6.3 Tap Water Quality Equation

Finally, the estimates of the equation for bacteriological pollution at the tap reveal some interesting findings. As might be expected, when water treatment is implemented by a municipality (basic, normal, mixed, or advanced), bacteriological pollution is significantly reduced.[15] We also find a positive and significant impact of nitrate contamination of surface water on bacteriological pollution at the tap. This is related to the fact that a majority of municipalities rely on surface water.

Additionally, we observe that temperature has a negative impact on tap water quality. So in municipalities where temperatures are higher, it is more difficult to provide high quality tap water, because the likelihood of drought and water scarcity problems is higher. Roibás et al. (2007) showed that during a severe drought in Seville, Spain, public authorities had to provide tap water that did not comply with minimum quality standards.

4 Conclusions

Preserving water quality is considered to be the primary objective of environmental policies around the world. There are some regions where quality problems are especially significant, due to pollution from economic activity. Pesticides, chemicals, and other pollutants have often contaminated surface and underground water bodies, leading to severe environmental and human health problems.

In this chapter, we have analysed the impact of water quality reductions on tap water demand. As we have said, there are not many papers that have analysed this specific issue in developed countries. We have contributed to this literature through a French case study involving a sample of municipalities located in the Southwest of the country (Adour–Garonne basin) to test the impact of water quality on residential water consumption. We have estimated a sophisticated model which allows us to identify interactions between water quality, quantity, and prices. Our findings suggest that quality of water at the tap (in particular, bacteriological quality) is a strong predictor of residential water use. We also show that French households adjust their tap water consumption to the quality of the raw water. They appear to be

[15] One may be concerned by a potential collinearity problem between the type of treatment and raw water quality. Hence, one may suspect the water quality variables (raw water quality and quality at the tap) to be collinear with the variables describing the type of water treatment. Although we expect more sophisticated treatments to be used when raw water is of lower quality, our data do not show such a pattern. No significant relationship was found between raw water quality and type of treatment. In fact, the type of treatment is highly driven by administrative constraints which depend on UDI size.

especially reactive to pesticide contamination of groundwater. These results call for joint monitoring and management of the water quality at the tap and in the environment (the rivers and groundwater).

Appendix

Table 16.5 Definition and description of variables

Variable name	Description	Source
WatCons	Average annual water consumption in m^3 per household (m^3/year)	IFEN-SCEES
Price	Average residential water price for a water consumption equal to 120 m^3 per year (euros/m^3)	IFEN-SCEES
IBR	Dummy equal to 1 if there is an increasing block rate	IFEN-SCEES
TapBactPol	Bacteriological pollution measured as the frequency of water analyses revealing bacteriological contamination at the tap (proportion of analyses which do not comply with bacteriological norms)	SISE-Eaux
BadSurfOrganic	Dummy equal to 1 if some contamination by organic matter has been reported in surface water	SIE-AEAG
BadSurfNitrates	Dummy equal to 1 if the average nitrate concentration in surface water is greater than 50 mg/L	SIE-AEAG
BadGwPesticides	Dummy equal to 1 if the average pesticide concentration in groundwater is greater than 0.01 mg/L	ADES
BadGwNitrates	Dummy equal to 1 if the average nitrate concentration in groundwater is greater than 50 mg/L	ADES
Groundwater	Dummy equal to 1 if the municipality uses groundwater	IFEN-SCEES
Income	Household taxable income (euros/year)	INSEE
Density	Population density (inhabitants/km^2)	INSEE
Pop0014s	Proportion of the population 0–14 years old	INSEE
Pop0003s	Proportion of the population 0–3 years old	INSEE
Single	Share of single households	INSEE
TouristicCapacity	Tourism capacity measured as the number of hotel or camping rooms divided by the number of permanent inhabitants	INSEE
Principal	Proportion of households living in their main residence	INSEE
House	Proportion of households living in a house	INSEE
OldHouse	Proportion of principal residences built before 1949	INSEE
RecentHouse	Proportion of principal residences built after 1990	INSEE
Temperature	Maximum temperature defined as the monthly average maximum temperature (degrees Celsius)	METEO-FRANCE
Customers	Number of connections to the water service	IFEN-SCEES
NetworkDensity	Customer network density defined as the number of customers per kilometre of network	IFEN-SCEES

(continued)

Table 16.5 (continued)

Variable name	Description	Source
ProtectAll	Dummy equal to 1 if all water withdrawal points used by the municipality are protected against potential pollution contaminations	IFEN-SCEES
TreatBasic	Dummy equal to 1 if the treatment applied to raw water is basic	IFEN-SCEES
TreatStandard	Dummy equal to 1 if the treatment applied to raw water is standard	IFEN-SCEES
TreatAdvanced	Dummy equal to 1 if the treatment applied to raw water is advanced	IFEN-SCEES
TreatMixed	Dummy equal to 1 if the treatment applied to raw water is mixed	IFEN-SCEES

IFEN-SCEES: survey on water utilities in France
SIE-AEAG: information system of the Adour–Garonne Water Agency
ADES: national groundwater database
INSEE: data provided by the French Institute of Statistics and Economic Studies
METEO-FRANCE: French climatic database

References

Abrahams, N. A., Hubbell, B. J., & Jordan, J. L. (2000). Joint production and averting expenditure measures of willingness-to-pay: Do water expenditures really measure avoidance costs? *American Journal of Agricultural Economics, 82*(2), 427–437.

Aini, M. S., Fakhrul-Razi, A., Mumtazah, O., & Meow Chen, J. C. (2007). Malaysian households' drinking water practices: A case study. *International Journal for Sustainable Development and World Ecology, 14*(5), 503–510.

Alberini, A., Eskeland, G. S., Krupnick, A., & McGranahan, G. (1996). Determinants of diarrhoeal disease in Jakarta. *Water Resources Research, 32*(7), 2259–2269.

Arbués, F., Garcia-Valiñas, M., & Martinez-Espiñeira, R. (2003). Estimation of residential water demand: A state-of-the-art review. *Journal of Socio-Economics, 32*(1), 81–102.

Bontemps, C., & Nauges, C. (2009). Carafe ou bouteille? Le rôle de l'environnement dans la décision du consommateur. *Economie et Prévision, 188*, 61–79.

Bukenya, J. O. (2008). Avoidance measures and household perceptions of water quality in Uganda. *Journal of African Business, 9*(2), 309–325.

Essex, S., Kent, M., & Newnham, R. (2004). Tourism development in Mallorca: Is water supply a constraint? *Journal of Sustainable Tourism, 12*(1), 4–28.

Ford, R. K., & Ziegler, J. A. (1981). Intrastate differences in residential water demand. *Annals of Regional Science, 15*(3), 20–30.

García-Valiñas, M. A. (2006). Analysing rationing policies: Drought and its effects on urban users welfare. *Applied Economics, 38*(8), 955–965.

Garcia-Valiñas, M. A., Martínez-Espiñeira, R., & González-Gómez, F. J. (2009). Does private management of water supply services really increase prices? An empirical analysis. *Urban Studies, 46*(4), 923–945.

Garcia-Valiñas, M. A., Martínez-Espiñeira, R., & González-Gómez, F. J. (2010). Affordability of residential water tariffs: Alternative measurement and explanatory factors in southern Spain. *Journal of Environmental Management, 91*(12), 2696–2706.

Garn, M. (1998). *Managing water as an economic good: The transition from supply-oriented to demand-responsive services*. Washington, DC: Water Supply and Sanitation Division, World Bank.

Genius, M., Hatzaki, E., Kouromichelaki, E. M., Kouvakis, G., Nikiforaki, S., & Tsagarakis, K. P. (2008). Evaluating consumers willingness to pay for improved potable water quality and quantity. *Water Resources Management, 22*, 1825–1834. doi:10.1007/s11269-008-9255-7.

Gössling, S. (2006). Tourism and water. In S. Gössling & C. M. Hall (Eds.), *Tourism & global environmental change* (pp. 180–194). London: Routledge.

Jalan, J., Somanathan, E., & Chaudhuri, S. (2009). Awareness and the demand for environmental quality: Survey evidence on drinking water in urban India. *Environment and Development Economics, 14*(6), 665–692.

Johnstone, N., & Serret, Y. (2012). Determinants of bottled and purified water consumption: Results based on OECD survey. *Water Policy, 14*(4), 668–679.

Martínez-Espiñeira, R. (2002). Residential water demand in the Northwest of Spain. *Environmental and Resource Economics, 21*(2), 161–187.

Monteiro, H. (2010). *Residential water demand in Portugal: Checking for efficiency-based justifications for increasing block tariffs*. ERC-ISCTE working paper 01/10, Lisbon.

Montginoul, M., & Rinaudo, J. D. (2011). Controlling households' drilling fever in France: An economic modeling approach. *Ecological Economics, 71*, 140–150.

Nauges, C., & Thomas, A. (2000). Privately-operated water utilities, municipal price negotiation, and estimation of residential water demand: The case of France. *Land Economics, 76*(1), 68–85.

Nauges, C., & Reynaud, A. (2001). Estimation de la demande domestique d'eau potable en France. *Revue Economique, 52*(1), 167–185.

Nauges, C., & Whittington, D. (2010). Estimation of water demand in developing countries: An overview. *The World Bank Research Observer, 25*(2), 263–294.

Ortalda, L., & Hatchuel, G. (2000). *La perception de la qualité de l'eau et la sensibilité de la population à l'environnement*. Paris: CREDOC.

Piper, S. (2003). Impact of water quality on municipal water price and residential water demand and implications for water supply benefits. *Water Resources Research, 39*, 1127–1136. doi:10.1029/2002WR001592.

Polyzou, E., Jones, N., Evangelinos, K. I., & Halvadakis, C. P. (2011). Willingness to pay for drinking water quality improvement and the influence of social capital. *Journal of Socio-Economics, 40*, 74–80.

Reynaud, A. (2010). *Doing better with less: Implementing peak-load pricing for managing residential water demand*. Mimeo, University Toulouse 1 Capitole.

Reynaud, A., Renzetti, S., & Villeneuve, M. (2005). Residential water demand with endogenous pricing: The Canadian case. *Water Resources Research, 41*(11), W11409.1–W11409.11.

Roibas, D., García-Valiñas, M. A., & Wall, A. (2007). Measuring welfare losses from interruption and pricing as responses to water shortages: An application to the case of Seville. *Environmental and Resource Economics, 38*(2), 231–243.

Schleich, J., & Hillenbrand, T. (2009). Determinants of residential water demand in Germany. *Ecological Economics, 68*(6), 1756–1769.

Shaw, W. D. (2005). Water resource economics and policy: An introduction. Edward Elgar Publishing, Cheltenham. ISBN 1-84376-917-4 364 pp.

Um, M. J., Kwak, S. J., & Kim, T. Y. (2002). Estimating willingness to pay for improved drinking water quality using averting behaviour method with perception measure. *Environmental and Resource Economics, 21*(3), 287–302.

United Nations (UN). (2002). *International covenant on economic, social and cultural rights*, General Comment No. 5.

Whitehead, J. C. (2006). Improving willingness to pay estimates for quality improvements through joint estimation with quality perceptions. *Southern Economic Journal, 73*(1), 100–111.

World Health Organization. (2011). *Pharmaceuticals in drinking-water*, WHO/HSE/WSH/11.05.

Worthington, A. C., & Hoffman, M. (2008). An empirical survey of residential water demand modeling. *Journal of Economic Surveys, 22*(5), 842–871.

Yongsung, C., Easter, K. W., McCann, L. M. J., & Homans, F. (2005). Are rural residents willing to pay enough to improve drinking water quality? *Journal of the American Water Resources Association, 41*(3), 729–740.

Zivin, J. G., Neidell, N., & Schlenker, W. (2011). Water quality violations and avoidance behaviour: Evidence from bottled water consumption. *American Economic Review: Papers and Proceedings, 101*(3), 448–453.

Chapter 17
A Simulation Model for Understanding the Consequences of Alternative Water and Wastewater Tariff Structures: A Case Study of Fayoum, Egypt

Céline Nauges, Dale Whittington, and Mohamed El-Alfy

1 Introduction

Households and firms in developing countries want piped water and wastewater services, but the high capital costs of this infrastructure often make it unaffordable. In many countries, donors and national governments step in to provide grants and concessional financing enables households to enjoy heavily subsidised piped water and wastewater for a period before the full cost has to be met.

With large subsidies, households and firms grow accustomed to low water bills and typically have little awareness of the real financial costs of piped services. At the same time, governments and donors may consider their grants and concessional financing to be one-off transfers which are needed at the introduction of piped water services to help low-income households and firms connect and access modern services. Over time, however, water and sanitation companies can find it

C. Nauges (✉)
School of Economics, University of Queensland, Brisbane St Lucia, QLD, Australia
e-mail: c.nauges@uq.edu.au

D. Whittington
School of Public Health, University of North Carolina at Chapel Hill,
Chapel Hill, NC 27599, USA

Department of Environmental Sciences & Engineering, Rosenau CB#7431,
The Manchester Business School, University of Manchester, Manchester, UK
e-mail: Dale_Whittington@unc.edu

M. El-Alfy
Ministry of Housing, Utilities and Urban Development, 1 Ismail Abaza Street,
Cairo, Egypt
e-mail: Alfy1958@hotmail.com

Q. Grafton et al. (eds.), *Understanding and Managing Urban Water in Transition*, Global Issues in Water Policy 15,
DOI 10.1007/978-94-017-9801-3_17

difficult or impossible to sustain the continual drain on their budgets that these subsidies require. As infrastructure depreciates and needs replacing, governments and donors commonly expect water users to pay more of the cost. At least some water customers will, however, remain poor, and nearly all customers develop feelings of entitlement to subsidised services whether or not they can afford them, and perceive tariff increases as unjust impositions rather than cost-recovery measures ensuring system financial sustainability. Because people invariably feel losses much more than equivalent gains (Kahnemann and Tversky 1979; Knetsch 2010), reform of water and wastewater tariffs in developing countries in order to recover financial costs is often politically explosive.

Most households in both urban and rural areas of Egypt now enjoy heavily subsidised water and wastewater services, and the quality and coverage of these services is high given the country's modest gross domestic product of about USD 2,000 per capita. The average water and wastewater price in Egypt is much lower than the price charged in countries with similar socio-economic, cultural, and climate characteristics: for household consumption of up to 15 m^3 per month, the average price is about 7 US cents (USD 0.068) per m^3, while it is USD 0.49, 0.90, and 0.37 per m^3 in Jordan, Morocco, and Tunisia, respectively.[1]

Even before the January 2011 revolution in Egypt, the Egyptian Water Regulatory Agency and the Egyptian Holding Company for Water and Wastewater (the institutions jointly responsible for the strategic direction and financing in the water and wastewater sector) struggled to implement tariff reforms. These institutions sought to design new tariffs that would balance four main objectives: (1) financial cost recovery, (2) economic efficiency, (3) fairness and equity, including affordability, and (4) simplicity. The government recognised that reform was needed since the tariffs in Egypt neither aimed at nor met these objectives. In general, most of the water and wastewater companies do not raise sufficient revenues to cover their operation and maintenance costs, their tariffs are quite complex, and the tariffs are unfair to customers using small quantities of water.

In this chapter we present a simulation model that was designed to assist the Egyptian Water Regulatory Agency to better navigate the tariff reform process. This model allows an analyst to simulate the effects of alternative tariff structures and prices on customer water use, which in turn affects utility revenues and costs. Sensitivity analyses test the effects of price and income elasticity assumptions on the performance of alternative tariffs. We apply this simulation model to individual customer records for an actual utility in Egypt, the Fayoum Water and Wastewater Company.

In Sect. 2 of the chapter, we describe the Fayoum Water and Wastewater Company and its current tariff structure. Section 3 describes our modelling strategy and analytical framework, including the two tariff options we analysed. Section 4 presents the results of our model simulations and our sensitivity analyses. In Sect. 5 we summarise our main findings and conclusions.

[1] The entire survey by Global Water International is available at http://www.globalwaterintel.com/tariff-survey-2009/

2 Description of the Fayoum Water and Wastewater Company

2.1 Number of Accounts, Revenues, and Costs

The Fayoum Water and Wastewater Company currently serves about 2.7 million people. The company has about 390,000 customers, all of whom have water service; about 124,000 customers (32 %) have connections to both piped water and sewer networks. The domestic customers with both water and sewer connections number about 116,000, and 8,000 non-domestic customers have both water and sewer connections. The annual operation and maintenance (O&M) costs in fiscal year (FY) 2009 were about 144 million Egyptian Pounds (LE) (or USD 27 m).[2]

The Fayoum water company's total revenues from all sources in FY 2009 were about LE 94 million, which covered 65 % of total O&M costs. Billing and payment records in the water company's database show about LE 54 m billed for core water and sewerage services and LE 46 m collected; of these amounts, LE 46 m was billed to domestic customers, and they paid LE 41 m. Billing records show that the government was billed for 13 million m^3 (13 M m^3), but did not pay directly for the services received. Instead, government bills are settled "outside the system". In this inter-governmental settlement of accounts, the company was paid about LE 15 m. Finally, commercial and industrial users paid LE 5 m for water services. Core water revenue was thus about LE 61 M (= 41 + 5 + 15).

Reported water production was 178 M m^3, and total sales from billing records were 140 M m^3, resulting in unaccounted-for water of 21 %. Domestic customers account for 79 % of total water sales, government accounts for 17 %, and commercial and industrial use accounts for 4 % of sales.

Table 17.1 summarises information for FY 2009 on the O&M costs, numbers of customers, and water sales to customers with water service and wastewater service for the Fayoum Water and Wastewater Company.

O&M costs amount to LE 83 m and LE 61 m for water supply and wastewater services, respectively (column *a*). There were 390,000 water connections and 124,000 sewer connections in FY 2009 (column *b*). Total water sales to Fayoum's customers with only water connections were 95 M m^3, and water sales to customers with both water and wastewater service were 45 M m^3 (column *c*).

Table 17.1 O&M costs for fiscal year 2009

	Total O&M cost, LE	Number of customers	Water sales, m^3
	a	*b*	*c*
Water service	83,000,000	390,000	95,000,000
Wastewater service	61,000,000	124,000	45,000,000

[2] The approximate exchange rate was LE 5.4 = USD 1.0 in mid-2010.

Assuming an economy-of-scale factor of 0.6[3] (i.e., a 10 % increase in production results in an approximate 6 % increase in cost), the data in Table 17.1 suggest the following relationship between the Fayoum Water and Wastewater Company's water O&M costs and water sales[4]:

$$\text{Water O\&M Costs} = 5.4 \times [\text{water sales to customers with water service}]^{0.6}. \qquad (17.1)$$

And similarly for wastewater O&M costs, we have[5]

$$\text{Wastewater O\&M cost} = 6.2 \times [\text{water sales to customers with sewer service}]^{0.6}. \qquad (17.2)$$

Thus, average O&M costs per cubic meter of water sold to customers are not constant. As water and wastewater tariffs increase during the reform process, in general customers will reduce consumption, and this will cause the utility's O&M costs to change. Our simulation model is designed to capture these effects.

2.2 *Current Tariff Scheme*

The Fayoum Water and Wastewater Company's tariff has a long and complicated list of customer categories. In addition to governmental institutions (including schools, mosques, and health centres), commercial businesses, and industries, there are some very specific categories such as youth clubs, private schools, gas stations, and coffee shops. The excessive number of customer categories makes the tariff structure complicated and difficult to understand, and there does not seem to be any rationale for the tariffs applied to these categories. Different customer classes currently face different prices in the current tariff, and even within the same customer class the prices are sometimes different. For example, restaurants and cafes are charged a minimum of about LE 10 each month whereas coffee shops are charged a minimum of LE 24.

---

[3] In an unpublished report to the Egyptian Water Regulatory Agency we estimated a stochastic cost frontier using cost data from 19 Egyptian utilities. We considered separate cost functions for water and wastewater and included as inputs the price of labour and the price of electricity. We estimated the economy-of-scale factor to be 0.68 (95 % confidence interval 0.52–0.84) for water services and 0.60 (95 % C.I. of 0.52–0.68) for wastewater services.

[4] The general cost function is O&M Costs = K × [water sales]$^{\gamma}$ where γ = economy-of-scale factor. The cost is 83 (from column *a* in Table 17.1) and sales are 95 (from column *c*). Inserting these values into Eq. 17.1 and solving yields K = 5.4.

[5] The amount of wastewater is approximated by the amount of water sold to customers with wastewater services.

Households' water bills are calculated using a two-part tariff. A volumetric charge is based on four blocks with increasing prices, and a fixed "charge" is the sum of several charges that can differ from one customer to another. All customers are charged using what is termed in Egypt a "simple block" tariff, which means that a customer's entire monthly consumption is charged at the price of the block in which the last cubic meter falls. This "simple block" tariff is best illustrated by example. In Fayoum, the first ("lifeline") block is 0–10 m^3, and the volumetric price is LE 0.23 per m^3; the second block (10–20 m^3) has a price of LE 0.45 per m^3. The fixed monthly charge is about LE 3. Thus, a customer using 10 m^3 per month will get a bill for LE 5.3 ($=3+0.23\times10$). However, if the customer's consumption increases to 11 m^3 in the following month, their bill will be LE 7.95 ($=3+0.45\times11$). A 10 % increase in consumption (from 10 to 11 m^3) therefore results in a 50 % increase in the water bill. This disproportionate increase in the bill is unrelated to the water company's costs. Eqs. 17.1 and 17.2 show that because of economies of scale, the increase in O&M costs for an increase in consumption of 10 % will be less than 10 %. It is unfair to have the household incur such a large increase in its water bill. Wastewater tariffs are commonly calculated as a percentage of the water tariff, which varies from about 35 % for residential users to 60–70 % for non-domestic users, which means that the unfairness affects wastewater as well as water.

Currently, households in Fayoum are charged for 10 m^3 if their consumption falls between 0 and 10 m^3 per month, which is called the "minimum charge". There are two problems with this practice. First, households have no financial incentive to reduce their usage below 10 m^3. Second, it is unfair to charge households for water they do not use. Unfortunately, the tariff is so lacking in transparency that many households trying to keep their water costs low by restricting their consumption to the lifeline block are probably not aware that there is no advantage in doing so.

According to billing records, domestic customers were billed an average price of 0.65 LE per m^3; and customers classified under small commercial activities were billed 1.2 LE per m^3. According to billing records, domestic customers paid 89 % of their bills, and non domestic customers paid 62 %. The overall collection rate was 85 %. If government payments "outside the system" are included, non-domestic customers paid 87 % of their bills.

Average water use per household in FY 2009 was about 12 m^3 per month, and the median domestic bill was about 6 LE per month (≈ USD 1). Despite the low price charged for water, the water use of many households was low. Some 60 % of domestic customers use less than 10 m^3 per month, which is in the lifeline block of the tariff. Water sold to domestic customers in the lifeline block represented about one third of total sales to the domestic population and 42 % of the total billed amount. Nearly 20 % of domestic accounts fall in the second block with average water use of 15 m^3 per month, so about 80 % of domestic users consume less than 20 m^3 per month.

Households with consumption in the lifeline block paid higher average prices than users in the higher blocks: those in the first block paid an average price of

0.90 LE per m^3, but the average price paid by households whose consumption fell in the third and fourth blocks was 0.54 and 0.59 LE per m^3, respectively. To the extent that poorer households use less water than richer households, the poor paid a higher average price. This occurred mainly because of the "minimum charge" regardless of the amount of water used.[6] Households with high water use (in the upper blocks) have lower collection rates than those in the lifeline block, which compounds the problem.

In summary, the current tariff in Fayoum does not achieve the government's objectives and thus needs to be revised. The tariff is overly complicated and lacks transparency. Customers do not understand their bills. The tariff is not generating sufficient revenue to cover financial costs, and none of the customer classes is paying the full financial cost of water. The existing tariff is not fair or equitable. Customers in the lifeline block of the tariff pay about LE 100 per year on average. However, customers in the highest block of the tariff pay just a little more, about LE 130 per year, and they use four times as much water.

3 Analytical Framework and Modelling Strategy

3.1 Model Objectives

The objective of our simulation exercise was to design a tariff that accomplishes several objectives, the first of which was financial. In this study we focus on recovery of O&M costs of the water company, i.e. we disregard the cost of replacement and rehabilitation of existing capital assets as well as the cost of future capital projects. The tariff should generate sufficient revenues to recover O&M costs by a specified future target date. In this chapter, a target date of 3 years is used. In the simulation model, O&M costs are endogenous because they depend on water sales. Moreover, O&M costs reflect economies of scale, and they are different for water and wastewater services. The model calculates the O&M cost of water services separately from the O&M cost of wastewater services, using Eqs. 17.1 and 17.2.

The second objective is "economic efficiency": as many customers as possible should face the marginal cost of water and wastewater supply, which we define for purposes of illustration as the sum of the opportunity cost of water withdrawn from other uses and the marginal cost of operation and maintenance. The third objective is "affordability", which we define as ensuring that the poor do not have to pay a water bill that is too high a percentage of their income. The fourth objective is "simplicity and transparency", by which we mean that a customer should easily be able to understand how their household's water bill is calculated.

[6] This fact – that an increasing block tariff does not actually help poorer households – has been noted in other studies. See, for example, Komives et al. (2005), Boland and Whittington (2000), and Whittington (1992).

3.2 *Alternative Tariff Structures to be Examined*

We analyse an increasing block tariff (IBT) because it is the type currently required by law and used in Egypt. To reduce the economic inefficiencies and inequities associated with the IBT, we reduce the number of blocks to two and retain a positive fixed charge (Whittington 1992; Boland and Whittington 2000; Griffin and Mjele 2011). We study two IBT structures: in the first, the lifeline block is 0–10 m^3/month, and in the second, the lifeline block is 0–5 m^3/month. In each tariff, the decision variables are the price of the first block, the price of the second block, and the size of the positive fixed charge.

In these tariffs, we eliminate the "minimum charge" and "simple block," using instead billing practices that are common in the global water industry. For example, if the first block is 0–10 m^3 per month and the price is LE 0.23 per m^3, then the volumetric charge for a customer that uses say 6 m^3 should be LE 1.38 (=0.23×6). If the second block is 10–20 m^3 with a price of LE 0.50 per m^3, a different customer using say 16 m^3 would pay a volumetric charge of LE 5.30, i.e. LE 2.30 for consumption in the first block plus LE 3.00 (=0.50×6) for 6 m^3 of consumption in the second block.

3.3 *Solution Procedure: Setting the Volumetric Price and Fixed Charge*

3.3.1 Water

The economic efficiency objective for the tariff requires that the volumetric price should be set at the Fayoum water company's marginal cost (MC) of operation. For an IBT with two blocks, the price for the first block is the lifeline rate, which includes both financial and political considerations for decision making. In the second block, we aim at setting the price equal to the MC as best it can be determined.

The MC of operation should include not only the marginal cost of providing water services but also the opportunity cost of taking water away from other sectors in the Egyptian economy (e.g., agriculture). The latter has been estimated to be about 1.4 LE/m^3 in the Egyptian Water Master Plan, and is assumed to be constant within the range of water sales under consideration (Ministry of Water Resources and Irrigation 2005). The marginal cost of operation is endogenous (it depends on water sales) and is roughly equal to the economy of scale factor (0.6) multiplied by the average water O&M cost derived from Eq. 17.1.[7]

[7] The economy-of-scale factor in Eqs. 17.1 and 17.2 is the ratio of marginal to average cost. The average cost of water service can be estimated from Eq. 17.1 using an exponent of −0.4 instead of 0.6 (i.e. by dividing Eq. 17.1 by sales).

Two other parameters of the tariff structure need to be chosen: the positive fixed charge and the volumetric price of the first block. Raising the fixed charge and keeping the volumetric price in block 1 low generates a higher, more stable revenue stream for the utility (since the amount generated by the fixed charge is known with certainty). However, a higher fixed charge is detrimental to low-consumption households (the higher the fixed charge, the higher will be the average price they pay). It is not known whether households with low water consumption are poorer than household with higher consumption; many factors besides income influence water use (e.g., family size, housing type, attitudes about water conservation). But to the extent that water use is positively correlated with income, fixed charges will tend to adversely affect low-income households. Also, if an objective is to send a signal on the economic value of water to customers, the volumetric price in block 1 should be raised instead of the fixed charge. In the simulation model, for the IBT we set the volumetric price in block 1 at 50 % of the volumetric price in block 2 (i.e., half of the MC) and search for the fixed charge that will fully recover O&M costs at the end of the transition period, year 3.

3.3.2 Wastewater

Fayoum's cost data for FY 2009 show that the annual O&M cost per connection with water and wastewater service is slightly more than twice the annual O&M cost per connection for water service only. Thus, when households connect to the sewerage system, their water bill should approximately double, or increase by even more than 100 % in order to recover financial costs. In order for the tariff to send the correct economic signal about the cost of wastewater services, we propose a 100 % surcharge for households with sewerage on top of the water volumetric charge (before inclusion of the opportunity cost of raw water), including households in the lifeline block of the IBT (in 2009, the wastewater surcharge was only 35 %). For simplicity, we assume that the fixed charge will be the same for households with and without a sewer connection.[8]

3.3.3 Customer Classes

We simplify the tariff structure by having only one domestic (residential) customer category plus two categories for non-domestic customers: Group 1, non-governmental worship places, charities, youth clubs, and schools; and Group 2, other non-domestic customers. The number of Fayoum customers in Group 1

[8] The estimation of the stochastic cost frontier on data from Egyptian utilities (see note 3 above) confirms that, for most utilities, the marginal cost of providing wastewater services is higher than the marginal cost of supplying water, so the wastewater surcharge should be at least 100 %. However, when working on tariff design with the Egyptian Water Regulatory Agency, we were asked to consider a maximum of 100 % for the wastewater surcharge.

is small compared to Group 2. These two non-domestic categories are a much simpler structure than the one currently in use; it reflects the objective of fairness that the existing tariff fails to achieve. Group 1 non-domestic customers pay the same volumetric price in the lifeline block of the domestic tariff for all the water they use, and Group 2 customers pay the same volumetric price in the second block in the domestic tariff.

3.4 *Assessing the Performance of the Tariff Options: Affordability Indices*

Our affordability indices measure water payments as the share of a household's total expenditures (or income). The data from the customer billing records for Fayoum do not contain information on income, total expenditure, household size, or other socioeconomic characteristics and thus do not permit estimates of the effects of tariff increases on poor households. Chemonics Egypt (2009)[9] used data from a 2007–08 survey of Egyptian households to estimate expenditure for several budget categories (including water supply) for households sorted by quintiles of income. There is a strong positive statistical correlation between water and sanitation expenditure and household total expenditure, which suggests that households with the lowest water bills tend to be the poorest.

We thus divide the domestic customers in Fayoum into 20 groups of equal size based on the average water use of each household. We then assume that the first group (the 5 % of households with lowest consumption) has total household expenditure equal to that of the first income strata in the Chemonics study. We repeat this process for the other 19 groups of domestic customers.

For each tariff scenario, we report the average affordability index in simulation year 3 for three groups of households: the *low-consumption group,* which includes the 20 % (quintile) of households with the lowest consumption; the *high-consumption group,* which includes the quintile of households with the highest consumption; and the *middle-consumption group,* which includes the remaining 60 % of households.

3.5 *Model Architecture*

The tariff simulations calculate the effects of tariff increases on each customer (both domestic and non-domestic) in the Fayoum water company's billing records for FY 2009, and then sum these individual customer effects to obtain totals. For each customer, the starting value is their average monthly water use in FY 2009. The model is run on a year-by-year basis (1-year steps), assuming that water prices,

[9] *Affordability assessment to support the development of a financing strategy for the water supply and sanitation sector in Egypt.* Final report, 161 pp.

fixed charges, and customer water usage are adjusted annually.[10] All monetary amounts are expressed in real (not nominal) terms.

The model simulates how each Fayoum customer adjusts their water use in each year of the 3-year simulation period to changes in water prices and income. Tariff reforms are implemented at the beginning of year 1; volumetric prices and the fixed charge are chosen to produce about the same total revenue as in the base year billing record. The model calculates the average price that each customer would be billed if their consumption in year 1 were the same as consumption the previous year (i.e. in the billing records), and then it calculates the percentage change in the average price from the base year to year 1. The percentage change in average price is then multiplied by the assumed price elasticity of demand (and the assumed percentage change in annual growth of income is multiplied by the assumed income elasticity of demand) to obtain the customer's change in consumption. The change in consumption is added to the previous year's consumption to get consumption in the current year. From that, the model estimates the bill in each year for each customer and the customer's payment, taking account of the collection rate for the customer's category.

This iterative process is repeated for all years over the simulation period. If the total revenue fails to meet the target, the tariff decision variables are changed by the user judgement and the simulation is repeated. Appendix 1 presents a detailed description of the simulation model, including model equations.

3.6 *Model Assumptions*

3.6.1 Customer Response to Prices

There has been a debate among water economists about the water price to which customers react when deciding how much water to use (for a summary of the literature on this question see Arbués et al. 2003). Economic theory suggests that customers react to the marginal price. However, because water tariff structures are often complicated and because water expenditure usually represents only a small portion of a household's budget, many economists consider it more realistic to assume that customers respond to the average price and not to the marginal price. The simulation model assumes that customers respond to change in average price.

3.6.2 Price Elasticity

It is not possible to estimate the price elasticity of demand for customers in Fayoum because our data are a single cross-section of customers facing the same tariff

[10] Customers are in fact billed monthly. However, the distribution of monthly records does not show any significant seasonal pattern (e.g. there is no significant increase in water use in the summer). For this reason, it is reasonable to use as a benchmark the average monthly water use for the entire fiscal year.

structure. However, studies in different countries and estimates from the literature suggest that the price elasticity of residential water demand is commonly in the range –0.3 to –0.6 (Nauges and Whittington 2010). In Egypt, the price of water is low and water expenditures are typically less than 1 % of household budgets (Chemonics Egypt 2009). We thus expect price elasticity to be low, probably smaller than –0.3.

The demand of households that use smaller quantities of water is probably more inelastic than that of households using larger quantities because large users have more options for reducing consumption if prices rise. For this reason, we assume a price elasticity of –0.2 for domestic customers using more than 10 m^3 per month per household, and –0.1 for domestic customers using less than 10 m^3 per month per household. We tested the sensitivity of simulation results to assumptions about price elasticity.

As the price of water increases, non-domestic customers may pay more attention to their water bills than before and may take steps to avoid wasting water, such as fixing leaks. Our tariff simulations assume a price elasticity of –0.2 for non-domestic customers.

3.6.3 Income Growth and Income Elasticity

Egypt's growth in gross domestic product was 7.2 % in 2008, 4.7 % in 2009, and 5.1 % in 2010 (World Bank).[11] In the model simulations, household income is assumed to grow in real terms by 5 % each year. The income elasticity of domestic customers' demand for water and sewerage is typically in the range 0.1–0.3 (Nauges and Whittington 2010). The model initially assumes an income elasticity of 0.2, which is then tested by a sensitivity analysis.

3.6.4 Threshold Domestic Water Use

We assume that households have a minimum water consumption of 5 m^3 per month, which means they will not reduce their consumption below this amount regardless of how high prices rise. The choice of the minimum threshold is *ad hoc* and based on our personal judgement. For greater accuracy the threshold should depend on the size of the household, but this information is unavailable in the customer billing records. No threshold is assumed for non-domestic customers.

3.6.5 Population Growth

The national master plan prepared by the Egyptian Holding Company for Water and Wastewater for the European Commission (May 2009) provides estimates of population growth by governorate in Egypt. Population in El Fayoum is estimated to

[11] Available at http://data.worldbank.org/indicator/NY.GDP.MKTP.KD.ZG

grow by 25 % over the next 10 years (from 2,961,431 in 2007–12 to 3,709,265 in 2017–22). Based on this forecast, we assume that the number of water connections (both domestic and non-domestic) will increase 2.5 % each year.

3.6.6 Sanitation Coverage

The Fayoum water company provided data on the number of new wastewater customers each year from FY 2006 through FY 2009. Projecting these data into the future, we estimate that new sewerage connections will increase 6 % per year during the simulation period. For modelling purposes, the domestic customers who connect to the sewer system are chosen at random each year from among those not already connected.[12] Documents provided by the water company reported that 75 % of non-domestic customers were connected to the sewerage system in FY 2009. However, billing records contain few actual bills for non-domestic customers, making it impossible to identify which ones already have (or do not have) sewer connections. Thus, for modelling the tariff, we assume that all non-domestic customers have sewer connections.

3.6.7 Collection Rate

The collection rate (payments divided by bills) can vary significantly from one customer to another. Instead of applying an individual collection rate for each customer in our tariff simulations, we use an average collection rate for groups of customers. We assume that operational efficiency and hence collection rate will remain the same over the 3-year planning period. More details on the assumed collection rates are given in Appendix 2.

4 Simulation Results

4.1 Tariff Option 1: Increasing Block Tariff with First Block 0–10 m³/month (IBT10)

The simulation outcome for an IBT with the first block 0–10 m^3/month (denoted IBT10) is shown in Table 17.2. The volumetric price in block 2 is 2.80 LE/m^3 for households with water and sewerage.[13] The volumetric price in block 1 is 50 % of

[12] This is unrealistic because in some cases there is no sewer line in a specific area. When one is installed, many households there may choose to connect. However, this assumption will not have much effect on our simulation results.

[13] The volumetric price in block 2 is chosen such that it is equal to the marginal cost, itself being a function of total water sales.

Table 17.2 Tariff option 1 (IBT10 with 2 blocks) – simulation results for year 3

Year 3 (target year)	Unit	Simulation outcome
Fixed charge[a]	LE/month	5.03
Volumetric price for water in block 1[b]	LE/m^3	0.70
Volumetric price for water and wastewater in block 1	LE/m^3	1.40
Volumetric price for water in block 2[c]	LE/m^3	1.40
Volumetric price for water and wastewater in block 2	LE/m^3	2.80
Total water sales	M m^3	80.9
Total water sales (domestic customers)	M m^3	64.6
Total water sales (non-domestic customers)	M m^3	16.3
Total O&M costs	M LE	132
Marginal cost of water and wastewater services, including opportunity cost of water	LE/m^3	2.80
Total collected payments	M LE	132

[a]This corresponds to a 34 % annual rate of increase from year 1 until year 3
[b]This corresponds to a 75 % annual rate of increase from year 1 until year 3
[c]This corresponds to a 87 % annual rate of increase from year 1 until year 3

the volumetric price in block 2, i.e. 0.70 LE/m^3 for households with only water service and 1.40 LE/m^3 for households with both water and sewerage. In order for collected payments from both domestic and non-domestic customers to cover O&M costs in year 3, the fixed charge has to be set at 5 LE/month, which would correspond to a 34 % annual rate of increase in the fixed charge from year 1 to year 3 (without taking inflation into account). With this tariff, our simulations indicate that about 81 M m^3 of water would be sold in year 3, of which 65 M m^3 would be used by the domestic population and 16 M m^3 would be used by the non-domestic population. Total collected payments are estimated to reach LE 132 m in year 3, which should cover O&M costs. In year 3, more than half of the domestic population (54 %) falls in the first block of the IBT, so less than half of the domestic consumers face the economically efficient price. The utility does not experience any decrease in revenue when the new tariff is put in place; the total amount of collected payments in year 1 is estimated at LE 55.4 m which is in the range of collected payments estimated for FY 2009.

In Table 17.3 we report the average consumption per household, the average bill, the average price, and the average affordability index in year 1 and in year 3 for low-, middle-, and high-consumption domestic customers.

Households in the third group (large consumers) consume on average 17 m^3/month in year 3, compared to 23 m^3/month in year 1. The decrease is smaller for households in the middle- and low-consumption groups because we have assumed a minimum monthly consumption of 5 m^3 per month per household and a lower price elasticity for households consuming less than 10 m^3 per month. In the group of low-consumption households, the average bill increases from 4.1 LE/month to 9.3 LE/month and the average share of the water bill in the average household's

Table 17.3 Impact of tariff option 1 (IBT10) on domestic customers in year 1 and year 3

		Year 1			Year 3		
		Low	Middle	Large	Low	Middle	Large
Average consumption per household	m³/month	5.5	10.8	23.1	5.2	8.9	17.1
Average bill	LE/month	4.1	6.3	13.3	9.3	13.9	29.3
Average price	LE/m³	0.75	0.59	0.58	1.80	1.57	1.71
Affordability index	%	1.2	0.5	0.2	2.7	0.9	0.4

total expenditure increases from 1.2 % in year 1 to 2.7 % in year 3. Despite the sharp increase in prices, water services remain affordable in year 3 for the low-income group (affordability indices from 2 % to 5 % are generally considered to be affordable). The average price is higher for the low-consumption group (LE 1.80/m³ in year 3) than for the high-consumption group (LE 1.71/m³).

4.2 *Tariff Option 2: Increasing Block Tariff With a First Block Set at 5 m³/month (IBT5)*

Table 17.4 presents the simulation results associated with the implementation of an increasing block tariff with a first block at 5 m³/month. In order to facilitate comparisons of IBT5 with IBT10, we also report the main results under implementation of an IBT10. The consequences for the domestic customers are shown in Table 17.5.

Tariff options 1 and 2 (IBT10 vs. IBT5) lead to the same level of water sales, total O&M costs, and collected payments in year 3. However, under the IBT5, only 9 % of the domestic customers fall into the first block, compared to 54 % in the case of an IBT10. So 91 % of the domestic customers are facing an economically efficient price if the size of the first block is reduced to 5 m³/month. The IBT5 provides small and poor households with low water bills, but the water use of larger households, even if they are poor, will not likely fall into the first block. As a consequence of the higher number of domestic customers falling in block 2, the fixed charge, which is required to generate revenues that cover O&M costs, with the IBT5 is 1.8 LE/month, while it was 5 LE/month with the IBT10. The lower fixed charge under IBT5 improves the affordability index for the group of low-consumption customers: the average share of water bill in total expenditure is now 2 % instead of 2.7 % with the IBT10 (Table 17.5). Also, the IBT5 could be seen as more "equitable" in the sense that the average price paid by households in the low-consumption group in year 3 (LE 1.47/m³) is smaller than the average price paid by households in the high-consumption group (LE 1.78/m³). The average consumption in the three groups of households is about the same whatever the size of the first block.

Table 17.4 Tariff options 1 and 2 (IBT10 and IBT5) – simulation results, year 3

Year 3 (target year)	Unit	IBT10	IBT5
Fixed charge[a]	LE/month	5.03	1.79
Volumetric price for water in block 1[b]	LE/m³	0.70	0.70
Volumetric price for water and wastewater in block 1	LE/m³	1.40	1.41
Volumetric price for water in block 2[c]	LE/m³	1.40	1.41
Volumetric price for water and wastewater in block 2	LE/m³	2.80	2.82
Total water sales	M m³	80.9	79.7
Total water sales (domestic customers)	M m³	64.6	63.4
Total water sales (non-domestic customers)	M m³	16.3	16.3
Total O&M costs	M LE	132	131
Marginal cost of water and wastewater services, including opportunity cost of water	LE/m³	2.80	1.41
Total collected payments	M LE	132	131

[a]This corresponds to a 34 % annual rate of increase in the case of IBT10 and a 20 % annual rate of decrease in the case of IBT5 from year 1 until year 3
[b]This corresponds to a 75 % annual rate of increase in the case of IBT10 and IBT5 from year 1 until year 3
[c]This corresponds to a 87 % annual rate of increase in the case of IBT10 and IBT5 from year 1 until year 3

Table 17.5 Impact of the IBT5 on the domestic customers, year 3

		IBT10			IBT5		
		Low	Middle	Large	Low	Middle	Large
Average consumption per HH	m³/month	5.2	8.9	17.1	5.2	8.5	17.0
Average bill	LE/month	9.3	13.9	29.3	7.6	13.3	30.4
Average price	LE/m³	1.80	1.57	1.71	1.47	1.57	1.78
Affordability index	%	2.7	0.9	0.4	2.2	0.9	0.4

4.3 *Sensitivity Analysis*

In this section we check the sensitivity of our simulation results to assumptions on price and income elasticity. In each case the sensitivity to an assumption is measured by the gap it generates between O&M costs and collected payments. We test how the main results change if the price elasticity for households using more than 10 m³ per month is –0.3 instead of –0.2. We still assume a price elasticity of –0.1 for households consuming less than 10 m³ per month. The outcome of the sensitivity analyses for the IBT10 and IBT5 are shown in Tables 17.6 and 17.7 respectively. Total water sales and collected payments in year 3 are quite sensitive to the assumed price elasticity of demand. With an IBT10, assuming a price elasticity of –0.3 instead of –0.2 leads to an increase in the share of domestic population falling in the first block from 54 % to 61 % (Table 17.6). A higher price responsiveness induces substantially lower water sales (from 81 M m³ to 69 M m³), and hence lower

Table 17.6 Sensitivity to assumptions on price and income elasticities, IBT 10

Year 3 (target year)	Unit	Base case assumptions[a]	Price elasticity: –0.3 for HH using more than 10 m^3 per month	Income elasticity: 0.3
Share of domestic population in first block	%	54	61	53
Total water sales	M m^3	80.9	69.8	81.9
Total water sales (domestic customers)	M m^3	64.6	56.5	65.6
Total water sales (non-domestic customers)	M m^3	16.3	13.2	16.3
Total O&M costs	M LE	132	120	133
Total collected payments	M LE	132	112	134

[a]Price elasticity set at –0.2 for households using more than 10 m^3/month and income elasticity set at +0.2

Table 17.7 Sensitivity to assumptions on price and income elasticities, IBT 5

Year 3 (target year)	Unit	Base case assumptions[a]	Price elasticity: –0.3 for HH using more than 10 m^3 per month	Income elasticity: 0.3
Share of domestic population in first block	%	9	9	9
Total water sales	M m^3	79.7	73.2	80.6
Total water sales (domestic customers)	M m^3	63.4	58.4	64.3
Total water sales (non-domestic customers)	M m^3	16.3	14.8	16.3
Total O&M costs	M LE	131	124	132
Total collected payments	M LE	131	118	132

[a]Price elasticity set at –0.2 for households using more than 10 m^3/month and income elasticity set at +0.2

collected payments. On the other hand, increasing the assumed income elasticity from +0.2 to +0.3 only slightly increases total water sales and total collected payments for both the IBT10 and IBT5 options.

5 Discussion

Our model simulations show that it is possible to modify the existing tariff structure used by the Fayoum Water Company and, over a 3-year planning period, increase average prices paid by customers in order to achieve O&M cost recovery targets. This financial objective can be achieved while meeting both affordability and economic efficiency objectives.

A two-block tariff with a positive fixed charge is simple to understand and transparent. The elimination of the simple block tariff and of the minimum charge reduces the burden imposed on small customers. Our results also show that reducing the size of the lifeline block at 5 instead of 10 m^3 per month could be an attractive alternative. If a lifeline block of 10 m^3 is used, and because volumetric prices and fixed charges must increase to meet revenue targets, the water use of households in the second block falls significantly, even though it is assumed that demand is very inelastic (a price elasticity of demand of –0.2 is assumed for households in the second block). The result is that more than 50 % of households are estimated to fall into the lifeline block. Also, because the volumetric charge in the lifeline block is low, to meet revenue targets a 10 m^3 lifeline block puts increasing pressure on the volumetric rate in the second block and on the fixed charge.

The results of the sensitivity analyses show that tariff design is complicated by uncertainty as to what will happen to both utilities and their customers if tariffs are increased substantially. There are several reasons why customers' responses to large tariff increases are uncertain and difficult to forecast. First, some customers may not be individually metered but instead share collective bills. Predicting how such groups will react to large increases in joint bills is especially difficult. Second, for customers that have individually-metered connections, price and income elasticities will be uncertain, and water utilities typically do not know the incomes or household sizes of their customers. In particular, when tariff structures are not transparent households simply may not understand how their water bills are calculated. From the utility's perspective, large tariff increases will reduce its water sales, which will in turn affect both its operating costs, net revenues, and capital expansion plans. Water regulatory agencies must take decisions in the face of such uncertainties about the consequences of their actions, but analytical tools such as the simulation model described in this chapter are still useful to help think systematically about future possibilities.

Appendix 1: Equations of the Simulation Model

We describe below how the model operates in year *t*. These equations correspond to the ones used to simulate an increasing two-block tariff with a positive fixed charge.

1 List of Variables and Parameters

AVPRICE	Average price (LE/m^3)
DAVPRICE	Change in average price (real rate)
COLLRATE	Collection rate for domestic customers
DFC	Annual (real) rate of increase for the fixed charge
DINCOME	Income growth (real rate)

DPRICELLB	Annual (real) rate of increase for the volumetric price in the lifeline block
DPRICEUPB	Annual (real) rate of increase for the volumetric price in the upper block
ELASINC	Income elasticity of demand
ELASPRI	Price elasticity of demand
FC	Positive fixed charge (LE/month)
MOCONS	Monthly consumption (m^3)
N_D	Total number of domestic customers
N_{ND}	Total number of non-domestic customers
NOFLAT	Number of apartments
PRICELLB	Volumetric price in the lifeline block in year 1 in LE/m^3
PRICEUPB	Volumetric price in the upper block in year 1 in LE/m^3
SANIT	A variable which takes the value 1 if the customer has a sewerage connection and 0 otherwise
SEWFEE	Sewerage surcharge
SIZELLB	Size of the lifeline block (m^3)
TOTSALES	Total water sales (m^3)
TOTCOLL	Total collected payments (LE)

2 List of Equations

2.1 *Fixed Charge and Prices in Year* t

At the beginning of year *t*, the tariff structure is readjusted as follows: the fixed charge (FC) in year *t* corresponds to the fixed charge in year *t*–1 to which we apply the annual rate of increase (DFC):

$$\text{FC}_t = \text{FC}_{t-1} \times (1 + \text{DFC}).$$

The volumetric price in the lifeline block (PRICELLB) in year *t* corresponds to the volumetric price in the lifeline block in year *t*–1 to which we apply the annual rate of increase (DPRICELLB):

$$\text{PRICELLB}_t = \text{PRICELLB}_{t-1} \times (1 + \text{DPRICELLB}).$$

The same applies to the volumetric price in the upper block (PRICEUB):

$$\text{PRICEUB}_t = \text{PRICEUB}_{t-1} \times (1 + \text{DPRICEUB}).$$

2.2 Description of the Customer's Behaviour

We calculate the *average price* that will be paid by each customer in year *t* if they consume in year *t* the same amount of water they consumed in year *t*–1. The average price is calculated as follows:

Domestic Customers

For household *i* in the lifeline block (i.e., $\text{MOCONS}_{i,t-1} \leq \text{SIZELLB}$) without sewerage connection:

$$\text{AVPRICE}_{i,t} = \left(\text{FC}_t + \text{PRICELLB}_t \times \text{MOCONS}_{i,t-1}\right) / \text{MOCONS}_{i,t-1}.$$

For household *i* in the lifeline block (i.e., $\text{MOCONS}_{i,t-1} \leq \text{SIZELLB}$) with sewerage connection:

$$\text{AVPRICE}_{i,t} = \left[\text{FC}_t + \text{PRICELLB}_t \times (1+\text{SEWFEE}) \times \text{MOCONS}_{i,t-1}\right] / \text{MOCONS}_{i,t-1}.$$

For household *i* in the upper block (i.e., $\text{MOCONS}_{i,t-1} > \text{SIZELLB}$) without sewerage connection:

$$\text{AVPRICE}_{i,t} = \begin{bmatrix} \text{FC}_t + \text{PRICELLB}_t \times \text{SIZELLB} + \text{PRICEUPB}_t \\ \times\left(\text{MOCONS}_{i,t-1} - \text{SIZELLB}\right) \end{bmatrix} / \text{MOCONS}_{i,t-1}$$

For household *i* in the upper block (i.e., $\text{MOCONS}_{i,t-1} > \text{SIZELLB}$) with sewerage connection:

$$\text{AVPRICE}_t = \begin{bmatrix} \text{FC}_t + \text{PRICELLB}_t \times (1+\text{SEWFEE}) \times \text{SIZELLB} \\ +\text{PRICEUPB}_t \times (1+\text{SEWFEE}) \\ \times\left(\text{MOCONS}_{i,t-1} - \text{SIZELLB}\right) \end{bmatrix} / \text{MOCONS}_{i,t-1}.$$

Non-domestic Customers

Non-domestic customers in Group 1 are charged at the price of the lower block for all cubic meters of water used and all customers are assumed to be connected to the sewerage system:

$$\text{AVPRICE}_{i,t} = \left[\text{FC}_t + \text{PRICELLB}_t \times (1+\text{SEWFEE}) \times \text{MOCONS}_{i,t-1}\right] / \text{MOCONS}_{i,t-1}.$$

Non-domestic customer *i* in Group 2 are charged at the price of the upper block for all cubic meters of water used and all customers are assumed to be connected to the sewerage system:

$$\text{AVPRICE}_{i,t} = \left[\text{FC}_t + \text{PRICEUPB}_t \times (1 + \text{SEWFEE}) \times \text{MOCONS}_{i,t-1}\right] / \text{MOCONS}_{i,t-1}.$$

From that, we can infer the *change in average price in year t* (DAVPRICE) faced by customer *i* as:

$$\text{DAVPRICE}_{i,t} = \left(\text{AVPRICE}_{i,t} - \text{AVPRICE}_{i,t-1}\right) / \text{AVPRICE}_{i,t-1}.$$

Hence the *predicted monthly water consumption* in year *t* is:

For domestic customer *i* using more than 5 m³/month/apartment:

$$\text{MOCONS}_{i,t} = \text{MOCONS}_{i,t-1} \times \begin{pmatrix} 1 + \text{ELASPRI} \times \text{DAVPRICE}_{i,t} \\ +\text{ELASINC} \times \text{DINCOME} \end{pmatrix}.$$

For domestic customer *i* using less than 5 m³/month/apartment, the minimum threshold applies:

$$\text{MOCONS}_{i,t} = 5 \times \text{NOFLAT}_i \times (1 + \text{ELASINC} \times \text{DINCOME}).$$

For non-domestic customer *i*:

$$\text{MOCONS}_{i,t} = \text{MOCONS}_{i,t-1} \times \left(1 + \text{ELASPRI} \times \text{DAVPRICE}_{i,t}\right).$$

The *monthly bill* for customer *i* is then calculated as follows:

Domestic Customers

For household *i* in the lifeline block (i.e., $\text{MOCONS}_{i,t} \le \text{SIZELLB}$) without sewerage connection:

$$\text{BILL}_{i,t} = \text{FC}_t + \text{PRICELLB}_t \times \text{MOCONS}_{i,t}.$$

For household *i* in the lifeline block (i.e., $\text{MOCONS}_t \le \text{SIZELLB}$) with sewerage connection:

$$\text{BILL}_t = \text{FC}_t + \text{PRICELLB}_t \times (1 + \text{SEWFEE}) \times \text{MOCONS}_t.$$

For household i in the upper block (i.e., $\mathrm{MOCONS}_{i,t} > \mathrm{SIZELLB}$) without sewerage connection:

$$\mathrm{BILL}_{i,t} = \mathrm{FC}_t + \mathrm{PRICELLB}_t \times \mathrm{SIZELLB} + \mathrm{PRICEUPB}_t \times \left(\mathrm{MOCONS}_{i,t} - \mathrm{SIZELLB}\right).$$

For household i in the upper block (i.e., $\mathrm{MOCONS}_{i,t} > \mathrm{SIZELLB}$) with sewerage connection:

$$\mathrm{BILL}_{i,t} = \mathrm{FC}_t + \mathrm{PRICELLB}_t \times (1 + \mathrm{SEWFEE}) \times \mathrm{SIZELLB} + \mathrm{PRICEUPB}_t \times (1 + \mathrm{SEWFEE}) \times \left(\mathrm{MOCONS}_{i,t} - \mathrm{SIZELLB}\right)$$

Non-domestic Customers

For non-domestic customers in Group 1, all cubic meters are charged at the price of the lower block and all customers are assumed to be connected to the sewerage system:

$$\mathrm{BILL}_{i,t} = \mathrm{FC}_t + \mathrm{PRICELLB}_t \times (1 + \mathrm{SEWFEE}) \times \mathrm{MOCONS}_{i,t}.$$

For non-domestic customers in Group 2, all cubic meters are charged at the price of the upper block and all customers are assumed to be connected to the sewerage system:

$$\mathrm{BILL}_{i,t} = \mathrm{FC}_t + \mathrm{PRICEUPB}_t \times (1 + \mathrm{SEWFEE}) \times \mathrm{MOCONS}_{i,t}.$$

2.3 *Total Water Sales, Collected Payments, and Target Revenue*

Total water sales in year t are calculated as the sum of water sold to all domestic customers (N_D) and water sold to all non-domestic customers (N_{ND}):

$$\mathrm{TOTSALES}_t = 12 \times \left(\sum_{i=1}^{N_D + N_{ND}} \mathrm{MOCONS}_{i,t} \right).$$

And *total collected payments* in year t are the sum of bills paid by domestic and non-domestic customers. The collection rate is applied to each domestic customer bill. The non-domestic customers pay their bills.

$$\mathrm{TOTCOLL}_t = 12 \times \left(\sum_{i=1}^{N_D} \mathrm{BILL}_{i,t} \times \mathrm{COLLRATE} + \sum_{i=1}^{N_{ND}} \mathrm{BILL}_{i,t} \right).$$

Finally, the target in year T (3 or 5 depending on the scenario) is defined as follows:

The target is to cover O&M cost in year T and is calculated as follows:

$$\text{TARGET1}_T = 5.4 \times \text{TOTSALES}_T^{\,0.6} + 6.2 \times [\text{TOTSALES}_T \times \text{SANIT}]^{0.6},$$

where $[\text{TOTSALES}_T \times \text{SANIT}]$ represents the total amount of water sales to customers with a sewerage connection.

Appendix 2: Assumptions on Collection Rates

The population of domestic customers in FY 2009 is divided into 10 bins of equal size (deciles), based on average monthly water use per household. The first bin (decile) has 10 % of the domestic population with the lowest average monthly water use per household, and the last bin (the tenth) has 10 % of the population with the highest average monthly water use per household. We calculate the average collection rate in each decile from Fayoum's billing records for FY 2009, and we then assume that the average collection rate in a decile applies to each customer in that decile. Fayoum's collection rates are shown below. We assume that collection rate remains constant over the 3-year simulation period.

Decile	Average collection rate (%)
1	92.1
2	96.1
3	97.0
4	95.6
5	93.9
6	88.1
7	95.6
8	95.5
9	93.9
10	69.3

References

Arbués, F., Garcia-Valiñas, M., & Martinez-Espiñeira, R. (2003). Estimation of residential water demand: A state-of-the-art review. *Journal of Socio-Economics, 32*, 81–102.

Boland, J., & Whittington, D. (2000). The political economy of increasing block water tariffs in developing countries. In A. Dinar (Ed.), *The political economy of water pricing reforms* (pp. 215–236). Oxford: Oxford University Press.

Chemonics Egypt. (2009). *Affordability assessment to support the development of a financing strategy for the water supply and sanitation sector in Egypt*. Final report prepared within the framework of the Mediterranean Component of the EU Water Initiative (MED EUWI) Egypt Country Dialogue on Water, February 2009, pp. 161.

Griffin, R. C., & Mjele, J. W. (2011). Distributing nature's bounty. *Ecological Economics, 72*, 116–128.

Kahnemann, D., & Tversky, A. (1979). Prospect theory: An analysis of decision under risk. *Econometrica, 47*(2), 263–292.

Knetsch, J. (2010). Values of gains and losses: Reference states and choice of measure. *Environmental and Resource Economics, 46*, 179–188.

Komives, K., Foster, V., Halpern, J., & Wodon, Q. (2005). *Water, electricity, and the poor: Who benefits from utility subsidies? Directions in development*. Washington, DC: World Bank.

Ministry of Water Resources and Irrigation. (2005). *Water for the future. National water resources plan 2017*. Cairo: MWRI.

Nauges, C., & Whittington, D. (2010). Estimation of water demand in developing countries: An overview. *World Bank Research Observer, 25*(2), 263–294.

Whittington, D. (1992). Possible adverse effects of increasing block water tariff in developing countries. *Economic Development and Cultural Change, 41*(1), 75–87.

Chapter 18
Managing Water Variability Issues

Hugh Sibly and Richard Tooth

1 Introduction

Despite the growing use of manufactured sources, urban centres still predominantly rely on climate-dependent water sources. In many cases, the availability of supply is subject to significant variability and uncertainty. Recently the additional concern has arisen that climate change will increase this uncertainty. This variability and uncertainty create a number of significant issues that need to be managed. This chapter reviews the issues and considers different approaches to managing them.

Urban water networks are characterised by relatively large infrastructure costs relative to operating costs. In common with many infrastructure industries, economic issues arise around the optimal pricing and timing of so-called 'lumpy' investments. A key feature of many urban water systems is the presence of large, cheap storages – a characteristic that is largely absent from other well-studied utility industries such as electricity and telecommunication networks. Storage has implications for both the pricing and timing of investments, and therefore how uncertainty in water availability is managed. These implications will be examined here.

Governance arrangements are also important in managing the uncertainty of water availability. It is often the case that urban water augmentation decisions, pricing, and operational decisions are centrally managed by government agencies

H. Sibly (✉)
Tasmanian School of Business and Economics, University of Tasmania,
Private Bag 84, Hobart, TAS 7001, Australia
e-mail: hugh.sibly@utas.edu.au

R. Tooth
Sapere Research Group Limited and Adjunct Professor with the School of Business
(Sydney Campus), Sapere Research Group, GPO Box 220, Sydney, NSW 2001, Australia
e-mail: rtooth@srgexpert.com

Q. Grafton et al. (eds.), *Understanding and Managing Urban Water in Transition*, Global Issues in Water Policy 15,
DOI 10.1007/978-94-017-9801-3_18

and/or government-owned organisations. However, markets offer a potential alternative approach, one where supply can more quickly adapt to changing conditions and needs. We consider the potential for markets to help deal with the issue of uncertainty.

An emerging issue in many jurisdictions is that of water security. When there is sufficient uncertainty, an urban centre can face the prospect of running out of water. This is a key driver of urban water policy in many regions. However, there is often a lack of clarity of how water security goals are determined and how security criteria are established. This chapter considers the government's and market's role in providing for water security. It is noted that an increase in diversity of supply may increase water security, although this diversity is likely to come at a cost.

This chapter has three major working sections. Section 2 examines the economics of bulk supply of urban water, including decisions over investment and pricing. Section 3 builds on this and discusses the issue of supply security. Then Sect. 4 examines issues of institutional design before we set out our final conclusions.

2 Economics of Augmentation and Use Decisions

2.1 Nature of Urban Water Supply

The majority of urban water systems rely extensively on water catchments. In many regions there is significant seasonal or annual uncertainty in the magnitude of inflows into these catchments, and so many urban water systems experience substantial variability in the available supply of water. Furthermore, the high cost (relative to its value) of transporting water means that urban water systems are highly localised, and dependent on a few water sources.

While supply from certain climate-dependent sources can be more or less predicted,[1] in general future supply is uncertain and has an unknown probability distribution.[2] A major concern associated with climate change is that the uncertainty of supply will progressively increase. This variability and uncertainty – coupled with demand uncertainty – creates a number of significant management issues.

So far as urban water is concerned, variability and uncertainty is primarily an issue to do with the provision and demand for bulk supply. Although uncertainty and variation in demand on other aspects of the supply chain can create challenges, these challenges tend to be relatively minor. For example, the distribution and reticulation network needs to be designed to meet seasonal peak demand requirements, which are not constant, but the incremental cost of providing additional capacity to meet higher peak demand is, in general, relatively small. In contrast, the

[1] In a number of systems inflows to water catchments follow broad seasonal patterns.

[2] Uncertainty in which the probability of the outcome is unknown is commonly referred to as Knightian uncertainty after Knight (1921).

network costs of meeting peak demand are major factors in other network utilities such as energy, transport, and telecommunications.

In addition to variability there are other important features of urban water. Urban water augmentation tends to involve 'lumpy' (i.e. large and infrequent) investments that have significant lead times (meaning they take a long time to develop). These factors create challenges for supply investment decisions.

The lumpiness of investment means that additional capacity needs to be created before it is fully required. An implication is that the relative merits of different supply alternatives will depend on projections of growth in demand; when demand growth is low, smaller investments may be more efficient even if their average cost per unit output is higher. When there is uncertainty as to future requirements (due to uncertainty in demand or available supplies) the issue of 'lumpiness' becomes more pronounced. Under uncertainty, there is a magnified risk that capacity will be unused.

When future needs are certain, a long lead time is of little consequence; investments with longer lead times just need to begin earlier. However when there is uncertainty (with regards to supply and/or demand), and investment is irreversible, lead times become important because conditions may change during the course of development and affect the investment's value. For example, a project that at one stage appeared necessary to supplement dwindling supplies may suddenly become unnecessary following a change in weather.

In terms of choice of investment, there are a number of implications of uncertainty. First, all else being equal, investments with shorter lead times are more valuable. Shorter lead times mean that investment can be made closer to when supply is required and thus when requirements are more certain. Second, flexible investments have greater value. Often the development process is not fixed, which provides the developer with options (known as *real* options) to re-optimise the design, and/or the development process, as conditions change. For example, a desalination plant has greater value if much of the time-consuming but inexpensive investment can be made early and the more expensive investment can be made later; this allows the developer to defer the expensive investment if conditions permit. The presence of real options adds complexity to investment decisions.

These issues are not unique to water. The problems of uncertain requirements and lumpiness of investments with long lead times are common in other infrastructure investments such as with energy, transport, and telecommunications. This combination of features, and the implications, as noted above, for timing and choices of investment has received substantial attention (see for example, Dixit and Pindyck 1994; Turvey 2000).

However, urban water differs from many other industries in that water can often be stored cheaply in large quantities. The services provided by transport and telecommunications must be used as they are provided, and whereas it is possible to store electricity the cost is prohibitive for general use. In contrast, climate-driven water sources are often captured in large storages within nearby water catchments. For example, in Sydney, Australia, the storage capacity of local water catchments is around 4 years of average demand.

The availability of storage has significant implications for urban water investment and management. First, storage can be used to help smooth over lumpy investments. Second, storage can be used to help manage uncertainty. These implications are explored in the following section.

2.2 Economics of Investment: No Storage

To analyse decisions around the type and timing of investment in urban water supply, it is useful to review the costs and pricing of urban water services. While urban water prices are usually administratively set, they are generally done with the dual aims of achieving cost recovery and efficiency. To meet these dual goals, a two-part tariff scheme, consisting of usage charges and fixed charges, is commonly implemented.[3]

If customers' fixed charges are sufficiently small so as not to materially affect connection decisions, then interest in the efficient use of the resource focuses only on usage charges. It is commonly acknowledged that efficient use of a resource is encouraged when the usage charge is set to marginal cost; however, debate has raged around how marginal cost should be defined.

To encourage efficient use, price may be set to short-run marginal cost (SRMC), which, when there is no storage available, is the cost of producing an additional unit of water. An issue with SRMC pricing is that when there are 'lumpy' investments the introduction of new capacity can have a dramatic effect on the price at which demand and supply are balanced. Two implications are that the price may be very volatile, and that the utility will under-recover for future investments. This effect is illustrated in Fig. 18.1 below. The market clearing price (the intersection of SRMC and demand) will be low when there is excess capacity, increase rapidly as demand grows and capacity becomes constrained, and then fall quickly when a new lumpy investment comes on line.[4]

Of particular concern is that the fall in price due to additional capacity creates a disincentive for investment.[5] This effect is illustrated in Fig. 18.2 below. Here, a price set to SRMC would be so low in the post-investment period that there is no producer surplus (i.e. return on investment) from which to fund new investment. Furthermore, due to the price drop, there is a large benefit transfer from producers to consumers associated with existing investments. Despite these adverse outcomes for suppliers, the investment may be socially beneficial if the net increase in producer and consumer surplus (areas B + C in Fig. 18.2) exceeds the cost of the investment.

[3] Other charges are imposed, most notably additional charges applied to new developments; however, these are extraneous to the analysis in this chapter.

[4] A further problem is that in some industries (e.g. landing slots at an airport) as capacity limits are reached congestion externalities start to kick in.

[5] This is the case when the private investor cannot charge a fixed fee to customers, as is likely to be the case when a private investor provides water to a publicly owned water corporation.

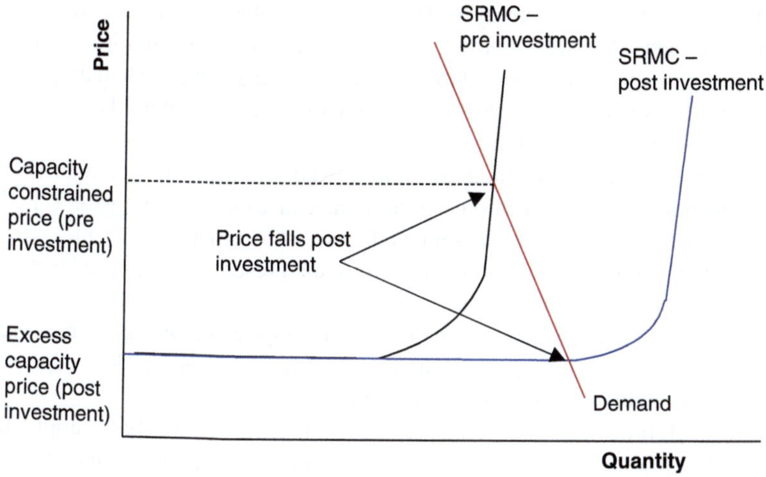

Fig. 18.1 Short-run marginal cost (SRMC) pricing, without storage, under lumpy investments

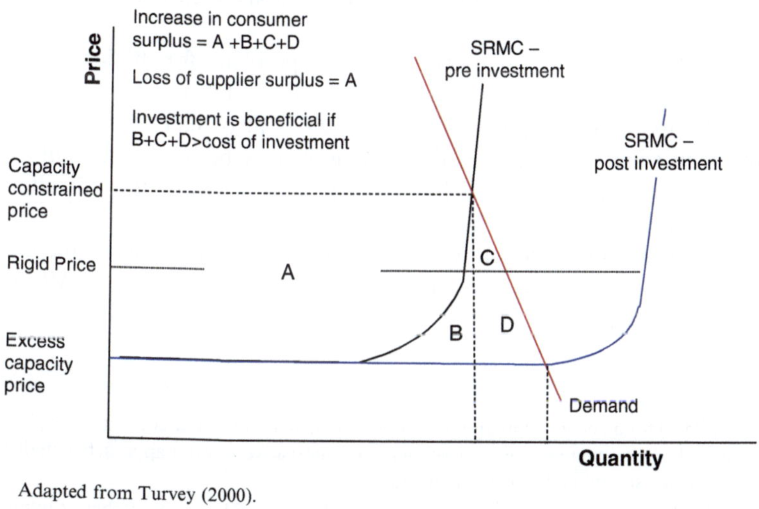

Fig. 18.2 SRMC pricing without storage under lumpy investments (Adapted from Turvey (2000))

The problem of lumpy investments prompted consideration of alternative price approaches and in particular pricing based on 'long-run marginal cost' (LRMC), a marginal cost value that incorporates capital costs of investment.

In determining LRMC a useful framework is to consider that, in an environment of continually growing demand, the effect of price on demand is primarily to change the timing of future investments. Thus a LRMC pricing approach may serve the dual objective of funding future investment as well as provide a signal for demand.[6]

[6] See OFWAT (2001) for a useful discussion on the role of LRMC pricing for water utilities in England and Wales.

With this framework, Turvey (1976) developed the 'present worth of system incremental cost' (PWSIC) method. This method involves estimating the optimal capital program over a long period to meet the demand and then calculating how the timing of the investment, and thus the present value of costs, would change with a permanent small change in demand.

However, the PWSIC method is complex and still leads to a saw-tooth price pattern. In practice, simpler, more approximate, methods are often used. Perhaps most common is an 'average incremental cost' (AIC) method that measures the present value of future investments divided by the present value of the incremental demand to be met by those investments.

While the relative merits of alternative pricing approaches has generated significant debate and discussion,[7] pricing based on LRMC has gained wide acceptance by regulators and influential bodies around the world.[8]

However significant challenges emerge when there is variability in supply. In the main, LRMC pricing is rigid; that is, the pricing path is set in advance and does not change with changing supply and demand conditions. Recent droughts in Australia and the US have resulted in severe supply shortages at prices set at LRMC. In response, governments have imposed water restrictions to curb demand, which are increasingly recognised as having problems.[9]

The rigid LRMC pricing approach also creates problems for efficient investment. A rigid pricing approach does not reward investors who invest in drought-proof sources of supply. The rigid pricing approach is also likely to result in over-investment in supply as it means that pricing cannot be used to curtail demand during severe shortages. Given the limited range of restrictions, the only alternative is to build additional supply earlier.

Both variability and uncertainty of water supply are key factors. If these factors did not exist then the bulk of the concerns relating to the rigid LRMC pricing approach would not exist.

---

[7] See Sibly (2006) for a debate on marginal cost pricing in an urban water context. Park (1989) uses simulation analysis to examine the relative merits of alternative pricing approaches under conditions of lumpy investment and growing demand.

[8] For example, the England and Wales regulator (see OFWAT 2001) and a number of regulators in Australia have endorsed the LRMC approach. The American Water Works Association (AWWA 2000, p. 120) endorses the LRMC, stating that "Economic theory suggests that water rates be set equal to long-run marginal costs to ensure an efficient allocation of water service."

[9] While restrictions have appeared to be successful in curbing demand, restrictions are an inefficient and inequitable approach relative to reasonable alternatives based on price. Demand restrictions have a number of problems. First, they are difficult to implement: they can only be reasonably applied to outdoor use and they are notoriously difficult to enforce and therefore they can only partially limit demand. Second, they are inefficient, as they restrict efficient allocation between different uses, users, and time periods. Third, restrictions have additional hidden consequences on business and society. These include loss of amenity due to loss of green spaces and costs spent by business and individuals in avoiding water restrictions. Fourth, relative to simple alternatives, they are highly regressive. In effect, when water is underpriced, the large users (who tend to be wealthier) receive a subsidy.

2.3 *Implications of Storage*

2.3.1 Implications Under Certainty

The provision of substantial storage capacity is a major feature of urban water systems that must deal with uncertain environmental water inflows. However, we first find it useful to consider the implications for urban water systems of having substantial storage capacity when there is certainty over future demand and supply.

Storage can be used to 'smooth over' the known (i.e. certain) variability in water inflows. For instance, in some regions annual river flows are known to be sufficient to provide for annual requirements; however sometimes, due to seasonal variations, river flows may be insufficient to supply daily requirements. In such cases storage capacity may be employed to smooth over the seasonal variation in supply.

Storages may also potentially be used to smooth over the excess capacity resulting from lumpy investments. To explain the impact of storage on the timing of lumpy investment, assume for simplicity that water production (e.g. inflows to catchments) and demand growth are constant and that investments (such as a desalination plant or recycling plant) are for an additional constant supply. The implications of storage for the timing of investment are illustrated in Fig. 18.3. The top half of the figure shows that, in the absence of storage, additional lumpy capacity is built prior to it being fully required. The bottom half of the figure shows how, with storage, the investment in capacity can be postponed by increasing stocks of water when capacity exceeds demand and then using the stored water to defer investment.

Storage has significant implications for marginal cost pricing. When water is being stored, the SRMC of using water is not the cost of additional production but rather the cost that reflects the opportunity cost of *not using* the water for future use.

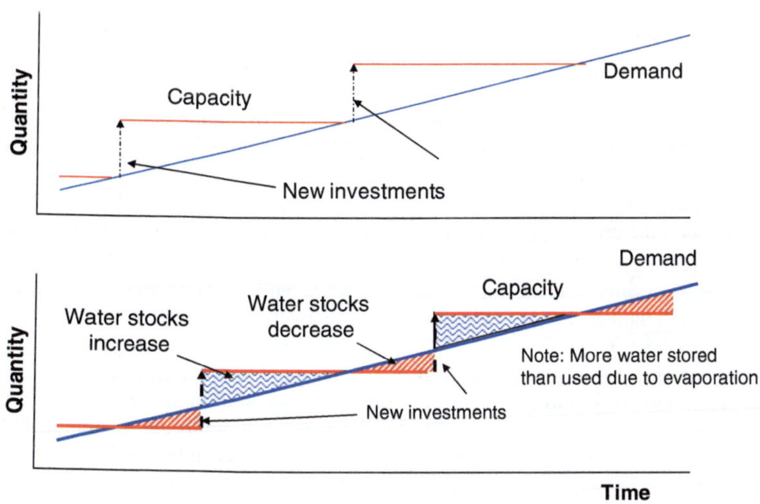

Fig. 18.3 The benefits of storage

This cost can be calculated as the present value of water in the future, less any marginal costs associated with storage. Thus to encourage an efficient (i.e. optimal) allocation of water over time, the bulk water price should incorporate this opportunity cost.

While water is being stored, an efficient pricing rule that governs the bulk water price[10] is one where the price grows at a rate reflecting the marginal cost of holding water, which is the combination of the financing cost (i.e. an interest rate) and the marginal costs of storage.[11]

The pricing growth implied by this price rule may be modest. If demand and supply (and related costs) are certain, then the appropriate discount rate is the risk-free interest rate.[12] Rates of storage loss vary according to climate and dam characteristics[13]; as a rough estimate, the average rate of storage loss in Sydney is less than 10 % and the marginal rate is less than 5 %.[14] Thus, assuming a marginal storage cost of 5 %, for example, the implied real price path may be an annual increase of less than 10 %. As this represents the bulk water price, the retail price growth may be substantially less.

This 'SRMC with storage' pricing rule differs significantly to pricing at SRMC when there is no storage. This is illustrated in Fig. 18.4. Under the 'SRMC with storage' pricing rule, while water is being stored the bulk water price rises continuously at a rate determined by the discount rate and the marginal rate of storage loss. In contrast, under 'SRMC pricing with no storage' the bulk water price might hardly move until capacity becomes a constraint, after which the price escalates at a rate driven largely by the elasticity of demand and the rate of demand growth.

The 'SRMC with storage' pricing path is more similar in form to that implied by the PSWIC method of calculating LRMC (which ignores storage). Under the PWSIC method the price calculated (based on a permanent demand) increases at close to the discount rate.[15] Thus the key difference between the two formulas is that, in the presence of storage, future price is discounted for storage loss. Both pricing rules yield similar price growth only if the marginal rate of storage loss is low.

---

[10] For the purposes of analysis in this chapter, and for ease of exposition, we focus on the bulk water price. Assuming pass-through of costs, the retail price will simply be an increment to the bulk water price, which is relatively stable and predictable.

[11] This is similar to Hotelling's price rule. It may be described as $P_{t+1} = P_t(1+r)/(1-m)$, where P_t denotes the price in period t, r is a constant discount rate, and m is the marginal rate of storage (see Appendix for the proof).

[12] At the time of writing, real risk-free interest rates are about 2 %. The cost of capital for utilities is higher, but still modest. For example, the real pre-tax weighted average cost of capital used in Australia is around 7 %.

[13] The marginal rate of storage loss will is likely to differ to from the average rate. Storage loss is primarily related to evaporation, and thus the marginal storage will primarily relate to the increase in evaporation from storing more water. This will depend on a number of factors including the shape of the dam. Assuming evaporation is proportional to surface area and that the walls of the dam have a positive slope, then the marginal storage rate will be positive.

[14] This is based on unpublished levels of storage and storage loss.

[15] That is, $P_{t+1} = P_t(1+r)$.

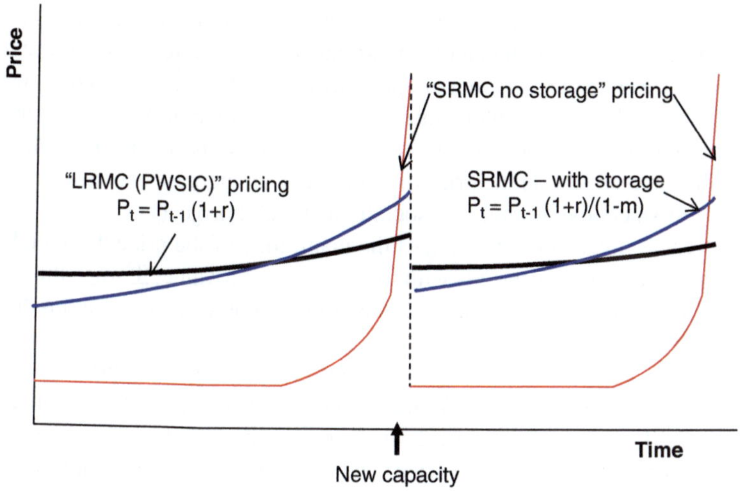

Fig. 18.4 Comparison of pricing paths

If demand is continually growing, then at some point the introduction of new capacity will become socially optimal. This new capacity puts a limit on the price growth. Assuming the capacity is sufficiently lumpy (i.e. adds significantly to the available supply), the new capacity lowers the marginal cost of supply.

Once storages are exhausted, the 'SRMC with storage' pricing rule does not apply and, quite simply, an SRMC pricing rule would follow 'SRMC without storage' pricing and equate short-term supply and demand and, due to demand growth, the price would continue to rise. However, under reasonable assumptions[16] it would be optimal to exhaust storages at the point when new lumpy capacity is installed.[17] Once new lumpy capacity is installed there is again excess capacity and the price to encourage efficient usage would drop. As illustrated in Fig. 18.4, the price drop will not be as severe as in the 'SRMC with no storage' pricing case.

The starting and ending price of this price path would be determined by the timing of the next augmentation; for example, a delay in the timing of the next augmentation necessitates a higher price to curb demand. A social planner would time investments so that the total consumer benefit arising from the infrastructure augmentation covers the cost of augmentation. That is, the social planner will undertake the project provided the discounted benefit of increased consumption outweighs the cost of investment.

[16] The reasonable assumptions are that the lumpy investment to be installed is not at significantly lower cost than future investment and there is no abnormal future growth in demand.

[17] If high price growth was expected, an inter-temporal optimisation would result in water being conserved for the period of higher prices. See Appendix for proof.

A private investor would only choose to undertake an investment that is timed so that demand has grown such that the investment cost is covered by the present value of the future revenue received (net of operating costs). As water demand is invariably inelastic, the bulk supply augmentation causes a substantial decrease in price. As illustrated in Fig. 18.2, a private investor in a new supply is not able to capture some of the value created. While the price drop provides an increase in consumer surplus, it creates a disincentive to invest in socially beneficial projects.

However, as noted, storage reduces the magnitude of the price drop, and allows water to be reallocated to periods with the highest water value. Storage thus helps to mitigate, but not eliminate, this issue of investors not being able to capture the value created.

Of note, the new investment also has an impact on consumption prior to the investment. The earlier the investment is made, the less water needs to be conserved in prior periods. Thus, bringing forward a new investment provides a benefit to consumers in prior periods by depressing prices and allowing greater consumption.

Another important implication of storage is that there is a relationship between 'SRMC under storage' and the LRMC that incorporates the capacity cost. Thus, an effect of storage is to reduce the distinction between SRMC and LRMC.

The potential for storage to address the problems of SRMC pricing associated with lumpy investments depends on the available storage capacity. The adequacy of available storage capacity in managing lumpy investments depends on the lumpiness of investment relative to demand growth. Under 'SRMC with storage' pricing, demand growth would be less than unconstrained demand growth; however the effect of price is likely to be small as the short-term price elasticity of bulk water is likely to be very small.[18]

These factors depend on the situation; however, some simple examples can provide a guide to the potential range. The total storage capacity requirements under certainty and constant demand growth are surprisingly light. Consider, for example, a situation in which capacity can be staggered (as in Fig. 18.3) and, for example, a constant marginal storage loss of 7 %, an annual net-demand growth of around 1.5 %,[19] and that new investment adds around 20 % additional capacity. Under this scenario a new lumpy investment is required after around 11 years and the water storage would peak at around the 5th year at less than 30 % of a year's aggregate demand.[20]

This requirement is small compared to the storage capacity of many urban centres that have large variance in supply; for example, Sydney's storage capacity is in

[18] Recently Abrams et al. (2012) estimated the short-run retail price elasticity of demand to be –0.05 for Sydney Water customers. The bulk water price elasticity would be lower in magnitude. However, using this price elasticity and a price growth of 10 %, the effect of price increases would be to dampen consumption growth by <0.5 % of unconstrained growth.

[19] For example, annual population growth in major urban centres in Australia is forecast to be around 1.5 % (see WSAA 2010). In general, per capita use of water will also tend to increase with rising incomes.

[20] This is estimated using a simple numerical simulation. Due to staggering of investment, initial demand is set at 90 % of capacity.

excess of 4 years of consumption. Whereas storage capacity is also needed to manage uncertainty, as discussed below, the large size of the storages suggests that the storages may be used to help smooth over the problem of lumpy investments. In locations where supply variability tends to be smaller (such as in England), storages are also smaller and the potential for smoothing appears less.

2.3.2 Introducing Uncertainty

For the purposes of examining supply uncertainty,[21] we assume the objective is to maximise the *expected* net benefit of water provision.

With this assumption, the introduction of variability and uncertainty makes little modification to the optimal pricing rule. While water is stored, the SRMC of water is equal to the opportunity cost, which is the *expected* present value of water in the future less the marginal costs associated with storage.[22]

Under uncertainty, the current price can be expressed as a function of the likelihood of different price scenarios in periods beyond the next period. Thus under this pricing rule, the SRMC of water, and thus the price of water, will depend on the expectations of demand and supply in future periods in addition to current level of water supply. For example, as the likelihood of a drought occurring increases, the expected future price will increase and require an increase in the current price. Similarly, prices would drop following a large fall of rain that significantly increased water stocks. All else being equal, the SRMC would shift positively with expected changes in current and future demand, and negatively with expected changes in current and future supply.

Numerical methods can be applied to estimate a pricing path that maximises a welfare objective, or, for example, ensures private return on investment.[23] Assumptions are required on the distribution of supply outcomes, cost of further investment, and the flexibility of demand and supply.

As has been demonstrated (Grafton and Ward 2011), the optimal pricing path can vary significantly over time as conditions change. The presence of uncertainty thus requires flexibility in pricing (and thus consumption) in response to uncertain outcomes. This is a key point of difference between pricing under certainty and uncertainty.

Another critical difference under uncertainty relates to the expected pricing path. Under uncertainty there is a risk that storages run out before new capacity can be installed. The cost of the short-term supply options may be significant, reflecting a

[21] Demand variability may be treated in a similar fashion. However for simplicity we restrict ourselves to supply variability.

[22] Thus the previous pricing rule can be modified to become $E_t P_{t+1} = P_t (1+r)/(1-m)$, where E_t denotes the expectation at time t. We might more formally express the expected future price $E_t P_{t+1}$ as a weighted average of prices that would occur in different states $\Theta_{i,t}$ with different probabilities $\varphi_{i,t}$. That is $E_t P_{t+1} = \sum_i \varphi_{i,t} P_{t+1} \mid \Theta_{i,t}$.

[23] For example, Grafton and Ward (2011) use a stochastic dynamic programming method to estimate a dynamically efficient water prices for Sydney to maximise consumer welfare.

convex short-term supply curve as illustrated in Fig. 18.2. Thus, under worst-case scenarios, the price may escalate significantly to balance demand with available short-term supplies. Under the pricing rule above, such future worst-case future scenarios will be reflected in the current price.

A further implication is that in almost all circumstances (barring this worst-case scenario), some level of buffer stock of water will be held, even as new (excess) capacity comes on-line. While a stock of water remains, prices between periods are linked and there would be no discrete shifts in the water price.

For a given level of demand, there will be an optimal level of buffer stock held in storage. All else being equal, we would expect that the level of buffer stock would increase with greater demand, greater uncertainty (in supply or demand), and reductions in or delays in the production of other supply sources. Furthermore, pricing would respond to the level of water stocks; if the stock is low, prices would increase to depress demand until stocks begin to rise (and vice versa).

The typical price path under uncertainty varies appreciably to that when there is certainty. Under uncertainty, the current price will reflect the discounted (for storage loss and financial discount rate) weighted average of these future price outcomes. Due to the convex shape of short-run supply, the set of future price scenarios will also be convex; that is, a reduction in supply will have a greater impact on price than a similar increase in supply. The convex nature of the price outcomes means that the expected price is greater than the median price, with the result that, despite the discount factors, the optimal price under average conditions may hardly move from one period to the next.[24] Furthermore, the nature of many storages is often such that they empty slowly (due to water use and storage losses) but rise quickly (following a downpour). This suggests that rapid price movements will be largely restricted to times when water stocks rise quickly from a low level (and thus prices would fall from a high level).

Similarly, under uncertainty the impact of new capacity on price will be less dramatic. Primarily this will be because the impact of a new supply source will be diluted over a long time period; a new source of capacity that comes on-line sometime in the future will impact on future expectations of future supply and thus have an impact on current prices. Importantly, the investment project does need to be committed to; rather the option of implementing the new supply would have an impact on future supply expectations and thus future and current prices.

Governments and/or consumers may also be risk-averse; that is they may value stability of prices. In such case they may seek to contract for future supply at pre-agreed prices. This would have the effect of reducing the elasticity of demand in response to future supply shortages. This in turn would result in higher current prices to accommodate these preferences.

---

[24] This can be demonstrated with a simple example. Assume that there are three possible inflow scenarios (0.5, 1, and 1.5 units) with equal probability (i.e. expected inflow is 1 unit). Assume SRMC pricing is used, and due to the convexity of the price set the future price in these circumstances is $4, $2, and $1.2 (i.e. expected value $2.4). Assume also that the combined discount factors (1+r)/(1–m)=1.2. Then the current price will be $2 =($2.4/1.2) and the future price given expected inflows will be $2.

3 Water Security

Security of supply is a critical issue in urban water management. The thought of a major urban centre running out of water is not a palatable one, but, given supply variability, one that must be considered.

Despite its importance, the issue of water security is generally not well defined. For the purposes of this chapter we define water security 'as the physical availability of supplies to satisfy demand at a given price'.[25]

Pricing is, of course, a key element of this water security definition. At higher prices demand is lower and alternative supplies become available. While raising prices can help manage water security by depressing demand and providing incentives for greater supply, this option is hardly used. There are a number of reasons for this. In some situations water use is not metered and so pricing is not an option. However, even where metering exists there is often large resistance to use pricing to help manage shortages.

The lack of pricing flexibility means that during shortages prices will be artificially low and rationing must occur to manage demand. This in turn necessitates central enforcement of rationing and central planning of water security targets. When price control and demand restrictions are used, there will be insufficient incentive for private investment in augmentation.

The lack of a market means that a central planner is required to set water security goals and determine the community's desire for water security. Goals are typically set with reference to a target frequency of entering demand restrictions and target levels of water shortage. To inform the community's desire for security, planners have sought to analyse the costs associated with restrictions and survey the community regarding its willingness to pay to avoid restrictions.

An important question in considering reform is whether there is a public interest in water security over and above private needs. *Prima facie*, there is no need for government involvement to manage security as private needs for security are reflected in the demand for water and considered in the decisions of suppliers. The greater the individual demand for private security, the less elastic the demand for water and the greater incentive for suppliers to ensure water supplies are available during a drought.

However, the need for water security may incorporate public as well as private interests – that is, demand for security over and above private demand. The issue of public demand for water security is currently masked by the public model of central planning for managing water. However if a market-based system is used (in which prices are determined by equating supply and demand), the question of public interest in water security becomes prominent.

The determination of security targets is an issue for other important goods. For example, it is recognised that that there are public costs of shortages in energy.

[25] This definition is adapted from a similar definition for energy security by Constantini and Gracceva (2004).

Notably, fuel supply shocks can lead to impacts on productivity and employment (Constantini and Gracceva 2004).

With regard to water, there are potentially a number of reasons for public interest. First, there are a range of potential external costs associated with water shortages. These include potential public health costs associated with using alternative water sources, and congestion costs associated with meeting supply.

Second, there are potential failures in the market responding adequately. For example, there will be metering challenges if prices are required to move rapidly. As has been discussed above, investors may, due to the lumpiness of investments, defer investment until after what is optimal for social welfare. Another key concern is that firms face regulatory risk – the risk that during a shortage governments may act in ways (e.g. by regulating prices) that prevent suppliers from recouping an adequate return for their investments. This concern reflects a government's desire for water security; it is hard to imagine that a state would let a major urban centre run out of water. Governments may also be concerned about the political implications of escalating water prices.

The extent of the public interest in security depends on the size of the supply shock. If the supply shock is small, and thus the associated price impact is small, then there would appear to be little interest in public security over private security. The extent to which this is the case will depend on the nature of the urban water system. In some systems the risk of significant price rises due to severe shortages is very low; for example, the risk will be low when there are manufactured supplies or a large diversity of supplies. In other systems where supply is heavily dependent on a single source, the risk of severe price shocks may be high.

In cases where it is deemed that government intervention in water security is required, there are alternative models that might be considered in which the level of supply is centrally determined but the requirements are met by markets. For example, governments might contract for a level of security of supply. Another approach used in electricity markets is to establish a capacity market whereby a central body determines a required level of supply (capacity) and sets obligations for participants which ensure that the required level of capacity is achieved.

4 Institutional Design

The complexity of investment under uncertainty raises the question of how best water supply investment should be governed.

There are a range of potential models.[26] Of particular interest to this chapter is the extent to which a central planning or markets-based approach can be used across the functions of water supply investment and management – functions that

[26] Models also differ significantly in terms of the level of vertical and horizontal (geographic) separation, and how functions are allocated across centralised bodies that include ministerial responsibility, government departments, economic regulators, and water utilities.

incorporate establishing security needs, forecasting and planning, investment selection, and the design, build, and operation of water sources.

At one extreme is a purely centralised approach in which all the functions, from forecasting and planning through to the design and build, are centralised. Complete centralisation is rare; typically some level of market competition is used in outsourcing some aspects of build and/or operation functions. Nevertheless, key functions of forecasting, planning, and investment selection are commonly centrally managed.

While simple in construct, there are a number of issues with the centralised planning approach, which have greater significance under uncertainty. The absence of markets necessitates administered pricing. Typically, administered pricing involves rigid prices – that is, prices do not fluctuate with changes in supply and demand.

A rigid price structure creates a challenge for investment planning when there is variability and uncertainty in supply sources. With an administered rigid price structure, it is more difficult to assess alternatives, as the price does not reflect the true value of the resource. In the absence of a price-based analysis, an alternative approach is to compare options based on their contribution to 'yield' – the level of water that may be drawn down while still meeting supply security requirements.

A related issue is that centralised water utilities may not have the motivation or the information to consider innovative alternative sources of water. Even when competition exists, the incumbent's proposed solutions are more likely to be selected – because the incumbent has an information advantage over alternative suppliers, who may offer innovative solutions not considered by the incumbent.

The problems with a centralised approach raise the question of whether liberalised markets might be used. In a market-based approach, prices are determined by market forces and reflect changes in supply and demand. Price changes give potential suppliers signals and incentives for investment, which in turn will help overcome variability and uncertainty in supply. Suppliers thus undertake their own forecasting and planning and make investment decisions based on knowledge of their own particular designs and their expectations of future prices.

There are a number of concerns with a market-based approach. These include concerns relating to private incentives to invest when augmentations are lumpy (discussed above) and whether these private incentives address security of supply. As has been discussed, the presence of storage can significantly mitigate the 'lumpy investment' issue. However, the security issue is likely to continue to be a factor. First, some government involvement is expected in assessments of water security and in making corrections when market failures occur. If water security is not explicitly addressed, potential suppliers may be hesitant to invest in fear of regulatory intervention during shortages. Second is the issue of whether suppliers sufficiently factor in gains from drought pricing. The timing and length of water shortages may not follow any known distribution, which in turn may discourage investment. For example, investors may be reluctant to finance a project for which a profitable return depends on a 1-in-20 year drought.

Nevertheless, a market-based approach appears possible. In some locations water is sourced directly from major river systems where water markets already exist. Sibly and Tooth (2008) show how the divisibility of water rights and the durability of water mean that a competitive market for urban water could be established from a single catchment.

The extent of the benefits associated with market competition will, in large part, depend on the nature of water supply investments. A centralised planning approach will be most effective when large water supply investments are required (e.g. dams and large-scale desalination). In such cases there will be limited alternatives and the challenges of selecting investments will be lessened.

There is, however, increasing interest in smaller decentralised water source solutions. Being smaller, they may be more flexible and quicker to develop and be more adaptable to the local area. For example, there may be opportunities for greater or more effective integration with local water uses, such as provision of water for the environment and waste disposal.

5 Conclusion

The availability of storage within urban water networks has important implications for managing both supply and investment. In all water networks, efficient use requires that price equals SRMC. This means efficient prices will vary with both the acquisition of new lumpy infrastructure and supply variability. However with storage, excess supply can be accumulated in periods of high inflows for future use, thereby helping to reduce the volatility that would otherwise be associated with SRMC pricing.

When water can be stored, the cost of current consumption (i.e. the SRMC) reflects the opportunity cost of reduced future supply. A key implication is that a pricing rule that encourages efficient use will mean that prices respond to both changes in current supply and demand and to expectations of the future. In such a way, flexible SRMC pricing can help to manage the problems of supply variability. We note that allowing for private trades in an urban water market provides a mechanism by which the opportunity cost of water is reflected in the prices at which it is traded.

Pricing also has important implications for managing security of supply. Current security of supply policies are generally based on a framework of rigid pricing and centrally planned investment. By remaining unresponsive to changes in demand and supply conditions, these policies not only create economic inefficiencies but are likely to exacerbate the problem of ensuring water supply security. A policy which, when flows are expected to be low, raises prices and triggers new investment in infrastructure will reduce the likelihood that at some future time water supply available to the community will be unacceptably low. Nonetheless, in order to remove or reduce the possibility that an urban area 'runs out of water' there is scope for a policy to encourage investment in a diversity of water sources, some of which may have relatively high average cost. Allowing for innovative private sector solutions to water provisioning will increase the diversity of water sources, adding to the level of water security.

Appendix

We assume the objective is to maximise the present value of expected consumer surplus, which for a given level of capacity is given by:

$$\sum_{t=0}^{\infty} E_0 \frac{B_t(C_t)}{(1+r)^t},$$

where C_t is consumption, $B_t(C_t)$ is consumer benefit, r is the discount rate, and E_t represents the expectations in period t.

Assuming no restrictions on, or rationing of, consumption, the marginal consumer benefit is equal to the price; that is $B_t' = P_t$.

Assume production is given by K_t and that the marginal storage loss is at a constant rate m. Then the change in the amount held in storage (S_t) is given by:

$$(S_{t+1} - S_t) + mS_t = K_t - C_t,$$

with the constraint that end-of-period storage cannot be negative and thus $C_t \le K_t + S_t(1-m)$.

Then over a period (t=0, …, T) that water is stored, it must be that:

$$S_{T+1} = \sum_{t=0}^{T} (K_t - C_t)(1-m)^{T-t} + S_0(1-m)^T$$

$$\sum_{t=1}^{T} C_t(1-m)^{T-t} = \sum_{t=0}^{T} K_t(1-m)^{T-t} + S_{T+1} + S_0(1-m)^T.$$

For a given capacity and starting and ending storage, the above equation then sets a constraint on C_t. Inter-temporal optimisation requires that the marginal consumer surplus from using water is equated over time. That is:

$$\frac{B_t'(C_t)}{(1+r)^t} = \lambda(1-m)^{T-t},$$

where λ is the marginal benefit of relaxing the consumption constraint.

Substituting $B_t'(C_t) = P_t$ and rearranging gives:

$$E_t P_{t+1} = P_t (1+r)/(1-m).$$

The PWSIC LRMC price is determined as the present value of an incremental increase in demand. Consider the case of a permanent increase in demand of ΔC in time t that results in bringing forward an infinitely lived investment with cost zK by a single period (i.e. from period T to period T–1). The PWSIC LRMC price at time t is thus:

$$\begin{aligned} P_t &= \frac{zK / \Delta C}{(1+r)^{T-t}}\left(\frac{1}{1+r} - 1\right) \\ &= P_{t-1}(1+r). \end{aligned}$$

Note this relationship depends on the assumptions of a permanent and constant demand increase and an infinitely lived asset.

References

Abrams, B., Kumaradevan, S., Sarafidis, V., & Spaninks, F. (2012). An econometric assessment of pricing Sydney's residential water use. *Economic Record, 88*(280), 89–105.

AWWA (American Water Works Association). (2000). *Principles of water rates, fees and charges. AWWA Manual M1* (5th ed.). Denver: AWWA.

Constantini, V., & Gracceva, F. (2004). *Social costs of energy disruptions* (INDES Working Paper No. 6).

Dixit, A. K., & Pindyck, R. S. (1994). *Investment under uncertainty*. Princeton: Princeton University Press.

Grafton, R. Q., & Ward, M. B. (2011). *Dynamically efficient urban water policy* (Crawford School Research Paper No. 10–13). Canberra: Australian National University.

Knight, F. H. (1921). *Risk, uncertainty, and profit*. Boston: Hart, Schaffner & Marx/Houghton Mifflin.

Ofwat (Water Services Regulation Authority). (2001). *The role of long run marginal costs in the provision and regulation of water services*. Annex B: Summary of previous Ofwat guidance on LRMC, MD170. http://www.ofwat.gov.uk/regulating/reporting/ltr_md170_lrmc

Park, R. E. (1989). *Incremental costs and efficient prices with lumpy capacity: The single product case*. Santa Monica: RAND Corporation.

Sibly, H. (2006). Urban water pricing. *Agenda, 13*(1), 17–30.

Sibly, H., & Tooth, R. (2008). Bringing competition to urban water supply. *Australian Journal of Agricultural and Resource Economics, 52*(3), 217–233.

Turvey, R. (1976). Analyzing the marginal cost of water supply. *Land Economics, 52*(2), 158–168.

Turvey, R. (2000). Infrastructure access pricing and lumpy investments. *Utilities Policy, 9*, 207–218.

WSAA (Water Services Association of Australia). (2010). *Implications of population growth in Australia on urban water resources* (Occasional Paper No. 25, July 2010).

Chapter 19
Volumetric Water Pricing, Social Surplus and Supply Augmentation

Quentin Grafton, Long Chu, Tom Kompas, and Michael Ward

> *Both cost and output have time dimensions, and both may be subject to uncertainty... cost analysis which is to be useful in decision-making needs to be historical dynamics, not comparative statics.*
>
> Ralph Turvey (1969, p. 287)

1 Introduction

Many of the world's urban centers have increasing populations and are close to their capacity to deliver residential water (Mays 2009). In locations such as the US South West, Spain and Southern Australia, the problems of meeting water demands are compounded by extended periods of low inflows into catchments and dams. Responding to this challenge involves balancing supply and demand over time and in ways that account for weather variability and the costs of supply augmentation.

Both non-price demand measures and dynamically efficient residential water pricing can help balance urban water supply and demand. While some non-price measures are effective at managing the quantity demanded by households (Michelsen et al. 1999), quantitative water restrictions are not welfare maximizing if they prevent the equalization of the marginal benefits of restricted and unrestricted water uses within households or the equalization of marginal benefits across households

This chapter is published with the kind permission of Elsevier. It is a reprint of a journal article with the same title and authors that was published in Water Resources and Economics 6 (2014), pp. 74–87.

Q. Grafton (✉)
Crawford School of Public Policy, The Australian National University, Canberra, Aust Capital Terr, Australia
e-mail: quentin.grafton@anu.edu.au

L. Chu • T. Kompas
Crawford School of Public Policy, The Australian National University, Building #132, Lennox Crossing, Canberra, ACT 2601, Australia

M. Ward
Monash Sustainability Institute, Monash University, Melbourne, VIC 3800, Australia
e-mail: michael.ward@monash.edu

Q. Grafton et al. (eds.), *Understanding and Managing Urban Water in Transition*, Global Issues in Water Policy 15,
DOI 10.1007/978-94-017-9801-3_19

(Hirshleifer et al. 1960, pp. 36–41).[1] Volumetric water pricing that fails to consider the effects of current prices on future quantity demanded and the decision of when to augment supply can also generate long-term welfare losses if it results in premature supply augmentation. By contrast, with dynamically efficient pricing it is possible to postpone investment in supply augmentation (Riordan 1971b; Gysi and Loucks 1971; Riordan 1984; Grafton and Kompas 2007) to the benefit of water consumers.

We address the problem of balancing supply and demand for urban water by deriving dynamically efficient volumetric prices for residential water using stochastic dynamic programming. In particular, we calculate efficient volumetric prices that account for weather variability; the costs of supply augmentation; and the quantity demanded using stochastic dynamic programming and supply, demand and pricing data from Sydney, Australia.

Until now, as far as we are aware, no one has modeled the dynamically efficient urban water price using actual consumption, supply and weather data from an urban center and for an actual supply augmentation. Our value add to the literature is not to show that dynamically efficient pricing is welfare enhancing, which is well known, but to indicate the likely size of the welfare losses in present value terms from premature supply augmentation in Sydney, Australia. The contribution is to estimate the present value of welfare losses from a supply augmentation that became operational in 2010 by calculating dynamically efficient volumetric prices with a net revenue constraint. The results from our modeling are important because globally investments in water infrastructure and supply augmentation are in the multiple of billions of dollars per year. Thus, if water utilities and water price regulators were to adopt a similar framework and analysis to what we use they might avoid premature investment in supply augmentation and, thus, generate potentially large welfare gains to water consumers.

In the following section, we review the theory and practice of efficient water pricing and explain why it is important to account for uncertainty in water supplies when setting the dynamically efficient price. In Sect. 3, we describe the problems with the regulatory pricing system employed in Sydney, Australia and its implications for the decision to invest in supply augmentation in the form of a desalination plant.

In Sect. 4, we use stochastic dynamic programming to determine the dynamically efficient volumetric price under a constraint that the water supplier continues to receive its existing net revenues from water sales and that these net revenues grow at a rate equal to the population increase in Sydney.[2] Our model generates a dynamically efficient volumetric price given the net revenue requirement that existed for the water supplier when the supply augmentation decision was made.

[1] Hirshleifer et al. (1960, p. 41) correctly noted more than 50 years ago that "…consumers served under identical cost conditions should be charged equal prices and the commodity should be supplied and priced in such a way that the price for each class of service should equal the marginal cost of serving that class."

[2] In 2008, the net real revenue requirement was approximately A$ 690 million dollars ($1.65/kL–$0.30/kL) *1.4 billion liters/day*365 days *without* the desalination plant. This amount is allowed to grow at 1 % per year in real terms or at a rate equal to the assumed population increase of Sydney over the programming horizon.

We calculate the net present value of the change in social surplus, including the desalination construction costs, with the supply augmentation that became operational in 2010 and compare volumetric prices, with and without supply augmentation, from 2008 to 2030 under two weather realizations. In Sect. 5 we discuss the general implications of our approach for optimal urban water supply augmentation and provide conclusions.

2 Efficient Water Pricing

The standard economic prescription for water pricing is to set the volumetric price of water equal to the marginal cost of supply, including all internal and external costs. Where marginal costs differ for groups or classes of users as a result of locational differences or level of service then the price charged should also differ by customer group (Hirshleifer et al. 1960, pp. 40–41).

Implementing marginal cost pricing for residential water consumers poses a number of challenges. Typically, marginal cost that is based on current water supply capacity is called short-run marginal cost while marginal cost that is based on augmentation to existing capacity is defined as long-run marginal cost. Although both types of marginal cost are well defined in a static framework, with a fixed capacity constraint, this is not the case if both supply and demand are variable and uncertain.

Methods have been developed to calculate the marginal costs associated with incremental increases to capacity (Turvey 1976), but these approaches assume no uncertainty in future supplies or homogeneous and infinite supplies of unprocessed water (Swallow and Marin 1988). If marginal cost is calculated on the basis of augmented capacity, however, then *when* this increased capacity is triggered, which depends on optimal volumetric pricing and uncertain future inflows to water catchments and storages, determines the actual cost.

2.1 *Volumetric Prices and Supply Augmentation*

The existing literature has long recognized the importance of pricing water to account for the costs of supply augmentation and that postponing investments in new capacity can be cost effective (Riordan 1971a, b; Gysi and Loucks 1971). In particular, Dandy et al. (1984) calculated the optimal time to augment water supply capacity in a hypothetical-deterministic setting with endogenously derived volumetric water prices while Timmins (2002) showed that if the volumetric price is set at level below marginal cost of supply, as practiced in some western US water utilities, there is a loss of social surplus from the too rapid extraction of water from aquifers. To account for differing marginal cost of supply, Renzetti (1992) proposed peak load pricing for municipal water prices whereby the price equates to long-run marginal cost at peak demands and to short-run marginal cost in off-peak periods.

To price water efficiently does not just depend on the cost of supply augmentation beyond the current capacity constraint which is typically how long-run marginal cost is calculated by water utilities. In addition, efficient pricing along with any non-price water conservation measures, should account for the variability over future inflows into existing storages and consideration of the *possibility* of either delaying or permanently avoiding supply augmentation through higher current prices (Riordan 1971a; Gysi and Loucks 1971) and non-price water conservation measures. In other words, given uncertainty about future water supply the long-run 'marginal cost' of supply is not exogenous, but is determined by the current and future volumetric water price. This is because the volumetric price determines quantity demanded and simultaneously the timing of the supply augmentation, or when the marginal cost of the next best incremental supply is triggered.

2.2 Natural Monopoly

A difficulty with marginal cost pricing is that many urban water suppliers are natural monopolies such that the average cost is decreasing in the amount supplied. Consequently, the short-run marginal cost pricing rule will not provide the water supplier with sufficient revenues to cover costs unless there is also a fixed cost component to the water tariff, or some subsidy paid to the water supplier. This problem is well recognized and, for this reason, almost all urban water utilities that impose volumetric water prices on consumers have a two-part tariff that includes a fixed charge component.

Setting the volumetric charge equal to the 'marginal cost of supply' and establishing the access fee at a level that allows the water supplier to break even is called the Coase Result (Coase 1946) by Train (1991, p. 197). According to Train (1991) it is a first-best short-run pricing scheme provided that the demand for water supply access is independent of the access fee. If, however, the demand for water supply access depends on the access fee then a first-best short-run pricing rule would be to set the access fee equal to the marginal cost of access and the volumetric price equal to the marginal cost of the volume of water supplied at the existing capacity. In this case, the water supplier would operate at a loss if it is a natural monopoly, and a second-best or Ramsey (1927) pricing rule may be considered. The Ramsey pricing rule, in theory, should set the price of the access fee and the volumetric price to account for the elasticity of demand of both the access charge and volumetric usage.[3]

[3] In the case of zero cross-price elasticities the price (access fee or volumetric charge) is raised more for the good that has the *lower* price elasticity of demand (Baumol and Bradford 1970). Where cross-price elasticities exist these should be subtracted from the own-price elasticities to account for the effect of one good's price on its demand *net* of the effect of the price change on the other good. Two possible services for which this might apply in the case of water consumption would be energy and water services where a rise in the water price may influence electricity consumption via the effect on the use some appliances (such as washing machines).

2.3 Block-Rate Pricing

An increasing block-rate pricing structure, as applied in many US cities, has been proposed because it promotes water conservation by imposing a higher marginal price on consumers who use the most water. Apart from being inefficient in that water consumers incur different marginal prices from water that has the same marginal cost, empirical studies indicate that the number of people in a household increases water consumption (Hanke and de Maré 1984; Lyman 1992). Consequently, with an increasing block tariff price structure large and poor households, who may have little discretionary water use, may pay a higher price for water than small and high-income households (Bithas 2008; Dahan and Nisan 2007). Whether or not high water users are worse off with an increasing block tariff relative to a uniform water price depends on whether scarcity values are included in the volumetric price charged to consumers. When scarcity-inclusive uniform prices are charged, low-income households may prefer a uniform scarcity-inclusive price rather than an increasing block tariff (Griffin and Mjelde 2011).

2.4 Regulated Prices

If water utilities and pricing regulators set volumetric water prices based on existing supply capacity then the prices charged to consumers are, typically, equal to the annualized cost of providing the most recent realized supply augmentation. If prices are determined on the costs of augmenting future supply then customers are, typically, charged the annualized supply cost associated with the next cheapest source of supply. In both cases, an allowance is given to account for operating costs and also the cost of the infrastructure or capital to deliver water to consumers.

In many jurisdictions the water supplier is either publicly owned and/or there are pricing rules imposed on private operators to protect water consumers from the abuse of market power. To avoid the misuse of market power there is sometimes contestability over which entity should be the designated water supplier. When access to the water distribution system is not preferentially assigned at a lower cost to an incumbent supplier, such as in France (Nauges and Thomas 2011), alternative suppliers can compete for long-run contracts of supply based on the existing distribution network. The use of private concessions to deliver water supplies is, however, not without its challenges (Lobina 2005).

In most countries water suppliers are regulated so that they are unable to charge their water consumers more than a given price or set of prices. In some cases, water suppliers are obliged to conform to a 'zero profit' constraint that prevents them from raising their prices in periods of drought if their costs have not increased (Howe 2005). While inefficient (Hirshleifer et al. 1960), many jurisdictions also impose 'postage stamp' pricing or a spatially uniform price regardless of differences in supply or delivery costs.

3 Regulatory Water Pricing: Sydney, Australia

Sydney is Australia's largest city with over four million residents. Its principal water supply is sourced from 11 dams that have been built in catchments west of the city. Water is delivered to households via a distribution network operated by a single supplier, the Sydney Water Corporation (SWC), which is a state-owned corporation that provides dividends to the Government of New South Wales. The maximum price that can charged by the SWC to its water consumers is set by the Independent Pricing and Regulatory Tribunal (IPART) of New South Wales that also sets the prices of energy and public transportation. To date, the maximum regulated volumetric price set by IPART has been the actual price charged by SWC.

Typically, IPART provides a determination of the maximum residential water prices in Sydney over a 4-year period. Maximum prices are set for both the access fee and the volumetric prices. Over the 2008–2012 period, coincident with the construction of the desalination plant, IPART allowed for a more than a 30 % increase in real terms in the average water bill of Sydney households. Almost half of this cost increase compensates SWC and its partners for the operating costs and annualized capital costs associated with a desalination plant that became operational in February 2010.[4] This desalination plant has a capacity to produce about 250 ML/day, about 15 % of Sydney's annual water consumption. The desalination plant cost A$ 1,918 million in 2008 dollars (IPART 2008) to establish and to construct.

IPART's price setting rule is to establish the maximum volumetric price equal to what it considers to be the long-run marginal cost of supply. To calculate this price it uses an Average Incremental Cost Method for what it considers to be the next best available supply augmentation. What IPART calls long-run marginal cost includes an allowance for the cost of the capital to construct the plant and its operating costs. A stylized representation of how it calculates the volumetric price charged to water consumers is given below:

$$\text{Average Incremental Cost} = \frac{\text{Least cost investment to equate demand and supply}\,(\$)}{\text{Incremental output from capacity expansion}\,(\text{kL})}.$$

IPART's regulatory pricing is typical of the procedures employed in other parts of Australia, and also in other countries (Howe 2005). It claims to base its pricing on long-run marginal cost. There is, however, no single or unique long-run marginal cost as claimed by IPART because it depends on the highly variable inflows of water into Sydney's catchments and dams (Grafton and Kompas 2007).

Sub-optimal volumetric pricing generates other welfare costs beyond investing in supply augmentation at appropriate scale or time. This is because to help balance current demand and supply with volumetric prices that are too low, and prior to any

[4]The announcement to build the desalination plant was made on 18 July 2007.

supply augmentation, water rationing in some form is typically imposed. In the case of Sydney, mandatory outdoor water restrictions were imposed on residential households in terms of how water could be used outdoors, at what times, and for what purposes. Rationing water with mandatory water restrictions was effective at reducing water demand. Nevertheless, water rationing resulted in estimated aggregate welfare losses equal to about A$ 250 million in 2008 dollars in 2004–2005 relative to a volumetric price that would have ensured the same level of demand and remitted revenue (in excess of supply costs) back to consumers in the form of lower access fees (Grafton and Ward 2008).

Before the decision was taken to build the desalination plant, IPART's regulated water prices did not include a scarcity component for the water in Sydney's dams despite the fact that Sydney had been suffering from a drought for about 3–4 years. As a result, the regulated maximum price and the price charged by SWC, was set at too low a level because a dynamically efficient price would have included a scarcity component. Consequently, the less-than-efficient price set by IPART had the effect of bringing forward the supply augmentation because the quantity demanded was greater given that IPART's price was lower than the efficient price. After the decision was made to invest in supply augmentation, the volumetric maximum price set by IPART included an allowance for the incremental costs (operational and fixed) associated with the desalination plant and became incorporated in IPART's price determination.

The key difficulty with IPART's pricing is that it fails to account for water supply uncertainty associated with variable inflows into Sydney's catchment and water storages. Inflow data indicates that the catchment has periods when inflows are below average for extended periods of time, and periods when the volume of water in storage can increase substantially following a major rainfall event.[5] The historical inflows into the catchment should be included in an economically optimal decision as to when to invest in supply augmentation *and* also the volumetric price charged to water consumers. For instance, if the historical inflows indicate that there is a high likelihood that a major drought will occur in the following 12 months, it may be economically efficient to raise the volumetric price and reduce the quantity demanded to balance it with current supply rather than prematurely invest in expensive supply augmentation.

Under current water metering in Sydney the price can change only once every quarter, but the practice is to change the price only once per year consistent with IPART's price determination. Smart meters that would allow households to observe their consumption by outlet and for the water utility to adjust prices to consumers instantaneously are not currently used in Sydney.

[5] For instance, based on historical inflow data from 1997 to 2007, 35 % of inflows occur in just 5 % of the months. In the month immediately preceding the announcement to build the desalination plant, water storages in the Sydney catchment increased from 37 % to 57 % of full capacity

4 Dynamically Efficient Water Pricing: Sydney, Australia

To calculate the economically efficient volumetric prices charged to Sydney households and the consumer losses associated with actual supply augmentation we use stochastic dynamic programming (Loucks and van Beek 2005; Ross 1983). This allows us to calculate volumetric prices that include a scarcity component and also account for the variability of inflows into Sydney's water storages *prior* to the decision to invest in desalination.

Ideally, the dynamically efficient water price should be set in accordance with the billing period which, in the case of Sydney, is quarterly. To provide a valid comparison to the actual volumetric price which is changed only annually we also impose an annual price change. To the extent that dam levels can change substantially over a quarter there could be additional welfare gains from quarterly rather than annual price changes.

The dynamic optimization problem we solve maximizes the expected present value of the total or aggregate social surplus (SS) from residential water consumption in Sydney over a 100 years planning horizon subject to an operational constraint that the water supplier receives a minimum amount of real net revenue sufficient to cover fixed costs and depreciation. Social surplus is defined as consumer surplus (CS) from water consumption less the marginal cost of the supplied water. Over the planning period the real net revenue constraint increases at a rate equal to the growth in Sydney's population, projected to be 1 % per year.

4.1 Model Specification

The state variable (s_t) is the amount of water in Sydney dam storage that is currently available for consumption and is defined by Eq. 19.1:

$$s_{t+1} = \min\left\{s_t + I(W_t) - d_t - O(s_t), D\right\} \tag{19.1}$$

where for period t, I(.) are dam inflows that are dependent on weather realizations W_t, d_t is the dam water supplied to Sydney's residents, D is the total dam capacity within the Sydney catchment and $O(s_t)$ is the quantity of water evaporated or released from the dams that equals total environmental releases from the dams (R) plus evaporation $Ev(s_t)$. All parameters and their assigned values are provided in Table 19.1.

The household water demand function for Sydney uses price elasticities of demand taken from a study undertaken for SWC by the Centre for International Economics (2010, p. 82). Total quantity demanded is decomposed into two parts: indoor and outdoor water quantity demanded each of which has a different price elasticity of demand as per Eq. 19.2:

$$q(p) = A_1 p^{-\varepsilon_1} + A_2 p^{-\varepsilon_1} \tag{19.2}$$

Table 19.1 Model parameters: definitions and values

Parameter values	Explanation	Sources
$D=2.6*10^6$(ML)	Dam capacity	CIE (2010)
$c_d=0.3*10^3*10^{-6}$(mil$/ML)	Marginal cost of supplying dam water is $0.3/kL	CIE (2010)
$c_u=0.6*10^3*10^{-6}$(mil$/ML)	Marginal cost of desalination water is $0.9/kL	CIE (2010)
$c_b=30*10^3*10^{-6}$(mil$/ML)	Base-case marginal cost of water soured using backstop technology is $30/kL	Authors' estimate
$R=44*365$(ML)	Environmental release (44 ML/day) is a minimum required.	CIE (2010)
$\varepsilon_1=0.216$	1 % increase in prices → 0.216 % decline in indoor quantity and 0.59 % outdoor water quantity. Outdoor water use is approximately 30 %	CIE (2010)
$\varepsilon_2=0.59$		
$A_1=0.0538$		
$A_2=0.0021$		
N	Number of houscholds (1.7 million and grows 1 % per year)	Australian Bureau of Statistics
$\rho=0.05$	Base-case real discount rate of 5 %	Authors' estimate
$T=100$	Assuming the plant, with upgrades, lasts for 100 years	Authors' estimate
I(W) and the distribution of W	Dam inflow and its weather-dependent distribution	Derived from actual data sourced from CIE (2010)
$Ev(s_t)$	Evaporation from dams (calculated using the fact that the evaporation in Sydney is about 1.733 m per year, i.e., approximately 5 % evaporation rate)	(http://readyreckoner.nceaprd.usq.edu.au/evaporationcalc.aspx) or (http://www.dpi.nsw.gov.au/archive/agriculture-today-stories/ag-today-archives/october-2011/calculate-losses-to-evaporation)

where the time subscripts are dropped for convenience, q and p are, respectively, household water consumption and the volumetric price charged to households in period t, A_1 and A_2 are parameters for indoor (1) and outdoor (2) water demand, and ε_1 and ε_2 are the price elasticities of demand for indoor and outdoor household water consumption.

To determine the net benefits of the Sydney desalination plant, we first calculate a baseline scenario that includes as water supply options only the water in Sydney's dams plus the option of a backstop technology (see Appendix). The dynamic optimization problem is to maximize the sum of the present value of the total social surplus in all years over the planning horizon if household water consumption from Sydney's dams and the backstop water each year (and consequently the volumetric price) are optimally determined.

The optimization framework with the desalination plant (see Appendix) is identical to the baseline optimization except for an additional control variable ($u_t \leq M$) included for water from a desalination plant subject to a capacity constraint given by M (defined as either 250 ML per day or 500 ML per day) and an extra net revenue requirement for the water supplier (ER) to cover the annualized construction cost and the annual maintenance cost of the desalination plant.

4.2 Backstop Technology and Weather Realizations

The backstop technology is included in both scenarios because of the very large consumer losses from not having water available to meet Sydney households' water demand, should the dams become empty. Thus, in addition to water available from Sydney's dams and the desalination plant, households are permitted the option of accessing water imported from outside the Sydney catchment, but at a much higher price. The costs associated with the backstop technology are not known as it would require an in-depth economic study of all possible water supply options outside of the catchment and could include building water pipelines from other locations or shipping water inter-state from Tasmania. The base-case backstop technology cost we assume represents a high-end estimate of the cost of supply and delivery of water from Tasmania by bulk tankers and is more than twice the estimated cost incurred by Barcelona when it imported bulk water by sea in 2008 during a severe drought (Burnett 2008). Our analysis is repeated assuming the marginal cost from the back-stop supply is equal to half its base-case value.

The solution depends on realization of the weather. We use data based on two time periods, a longer and wet period (1919–2008) and a shorter and drier period (1969–2008). The data for both were obtained from CIE (2010). The drier the period, all else equal, the greater is the net benefit associated with a desalination plant because the fewer are the periods where the expensive backstop technology needs to be utilized.

4.3 Method of Solution

The welfare effects of the desalination plant are determined by comparing the net benefit under the two scenarios (baseline and desalination). We solve the two dynamic optimization problems in three stages. In the first step we solve for the value function at the end point of the planned lifetime of the plant ($t = 100$, equivalent to calendar year 2110) using the standard value-function-iteration algorithm for a stationary Bellman equation. In the second step, and starting with the value function obtained from the first step $V_{100}(.)$ we solve the time-dependent Bellman equation using the backward induction procedure to obtain the value functions at each year during the planning horizon ($t = 99, \ldots .0$). At the conclusion of the backward induction process, we obtain two value functions at time t=0 (i.e., calendar year 2010), namely $V_0^{baseline}(.)$ and $V_0^{desal}(.)$.

A comparison of the two value functions is sufficient to evaluate the net benefit of the desalination plant when it became operational in 2010 or afterwards, but not before this date. The decision to build the plant was made in 2007, thus, in the third step we go back a further 2 years until 2008, but *without* the desalination water control variable given that the desalination was not yet operational. At the termination of the three steps, we obtain two value functions, both measured in Australian 2008 dollars. The difference in the two values is the change in the social surplus as a result of the decision to build the desalination plant in 2007.

4.4 Welfare Losses from Premature Supply Augmentation

The sensitivity of the solution of when to optimally build the desalination plant varies according to several factors including the marginal supply cost of the backstop technology and historical weather and inflows. Results in present value terms are presented in Tables 19.2 (marginal cost of backstop water of \$30/kl) and 19.3 (marginal cost of backstop water of \$15/kl) for two possible sizes of the desalination plant: the 250 ML/day which is the current capacity of the plant and 500 ML/day which is the maximum capacity the plant could be enlarged to, if required.

Two sets of results are provided: a drier and medium-term weather realization associated with a shorter period of data (1969–2008) and a wetter weather realization associated with a long-term time-series (1919–2008). In all results the construction and establishment costs of the desalination plant are fixed at the actual cost in A\$ of 1,918 million and the cost of increasing the capacity of the plant from 250 ML/day to 500 ML/day estimated at A\$ 1,020 million (IPART 2008). The change in social surplus is the difference in the value function with

Table 19.2 Social surplus and net present value (A\$ 2008 million) of the Sydney desalination plant assuming the marginal cost of backstop water is \$30/kL

Cost and Benefit components (million 2007–08 dollars)	Long-term (wetter) weather distribution		Medium-term (drier) weather distribution	
	250 ML/day	500 ML/day	250 ML/day	500 ML/day
Construction cost of the desalination plant	−1,918	−2,938	−1,918	−2,938
Change in social surplus as a result of desalination plant	−2,083	−3,159	−1,283	−3,159
Net present value benefit of desalination plant	−4,001	−6,097	−3,201	−4,997

Notes

1. The minus sign before a number implies a net cost or loss
2. Construction and capital costs for a desalination plant with a capacity of 250 ML/day and 500 ML/day obtained from IPART (2008, p. 218)

Table 19.3 Social surplus and net present value (A$ 2008 million) of the Sydney desalination plant assuming the marginal cost of backstop water is $15/kL

Cost and Benefit components (million 2007–08 dollars)	Long-term (wetter) weather distribution		Medium-term (drier) weather distribution	
	250 ML/day	500 ML/day	250 ML/day	500 ML/day
Construction cost of the desalination plant	–1,918	–2,938	–1,918	–2,938
Change in social surplus as a result of desalination plant	–2,547	–4,491	–1,904	–3,583
Net present value benefit of desalination plant	–4,465	–7,429	–3,822	–6,521

Notes
1. The minus sign before a number implies a net cost or loss
2. Construction and capital costs for a desalination plant with a capacity of 250 ML/day and 500 ML/day obtained from IPART (2008, p. 218)

and without the desalination plant, but *not* including the construction/establishment costs of the desalination plant.

The net present value (NPV) of the change in welfare is the sum of the establishment/construction cost of the desalination plant plus the change in the social surplus. The loss in social surplus is a result of the higher price paid by water consumers from paying a higher marginal cost of water delivered from the desalination plant earlier than is necessary and the associated deadweight loss. The overall welfare loss is a greater amount because of the capital costs incurred in constructing the desalination plant. In NPV terms under the base case assumptions about the discount rate (5 %) and operational life of the plant (100 years), and assuming a drier weather realization, the loss in welfare is A$ 3,201 million or an amount per household of about A$ 1,900.[6] Under the wetter weather realization the welfare loss is even greater.

In both weather realizations (drier and wetter) the NPV of the desalination plant is negative under the base case assumptions. In the drier-weather realization the NPV is greater (or closer to zero) than the wetter-weather realization because eventually, in 2061 with an assumed population growth of 1 %/year, it becomes worthwhile to operate a desalination plant under the medium-term weather realization. Thus, beyond 2061 the benefit of having the plant partially offsets the intervening period when the desalination plant generated a negative social surplus. In the wetter scenario (long-term weather realization) it is becomes worthwhile to operate a desalination plant in 2084.

Figure 19.1 presents, under two different weather realizations, calculated volumetric water prices in 2008 Australian dollars per kL that include: (1) the

[6]Taking the estimate of the net present value loss terms of $A 3,201 million (medium-term weather realization, a 250 ML/day plant, 5 % discount rate and a 100 operational plant life) and dividing by the number of occupied residential dwellings in Sydney of 1.7 million in 2011, the net present value of the welfare loss is $A 1,883 per household.

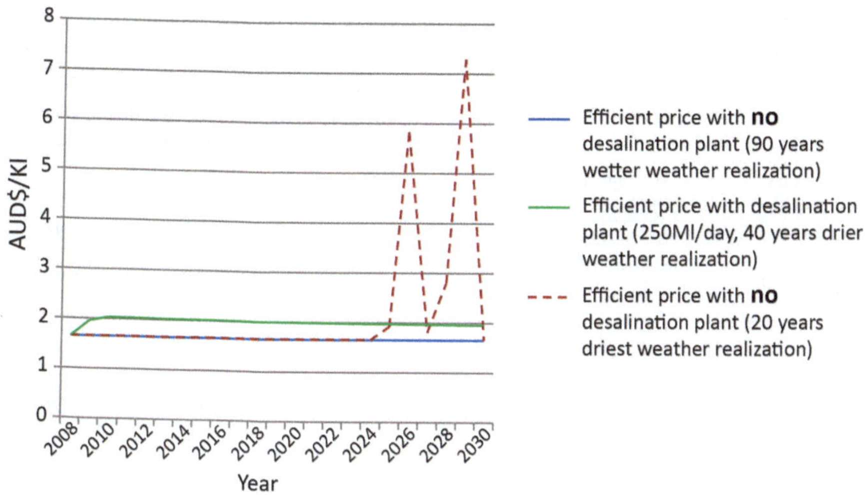

Fig. 19.1 Volumetric water prices ($A2008) for different weather realizations

dynamically efficient price *with* the desalination plant and net revenue constraint; and (2) the dynamically efficient price *without* the desalination plant including a net revenue constraint. The key comparison in Fig. 19.1 is between the dynamically efficient prices with and without the desalination plant. In the absence of the desalination plant, water consumers do not need to fund the operating costs of the desalination plant nor do they need to incur the annual fixed costs of $111 million/year (Sydney Water 2011) that includes $15 million for maintenance of the plant and $96 million to cover the annualized capital costs that are built into the actual price determination by IPART. The saving to Sydney households from not building the desalination plant and setting volumetric price efficiently is, in real dollars, about $ 0.30/kL using the long-term and wetter weather realization.

If there were to be a drier weather realization, and supposing the driest twenty years recorded in the weather data in terms of dam inflows became the norm, the dynamically efficient water price without a desalination plant would eventually exceed the volumetric price with the desalination plant in 2026. Under this driest 20-years weather realization the water level in Sydney's dams becomes so low that the expensive backstop technology water source is required to satisfy the quantity demanded. It is at this point, in 2027, that it becomes economically efficient to operate a 250 ML/day desalination plant. The economically efficient time to begin operating the desalination plant under the driest 20 years, the drier (1969–2008) and the wetter (1919–2008) weather realizations are provided in Table 19.4 under the assumption that Sydney's population grows at 1 %/year over the next 100 years.

Table 19.4 Economically efficient time to start operating a 250 ML/day desalination plan under three weather realizations

		Long-term (wetter) weather distribution	Medium-term (drier) weather distribution	20-year driest period
Summary of climate characteristics	Average annual inflow (GL)	1186.9	960.8	715.4
	SE of the annual inflow (GL)	845	604	597
Optimal construction time and supply level	Year	2084	2061	2027
	Number of households at the optimal construction time (million)	3.55	2.82	2.01
	Average supply capacity of the dam system at the optimal time (L/household/day)[1]	803	790	775

[1]Capacity is calculated after netting out overflow spill, evaporation and required environmental releases. Sydney's population is assumed to grow at 1 %/year over the entire planning period

The results are sensitive to assumptions made about the discount rate and also the longevity of the desalination plant. The longer the plant remains operational the greater the net benefit of the plant for a fixed construction cost. The influence of the discount rate on the results is not monotonic in the case when the life of the plant is of a short duration. This is because if the plant has only a short duration and is not needed during its operational period then a smaller discount rate will reduce its desirability because the benefit of not having the plant (a lower volumetric price) further into the future is more highly valued. By contrast, when the plant has long operational life a smaller discount rate will increase its desirability because over a longer operational life the plant gets used and the value of the plant when used in the future is more highly valued with a lower discount rate.

Sensitivity results in terms of the NPV of the plant that includes both the social surplus and construction costs are shown in Table 19.5 for 250 ML/day plant and Table 19.6 for a 500 ML/day plant where the base case values are a discount rate of 5 % and a plant life of 100 years. Both tables show that under the assumed discount rates (2 %, 5 % and 8 %) that the desalination plant generates a negative net present value under the two weather realizations should the plant have an operational life of 50 years or less. In the case where the plant life has an operational life of 70 years or more, and for a discount rate of 2 %, the net present value is positive under a medium (drier) weather realization. Assuming a long-term (wetter) weather realization the desalination plant has a positive net present value if it has an operational life of 100 years with a discount rate of 2 %.

Table 19.5 Net present value (A$ 2008 million) of the Sydney desalination plant assuming the marginal cost of backstop water is $30/kL for different discount rates, plant life and weather realizations (plant capacity 250 ML/day)

	Desalination plant's operational life	Annual discount rate		
		ρ=2%	ρ=5%	ρ=8%
Medium-term (drier) weather realization	T=30	-5,477	-5,283	-5,146
	T=50	-3,540	-4,717	-4,974
	T=70	+727	-3,950	-4,813
	T=100	+8,310	-3,201	-4,758
Long-term (wetter) weather realization	T=30	-5,886	-5,507	-5,277
	T=50	-4,969	-5,233	-5,209
	T=70	-1,882	-4,695	-5,106
	T=100	+4,555	-4,001	-5,048

Notes
1. Shaded cell values represent base case assumptions about the operational life of the desalination plant and the discount rate

Table 19.6 Net present value (A$ 2008 million) of the Sydney desalination plant assuming the marginal cost of backstop water is $30/kL for different discount rates, plant life and weather realizations (plant capacity 500 ML/day)

	Desalination plant's operational life	Annual discount rate		
		ρ =2%	ρ =5%	ρ =8%
Medium-term (drier) weather realization	T=30	-8,372	-8,015	-7,791
	T=50	-6,256	-7,436	-7,628
	T=70	+267	-6,812	-7,422
	T=100	+13,834	-4,997	-7,302
Long-term (wetter) weather realization	T=30	-8,804	-8,249	-7,926
	T=50	-8,003	-8,061	-7,886
	T=70	-3,532	-7,306	-7,751
	T=100	+7,842	-6,097	-7,653

Notes
1. Shaded cell values represent base case assumptions about the operational life of the desalination plant and the discount rate

5 Implications and Conclusions

Using data from Sydney, Australia we show that the decision to build a desalination plant in July 2007 was taken prematurely. According to our modeling, and under base case assumptions (5 % discount rate and 100 operational plant life), the NPV from lower volumetric water prices from *not* building a desalination plant that began operation in 2010 and not incurring the plant's construction cost is worth more than three billion dollars in total with a drier weather realization, and more than four billion dollars with a wetter weather realization. In large measure because of the decision to prematurely build the desalination plant, the regulated volumetric price increased by about 50 % between 2007 and 2010. By contrast, with dynamically efficient water prices and *if* the desalination plant had not been built volumetric prices paid by Sydney households would, in real terms, average about $0.30/kL less than current and projected prices.

A possible explanation as to why the decision to invest in the desalination plant was taken in July 2007 by the State Water Minister is that the government, at the time, was perceived to not be effectively managing reductions in water storages associated with a drought and there was growing animosity about the use of mandatory water restrictions in Sydney. By building a desalination plant, sooner rather than later, the State government sought to be seen as decisive in its actions and also to bring forward the date it could relax its unpopular mandatory water restrictions.

While our model is specific to Sydney, the results are of general interest because, as far as we are aware, there are no water utilities or water regulators that have implemented dynamically efficient volumetric water pricing or used it to determine optimal supply augmentation. Given that global expenditures in water infrastructure are estimated to be some $US 75 billion per year, our results suggest that there could currently be very large welfare losses world-wide from inefficient volumetric water pricing and premature water supply augmentation.

Appendix

Baseline Scenario Dynamic Optimization Problem

$$V_0^{baseline} = \max_{\left\langle \substack{d_t, b_t, Q_t = d_t + b_t \\ d_t \le s_t \\ p_t Q_t - (d_t c_d + b_t c_b) \ge FC_t^{baseline}} \right\rangle} E_W \sum_{t=0}^{T-1} \left(\frac{1}{1+\rho} \right)^t SS_t$$ which is associated with the

Bellman equation $$V_t^{baseline}(s_t) = \max_{\left\langle \substack{d_t, b_t, Q_t = d_t + b_t \\ d_t \le s_t \\ p_t Q_t - (d_t c_d + b_t c_b) \ge FC_t^{baseline}} \right\rangle} \left\{ SS_t + \frac{1}{1+\rho} E_{W+1} \left(V_{t+1}^{baseline}(s_{t+1}) \right) \right\}$$

subject to the water stock accounting rule where the E_w operator specifes the expected value over all realization of weather W and the social surplus defined as the net of the change in the consumer surplus and the marginal cost of the supplied water, compared to the empty-dam scenario: $SS_t = N_t \int_{p_t}^{c_b} q(p)dp - \left(\left[d_t c_d + b_t c_b\right] - q^* c_b\right)$

where N_t is the number of Sydney households, ρ is the discount rate, c_d is the marginal cost of supplying water from Sydney's dams, q* is the quantity of water supplied from the backstop technology if there were no dam water, and c_b is the marginal cost in ML of water supplied by the backstop technology.

Desalination Scenario Dynamic Optimization Problem

$$V_0^{desal} = \max_{\left\langle \begin{array}{l} d_t, b_t, u_t, Q_t = d_t + b_t + u_t \\ d_t \le s_t \\ p_t Q_t - (d_t c_d + b_t c_b + u_t c_u) \ge FC_t^{baseline} + ER \\ u_t \le M \end{array} \right\rangle} E_W \sum_{t=0}^{T-1} \left(\frac{1}{1+\rho}\right)^t SS_t$$

which is associated with the Bellman equation

$$V_t^{desal}(s_t) = \max_{\left\langle \begin{array}{l} d_t, b_t, u_t, Q_t = d_t + b_t + u_t \\ d_t \le s_t \\ p_t Q_t - (d_t c_d + b_t c_b + u_t c_u) \ge FC_t^{baseline} + ER \\ u_t \le M \end{array} \right\rangle} \left\{ SS_t + \frac{1}{1+\rho} E_{W+1}\left(V_{t+1}^{desal}(s_{t+1})\right) \right\}$$

where $SS_t = N_t \int_{p_t}^{c_b} q(p)dp - \left(\left[d_t c_d + b_t c_b + u_t c_u\right] - q^* c_b\right)$ and c_u is the marginal cost of the desalination water.

References

Baumol, W. J., & Bradford, D. F. (1970). Optimal departures from marginal cost pricing. *The American Economic Review, 60*(3), 265–283.

Bithas, K. (2008). The sustainable residential water Use: Sustainability, efficiency and social equity. *The European Experience, Ecological Economics, 68*, 221–229.

Burnett, V. (2008, June 13). Thirsty Barcelona imports drinking water. *The New York Times*. http://www.nytimes.com/2008/05/13/world/europe/13iht-spain.4.12853501.htm

Centre for International Economics. (2010). *Review of operating regime for Sydney water's desalination plant: Canberra*. http://sydneydesal.com.au/images/documents/desalination-operating-rules/desalination-operating-rules-cie-report.pdf Accessed 22 Apr 2014.

Coase, R. (1946). The marginal cost controversy. *Economica, 13*, 169–189.

Dahan, M., & Nasan, U. (2007). Unintended consequences of increasing block tariffs pricing policy in urban water. *Water Resources Research, 43*, 3. doi:10.1029/2005WR004493.

Dandy, G. C., McBean, E. A., & Hutchinson, B. G. (1984). A model for constrained optimum water pricing and capacity expansion. *Water Resources Research, 20*(5), 511–520.

Grafton, R. Q., & Kompas, T. (2007). Pricing Sydney water. *Australian Journal of Agricultural and Resource Economics, 51*, 227–241.

Grafton, R. Q., & Ward, M. (2008). Prices versus rationing: Marshallian surplus and mandatory water restrictions. *The Economic Record, 84*, S57–S65.

Griffin, R. C., & Mjelde, J. W. (2011). Distributing water's bounty. *Ecological Economics, 72*, 116–128.

Gysi, M., & Loucks, D. P. (1971). Some long Run effects of water-pricing policies. *Water Resources Research, 7*(6), 1371–1382.

Hanke, S. H., & de Maré, L. (1984). Municipal water demands. In J. Kindler & C. S. Russell (Eds.), In collaboration with B. T. Bower, I. Gouevsky, D.R. Maidment, & W. R. D. Sewell, *Modeling water demands* (pp. 149–169). London: Academic.

Hirshleifer, J., DeHaven, J. C., & Milliman, J. W. (1960). *Water supply economics, technology and policy*. Chicago: University of Chicago Press.

Howe, C. (2005). The functions, impacts and effectiveness of water pricing: Evidence from the united states and Canada. *Water Resources Development, 21*(1), 43–53.

Independent Pricing and Regulatory Tribunal (IPART). (2008). *Review of prices for Sydney Water Corporation's water, sewage and storm water and other services from 1 July 2008.* Water – Determination and final report. www.ipart.nsw.gov.au/Home/Industries/Water/Reviews/Metro_Pricing/Investigation_into_Prices_for_Water_Sewerage_and_Stormwater_services_provided_by_Sydney_Water_Corporation_-_From_1_July_2008/16_Jun_2008_-_Final_Determination_and_Report/Final_Report_and_Determination_-_Review_of_prices_for_Sydney_Water_Corporations_water_sewerage_stormwater_and_other_services_-_June_2008. Accessed 26 Apr 2014.

Lobina, E. (2005). Problems with private water concessions: A review of experiences and analysis of dynamics. *Water Resources Development, 21*(1), 55–87.

Loucks, D. P., & van Beek E., (with contributions from J. R. Stedinger, J. P.M. Dijkman & M. T. Villars). (2005). *Water resources systems planning and management.* Paris: UNESCO.

Lyman, R. A. (1992). Peak and Off-peak residential water demand. *Water Resources Research, 28*(9), 2159–2162.

Mays, L. W. (Ed.). (2009). *Integrated urban water management: Arid and semi-arid regions.* Leiden: UNESCO Publishing/Taylor & Francis.

Michelsen, A., McGuicken, J. T., & Stumpf, D. (1999). Nonprice water conservation programs as a demand management tool. *Journal of the American Water Resources Association, 35*, 593–602.

Nauges, C., & Thomas A. (2011). Pricing of urban water services: The case of France. In R. Q. Grafton & K. Hussey (Eds.), *Water resources, planning and management: Challenges and solutions*. Cambridge, UK: Cambridge University Press.

Ramsey, F. (1927). A contribution to the theory of taxation. *Economic Journal, 37*(1), 47–61.

Renzetti, S. (1992). Evaluating the welfare effects of reforming municipal water prices. *Journal of Environmental Economics and Management, 22*, 147–163.

Riordan, C. (1971a). General multistage marginal cost dynamic programming model for the optimization of a class of investment-pricing decisions. *Water Resources Research, 7*(2), 245–253.

Riordan, C. (1971b). Multistage marginal cost model of investment pricing decisions: Applications to urban water supply treatment facilities. *Water Resources Research, 7*(3), 463–478.

Riordan, M. (1984). On delegating price authority to a regulated firm. *Rand journal of Economics, 15*(1), 108–115.

Ross, S. M. (1983). *Introduction to stochastic dynamic programming*. London: Academic.

Swallow, S. K., & Marin, C. M. (1988). Long Run price inflexibility and efficiency loss for municipal water supply. *Journal of Environmental Economics and Management, 15*, 233–247.

Sydney Water. (2011). *Sydney water submission to IPART. 2012 pricing determination.* http://www.google.com.au/url?sa=t&rct=j&q=&esrc=s&frm=1&source=web&cd=1&ved=0CCsQFjAA&url=http%3A%2F%2Fwww.ipart.nsw.gov.au%2Ffiles%2Ff5cf61e1-7981-4d16-b567-9f9101196200%2FSubmission_-_Review_of_Prices_for_Sydney_Water_Corporations_water_sewerage_stormwater_and_other_services_from_July_2012_-_Sydney_Water_-_Kevin_Young_-_16_September_2011-_Website_Document.pdf&ei=_kVXU9T2GI_j8AWJ8oKoDQ&usg=AFQjCNFVpjVMTxDQ-iPQfHuo13Mx_-wwdQ&sig2=D8Dxpq-j4XD9sLLJa-NZ2g&bvm=bv.65177938,d.dGc. Accessed on 23 Apr 2014.

Timmins, C. (2002). Measuring the dynamic efficiency costs of Regulators' preferences: Municipal water utilities in the arid west. *Econometrica, 70*(2), 603–629.
Train, K. E. (1991). *The economic theory of natural monopoly*. Cambridge, MA: MIT Press.
Turvey, R. (1969). Marginal cost. *The Economic Journal, 79*(314), 282–299.
Turvey, R. (1976). Analyzing the marginal cost of water supply. *Land Economics, 52*(2), 158–168.

Part III
Water Governance and Integrated Management

Chapter 20
Governance and Regulation of the Urban Water Sector: *Quoi de Neuf*?

Walter Reinhardt and Lætitia Guérin-Schneider

1 Introduction

This chapter examines three broad themes of change in governance and regulation which have emerged in recent decades in the urban water supply, treatment, and wastewater sector – collectively described here as the 'urban water sector'. In brief, these themes are the *devolution* in management and control, increased *sophistication in regulation*, and the re-emergence of *social and environmental concerns*.

The themes of change analysed here are necessarily general, reflecting the variety of forms possible under the particular conditions present in different locations. The themes are drawn from observation of the urban water sector in the developed world, which we take to be Europe, North America, parts of Asia, and the Pacific. Examples and exceptions are included where possible. The experience at each location differs, reflecting particular economic, legal, social, cultural, and environmental conditions.

We provide case studies from Australian and French experience to illustrate how the themes of devolution, sophistication, and social and environmental concerns have evolved in practice. Through the Australian case studies, particularly that of Melbourne in the State of Victoria, we observe an interaction between governance change and extreme environmental stress (drought). French case studies provide

W. Reinhardt (✉)
Fenner School of Environment and Society, The Australian National University,
Canberra, ACT 0200, Australia
e-mail: walter.reinhardt@anu.edu.au

L. Guérin-Schneider
Centre de Montpellier, Institut National de Recherche en Sciences
et Technologies pour l'Environnement et l'Agriculture,
361 rue Jean-François Breton, BP 5095, Montpellier 34196, Cedex 5, France
e-mail: laetitia.guerin-schneider@irstea.fr

Q. Grafton et al. (eds.), *Understanding and Managing Urban Water in Transition*, Global Issues in Water Policy 15,
DOI 10.1007/978-94-017-9801-3_20

insight into regulatory evolution in a country with longstanding private sector participation. Together these case studies provide an illuminating exploration of how these themes have manifested under a variety of institutional, regulatory, and environmental circumstances.

Our analysis begins with the premise that government has a strong role in directing the provision of urban water and wastewater services. If its role is not in the ownership and management of assets and delivery services, then there is certainly a role for government in the regulation of services provided by the private sector. More than other network industries such as electricity or telecommunications, the urban water network is a monopoly par excellence (Littlechild 1988; Abbott and Cohen 2010). Government involvement ensures that the quantity and quality of services provided are socially optimal and meet the public interest (Bozeman 2007; PC 2011). Because water is essential for life, the security and quality of its provision are necessary for population health as well as economic development.

Although there is an acknowledged role for government in the urban water sector, changes in what that role is, and how it is performed, are the underlying themes analysed in this chapter. The first theme we examine, *devolution*, is a trend towards reduced direct government ownership, management, and control of water sector assets and services. As governance of the water sector has changed, this has necessitated changes in the regulatory tools of government. The next theme, *sophistication in regulation*, examines the evolution of a greater variety and complexity of regulatory tools used by governments in the water sector. Overlaying changes in governance and regulation, the re-emergence of *social and environmental concerns* have given new dimensions to the activities of the urban water sector.

Particular aspects of the themes contained here can be analysed more comprehensively in isolation, and therefore we draw on the scholarly works of others who have done so. To explore broad themes in a short chapter it is necessary to only outline particular aspects of urban water sector governance and regulation. The cost of this approach in terms of details is outweighed by the benefit of a more integrative perspective. Through analysis and synthesis of these three major themes, we will show how the themes interact and attempt to provide a more meaningful understanding of each in the context of the others.

2 Devolution

One of the most significant changes in the urban water sector over the past four decades has been the reduction of direct government involvement in the provision of service. 'Direct government' is used in this analysis to describe elected officials and the offices and departments that directly report to them.

We define 'devolution' to be the process whereby some or all management and control functions are delegated from direct government to other autonomous or semi-autonomous bodies. Recipients of delegated functions can be lower levels of government, independent government agencies, or the private sector. A strict interpretation of devolution would imply delegation to a lower level of government only (such as that used by Wilkins 2003); however we use a broader definition that includes delegation to external organisations such as privately owned firms ('the private sector') or publicly owned, autonomously managed companies. Most devolution is considered temporary, in that it is either of fixed term or made under a legislative process that can be repealed or amended as desired.

Devolution is a distinct and separate process from privatisation, and is a term often conflated with private sector participation. Bakker (2010) uses a broader definition of privatisation to describe redistribution of governance to non-state actors; this definition might approximate our definition of devolution if not for the fact that we see governance being redistributed to other state actors too. Private sector participation in water service provision can be done through contractual arrangements without necessarily redistributing governance. As our analysis illustrates – and is reflected in some degree of literature consensus (such as Bel and Warner 2008; Lobina and Hall 2008; and Bakker 2010) – the merits of privatisation and private sector participation can only be evaluated by understanding their local governance and institutional contexts.

Two global driving forces for devolution have been the spread of neoliberalism and the development of 'New Public Management' principles of public administration. In the late 1970s the neoliberal perspective on the appropriate role and size of government in public service provision found political support in both the UK (under Thatcher) and the US (under Reagan). There are a multiple strands of neoliberal reform that have been applied in varying ways to the water sector, namely privatisation, marketisation, deregulation, reregulation, commercialisation, and corporatisation (Bakker 2007). New Public Management, as a defined approach to government, can be described as adopting principles of compartmentalisation around service functions, accountability, arm's length regulation, competition, and support of private sector management styles (Hood 1991, 1995).

Together these global forces transformed conceptions of service provision by government and led to significant changes in the size and shape of governments for much of the developed and developing world, including of course, government service provision in the water sector. Figure 20.1 illustrates a timeline of major institutional changes in the governance of urban water in a number of global cities over the period 1980–2010.

In practice, the way devolution manifests depends on the initial governance framework and the aspects of management and control which are delegated. Governance and responsibility for urban water services may be held at a national, state, local, or other level of government. Aspects of management and control for devolution may be grouped based on a particular process or location (such as the operation of a water treatment plant), across a particular system (such as maintenance of the sewerage system), at a particular stage in a project lifecycle (such as construction of new infrastructure), or in connection with particular groups or interests (such as interfacing with retail customers). Potentially, there is an endless variety of functions and processes which can be devolved, separately or in combination.

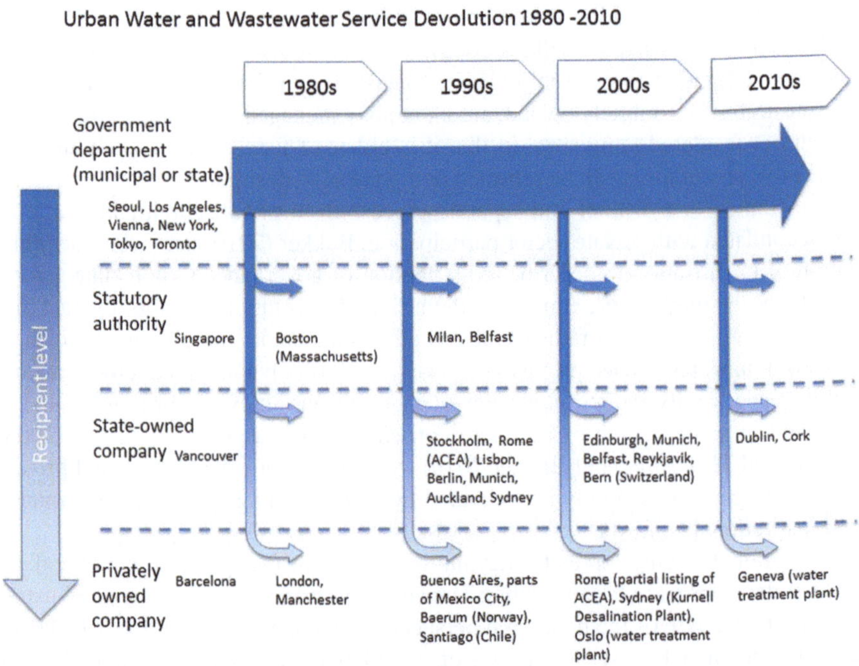

Fig. 20.1 Select examples from global cities of the changes in urban water governance that we have described as 'devolution'. For each of the locations listed the management of urban water services has been delegated to other bodies along a spectrum from statutory authorities to privately owned companies. The locations shown prior to the start of the timeline (1980) are those whose management has not changed

2.1 *Australian Experience*

From the earliest times, state and territory governments (or their colonial antecedents) have been responsible for providing urban water services in Australia. It was a natural consequence that this level of government should concentrate ownership and management, since it held constitutional jurisdiction over water resources and, at the end of the nineteenth century, it was the only level of government in the country with access to sufficient capital necessary to finance construction (Lloyd et al. 1992, p. 13).

By the mid-1990s, Australian governments, led by the Commonwealth, were focusing on liberalising the Australian economy and increasing productivity. The public water boards were among the first agencies of government to be targeted for reform. This reflected the interest in devolving government services and increasing competition, as identified by the 1995 National Competition Principles (COAG 1995). The primary responses were commercialisation, contracting out, and corporatisation of urban water services (Schott et al. 2008).

The congenital supremacy of the states and territories over Australian water resources has meant that national standardisation and oversight of water sector institutional development and reform has been late and light. After the initial COAG reform attempts in the 1990s, a diversity of institutional structures still exist (Abbott and Cohen 2010). A uniform industry structure may not be desirable, but there remains much research and industry interest in improving productivity and efficiency through institutional reform and increased competition (Crase et al. 2008; Schott et al. 2008; Dollery and Crase 2010). The institutional reform process continues today, nearly 20 years on, even as 'reform fatigue' shows (McKay 2005; PC 2011).

The State of Victoria led Australia in corporatising many government-operated services, even in advance of the National Competition Principles. The Kennett government passed legislation in 1992 to corporatise the water boards, in a never-completed process of privatisation of water retailers (Alford and O'Neill 1994). The Melbourne and Metropolitan Board of Works (MMBW) was broken up into three retail water corporations (South East Water, Yarra Valley Water, and City West Water) and one bulk water supply and treatment corporation (Melbourne Water). In the legislation underpinning the break-up of MMBW and the formation of Melbourne Water, there is explicit guidance that Melbourne Water is not to be an agent of the government for contracting purposes, unlike the former MMBW was (Government of Victoria 1992).

The change in Victorian water governance served as a testing ground for reforms in other Australian states. Not all Australian states vertically separated bulk water supply from water retailing, and some separated waste water from water supply. Table 20.1 shows the emergent structural separation of urban water services in various Australian capital cities as of 2011. With the formation in 2002 of the Northern Territory Power and Water Corporation, all of the Australian capital cities now have corporatised water services. Some states have not gone as far as Victoria in corporatising water services in rural communities, and local government still directly provides water services in much of Tasmania, Queensland, Western Australia, and New South Wales. Even after a few decades of flirtation with privatisation, the urban water sector in all major Australian cities remains firmly within state or territory government ownership and (devolved) management (Edwards 2008).

While ultimate ownership, management, and control of water services has remained with Australian state governments, a substantial proportion of services and capital works has been contracted out to the private sector (Schott et al. 2008). For the water corporations of Sydney and Melbourne, over 90 % of capital expenditure in 2010–2011 was outsourced to the private sector; over the 2006–2011 period, 80 % of Sydney Water's expenditure was for services provided by the private sector (PC 2011, p. 113). The sale of existing state water assets may not have occurred, but the private sector has had an increased role in performing a variety of capital works on behalf of state-owned water corporations.

While general trends of devolution across the Australian urban water sector are visible, it is important to note that trends belie actual paths. The temporary return of direct government management in the sector is illustrated by the Victorian

Table 20.1 Corporations and government offices undertaking functions of urban water management in Australian states 2010–2011

Jurisdiction	Bulk supply	Water retail	Wastewater retail	Stormwater
New South Wales	Sydney Catchment Authority	Sydney Water	Sydney Water	Sydney Water
	Sydney Desalinisation Pty Ltd (subsidiary of Sydney Water)	Hunter Water	Hunter Water	Hunter Water
		Gosford Wyong Joint Water Authority	Gosford Wyong Joint Water Authority	Gosford Wyong Joint Water Authority
	State Water Corporation	105 local water utilities	105 local water utilities	Local governments
Victoria	Melbourne Water	Yarra Valley Water	Yarra Valley Water	Melbourne Water
	13 regional urban water utilities	South East Water	South East Water	Local governments
		City West Water	City West Water	
		13 regional urban water businesses	13 regional urban water businesses	
Queensland	Seqwater	Queensland Urban Utilities	Queensland Urban Utilities	73 local governments
	SunWater	Allconnex Water	Allconnex Water	
	Local government-owned providers	Unitywater	Unitywater	
		71 local water utilities	71 local water utilities	
South Australia	SA Water	SA Water	SA Water	Local governments
		Small local government providers	Small local government providers	Natural Resource Management Boards
Western Australia	Water Corporation	Water Corporation	Water Corporation	Water Corporation
	Busselton Water	Busselton Water	Hamersley Iron Pty Ltd	Local governments
	Aqwest Water (Bunbury)	Aqwest Water (Bunbury)	Some local government providers	
	local government providers	Hamersley Iron Pty Ltd		
Tasmania	Southern Water	Southern Water	Southern Water	8 local government drainage trusts
	Ben Lomond Water	Ben Lomond Water	Ben Lomond Water	
	Cradle Mountain Water	Cradle Mountain Water	Cradle Mountain Water	
Northern Territory	Power and Water Corporation	Power and Water Corporation	Power and Water Corporation	Department of Lands and Planning
				Local governments
Australian Capital Territory	ACTEW	ACTEW	ACTEW	Roads ACT

Source: PC (2011, p. 44) (Productivity Commission report), used with permission. With the exception of Hamersley Iron Pty Ltd, all listed corporations are government owned; Hamersley Iron is a private company that provides water and wastewater services to Dampier, Paraburdoo, and Tom Price, Western Australia

government's response to water shortage in Melbourne (see box below). In a near-opposite response to the same millennium drought, the Queensland government vested water resource planning for the most populous part of the state in an independent statutory authority, devolved from direct government control (Spiller 2008). The Queensland Water Commission was created in 2006 near the peak of the drought and was abolished in 2013. Its policy and planning roles were then subsumed by the Queensland government's Department of Energy and Water. Centralisation of management across a greater geographic area and range of water resources has been a common response to water shortage in Australia (Searle and Head 2011). However, as these examples show, water shortage does not necessarily predicate greater direct government management and control.

Governance Responses to a Water Shortage in Melbourne, Victoria

The 'millennium drought' over south-east Australia was the driest 13 year period since modern records began. Starting in 1996, the average reduction in annual rainfall was approximately 11.4 % (CSIRO 2010). This reduction was a major challenge for Melbourne, Australia's second most populous city, as it receives over 90 % of its water from rain-fed catchments to supply an average annual urban consumption of approximately 450 gigalitres (GL). Historically, these catchments received an average of 590 GL per year in inflows; however, over the 1996–2007 period, inflows shrank to an average of 387 GL (Melbourne Water 2008, 2010) and the water security of the city was threatened (Ker 2009).

From their creation in 1992 and up until 2007, the Office of Water within the then Department of Sustainability and Environment (DSE) had encouraged the state-owned, independent water corporations to plan and implement water strategies to meet Melbourne's water demand–supply balance (Melbourne Water 2006). With the water crisis at hand, and the inadequacy of response apparent, the planning and management of Melbourne's water resources were returned to the Victorian Minister for Water and DSE (Searle and Head 2011). As noted by the regulator, the Victorian Competition and Efficiency Commission (2008, p. 174):

> … responsibility has effectively moved from Melbourne Water and the retailers to DSE … This change in roles reflects a concern that managing the risk of significantly reduced water inflows involves policy choices for which the Government will be regarded as accountable.

Through return of planning and management functions, the Victorian government was able to assert more direct control over water supply planning than it had previously been able to do under the devolved governance structure. Their policy response was a new water strategy for the region, *Our water our*

(continued)

future: The next stage of the government's plan (DSE 2007). The 2007 strategy contained a number of major supply augmentation projects including construction of a new desalination plant and new pipe connections to distant catchments. These were to supply an additional 240 GL by 2011 (DSE 2007), enough to supply half of Melbourne's existing water needs. The original estimated cost of the newly planned supply augmentation projects was A$4.9 bn, approximately $1,225 per capita, but it is understood that actual costs were well above budget.

In 2010, 2 years before the desalination plant was due to commence operation, Melbourne received its highest rainfall in decades. The desalination plant was commissioned in 2012 but has since remained on standby. The decision to invest in such a large and expensive supply augmentation has been roundly criticised since (Barnett and O'Neill 2010; PC 2011).

2.2 French Experience

France has had long experience with private sector participation in the water sector. The first concession was introduced in Paris in 1777: King Louis the XVIth granted the Périer brothers the monopoly right to construct and operate the first water network to serve houses, not just public fountains. Though this first experience ended with bankruptcy (Guérin-Schneider 2011), the private sector's varying participation in public water service provision in France had been established.

Although they retain ultimate ownership and control, the 36,000 individual French municipalities have the capacity to devolve aspects of management and control to the private sector. The diversity of management forms is high, as illustrated in Table 20.2. In practice, there are a large number of intermediate situations (for example, lease contracts – *affermage* – with a concession clause including limited investments) where operations and management roles are split between the local government and the private sector.

Contrasting Australian state-level governance of water services, French municipalities at a local level have always retained responsibility for the provision of water services. However, like in Australia, investment from a higher and more financially capable level of government was critical for the construction of the first urban water networks, as well as in the recovery following the Second World War, after which the French national government maintained a guiding role in water service management. For instance, up until 1982, a standardised contract was required by the national government for *affermage* contracts between local authorities and the private sector. From 1952 to 1986, under an anti-inflation policy, the national government also regulated tariffs, limiting possible increase. In 1982–1983, decentralisation laws launched a devolution process, in the sense of giving more responsibilities to

Table 20.2 Management and contracting types for French urban water services

		Direct public management (*régie*)	Public procurement	Public service delegation (*délégation*)	
Type of contracts		N/A	Management contract	Lease contract (*affermage*)	Concession contract (*concession*)
Approximate duration		N/A	3–5 years	10–12 years	25–30 years
Distribution of functions	Technical and commercial exploitation	Local authority	Private operator	Private operator	Private operator
	Maintenance and replacement of infrastructure	Local authority	Local authority	Private operator	Private operator
	Funding of upkeep of infrastructure	Local authority	Local authority	Local authority	Private operator
	Owner of the infrastructure	Local authority	Local authority	Local authority	Local authority
Payment of the operator		By users	By the local authority, fixed part with incentive remuneration	By users	By users

Source: Guérin-Schneider et al. (2014). Used with permission

local government.[1] From 1983, municipalities became free to tailor their contracts and in 1986 retail prices were deregulated.

The high number of services shown in Table 20.3 – about 31,000 – is due to the extreme fragmentation of French municipalities. Some municipalities can form a so-called 'intercommunality' to combine their means, such as by sharing operation of a combined water network. Conversely, in a given territory, the responsibility for water production, water distribution, wastewater collection, and wastewater treatment can be shared among different public authorities. For instance, production and collection would depend on two different intercommunalities while the distribution and collection would remain municipal: that would count for four services in Table 20.3. The break-down between global and partial services is given in Table 20.4.

Three key factors have encouraged private sector participation in the French water sector. Firstly, in the short term, contracting with the private sector allows local municipalities more financial flexibility than if they provide the water infrastructure or services themselves. Contracting with the private sector allows municipalities to transfer the financial burden to an external organisation – and finally to water

[1] Law No. 82–213 of 2 March 1982 on the rights and freedoms of the communes, departments and regions, followed by other laws in 1983 (No. 83–8 and No. 83–663).

Table 20.3 French urban water service management in 2009

		Délégation contract	*Régie* (direct management)	Total[1]
Drinking water	Number of services	4,470	9,520	13,990
	Fraction of population served	62 %	38 %	100 %
Sewerage	Number of services	4,509	12,847	17,356
	Fraction of population served	44 %	56 %	100 %

Source: Office National de l'Eau et des Milieux Aquatiques (ONEMA)
[1]Note the total number of services counted in Table 20.3 is less than the services counted in Table 20.4 due to a number of municipalities withholding management information

Table 20.4 Organisation of technical responsibilities for French urban water services in 2009

Number of services	Global service	Partial service	Total
Drinking water[a]	12,335	1,704	14,039
Sewerage[b]	12,843	4,524	17,367

Source: Office National de l'Eau et des Milieux Aquatiques (ONEMA)
[a]Global service: production, transfer, and distribution of drinking water
[b]Global service: collection, transfer, and treatment of waste water. A partial sewerage service provides only some of the elements of a global service

customers – rather than borrowing for themselves. This allows local municipalities to improve their debt ratios and preserve more flexibility to borrow for other public service needs and responsibilities. Secondly, through private sector participation, municipalities can gain access to the technical skills and organisational capacities of large private corporations which may otherwise be lacking in small municipal administrations. Thirdly, under certain forms of management delegation, private corporations may be authorised to collect water rates from retail customers directly. The French legal interpretation of water and sewerage as an 'industrial and commercial public service' encourages true cost recovery, and in practice allows private companies to bill water consumers directly rather than relying on municipal governments for payment. As long as water use continues to increase, the sector is considered as stable and profitable, encouraging the participation of private operators (Lorrain 1998).

The balance between the different management modes in France has varied over time. From the Second World War to the beginning of the 1970s, the *régie* (direct public management) was dominant. The delegation of services to the private sector became dominant in the water supply sector through to the 1990s as French municipalities responded to higher European Union water quality standards by employing private sector investment and technical skills. A public outcry during the 1990s over increased water prices (in part, a result of implementation of EU water quality standards) focused on the lack of accountability in private management, which led to strengthening of national regulations and the regrowth of *régie* management. Despite the return of *régie*, the majority of the population still has private sector provision of water supply (see Table 20.3). The composition of urban water services is likely to remain dynamic because of increased interest and competition from the private sector (Guérin-Schneider et al. 2014).

While the French have a long history of private sector participation, they have only recently used state-owned corporations as a recipient of delegated water services. The national government recently enacted law to enable municipal governments to establish state-owned water corporations (Law no. 2010–559 of 28 May 2010). The new so-called 'Sociétés Publiques Locales' are state-owned companies governed under private law. They can operate water or sewerage utilities on the territory of the shareholder municipalities and no competition is required (allowing municipalities to retain services 'in house'). As of early 2012, only a few municipalities had chosen this form of devolution for water services.

2.3 Synthesis

Devolution of urban water management and operation inherently requires a change in governance and institutional configuration. As we note, the way that this format presents itself will depend on the initial governance and institutional forms and the aspects of government which are devolved. In our case study of Australia, we identify this theme from the granting of responsibility for urban water management to water corporations. In France, it has come from the national government granting more responsibility to local government.

Comprised within the process of devolution is the message from government that water services are to be made more economically efficient. The newly created institutional structures provide incentives for this. Evidence from Australia exists in the creation of corporations to provide water services, which are responsible to both the Minister for Water and the Treasurer. France, with a long history of private sector participation in the water sector, has had in place institutional structures to devolve water service provision from direct government in a manner that encourages economic efficiency (through *affermage* or concessions).

The nature, extent, and merits of private sector participation as a result of devolution will depend on the particular circumstances in each location. Using case study experience, we find an increased role of the private sector in certain functions of the water sector but for different reasons. French municipalities found private sector participation politically and technically expedient, whereas Australia has looked to the private sector as a means to raise productivity through competition.

In observing Australia and France, we note that from different starting points of private sector participation in the water sector, both countries have enabled corporatisation of public water asset operations under private law. The use of corporate vehicles for water services has important implications for exercise of government control and regulation. This evolution and development is the focus of the next section, sophistication in regulation.

3 Sophistication in Regulation

The second theme we observe is the evolution and increasing sophistication of regulation in the urban water sector. As governments have devolved aspects of management and control, there has been a growing need for governments to transition from direct control to more nuanced tools for guiding and regulating the provision of water services. Describing this move as 're-regulation', Bakker (2005) takes a narrow view of the variety of policy instruments that have been implemented. Another conception of this trend is 'free market environmentalism' (Anderson and Leal 2001), but again this takes a skewed, market-based, view of policy innovation. Our view of sophistication in regulation sees it as a broader attempt by governments to achieve political and community desires through alternative means of influence.

Part of the evolution of influence is a re-conceptualisation of 'regulation' in theory and practice. The cleaving of policy making, service provision, and regulation was a feature of New Public Management, an effort to make public service and monopoly provision more accountable and transparent (Hood 1991). It has been pithily described as the separation of 'steering' from 'rowing' (Osborne and Gaebler 1992). In legal terms, regulation has been defined as "a government activity that is intended to affect directly the behaviour of private sector agents in order to align them with the 'public interest'" (Chang 1997). However, as shown in subsequent sections, the recipient of regulation may be a state-owned entity incorporated in the private sector or at a lower level of government. Thus we define regulation as *a government activity that is intended to directly affect the behaviour of public and private sector agents in order to align them with the public interest.*

As established in the first part of this chapter on devolution (Sect. 2), recent decades have witnessed a reduction in direct government management and control in the water sector. The perception of direct or sole government control in the operations of the urban water sector is generally not justified. There are a multitude of actors beyond a minister's office that influence the direction and nature of water service provision. The instrumental approach to governance and regulation recognises that, beyond direct instruction, there are a range of approaches possible to influence people and events (Salamon and Lund 1989; Gunningham et al. 1998). Occasionally, we find explicit recognition of policy innovation in water sector operations, such as in Quebec's water strategy (Government of Quebec 2002). However it is policy and regulatory theorists who have provided the framework for analysis.

The policy classification framework developed by Freiberg (2010, p. 85) serve as a useful foundation for analysing the policy tools applied by governments to manage the urban water sector. The policy instruments classified by Freiberg include economic, transactional, authority, structural, informational, and legal instruments. However, given continued government ownership in much of the water sector, the Freiberg policy tool classification system does not consider a government's ability to command, through ministerial direction, state-owned corporations to act (Thynne 2011). This may reflect different perspectives on regulation taken by scholars of law and of policy. In the analysis here, an additional 'command' tool is added to the

Freiberg policy classification framework, thereby completing, we believe, the classification of government policy tools used in the urban water sector. The categories, explanations, and examples are listed in Table 20.5.

A critical and common feature of the management of water services provided by companies governed under private law is the use of contracts. Water companies incorporated under private law require the use of contracts, which in turn, define relationships in terms of transaction outcomes and expectations. This constitutes transactional regulation under the typology we adopt here. However, using contracts to integrate public interests (such as equity, fair process, and adaptability) with private incentives (such as economic efficiency) is inherently difficult (Collins 1999,

Table 20.5 Government policy instruments for directing water services

Class	Instrument and example
Command	Direct instruction from minister office or central government office to undertake an action using government resources. The instruction in 2007 by Hon. Tim Holding, the Victorian government's Water Minister, for Melbourne Water to build a water transfer between the Murray–Darling Basin and Melbourne is an example
Economic	Actions that create a new market or influence an existing market for a good or service. An example would be the decision by the Victorian government in 2007 to establish a competitive tender for the right to build, own, and operate a desalination plant to supply water to Melbourne
Transactional	Actions which specify the delivery of service in return for payment, or which regulate the form of contract. The former would include government grants or contracts for delivery of water services (such as the delivery of water from a privately owned desalination plant). The latter would include regulating the form and content of contracts between third parties (such as requiring Victorian utilities to allow for hardship terms in customer water bills)
Authority	Actions that grant authority, such as licensing, certification, or permissions. As applied to the water sector, French municipalities may grant the right to private companies to bill water customers
Structural	Policy actions which seek to change the circumstances or environment of decision making so to avoid or reduce harm. In this case, structural separation may include the institutional separation of water supply management from water retail or waste water treatment
Informational	Actions which require disclosure of information, either as publicly released performance indicators, credit ratings, or delivery of specific information at specific times. The publicly available, national performance indicators of France are an example
Legal	Legislative action by governments that proscribe specific outcomes on threat of civil or criminal penalty enforced by courts. This may include primary legislation made by governments or the delegation of standards and rule-making to other authorities (such as health departments). For the water sector, this would include French water quality standards for water supplies, on threat of fine or civil punishment to the operator

Adapted from Freiberg (2010), and Thynne (2011)

p. 305). Global experience in grappling with this challenge has led to innovative governance and regulatory forms with intertwined aspects of public governance and private property (Godden 2008). We note that the French case of *intuitu personae* (see inset) provides an interesting adaptation.

One of the more common means of achieving influence, accountability, and transparency has been the creation of independent regulatory bodies to oversee certain water sector functions. As mentioned previously, the urban water sector is a natural monopoly and thus warrants government oversight. Governments have established external regulators to measure and manage performance in a variety of urban water sector functions. Regulators of health and water quality, economic performance and pricing, and environmental standards abound. Many countries, regions, and even cities have created regulatory authorities which undertake monitoring and regulatory functions ranging from economic performance (Littlechild 1988) through to public accountability and probity (Lovett 2010).

The advantage of independent regulation in the water sector is that politically motivated decisions and poor performance are reviewed transparently and therefore, if unsound, are likely to be avoided or corrected (PC 2011). The disadvantage is that regulation by independent bodies may be overly burdensome, ineffectual if not enforced, and inconsistently applied (NWC 2011). For example, the Australian Productivity Commission recommended a degree of self-regulation within the nation's water sector because of a lower societal cost than heavy-handed external regulation (PC 2011, p. 295).

Descriptions of government attempts to influence the urban water sector will differ depending on location and government priorities. The use of the policy instrument typology in Table 20.5 gives us a framework to encapsulate the variety of approaches found across the developed world. In the next subsections on Australian and French experiences, we provide specific examples of how governments have experimented and developed their policy instruments for management of the urban water sector.

3.1 Australian Experience

The complete separation of policy making, service delivery, and regulation is an apparent goal of Australian state and territory governments. Much remains incomplete, even after 15 years of devolved and corporatised water services. This has not necessarily prevented a proliferation of government attempts to regulate the water sector, and we use the example of Melbourne Water to show the variety of approaches used.

Following the corporatisation of the Melbourne Water through the *Water Industry Act 1994* (Government of Victoria 1994), and subsequent legislative evolution, the directors of the water utility are accountable to both the Treasurer and the Water Minister. The Minister has the ability to appoint directors and introduce government legislation on water matters. Monitoring and regulating the performance of Melbourne Water are the Victorian government's Department of Health, the state economic regulator (the Essential Services Commission), the state environmental regulator (the Environmental Protection Agency), the Commonwealth government's water, health, and environment agencies, the local governments in whose area Melbourne Water operates, and, of course, the three Melbourne water retailers as customers. Melbourne Water is also required to provide public annual corporate and sustainability reports.

The creation of property, procedural, or service rights has been used in Melbourne as an alternative means of improving water services to residents. Residents can seek redress against water retailers through the Victorian Civil and Administrative Tribunal (VCAT) and representation from the Energy and Water Ombudsman. The Energy and Water Ombudsman was established in 1995, with ongoing funding from energy and water retailers in Victoria (EWOV 2010). The first modern use of a parliamentary ombudsman originated in 1809 in Sweden, but Victoria claims the first energy and water ombudsman.

In addition to state-based reporting of performance, the National Water Commission reports annual performance of water services in Australian capital cities (NWC 2012) and Commonwealth government agencies such as the Productivity Commission will occasionally undertake a review (PC 2011). These national agencies do not have approval or regulatory power over the investment decisions of state governments, but they do provide national public benchmarking and scrutiny.

3.2 French Experience

Similar to the national reporting process in Australia, France has adopted benchmarking and performance reporting through a national office, based on the *Système d'Information sur les Services Publics d'Eau et d'Assainissement* (SISPEA) (Information System on Water and Sewerage Public Services). The legal foundation of the national office was laid in 2006, but operations did not commence until a few years later. The first report was published in 2012 (ONEMA 2012). The evolution of this office will be detailed in the following section; however, compared to the National Water Commission in Australia the performance indicators used are more comprehensive in scope (they include economic, social, and environmental indicators) and technically more detailed. Health and environmental regulation occur at the *préfet* level, but local municipalities have wide scope to use available policy instruments to regulate and influence water service provision from external parties.

France provides an interesting policy lesson in how governments contract for service delivery in a structurally separated water industry under private law. The original concept of *intuitu personae* (see inset) recognises that no contract is complete, regardless of the terms submitted in the tender. The personal choice granted to the mayor allows for, as a factor in selecting a private partner, their relationship and the perception of amenability. Amenability to change in the contract terms is a manifestation of the French public service principle of adaptability. Though occasionally tested, these principles have been maintained beyond the advent of neoliberalism and New Public Management.

Intuitu Personae

The French *intuitu personae* system of public procurement of water services from private sector providers is distinct from the many other systems found in developed nations.

In much of the developed world, public service procurement processes require highly detailed tender criteria and contracting agreements where expectations and obligations are specified explicitly ex ante (that is, prior to agreement). In direct contrast to this, the French employ *intuitu personae* during the tendering process, which allows, to some extent, the use of open tender criteria. In effect, this gives the mayor of the municipality greater freedom to decide which company to contract with. The mayor of the municipality does not need to establish tender criteria with the same level of precision as would accompany other public procurement processes specified under European public procurement legislation.[2] The mayor can choose the preferred private company after a closed negotiation phase and provide a public written report afterwards. This process developed from the original principles in the French water sector of adaptability and trust between government and private companies.

In recent decades the *intuitu personae* process of French water service tendering has been subject to increased domestic and EU scrutiny. Following the public opposition to water price increases in the early 1990s, a national debate on competition and regulation was opened and transparency of contracting became the focus. A new national law (the Sapin Law, No. 1993–122) was introduced in 1993 to increase transparency in contracting without removing *intuitu personae*. However, the improvement in competition and water prices has not been clear enough to close the debate. If consumers in a majority of larger cities obtained better water prices, others, mainly in smaller cities, faced price increases. The market has remained oligopolistic with only three major private companies, and one-third of tendering processes receive only one bidder (Brunet et al. 2003; Guérin-Schneider et al. 2003).

The closed contract negotiation process of French water service tendering is likely to be challenged by future European Union directives. In December 2011, the European Commission[3] proposed a draft directive on concessions which would reinforce transparency obligations. The award criteria were to be defined ex ante, and could not be changed during negotiations. Many southern European countries, not only France, were reluctant to accept this directive requirement. For the moment, water services have been excluded from increased transparency obligations in the directive voted in the European Parliament (Directive 2014/23/EU).

[2] Directives 2004/18/EC, 2004/17/EC, and 2009/85/EC. A directive is a European law that all member states must adopt in their countries.

[3] The European Commission (EC) is an executive body of the European Union. It is composed of one appointed commissioner per member state.

3.3 Synthesis

Governments across the developed world have been experimenting with alternative policy means, beyond direct command and control, for influencing the urban water sector.

As we have noted in this section, the proliferation of policy instruments stems from the devolution of direct government management and operation of the water sector. Other observers have viewed this proliferation as a consequence of privatisation and the adoption of free market environmentalism (Anderson and Leal 2001; Bakker 2005), which we disagree with when noting that the same proliferation of policy instruments has occurred where governments have delegated responsibility to lower levels of government and state-owned corporations.

Australian and French case studies are used to illustrate the experimentation and policy innovation we see across much of the developed world. It is evident that there remain significant challenges and opportunities for development of policy instruments to guide the urban water sector. The physical nature and monopoly qualities of water service provision have not evolved to the same degree as other public service monopolies, such as telecommunications and electricity. Both Australia and France have grappled with issues of public interest and private incentives in the context of the water sector and have responded differently. Australia states and cities have created a variety of regulators to monitor performance of water utilities, but not major supply augmentation decisions, while France has employed *intuitu personae* for adaptability when contracting with the private sector for public water services.

Both Australia and France have developed public reporting systems which have social and environmental indicators. The following section details the emergence of social and environmental concerns which led to this reporting system.

4 Social and Environmental Concerns

The third theme that we identify is the emergence of broader social and environmental concerns. Social concerns are intrinsically present in the water sector, as the provision of safe and secure water supplies and sanitation is a major public health target (Goubert 1986). However, social and environmental concerns acquired a new dimension with the growth of international consciousness about sustainable development (UN 1992) and as a backlash against neoliberalism and its effects on environment and society (Bakker 2005). This theme emerged during the 1990s in both developed and developing nations, although in this analysis we maintain our focus on the developed world.

Urban water supply is indelibly marked with social concern through its original conception as a public health activity (Goubert 1986). Social concerns have re-emerged more recently in response to broader declines in social equity in western society since the 1970s (OECD 2011) and from research and political interest in social cohesion. In the water sector this has appeared through calls for greater research into social 'structuration' of water use (Syme 2008) and greater participation of citizen-consumers in urban water management (Wong and Brown 2009). Commonly recognised government responses have included the incorporation of community consultation in water planning and management, participatory governance, and attempts to align decision-making scales to catchment scales. The 'water as a human right' catch-cry has been used to lobby legislatures to embed social aspects in water management, with varying success (Bakker 2007).

Environmental concerns in society at large have emerged from a variety of sources, but importantly there has been shared recognition of the problem (particularly at Rio in 1992) and a shared understanding of the necessity to act (Dryzek 1997). In the water sector these concerns have manifested as environmental principles or aspirations, triple bottom line reporting, recognition of environmental services and other water 'users', technical design features such as water sensitive urban design (WSUD), and the allocation of water stocks or flows for environmental purposes.

When we see concurrent rises in social and environmental concerns, we also observe a conflation of the concerns under broad ambiguous titles such as 'sustainability', 'liveability', and 'water sensitive' cities, designs, or resource management. This is not to discount the value of multipurpose infrastructure and programs, but it does highlight the formation of coalitions of interest groups agitating for change in how water is managed in cities (Bakker 2010). Translation of rhetoric into action and change in urban water is not as smooth as many would like, although there are notable champion cities such as Singapore (Brown and Farrelly 2009).

4.1 Australian Experience

Melbourne Water offers a case study of the emergence of social and environmental concerns in Australia. The Melbourne water corporations were established in the early 1990s without sustainability principles, nor social or environmental goals (Government of Victoria 1992, 1994). Mounting national and international interest in environmental and social issues through the 1990s, led by such events as the Rio Earth Summit in 1992, allowed a 2002 Australian government senate enquiry into urban water to conclude: *...management solutions must also be based on the three parameters of environmental, social and economic sustainability* (Commonwealth of Australia 2002, p. xiii) and subsequently included sustainability and urban water as part of a nationally coordinated attempt at water reform (Hussey and Dovers 2006). By 2007 the legislative foundations for Victorian water corporations were amended to include these as 'sustainable management principles' in operations (Government of Victoria 2007) which were subsequently translated into corporate principles for public water companies. For the directors of Melbourne Water, this includes a legal obligation to evaluate actions and report annual performance against (social) *relationships, integrated water management*, and *environmental* *stewardship*, alongside *service delivery* and *financial sustainability* (Melbourne Water 2012). These principles of operation have been used to justify a variety of activities which are not strictly in the provision of water supply and treatment, such as school education programs and renovating an 1893 heritage-listed sewer for community use (Melbourne Water 2013, p. 6).

Environmental and social concerns have evolved following the millennium drought and the election of a conservative state government in 2010. Melbourne's water policy has swung away from large infrastructure solutions and towards decentralised supply under the aegis of water sensitive urban design. A new statutory body, the Office of Living Victoria, has been established to oversee water planning and policy with the purpose of promoting 'liveability'. A newly released paper, *Melbourne's Water Future* (OLV 2013), proposes changes in stormwater use and recycling, building codes, and community participation in the water sector.

Competing social, environmental, and economic principles can be a source of conflict for water service providers. In the case of Melbourne Water, the millennium drought caused the corporation to choose between retaining water for urban use or releasing water for environmental flows in the Yarra River. In doing so, the principles of secure urban water supply and environmental health were brought into conflict, a challenge noted by other Australian water authorities with similar conflicts of policy making and service delivery (WAWA 1995). The Productivity Commission in their recent inquiry into the Australian water sector has challenged the efficiency of requiring a water service provider to perform these functions (PC 2011). In essence, this review recommended water corporations retreat from the role of service provider (delivering a suite of environmental, social, and economic outcomes) to become a focused commodity supplier.

4.2 French Experience

Despite its reputation for centralised and regulation-reliant government policy, France has recently started using the national office (SISPEA) for public reporting of environmental and social outcomes in the water sector. This comes on top of European and national legislation prescribing specific technical standards, implemented due to public concerns about the quality of water and wastewater treatment. Social concerns about equity of treatment and continuity have been part of public law doctrine for a long time.[4]

France's response to the rise of environmental concerns deserves an explanation of the history of environmental regulation in the European Union. European Union

[4] In France, as in many Mediterranean countries influenced by Roman law, public and private sectors are subject to separate legislations (public law versus private law). The doctrine of public law relies first on this specific legislation and second on the jurisprudence of the *Conseil d'Etat*, a specific court of appeal for public law. *Equity of treatment* and *continuity* are two fundamental principles of public law deriving from this doctrine.

legislation in the 1980s significantly increased France's water quality and sanitation obligations. As referred to previously, new technical requirements and large investments led to subsequent increases in the water prices paid by households. Increases in water prices provoked large public outcry and debate about water sector regulation through the 1990s. In response to the outcry, the left-wing French government planned to introduce a national regulatory authority with coercive powers over water services. The main objective was to reduce information asymmetry between private operators and local authorities, and to improve local regulation of the price and quality of services. The main issue was transparency and pricing, rather than environmental outcomes. A proposed national regulator was meant to provide local authorities with a national expertise and assist in performance monitoring using a range of indicators on service operation, including an environmental dimension. A political change in 2002 stopped this project of national regulation coming to fruition. However, the concept of a national office for performance reporting (based on SISPEA) was maintained in the 2006 water law. The justification was to create a national information and benchmarking system for self regulation, rather than generate more national regulation. Environmental and social concerns were used to justify the reinstatement of public reporting, including performance indicators. The reporting was made compulsory at the local level, but the transmission of performance indicators to SISPEA remained optional.

It is interesting to analyse this shift. The list of publicly reported indicators is almost the same as those originally proposed for regulation, but the regulatory system (regulation of operators' performance versus information to users) and the role of the national authority (strong support to local regulation versus informing consumers and citizens) has changed. The hybridisation of the two regulatory systems (performance regulation by local authorities and sustainability information for users) has led to imperfection both in regulation of operational aspects and in achieving the desired social and environmental outcomes (Canneva and Guérin-Schneider 2011a). Some publicly reported indicators were selected for the use of experts, and thus are difficult for much of the public to understand. Local authorities are less than conscientious in their data collection for public indicators, and fail to see the importance of a national reporting office which does not assist in the enforcement of regulation (Canneva and Guérin-Schneider 2011b). This frustrated process reflects a disjoint between the two uses of French performance indicators. They were established as a tool for regulatory control over operators, but subsequent implementation has been as a communication tool to restore the environmental and social reputation of operators.

4.3 Synthesis

With the advent of environmental and social concerns in the water sector, Australia (as exemplified in the State of Victoria) and France have applied similar information regulation and legal tools of government to influence the performance of the water sector. The French responded with increased public surveillance of water utility function, requiring publication of non-financial performance indicators including environmental and social ones. Australia's state ownership of water corporations has enabled jurisdictions to establish requirements, through amendments to legislation, for water utilities to have environmental and social principles which they must report and report on.

This chapter has identified a conflict between inclusion of environmental and social goals and goals of efficient service delivery. This conflict has been noted by previous commentators, particularly from the economic field. We have explored different policy instruments that governments have used to promote environmental and social performance in the water sector.

5 Water Sector Governance and Regulation: Continuously Fluid?

Governance and regulation in the urban water sector is changing in response to a variety of factors. This chapter has reviewed three major themes of change over recent decades and their respective drivers: devolution, sophistication in regulation, and the emergence of social and environmental concerns.

Devolution in direct government management and control in the urban water sector has been a significant and on-going trend. Different initial starting points and interpretations have yielded a variety of structural forms and responses. Private sector participation has increased for a variety of reasons beyond economic efficiency, as French and Australian experience illustrates.

The necessary effect of devolution has been the evolution of government instruments for regulating the urban water sector. Rarely can elected officials now issue instructions for urban water management and operations. A surprising variety of government policy instruments are now used to influence the provision of urban water services by autonomous and semiautonomous bodies. As we argue in this section, much of this expansion in the tools of government is due to the provision of public services through private law incorporated corporations. The similarly wide and varied use of policy instruments in the French and Australian water sectors, despite their different levels of private sector participation, validates this point. The French use of *intuitu personae* provides an illuminating case of public procurement of services from the private sector while attempting to maintain public service principles.

Environmental and social concerns in the water sector have re-emerged from a variety of sources. Integrating these concerns with economic efficiency and service standards has been a challenge for the water sector. We have observed trade-offs between such conflicting concerns and obligations in the Australian urban water sector. We find the use of sustainability indicators to make up for a lack of a greater regulation, as in France, while environmental and social corporate principles have been used in Australia to justify water corporation activities that are not strictly for the supply of water to urban areas.

Necessarily, this chapter has only been able to outline certain aspects of urban water sector governance and regulation. However, by analysing these themes we have illustrated how they interact and given a wide integration of the changes in urban water sector governance and regulation. For example, we have shown how devolution has led to a greater need for, and innovation in, the use of policy instruments to guide the provision of urban water services. We have explored the evolution of a diverse array of policy instruments in their application to emergent social and environmental concerns. This type of analysis is uncommon in the context of academic enquiry into the urban water sector.

Looking ahead, there are other emerging issues for governance and regulation in the urban water sector. Rising costs of energy, impacts of climate change, and the necessity to replace aging infrastructure are issues with potentially large impacts on the urban water sector globally. We expect that the governance and regulation of the urban water sector will continue to be as fluid as the resource it seeks to manage.

References

Abbott, M., & Cohen, B. (2010). Industry structure issues in the water and wastewater sectors in Australia. *Economic Papers, 29*, 48–63.

Alford, J., & O'Neill, D. (Eds.). (1994). *The contract state: Public management and the Kennett government*. Geelong: Deakin University Press.

Anderson, T., & Leal, T. (2001). *Free market environmentalism*. New York: Palgrave.

Bakker, K. (2005). Neoliberalizing nature? Market environmentalism in water supply in England and Wales. *Annals of the Association of American Geographers, 95*, 542–565.

Bakker, K. (2007). The "commons" versus the "commodity": Alter-globalization, anti-privatization and the human right to water in the global south. *Antipode, 39*, 430–455.

Bakker, K. (2010). *Privatizing water: Governance failure and the world's urban water crisis*. Ithaca: Cornell University Press.

Barnett, J., & O'Neill, S. (2010). Maladaptation. *Global Environmental Change, 20*, 211–213.

Bel, G., & Warner, M. (2008). Does privatization of solid waste and water services reduce costs? a review of empirical studies. *Resources, Conservation and Recycling, 52*, 1337–1348.

Bozeman, B. (2007). *Public values and public interest: Counterbalancing economic individualism*. Washington, DC: Georgetown University Press.

Brown, R., & Farrelly, M. (2009). Delivering sustainable urban water management: A review of the hurdles we face. *Water Science and Technology, 59*, 839.

Brunet, E., Guérin-Schneider, L., & Bonnet, F. (2003). Impact of a new legislation on the water market and competition in France. *Water Science and Technologie: Water Supply, 3*, 389–394.

Canneva, G., & Guérin-Schneider, L. (2011a). La construction des indicateurs de performance des services d'eau en France: mesurer le développement durable ? *Natures Sciences et Sociétés, 19*, 213–223.

Canneva, G., & Guérin-Schneider, L. (2011b). National monitoring of water utility performance in France. *Water Science and Technologie: Water Supply, 11*, 745–753.

Chang, H.-J. (1997). The economics and politics of regulation. *Cambridge Journal of Economics, 21*, 703–728.

COAG. (1995). *National competition principles*. Hobart: Council of Australian Governments.

Collins, H. (1999). *Regulating contracts*. Oxford: Oxford University Press.

Commonwealth of Australia. (2002). *The value of water: Inquiry into Australia's management of urban water.* Report of the Senate Environment, Communications, Information Technology and the Arts References committee. Canberra

Crase, L., O'Keefe, S., & Dollery, B. (2008). Can urban water markets work? Some concerns. *Agenda, 15*(3), 73–82.

CSIRO. (2010). *Climate variability and change in south-eastern Australia: A synthesis of findings from phase 1 of the south eastern Australian climate initiative (SEACI).* Canberra: CSIRO.

Dollery, B., & Crase, L. (2010). Industry structure issues in the water and wastewater sectors in Australia: A comment on Abbott and Cohen. *Economic Papers, 29*, 365–367.

Dryzek, J. (1997). *The politics of the earth: Environmental discourses.* Oxford: Oxford University Press.

DSE. (2007). *Our water our future: The next stage of the government's plan.* Melbourne: Department of Sustainability and Environment, Government of Victoria.

Edwards, G. (2008). Urban water management. In L. Crase (Ed.), *Water policy in Australia: The impact of change and uncertainty.* Washington: Resources for the Future.

EWOV. (2010). *Constitution of energy and water ombudsman (Victoria) Ltd.* Melbourne: Energy and Water Ombudsman (Victoria) Ltd.

Freiberg, A. (2010). *The tools of regulation.* Sydney: The Federation Press.

Godden, L. (2008). Property in urban water: Private rights and public governance. In P. Troy (Ed.), *Troubled waters: Confronting the water crisis in Australia's cities.* Canberra: ANU E-Press.

Goubert, J.-P. (1986). *The conquest of water: The advent of health in an industrial age.* Oxford: Polity Press.

Government of Quebec. (2002). *Water. Our life. Our future.* Montreal: Government of Quebec.

Government of Victoria. (1992). *Melbourne Water Corporation Act 1992.* Melbourne.

Government of Victoria. (1994). *Water Industry Act 1994.* Melbourne.

Government of Victoria. (2007). *Water Act 1989 (version 85).* Melbourne.

Guérin-Schneider, L. (2011). Histoires des services publics d'eau potable et d'assainissement: entre stabilité et reconfiguration. In G. Bouleau & L. Guérin-Schneider (Eds.), *Des tuyaux et des hommes. Les réseaux d'eau en France.* Versailles: NSS Dialogues and Editions Quae.

Guérin-Schneider, L., Bonnet, F., & Breuil, L. (2003). Dix ans de loi Sapin dans les services d'eau et d'assainissement: évolutions et perspectives du modèle de délégation à la française. *Responsabilité & Environnement Annales des Mines, 31*, 44–57.

Guérin-Schneider, L., Breuil, L., & Lupton, S. (2014). Liberalisation of water services in Europe: The end of the French water exception? In G. Schneier-Madanes (Ed.), *Globalized water, a question of governance.* New York: Springer.

Gunningham, N., Grabosky, P., & Sinclair, D. (Eds.). (1998). *Smart regulation.* New York: Oxford University Press.

Hood, C. (1991). A public management for all seasons? *Public Administration, 69*, 3–19.

Hood, C. (1995). The 'new public management' in the 1980s: Variations on a theme. *Accounting, Organizations and Society, 20*, 93–109.

Hussey, K., & Dovers, S. (2006). Trajectories in Australian water policy. *Journal of Contemporary Water Research & Education, 135*, 36–50.

Ker, P. (2009). Dam water levels dwindling to historic low. *The Age*, 15 April.

Littlechild, S. (1988). Economic regulation of privatised water authorities and some further reflections. *Oxford Review of Economic Policy, 4*, 40–68.

Lloyd, C., Troy, P., & Schreiner, S. (1992). *For the public health: The hunter district water board 1892–1992.* Melbourne: Longman Cheshire.

Lobina, E., & Hall, D. (2008). The comparative advantage of the public sector in the development of urban water supply. *Progress in Development Studies, 8*, 85–101.

Lorrain, D. (1998). Le régulateur, le service public, le marché et la firme. *Flux, 31–32*, 13–23.

Lovett, I. (2010). City council tightens reins on LADWP. *Park Labrea News Beverly Press*, 11 April.

McKay, J. (2005). Water institutional reforms in Australia. *Water Policy, 7*, 35–52.

Melbourne Water. (2006). *Water supply–demand strategy for Melbourne 2006–2055*. Melbourne: State of Victoria.
Melbourne Water. (2008). *Annual report 2007/08*. Melbourne: Melbourne Water.
Melbourne Water. (2010). *Annual report 2009/10*. Melbourne: Melbourne Water.
Melbourne Water. (2012). *Annual report 2011/12*. Melbourne: Melbourne Water.
Melbourne Water. (2013). *Melbourne water's submission to Melbourne's water future*. Green paper. Melbourne: Melbourne Water.
NWC. (2011). *Urban water in Australia: Future directions*. Canberra: National Water Commission.
NWC. (2012). *National performance report 2011–12: Urban water utilities*. Canberra: National Water Commission.
OECD. (2011). *Divided we stand: Why inequality keeps rising*. Paris: OECD.
OLV. (2013). *Melbourne's water future*. Consultation paper. Melbourne: Office of Living Victoria.
ONEMA. (2012). *Observatoire des services publics d'eau et d'assainissement. Panorama des services et de leurs performances*. Paris: ONEMA.
Osborne, D., & Gaebler, T. (1992). *Reinventing government: How the entrepreneurial spirit is transforming the public sector*. Reading: Addison-Wesley.
PC. (2011). *Australia's urban water sector*. Report No. 55, Final inquiry report. Canberra: Productivity Commission.
Salamon, L. M., & Lund, M. (1989). *Beyond privatization: The tools of government action*. Washington, DC: Urban Institute Press.
Schott, K., Wilson, S., & Walkom, S. (2008). Urban water reform: An industry perspective. *Australian Economic Review, 41*, 413–419.
Searle, G., & Head, B. (2011). *Urban drought, infrastructure crisis, and governance centralisation in Sydney, Melbourne and SE Queensland.* Paper presented to State of Australian Cities conference, Melbourne.
Spiller, D. (2008). Water for today, water for tomorrow: Establishment and operation of the SEQ water grid. *Australian Economic Review, 41*, 420–427.
Syme, G. (2008). Sustainability in urban water futures. In P. Troy (Ed.), *Troubled waters: Confronting the water crisis in Australia's cities*. Canberra: ANU E-Press.
Thynne, I. (2011). Ownership as an instrument of policy and understanding in the public sphere: Trends and research agenda. *Policy Studies, 32*, 183–197.
UN. (1992, June). *United Nations Conference on Environment and Development (UNCED)*. Rio de Janeiro: United Nations.
Victorian Competition and Efficiency Commission. (2008). *Water ways: Inquiry into the reform of the metropolitan retail water sector*. Melbourne: State of Victoria.
WAWA. (1995). *A water supply strategy for Perth and Mandurah to 2021 (with a focus on 2010)*. Perth: Water Authority of Western Australia.
Wilkins, J. K. (2003). Conceptual and practical considerations in alternative service delivery. *International Review of Administrative Sciences, 69*, 173–189.
Wong, T. H. F., & Brown, R. R. (2009). The water sensitive city: Principles for practice. *Water Science and Technology, 60*, 673–682.

Chapter 21
Public–Private Partnerships and Their Ownership in the Urban Water Sector

Arnaud Reynaud

1 Introduction

Water services, including sanitation, are essential to life and health, economic development, and human dignity. As a result, those services have either been the responsibility of the public or the private sector, but in both cases under strict public regulation. For a long time, the debate about private participation in the water industry has been limited to the question of privatisation, as was done for example in the United Kingdom in 1989. However, in recent years, there has been increased interest in alternative organisational arrangements for the provision of water services involving private participation at various levels (Chong et al. 2006). Such public–private partnerships (PPPs) are increasingly being considered viable alternatives to the full privatisation of the water industry (Bel and Warner 2010). Formally, the concept of PPP is not well defined within the economics literature, but it refers to any kind of contractual arrangement by which a public authority assigns to a private operator the fulfilment of a mission of public interest.

As will be discussed in this chapter, across the public and the private parties to a contract there are a wide range of PPPs differing in their allocation of decision prerogatives, risks, and revenues. Delivery of water services through PPPs is not new. For instance, the first well-documented case of a PPP in France dates back to 1776 when the city of Paris decided to allocate to "Sieurs Perrier" a 15-year exclusive concession to the water supply system (Duroy 1996). Two years later, Louis XVI gave the Perrier brothers the exclusive right "to build and establish all facilities (pumping machines, pipes, etc.) required to bring water from the Seine to all

A. Reynaud (✉)
Toulouse School of Economics (LERNA), Université Toulouse 1 Capitole,
21 Allée de Brienne, 31000 Toulouse, France
e-mail: areynaud@toulouse.inra.fr

Q. Grafton et al. (eds.), *Understanding and Managing Urban Water in Transition*, Global Issues in Water Policy 15,
DOI 10.1007/978-94-017-9801-3_21

Parisian districts and suburbs". The need to have a well-specified contract between the public authority and the private operator was already noted at the time. Hence, a contractual clause concerning water affordability to the poor households was already included in the concession contract: the Louis XVI text stipulated that "The Sieurs Perrier must build some water fountains in order to guarantee a moderate price for poor households".

In this chapter we propose to summarise the experience of PPPs in the provision of urban water and sanitation services. We start by discussing the various forms of organisational arrangements between fully public provision of services and complete privatisation. We then discuss the main arguments in favour of PPPs while stressing some possible limitations of this kind of organisational arrangement. We then conclude by summarising the empirical literature having tried to establish a link between PPPs and performance of water services.

2 Models of PPP in the Urban Water Sector

As mentioned previously, PPPs are not formally well defined within the economics literature and, in their widest sense, they embrace a variety of possible relationships between a public and a private contractor. The concept of PPP refers in fact to any kind of contractual arrangement (covering usually a long time period) by which a public authority (state, province, city, etc.) assigns to a private operator the fulfilment of a mission of public interest.[1] The private partner may then bear substantial risks and may even raise private finance. Its revenue usually derives from a combination of public authority payments and user fees. A wide range of contractual forms may be adopted by the public sector in establishing partnerships with the private sector, ranging from those where there is a high level of public sector participation to those where there is very little involvement.

Generally, the spectrum of contractual options ranges from "build–own–operate–transfer" (BOOT) contracts or concession contracts, to management and service contracts (see Table 21.1). In the specific case of the water sector, we can make the distinction between the following various forms of PPPs.

A *BOOT contract* (build, own, operate, transfer) corresponds to a situation in which the private sector contractor builds a facility in accordance with a design prepared by the public sector. The private sector contractor finances the construction of the facility and legal ownership of the facility rests with the contractor until the end of the contract term. The private sector contractor is responsible for operating the facility for the contract term and bears all the financial risk. Sometimes, the contract also includes the design of the facilities to be afterward operated (design, build, own, operate, transfer, or DBOOT). The private contractor's revenue can be

[1] Some authors have adopted a more restrictive definition for PPPs. Maskin and Tirole (2008) define a PPP as a situation in which the public authority typically engages its partner "both to develop the project and to operate and service it".

Table 21.1 Allocation of risks and responsibilities in PPPs

	Asset ownership	Operating and maintenance	Capital investment	Commercial risk	Contract duration
BOOT	Private	Private	Private	Private	20–30 years
BOT	Public	Private	Private	Private	20–30 years
Concession	Public	Private	Private	Private	25–30 years
Lease contract (affermage)	Public	Private	Public	Public/private	8–15 years
Management contract	Public	Private	Public	Public	3–5 years
Service contract	Public	Public/private	Public	Public	1–5 years

Source: Adapted from Walker and Marr (2002)

based on a variety of arrangements, ranging from a fixed annual fee (i.e., flat rate) to quantity-based pricing (i.e., unit or block rates). This type of PPP is well established for the water industry in Italy, Spain, and Portugal.

A *BOT contract* (build, operate, transfer) corresponds to a situation in which the contractor builds (and sometimes designs) the facility, operates it for a specified period, and then hands it over to the public water undertaking in good condition. The contractor may not always provide finance, in which case he does not own the facility as he does in the case of a BOOT scheme. Such arrangements have proved popular in Australia, especially for water utilities that do not have a financing problem.

A *concession contract* is similar to a BOOT contract including the design of facilities, except that ownership of the asset remains public. The firm has the control of operation and management, as well as financial responsibility for investment in the expansion and maintenance of water infrastructure. The firm generally collects fees directly from customers, with profits dependent on the efficiency of water service management. The tariff is defined in the concession contract, which also includes some rules on how it may be changed over time. Typically, the duration is 20–30 years. Concession contracts have been used for a long time in the water industry in France, and in the late 1980s in the developing world, to attract private investment in infrastructure sectors (Guasch et al. 2008).

A *lease contract* ("affermage") is very similar to a concession contract except that public authorities remain responsible for the expansion and maintenance of water infrastructure. Under a lease contract, the private contractor obtains its revenue through the water bills paid by consumers. As a result, the commercial risk is shared between the public and the private contractors. Duration is typically 8–15 years. At the end of a leasing contract, all system assets and non-system assets required to deliver the services are handed over to the public water undertaking in good repair. Lease contracts are very popular in the water sector in France. However, since private contractor revenues come from customer payments, under a lease contract the question of tariff setting is crucial. Lease contracts may require structuring and revising complex price schemes.

A *management contract* corresponds to a situation in which the private firm assumes full day-to-day operational control of a water service. Ownership, investment in infrastructure, and other capital expenses remain in the public hand. Although management contracts generally involve a transfer of managerial responsibility to the private sector, the private contractor does not typically bear much of the associated commercial risk. The private firm receives a fee (often flat) for the service provided, the fee being in accordance with the conditions laid down in the contract. Typical duration is 3–5 years. Management contracts are most likely to be useful where the primary objective is to rapidly enhance the technical capacity and efficiency of a utility to perform specific tasks.

A *service contract* (outsourcing) corresponds to a situation in which only certain aspects of the operation of the facilities are delegated to third parties (for example the maintenance of the facilities). These contracts generally have a term of 1–5 years. The simplest form of service contract involves the payment of a fee for a service such as meter installation. Public authorities generally use competitive bidding procedures to award service contracts.

Divestiture and *joint ventures* are another form of possible organisation of the water sector involving private participation. Depending upon the author, these kinds of arrangements are generally not formally included as PPPs. A complete divestiture involves the selling of all of the public water undertaking to the private sector. The private sector then has the full responsibility for operations, maintenance, and investment, as well as the ownership of the assets. The role of public authorities is then to regulate the private firm, in order to ensure that the services provided by the private sector meet the expectations of the government and the customers. A typical example of complete divestiture for the water sector is the United Kingdom. Partial divestitures (sometimes described as joint ventures) limit the sale to a percentage, providing the private sector limited ownership and control over the assets. This model has been used in Germany and Chile. Partial divestiture is an option that is politically attractive, but regulation and the allocation of risks between the private and public sectors can be problematic and must be carefully designed.

Globally, it is very difficult to assess any trend in the use of any specific PPP, first because no consolidated data at this level exists, and second due to the very strong regional specificities of PPPs. Hence, PPPs in Latin America have mainly followed the concession model, whereas many water PPPs in Sub-Saharan Africa have adopted the "affermage" model. If the structure of PPPs in developed countries has remained relatively stable over the last decade, two important changes should be noticed for low- and middle-income countries. First, whereas for a long time PPPs where signed with a few multinational firms, in the last decade more than half of water PPPs have been contracted by private firms originating from low- and middle-income countries. Second, the last decade has also viewed the opening of China to private participation in water infrastructure and its emergence as the first water PPP market among low- and middle-income countries.

3 Rationale and Limits to PPPs in the Water Sector

3.1 Rationale for PPPs in the Water Sector

The various forms of PPPs presented in Table 21.1 differ in terms of whether the initial investment is undertaken by public or private contractors, what the level of involvement of the private contractor is in the management of the water service, and the way risks are shared between the two contracting parties. The rationale for using any PPP then depends upon the weight put by public authorities on each of these three characteristics.

The first rationale for using a PPP is based on the idea that the partnership will provide both infrastructures and services at a lower cost (Vining and Boardman 2008). Two main arguments have been proposed to support this cost-efficiency argument. First, private firms may benefit from scale,[2] scope,[3] or learning economies because they are more specialised, larger, and have more experience in construction and operation. Second, they face stronger incentives to minimise their costs and to innovate, although this may depend on the level of competition and the economic regulations in place.[4] In the next section, we survey the empirical literature that has tried to establish a relationship between PPPs and cost efficiency of water services.

The second argument for using PPPs is the minimisation of on-budget government expenditures and/or the desire not to increase current debt levels (Vining and Boardman 2008). Hence, under BOOT, BOT, and concession contracts the capital investment is made by the private contractor. This argument is particularly important in the water sector since the water industry is known to be a very capital-intensive industry,[5] with a high proportion of fixed assets (water and sewerage pipes, water and wastewater treatment plants) having very long lives (up to 100 years).

The third rationale relates to the government's desire to reduce the risk associated with its financial exposure to construction costs, maintenance costs, and variability of revenues. This argument relies on the fact that the private partner is usually engaged in similar projects simultaneously and can therefore spread the risks (Perold 2004).

[2] In the light of their review of the literature, González-Gómez and García-Rubio (2008) conclude that although there are important economies of scale in the urban water sector, they are however not unlimited. Saal et al. (2007) conclude for instance that the excessive size of the water service companies created in England and Wales after industry reforms in 1989 has had a negative effect on the growth of productivity over the subsequent years.

[3] For instance, in the UK case Hunt and Lynk (1995) report cost efficiency gains for utilities jointly operating water distribution and sewerage services.

[4] See for instance Averch and Johnson (1962) for some perverse cost-minimising effects of rate-of-return regulations.

[5] The capital-to-revenue ratio is between 10 and 12. The water industry is 3–4 times more capital intensive than the electric industry and 5–6 times more capital intensive than the railroad industry (Jordan 1998).

Two political arguments in favour of PPPs have also been proposed. First, a PPP might be an efficient setting when public authorities do not behave in a benevolent way. This might be the case, for instance, if public decisions are influenced or controlled by interest groups.[6] Recently, Picazo-Tadeo et al. (2009) suggested that public ownership in the water sector results in greater difficulties in adapting staff to the real needs of the company, partly because of the greater degree of unionisation in the public sector and partly because local governments tend to avoid confrontations that could have political and social consequences. The second political argument favouring PPPs is that with a private-sector operator in charge it is politically easier to impose unpopular decisions such as higher user-fees (Vining and Boardman 2008).

3.2 Some Limits to the Use of PPPs in the Water Sector

The first problem with using PPPs in the water industry is the issue of asymmetric information between public authorities and private firms. Due to information asymmetry, public authorities are not always in a good position to efficiently control water service providers. Sources of asymmetric information may be the lack of technical expertise (which is likely to be especially true for small communities), the unbalanced negotiation power between the private firm and the public authority, or the lack of financial resources for implementing efficient regulations. The problem is that the water industry is characterised by oligopolistic markets with a limited number of large firms. This raises some ex-ante competition issues such as the possible lack of bidders, the existence of restrictive agreements between bidders (bid rigging, market allocation, etc.), or even possible abuses of dominant bidders. The situation also means that threats of non-renewal of a contract by a public authority are weaker, even though such threats have been shown to be an important way of disciplining private firms.

Transactions costs are another important limit to the use of PPP. Transaction costs are directly and positively linked to difficulties in measuring the performance and quality of a service and with the unpredictable nature of a task. In such situations, the costs of designing and enforcing contracts are high. According to Pérard (2009, p. 202), "because of asymmetries of information, the cost of bid evaluation and due diligence process is extremely important for infrastructure projects and more particularly for water supply". In the water sector, these costs have been estimated at around 3–5 % of the total project costs (Klein 1996a, b).[7] Working on a sample of French water utilities under various forms of PPPs, Chong et al. (2006)

[6] There is substantial evidence that politicians' project choices are influenced significantly by the desire to please constituencies and by budgetary constraints (Levin and Tadelis 2010).

[7] Klein (1996a, b) reviews 33 infrastructure projects of the World Bank. In these projects transaction costs represent on average 3–5 % of the total project cost, but they can reach up to 12 % for some projects.

concluded that their results are consistent with a theory in which high transaction costs make the use of PPPs inefficient.

4 Performance of PPPs: Some Elements from the Empirical Literature

To more formally assess the efficiency of PPPs in providing water services at low cost, a prerequisite is to be able to measure the performance of urban water utilities. This is a challenging issue since performance is typically multidimensional. In this section we limit the discussion to three dimensions of water service performance, namely technical and cost efficiency, price, and access to water and sanitation services.

4.1 Technical and Cost Efficiency

We start by discussing the results of empirical studies which analysed the cost efficiency of PPPs in the water sector. Since PPPs have mainly emerged as an alternative to full privatisation only in the last 10 years, most of the existing work relies on a comparison between the cost efficiency of private and public services. The interested reader may also refer to Bel and Warner (2008) who reviewed all the econometric studies published between 1976 and 2006 which compared the costs of public and private water services.

Comparisons of cost efficiency of public and private water utilities have been conducted in two different ways. Some studies have been conducted in countries where both forms of ownership in management coexist (US, France, Brazil). The second approach has been to compare the efficiency of water services before and after privatisation (England, Wales). In Table 21.2, we report the results concerning technical and cost efficiency of public and private water services from the main empirical published studies.

The issue of efficiency of public versus private ownership has been long debated in the US since the late 1970s. The results of a study by Feigenbaum and Teeples (1983) on a sample of US water services suggest that there are no significant differences in cost of service between private and public firms. An important contribution of their work is to show that the empirical analysis of ownership and efficiency is highly sensitive to the cost specification. Raffiee et al. (1993) reach a different conclusion. Using a sample of 238 public water utilities and 33 private ones observed in 1989, they conclude in favour of private management. Bhattacharyya et al. (1994) use a survey conducted by the American Water Works Association and provide some evidence that public water utilities are on average more efficient than private ones. Using the same data set, but adopting a stochastic frontier approach, Bhattacharyya et al. (1995) show that small privately owned water utilities are

Table 21.2 Selected studies comparing the efficiency of public and private water services

Study	Area	Main result	Comments
Feigenbaum and Teeples (1983)	US	No difference	Hedonic cost model
Raffiee et al. (1993)	US	Private	Cost function estimated on 271 firms
Lynk (1993)	England and Wales	Public	Comparison before and after privatisation
Bhattacharyya et al. (1994)	US	Public	Translog cost function estimated on 225 public and 32 private water utilities
Bhattacharyya et al. (1995)	US	Public	Results depend upon the utility size
Saal and Parker (2000)	England and Wales	No difference	No impact on productivity of the privatisation in 1989
Estache and Rossi (2002)	Asia	No difference	Cost frontier approach on 50 firms in 19 different countries. No systematic relation between costs and production
Estache and Kouassi (2002)	Africa	Private	Stochastic frontier analysis
Seroa da Motta and Moreira (2006)	Brazil	No difference	Data envelopment analysis and stochastic frontier analysis
Saal and Parker (2004)	England and Wales	No difference	No impact on productivity of the privatisation in 1989
Picazo-Tadeo et al. (2007)	Spain	Private	Private firms outperform public utilities in their management of labour
Kirkpatrick et al. (2006)	Africa	No difference	Data envelopment analysis and stochastic frontier analysis on 76 African countries
Da Silva et al. (2007)	Brazil	No difference	Stochastic cost frontier estimated on 279 firms in year 2002
Munisamy (2009)	Malaysia	Private	Private entities outperform public entities in terms of technical efficiency
Romano and Guerrini (2011)	Italy	Public	Public owned companies purchase and use inputs in a better way compared with mixed ownership firms

nevertheless comparatively more efficient.[8] Estache and Rossi (2002) estimated a stochastic cost frontier on a sample of Asian and Pacific regional water companies, and found that efficiency was not significantly different for private and public firms. Seroa da Motta and Moreira (2006) focus on the productivity level of the sanitation sector in Brazil using a data envelopment analysis approach. They conclude that ownership does not matter for productivity gains in the sanitation sector. Since the private sector represents only 10 % of municipalities, their results should be taken with caution. Picazo-Tadeo et al. (2007) consider a sample of water utilities located in Spain (Andalusia). They did not find any difference in the average technical efficiency scores for privately and publicly own water services. However, they

[8]The stochastic cost frontier approach is attractive since it allows one to distinguish the impact on the firm's output of variation in technical efficiency from external stochastic error.

report that privately owned companies outperform public utilities in their management of labour. Kirkpatrick et al. (2006), considering African water utilities, also address the issue of ownership and its effect on performance of the sector. Their results show that there is no difference in the performance between privately owned utilities and publicly owned ones. Da Silva et al. (2007) assessed the cost efficiencies of Brazilian public and private water supply companies using a stochastic frontier approach. Statistical inference leads to the conclusion that there is no evidence that private firms and public firms are significantly different in terms of efficiency. Munisamy (2009) applied a data envelopment analysis approach on a sample of 11 Malaysian water utilities. The main result was that private water services outperform public ones in terms of technical efficiency. The private sector had an average overall technical efficiency score of 86 %, while the public sector efficiency score was 70 %. The source of inefficiency in the private sector is mainly its scale, while inefficiency in the public sector is related both to scale and to technically inefficiency. Lastly, Romano and Guerrini (2011) studied a sample of 43 publicly and privately own Italian water utility companies using a data envelopment analysis approach. Their main findings was that publicly owned firms have the highest efficiency scores, implying that publicly owned companies purchase and employ inputs in a much better way compared with mixed ownership firms.

A number of studies have investigated the change in efficiency after privatisation of this industry (particularly after it happened in England and in Wales). In most cases, the evidence showed limited effects of privatisation. Lynk (1993) reported that productivity increases following privatisation appeared to be less than the improvements in productivity that could be traced to the period immediately prior to privatisation. Similar results were obtained by Saal and Parker (2000); comparing the productivity of water services in England and Wales over time, they found that total productivity did not improve relative to the pre-privatisation period.

The general message from Table 21.2 is that there is no definitive evidence of private utilities outperforming public ones in terms of cost or technical efficiency. Rather than focusing on public management versus private management, the debate should be redirected on the form of regulation to be implemented when water services are privately operated or owned, and on changes in legislation that could improve technical and cost efficiency. Aubert and Reynaud (2005) showed for instance that the cost efficiency of US water utilities is directly and significantly affected by the type of economic regulation implemented by public authorities. In England and Wales, Saal and Parker (2004) reported evidence of a small increase in the rate of total factor productivity growth in following a substantial tightening of regulations in 1995. To improve cost efficiency, efforts should be made to promote competition (although direct competition is limited due to the cost structure in the industry), and to reduce the level of asymmetrical information between water services and public authorities in charge of regulation and control.

Table 21.3 Selected studies comparing prices of public and private water services

Study	Area	Main result	Comments
Carpentier et al. (2006)	France	Higher price for private	Treatment effect approach on a large sample of French water services
Andrés et al. (2008)	Latin America	Higher price for private	Performance of 49 water utilities in seven Latin American countries
Gassner et al. (2008)	Developing countries	No difference	977 water utilities located in various developing countries
Martínez-Espiñeira et al. (2009)	Spain	Higher price for private	Treatment effect approach on a sample of 53 urban municipalities
Ruester and Zschille (2010)	Germany	Higher price for private	765 German water suppliers

4.2 *Water Prices*

Assessing the impact of PPPs on the price of water is important since price is the main driver of water service affordability. As a result, a number of case studies have been conducted to assess the relationship between PPPs and water price. The consensus that seems to emerge is that prices often rise following introduction of PPPs.[9] In the absence of counterfactuals, the causal relationship between PPP and increase in water price is difficult to establish. A few published papers have adopted some econometric approaches to assess the differences in water price between privately and publicly managed water services. We report in Table 21.3 the mains insights from this literature.

Carpentier et al. (2006) have provided the first econometric analysis of water price in France with a special focus on the comparison between public and private pricing schemes. They report that the average domestic water price in 1998 was 33 % higher in the case of private management than in the case of a public one. However, Carpentier et al. (2006) show that French local authorities tend to delegate water services in cases of difficult operational conditions (complex water network, large seasonal population, high level of capital investment). Taking into account these more complex operational conditions, the water price is in fact 15 % higher in the case of a private management for small local authorities (less than 10,000 inhabitants) and only 5 % for large authorities (population greater than 10,000 inhabitants). A direct interpretation of this price differential remains difficult since the accounting practices for asset depreciation are not strictly the same in the case of public and private services. Andrès et al. (2008) compare the performance of 49 water utilities in seven Latin American countries, before and after the introduction of a private operator. They report that the introduction of a private operator usually resulted in tariff increases, which is in fact not surprising, because most Latin

[9] For Chile, Bitran and Valenzuela (2003) report for instance that water and sewerage rates increased by 40 % for privatised utilities compared to 20 % for non-privatised ones.

American utilities had tariffs below cost-recovery levels in the early 1990s. Gassner et al. (2008) used a large sample made of 977 water utilities located in various developing countries, of which 141 had some form of private participation. They conclude that there was no significant difference in tariff levels between PPPs and comparable public operators over the same period. Martínez-Espiñeira et al. (2009) explain differences in the average price of domestic water supply services in Spain, with special attention on the effects of privatisation on price. They used a treatment effect approach on a sample of 53 major urban municipalities. Taking into account the fact that municipalities do not randomly distribute themselves between public and private (one group having strictly public ownership and management and another where all or part of the service has been delegated to a private firm), they find that there is a significant positive effect of privatisation on water price.

Most recently, Ruester and Zschille (2010) investigated the impact of governance choice on firm performance using a database of 765 German water suppliers. They analysed the relationship between organisational form (i.e., private sector participation in water supply versus public service provision) and retail prices, controlling for economies of scale as well as technical and structural characteristics of the suppliers. They showed that private sector participation results in higher prices.

The general result from Table 21.3 is that private sector participation seems to be associated with higher prices. One should stress however that this correlation cannot be interpreted as causal, especially due to self-selection problems (Carpentier et al. 2006; Ruester and Zschille 2010). One possible explanation of higher prices in the case of private management could be the weak enforcement of regulations (Guasch et al. 2008), since quite often the economic regulation of delegated services is carried out directly by local communities.[10] First, due to information asymmetry, local communities are not always in a good position to exercise efficient control of water service providers. Due to long periods of water service delegation, some communities lack the technical expertise to evaluate the delegation contract. This is especially true for small communities with limited financial resources. Second, the main sanction available (non-renewal of the contract) may not be credible. Third, ex-ante competition between private operators, via the bidding process for the delegation contract, can be limited due the oligopolistic nature of this industry. The debate therefore should focus not directly on the price differential between private and public water services but on the factors contributing to this result.

4.3 *Access to Water and Sanitation Services*

As one might expect, most existing studies of access to water and sanitation services have focused on developing countries, even though in 2007 only 7 % of urban populations in such countries were served by formal PPPs (Marin 2009). In Table 21.4, we summarise the main findings of econometric studies comparing the performance

[10]This is for instance the case in France.

Table 21.4 Selected studies assessing the impact of PPPs on access to water services

Study	Area	Main result	Comments
Gomez-Lobo and Melendez (2007)	Colombia	Neutral impact of private	Neutral impact of private on access to piped water. Positive impact for sewerage connections
Gassner et al. (2008)	Developing countries	Positive impact of private in terms of growth in the number of connections	Sample of 977 water utilities located in various developing countries
Barrera-Osioro et al. (2009)	Colombia	Positive impact of private in urban areas but negative effect in rural areas	Sample of 82 municipalities. Pre and post privatisation data set
Clarke et al. (2009)	Argentina, Bolivia, and Brazil	Positive impact of private	The share of households connected to piped water and sewerage improves following the introduction of PPPs, but not in a way different to control regions

of public and private utilities in terms of the connection rate to water and sewerage services. The interested reader should also refer to Marin (2009) for a survey of PPPs in the water sector of developing countries.

Gomez-Lobo and Melendez (2007) used a living standards household survey in Colombia to assess the impact of private participation on the population's connection rate to water and sewerage services. They found that there was either no effect on water connection rates or a positive effect if the largest municipalities were dropped from the sample. Their results are less ambiguous for sewerage connections since the effects of private participation on sewerage connections are in almost all cases positive and statistically significant, especially among poorer communities. In Latin America, Andrés et al. (2008) found that introducing private participation did improve coverage, but they describe only a transitory improvement (1 year before and 2 years after the takeover). Gassner et al. (2008), using a sample of 977 water utilities in developing countries, found that private operators performed better when performance was measured by growth in the number of connections, both during the transition period to private operations and after. However, their results for residential coverage are inconclusive. Barrera et al. (2009) studied the effects of water privatisation on consumer welfare in Colombia. In terms of access, they found there was no evidence of a positive impact from privatisation, even after they controlled for the technical capacity of the municipalities. However, they did find that the effects of privatisation were not homogeneous across income and across areas. In particular, in rural areas privatisation had strong negative effects on access to water. Clarke et al. (2009) used household-level data to explore the effects of privatisation on water and sewerage coverage in Latin America (Argentina, Bolivia, and Brazil). Their analysis revealed that, in general, the share of households connected to piped water and sewerage improved following the introduction of PPPs, a result consistent with the case study literature also reported in that paper.

The same authors mention, however, that the share of households connected improved by the same amount in control regions, suggesting that private sector participation, *per se*, may not have been responsible for the improvements. Results were similar when they looked only at the poorest households.

5 Conclusion

Over the last decades debate on ownership and PPPs in the water industry has been dominated by dogmatic arguments. The main conclusion of this chapter is that there is little empirical justification for a general presumption in favour of a type of ownership or PPP. Case-by-case evaluations of the various trade-offs need to be conducted. In fact, when trying to relate the performance of a water utility with its ownership or with the form of its PPP, studies simply provide an indirect test of the trade-off between two sources of inefficiency in publicly owned and privately owned water utilities – inefficiencies from attenuation of property rights in publicly owned water services, and inefficiencies from conflicts between the regulator's and the shareholders' goals in privately owned utilities. This result is in line with some previous surveys assessing the efficiency of PPPs in other industries. For urban infrastructure, Koppenjan and Enserink (2009) conclude that proper governance practices are required to reconcile private sector participation with the objective of increasing sustainability of the urban environment. A similar conclusion emerges from Reich (2002), who has summarised the existing literature on PPPs in the health sector, and from Falch and Henten (2008) who discuss the role played by PPPs in the telecommunication sector.

From the above discussion, no clear and definitive answer for the impact of PPPs on water utility performance and consumer benefit seems to emerge. Following Vickers and Yarrow (1988), we conclude that "There is little empirical justification for a general presumption in favour of either type of ownership, and case-by-case evaluation of the various trade-offs is therefore in order". Hence, the impact of PPP on water and sanitation utility performance will depend mainly on local conditions, including efficiency of regulation.

References

Andrés, L., Guasch, J. L., Haven, T., & Foster, V. (2008). *The impact of private sector participation in infrastructure: Lights, shadows and the road ahead* (Latin American Development Forum Series). Washington, DC: World Bank.

Aubert, C., & Reynaud, A. (2005). The impact of regulation on cost efficiency: An empirical analysis of Wisconsin water utilities. *Journal of Productivity Analysis, 23*, 383–409.

Averch, H., & Johnson, L. L. (1962). Behavior of the firm under regulatory constraint. *The American Economic Review, 52*(5), 1052–1069.

Barrera-Osorio, F., Olivera, M., & Ospino, C. (2009). Does society win or lose as a result of privatisation? The case of water sector privatisation in Colombia. *Economica, 76*(304), 649–674.
Bel, G., & Warner, M. (2008). Does privatisation of solid waste and water services reduce costs? A review of empirical studies. *Resources Conservation and Recycling, 52*(12), 1337–1348.
Bel, G., & Warner, M. (2010). Partial privatisation in local services delivery: An empirical analysis on the choice of mixed firms. *Local Government Studies, 36*(1), 129–149.
Bhattacharyya, A., Parker, E., & Raffiee, K. (1994). An examination of the effect of ownership on the relative efficiency of public and private water utilities. *Land Economics, 70*(2), 197–209.
Bhattacharyya, A., Harris, T. R., Narayanan, R., & Raffiee, K. (1995). Specification and estimation of the effects of ownership on the economic efficiency of the water utilities. *Regional Science and Urban Economics, 25*, 759–784.
Bitran, G. A, & Valenzuela, E. O. (2003). *Water services in Chile: Comparing private and public performance*. Public policy for the private sector 255. Washington, DC.: World Bank.
Carpentier, A., Nauges, C., Reynaud, A., & Thomas, A. (2006). Effets de la délégation sur le prix de l'eau potable. Une analyse à partir des résultats de la littérature sur les effets de traitement. *Economie Prévision, 174*(3), 1–19.
Chong, E., Huet, F., Saussier, S., & Steiner, F. (2006). Public–private partnerships and prices: Evidence from water distribution in France. *Review of Industrial Organization, 29*(1/2), 149–169.
Clarke, G. R. G., Kosec, K., & Wallsten, S. (2009). Has private participation in water and sewerage improved coverage? *Journal of International Development, 21*(3), 327–361.
Da Silva e Souza, G., De Faria, R. C., & Moreira, T. B. S. (2007). Estimating the relative efficiency of Brazilian publicly and privately owned water utilities: A stochastic cost frontier approach. *Journal of the American Water Resources Association, 43*, 1237–1244.
Duroy, S. (1996). *La distribution d'eau potable en France, contribution à l'étude d'un service public local*. Paris: Librairie générale de droit et de jurisprudence.
Estache, A., & Kouassi, E. (2002). *Sector organization, governance, and the inefficiency of African water utilities*. Policy research working paper 2890. Washington, DC: World Bank.
Estache, A., & Rossi, M. (2002). How different is the efficiency of public and private water companies in Asia? *World Bank Economic Review, 16*(1), 139–148.
Falch, M., & Henten, A. (2008). Investment dimensions in a universal service perspective: Next generation networks, alternative funding mechanisms and public–private partnerships. *Info, 10*(5/6), 33–45.
Feigenbaum, S., & Teeples, R. (1983). Public versus private water delivery: A hedonic cost approach. *The Review of Economics and Statistics, 65*, 672–678.
Gassner, K., Popov, A., & Pushak, N. (2008). *Does private sector participation improve performance in electricity and water distribution? An empirical assessment in developing and transition countries* (PPIAF trends and policies series). Washington, DC: PPIAF.
Gomez-Lobo, A., & Melendez, M. (2007). Social policies and private sector participation in water supply: The case of Colombia. In N. Prasad (Ed.), *Social policies and private sector participation in water supply: Beyond regulation*. Houndmills: Palgrave Macmillan.
González-Gómez, F., & García-Rubio, M. A. (2008). Efficiency in the management of urban water services. What have we learned after four decades of research? *Revista de Economía Pública, 185*(2), 39–67.
Guasch, J. L., Laffont, J. J., & Straub, S. (2008). Renegotiation of concession contracts in Latin America: Evidence from the water and transport sectors. *International Journal of Industrial Organization (special issue on PPPs), 26*, 421–442.
Hunt, L. C., & Lynk, E. L. (1995). Privatisation and efficiency in the UK water industry: An empirical analysis. *Oxford Bulletin of Economics and Statistics, 57*(3), 371–388.
Jordan, J. (1998). An introduction to water: Economics concepts, water supply and water use. In *Georgia water series issue 1*. Athens: University of Georgia.
Kirkpatrick, C., Parker, D., & Zhang, Y. F. (2006). An empirical analysis of state and private sector provision of water services in Africa. *World Bank Economic Review, 20*(1), 143–163.

Klein, M. (1996a). *Competition in network industry*. Policy research working paper 1591. Washington, DC: World Bank.
Klein, M. (1996b). *Economic regulation of water companies*. Policy research working paper 1649. Washington, DC: World Bank.
Koppenjan, J. F. M., & Enserink, B. (2009). Public–private partnerships in urban infrastructures: Reconciling private sector participation and sustainability. *Public Administration Review, 69*(2), 284–296.
Levin, J., & Tadelis, S. (2010). Contracting for government services: Theory and evidence from U.S. cities. *The Journal of Industrial Economics, 58*(3), 507–541.
Lynk, E. (1993). Privatisation, joint production and the comparative efficiencies of private and public ownership: The UK water industry case. *Fiscal Studies, 14*, 98–116.
Marin, P. (2009). *Public–private partnerships for urban water utilities: A review of experiences in developing countries* (Trends and policy options series, no. 8). Washington, DC: PPIAF.
Martinez Espiñeira, R., García Valiñas, M. A., & González Gómez, F. (2009). Does private management of water supply services really increase prices? An empirical analysis in Spain. *Urban Studies, 46*(4), 923–945.
Maskin, E., & Tirole, J. (2008). Public–private partnerships and government spending limits. *International Journal of Industrial Organization, 26*(2), 412–420.
Munisamy, S. (2009). Efficiency and ownership in water supply: Evidence from Malaysia. *International Review of Business Research Papers, 5*(6), 148–260.
Pérard, E. (2009). Water supply: Public or private? An approach based on cost of funds, transaction costs, efficiency and political costs. *Policy and Society, 27*, 193–219.
Perold, A. (2004). The capital asset pricing model. *The Journal of Economic Perspectives, 18*(3), 3–24.
Picazo-Tadeo, A. J., Sáez-Fernández, F. J., & González-Gómez, F. (2007). The role of environmental factors in water utilities' technical efficiency Empirical evidence from Spanish companies. *The Journal of Applied Economics*. doi:10.1080/00036840601007310.
Picazo-Tadeo, A. J., González-Gómez, F., & Sáez-Fernández, F. J. (2009). Accounting for operating environments in measuring water utilities' managerial efficiency. *The Service Industries Journal, 29*, 761–773.
Raffiee, K., Narayanan, R., Harris, T. R., Lambert, D., & Collins, J. M. (1993). Cost analysis of water utilities: A goodness-of-fit approach. *The Atlantic Economic Journal, 21*(3), 18–29.
Reich, M. R. (Ed.). (2002). *Public–private partnerships for public health. Harvard series on population and international health*. Cambridge, MA: Harvard University Press. ISBN 0-674-00865-0.
Romano, G., & Guerrini, A. (2011). Measuring and comparing the efficiency of water utility companies: A data envelopment analysis approach. *Utilities Policy, 19*(3), 202–209.
Ruester, S., & Zschille, M. (2010). The impact of governance structure on firm performance: An application to the German water distribution sector. *Utilities Policy, 18*(3), 154–162.
Saal, D., & Parker, D. (2000). The impact of privatisation and regulation on the water and sewerage industry in England and Wales: A translog cost function model. *Managerial and Decision Economics, 21*, 253–268.
Saal, D., & Parker, D. (2004). The comparative impact of privatisation and regulation on productivity growth in the English and Welsh water and sewerage industry, 1985–99. *International Journal of Regulation and Governance, 4*(2), 139–170.
Saal, D. S., Parker, D., & Weyman-Jones, T. G. (2007). Determining the contribution of technical change, efficiency change and scale change to productivity growth in the privatized English and Welsh water and sewerage industry: 1985–2000. *Journal of Productivity Analysis, 28*, 127–139.
Seroa da Motta, R., & Moreira, A. (2006). Efficiency and regulation in the sanitation sector in Brazil. *Utilities Policy, 14*, 185–195.
Vickers, J., & Yarrow, G. (1988). *Privatisation: An economic analysis* (MIT Press Series on the Regulation of Economic Activity, Vol. 18). Cambridge: MIT Press.
Vining, A., & Boardman, A. (2008). Public–private partnerships in Canada: Theory and evidence. *Canadian Public Administration, 51*(1), 9–44.
Walker, S. & Marr, S. (2002). Study on the application of the competition rules to the water sector in the European Community. Report for the European Commission – Competition Directorate General. 139 pages.

Chapter 22
Issues of Governance, Policy, and Law in Managing Urban–Rural and Groundwater–Surface Water Connections

Rebecca Nelson, Heather N. Bischel, Richard G. Luthy, and Barton H. Thompson Jr.

1 Introduction

Historically, water law has taken a 'silo' view of water management, treating groundwater as separate from surface water, and often overlooking potential connections between urban or agricultural water and wastewater systems. This chapter investigates two major aspects. First, how modern water laws and institutions can facilitate conjunctive management of groundwater and surface water sources, both by recognising the impacts that pumping groundwater has on surface water and, in

R. Nelson (✉)
Stanford Woods Institute for the Environment, Stanford University, Stanford, CA 94305, USA

United States Studies Centre at the University of Sydney, Institute Building H03/City Road, Sydney, NSW 2006, Australia
e-mail: rebeccalouisenelson@gmail.com

H.N. Bischel
École Polytechnique Fédérale de Lausanne, EPFL ENAC IIE LCE, CM 1 119 (Centre Midi) Station 2, CH-1015 Lausanne, Switzerland
e-mail: heather.bischel@epfl.ch

R.G. Luthy
Department of Civil and Environmental Engineering, Yang & Yamazaki Environment and Energy Building, 473 Via Ortega - Room 191, Stanford, CA 94305-4020, USA
e-mail: luthy@stanford.edu

B.H. Thompson Jr.
Law School and Stanford Woods Institute for the Environment, Stanford University, Stanford, CA 94305, USA
e-mail: buzzt@law.stanford.edu

Q. Grafton et al. (eds.), *Understanding and Managing Urban Water in Transition*, Global Issues in Water Policy 15,
DOI 10.1007/978-94-017-9801-3_22

the context of urban supply, by providing frameworks for undertaking managed aquifer recharge. Second, it explores how law can support interchange between urban and rural water supplies, both through water trading and exchange as well as through facilitating agricultural use of urban wastewater.

These interactions are examined as they occur in a variety of communities and over a diversity of laws and projects in California. Each section of the chapter outlines California's legal and regulatory approach, examples of innovative water programs and projects that illustrate that approach, and key issues that can arise in designing and implementing such innovations. These approaches and programs can provide inspiration to other jurisdictions that share key characteristics with California, such as: complex and highly managed natural and engineered hydrological systems; intricate hydro-political arrangements with cultural preferences for local rather than state-centralised water management; evolving technology; and challenges of population growth, water scarcity, and heightened recognition of environmental demands.

A common theme – the importance and value of more integrated water planning – emerges from the eight major examples of developed and emerging innovative approaches and projects presented (see Fig. 22.1). The water planning paradigm, especially where it counsels significant local involvement, has become accepted internationally over the past several decades as a way of ensuring rational, proactive water management, particularly in areas with great water supply variability, like California (Caponera 2007). This paradigm contrasts with an historical approach in which projects were implemented (and natural resources managed) without coordination, which invited potential for conflict and inefficient allocation of resources.

In California, statute-based groundwater management plans (GWMPs) and integrated regional water management plans (IRWMPs) can help water agencies to recognise and manage interactions among substantive elements of water systems. Importantly, they can also help to forge institutional connections. This is evident even though water agencies are compelled to adopt neither IRWMPs nor GWMPs. Accordingly, in addition to demonstrating how law can facilitate particular interactions, the chapter underscores the value of a planning approach to tackling the complexity of integrated water management in urban settings.

2 Interactions Between Groundwater and Surface Water

2.1 *Background*

Though it is rarely apparent to the naked eye, water below the ground surface is frequently connected to water in rivers, lakes, wetlands, and infiltration areas at the surface. Pumping water from aquifers can therefore 'pull in' water from, and deplete, connected surface water bodies over days to decades, depending on local geological conditions (Winter et al. 1998). Further, greater impermeable paved

Fig. 22.1 Locations discussed as examples of interactions between groundwater and surface water, or urban and rural water systems

spaces in urban areas prevent the infiltration of rainfall to aquifers and can decrease the baseflow component of streamflow by lowering urban water tables (Otto et al. 2002).

In planning, developing, and monitoring their water supplies, urban water suppliers need to be aware of these connections to protect their own and others' water sources from unanticipated depletion caused by pumping. These issues are especially important in California, where municipal water utilities source around one-fifth of their water from groundwater (Kenny and USGS 2009) and where groundwater pumping is identified as the largest single source of future water supplies (Hanak 2005). It is estimated that groundwater contributes 27–60 % of total annual stream flows (Howard and Merrifield 2010).

Traditionally, most western U.S. water law and policy has largely ignored the potential for groundwater pumping to deplete streamflow. Although California law retains a traditional preference for recognising this interaction to a limited degree, the Carmel River case, described in Sect. 2.2.2, shows that even this narrow legal recognition can significantly affect urban water suppliers. The state pursues more comprehensive recognition of groundwater–surface water and groundwater–infiltration area interactions by encouraging local water suppliers who use groundwater, such as the Zone 7 Water Agency introduced in Sect. 2.2.3, to voluntarily adopt plans that recognise these connections.

In addition to managing the potential for wells to negatively impact streams, and the potential for paved areas to reduce recharge, water resource planning approaches can identify valuable opportunities to use the immense storage capacity of subterranean aquifers to store surface waters. Precipitation and infiltration to groundwater from rivers and streams naturally replenish groundwater resources (de Vries and Simmers 2002), as does incidental infiltration from human activities (Bouwer 2002). The natural storage capacity of groundwater aquifers can also be harnessed intentionally using managed aquifer recharge (MAR), providing "engineered" storage for future use.

The urban environment can provide diverse water sources for MAR including high quality drinking water, captured stormwater, and recycled water (a term used interchangeably here with reclaimed water). This surface water, placed in basins, furrows, or ditches with permeable surface soils, infiltrates into these underground storage systems, artificially recharging groundwater. When permeable surface soils are not available, direct injection of surface water through wells can bypass this constraint (Bouwer 2002). Alternatively, agencies can conduct indirect, or 'in lieu', groundwater storage by foregoing pumping groundwater, instead using excess surface water or re-operating reservoirs. They account for and later access the resulting 'stored' groundwater when insufficient surface water is available (Purkey and Mansfield 2002). This method also has the advantage of avoiding costs and technical difficulties that may be involved in direct MAR.

MAR complements and offers some advantages over traditional surface water storage behind dams. For example, it saves water by reducing evaporation during storage; avoids the environmental impacts involved in enlarging or constructing new surface reservoirs; and, by raising water tables, reduces pumping costs. Further, the treatment capacity of soils can enhance the quality of injected sources through soil-aquifer treatment (Bekelea et al. 2011). Recharging aquifers can also reduce seawater intrusion and land subsidence, and facilitate water conveyance (Bouwer 2002).

Reclaimed urban wastewater is increasingly used for MAR as a seawater intrusion barrier and to supplement drinking water supplies (NRC 2012). When used for drinking water supplies, the practice is commonly termed indirect potable reuse (IPR) because the recycled water passes through the aquifer prior to extraction for consumption. IPR has been practiced for over 30 years, with projects currently in place in the U.S.A., Europe, and Singapore (Rodriguez et al. 2009), though it remains relatively rare in planned water management. In the U.S.A., where environmental buffers between the recycled water source and the potable reuse end point are important features of projects constructed over the last several decades, the most common practices involve surface spreading of recycled water using recharge basins and direct injection into the subsurface (NRC 2012). Both processes involve the water passing through the subsurface, typically with greater than 6 months retention time, prior to withdrawal.

In California, MAR projects using treated wastewater are well established. Orange County Water District operates a widely lauded facility using treated wastewater, as Sect. 2.3.2 describes. However, the state lacks a comprehensive legal framework for these activities, in particular for 'water banking' programs that involve multiple parties engaged in recharging, storing, withdrawing, and selling water. Accordingly, Californian water agencies have developed complex legal arrangements to enable water suppliers to bank and later recover water in subsurface facilities operated by others, as explained in the Kern Water Bank case described in Sect. 2.3.3.

2.2 *Managing the Impacts of Pumping Groundwater on Surface Waters*

2.2.1 The Californian Regulatory Approach: From Subterranean Streams to Groundwater Management Plans

The potential for groundwater pumping to deplete connected streams arises when a decision-maker considers whether to grant (or allow a modification to) a permit for groundwater extraction, or whether to curtail an existing groundwater use that is interfering with a surface water use that has a higher legal priority. The water laws of many western U.S. states recognise that groundwater is connected to surface water, but they differ radically in the degree of connection that they recognise (in the context of making permitting and curtailment decisions). Some states, like Washington, adopt the so-called 'one molecule rule' under which any ascertainable impact of pumping groundwater on surface water is legally recognised (*Postema v. Pollution Control Hearings Board* 2000). Any impact above this low level can justify imposing conditions on a permit, or restricting pumping that is later found to cause interference with an existing surface water right. At the opposite end of the spectrum, California restricts the pumping of only groundwater that is very closely connected to surface water. California law deems "subterranean streams flowing through known and definite channels" – an antiquated legal notion rooted in historical misunderstandings of the nature of groundwater – to be equivalent to surface water (North Gualala Water Company v. SWRCB 2006), and to require a permit from the State Water Resources Control Board (SWRCB) for extraction. Such pumping may be restricted if it interferes with pre-existing surface water rights. By contrast, a well that taps other ('percolating') groundwater requires no permit from the SWRCB, and in the absence of rare local rules, pumping from such wells escapes regulation (Little Hoover Commission 2010). Rather, where groundwater–surface water connections are too weak to warrant the 'subterranean stream' label, the state deals with these connections by encouraging local agencies to consider them under groundwater management plans (GWMPs).

The Ground Water Management Act – California Water Code (CWC) §§10,750–10,767 – establishes the state's framework for local water agencies to adopt GWMPs. Some GWMPs adopted by urban water agencies operate at a regional level through collaborations between agencies; others address only the local issues of a single agency (Nelson 2011), though they may link communities and water managers through advisory committees appointed to assist in the preparation of a GWMP (CWC §10753.4), as demonstrated in the Alta Irrigation District example outlined in Sect. 3.2.3.

A GWMP may cover a range of matters, including mitigating conditions of overdraft, replenishing extracted groundwater, monitoring groundwater, facilitating conjunctive use operations, constructing and operating groundwater recharge projects, and identifying recharge areas (CWC §10753.8). Importantly, to receive state funding, a GWMP must also include a component relating to monitoring and managing groundwater–surface water interaction (CWC §10753.7(a)(1)). Although the power of GWMPs is limited – they are not mandatory, do not have binding legal effect, and do not affect water rights – this component encourages water utilities to consider whether their groundwater pumping may adversely affect streams. Given California law's traditional reluctance to recognise such effects, the state's planning-based approach to this issue represents a significant policy step.

GWMPs may also be used to link overland flow, land use, and groundwater, which, in traditional legal frameworks, has proven to be problematic – particularly in relation to the influence of paved areas on recharge (Thompson 2011; Davies 2007). Other connections between groundwater and land use are comparatively better regulated: groundwater protection programs developed jointly by states and the USEPA consider the impacts of land uses on groundwater quality, and 'assured water supply' laws link land use and water demand by requiring developers to prove that adequate water is available for development, before construction begins (Thompson 2011; Davies 2007).

California links land use, overland flow, and groundwater recharge through a creative addition to its groundwater planning framework. From 2013, California requires GWMPs to map recharge areas in the subject groundwater basin, and describe how these areas replenish the basin (CWC §10753.7(a)(1), (4)). Local agencies may request state funds for recharge mapping. Though agencies without a GWMP need not carry out recharge mapping, and no agency is mandated to take any specific action to protect important recharge areas, the new mapping requirement will provide a necessary tool for cities and water supply agencies to take protective action. Section 2.2.3 describes the early adoption of this mapping approach by the Zone 7 Water Agency in northern California.

2.2.2 Curtailing Groundwater Pumping to Protect Surface Waters: California American Water and the Carmel River

Drawing the legal line between a connected 'subterranean stream' and supposedly unconnected 'percolating groundwater' can profoundly impact Californian urban water providers that install wells adjacent to a stream in order to avoid the costs of treating lower-quality water diverted directly from the stream (e.g., North Gualala Water Company v. SWRCB 2006). Compared to pumping further away from the stream, this practice greatly increases the probability of depleting the stream. This supply strategy was the root of 'cease and desist' orders that the SWRCB issued in 1995 and 2009 to California American Water (Cal-Am), a private water utility that supplies urban water needs on the Monterey Peninsula (SWRCB 2009) using 69 % groundwater.

In 1995, the SWRCB found that flows in the Carmel River ceased for 5–6 months of the year due to Cal-Am pumping from a subterranean stream associated with the Carmel River, causing damage to public trust resources – riparian habitat, associated wildlife, and fisheries. Legally, Cal-Am required a SWRCB permit to pump 10,730 acre-feet/yr (0.42 m^3/s) from the wells, which it lacked. The SWRCB ordered Cal-Am to cease its unlawful pumping by obtaining other source water, to implement an urban water conservation plan, and, in the interim, to mitigate the effects of its pumping (for example, by favoring pumping from its most downstream wells to maximise fish habitat; SWRCB 1995). In 2009, the SWRCB found that Cal-Am had not diligently implemented the required actions and had continued to pump without a valid water right, causing ongoing damage. It ordered Cal-Am to cease its unlawful pumping by the end of 2016, in eight annual phases that involve undertaking a series of demand management and small supply augmentation measures, such as MAR and desalination (SWRCB 2009).

The significant nature of the reductions required in the Carmel River case, and the repeated state action, show that although California law recognises the impacts of pumping groundwater on rivers in only a narrow set of circumstances; this mechanism can powerfully protect affected rivers, even when core urban water supplies are at issue. Comprehensive water planning approaches that address groundwater–surface water interactions could avoid such costly scenarios by facilitating pre-emptive consideration of potential stream impacts.

2.2.3 Managing Surface Water–Land use–Groundwater Links Under Groundwater Management Plans

In contrast to the legal mandates that apply to subterranean streams, GWMPs represent a voluntary, albeit statute-supported, approach to recognising groundwater–surface water interaction. First, they facilitate goal-setting, monitoring, and quantification in relation to stream–aquifer interactions (Nelson 2011). Second – and the focus of this section – new recharge mapping requirements encourage agencies to collect basic information to protect recharge areas and maintain links between land use, overland flows, and groundwater. Innovative tools in leading existing GWMPs, like that of the Zone 7 Water Agency (Zone 7), demonstrate how local agencies could use the new recharge maps.

Zone 7 is an independently governed sub-district of the Alameda County Flood Control and Water Conservation District in northern California. It is a flood manager and water wholesaler for three cities overlying the Livermore–Amador Valley basin: Livermore, Pleasanton, and Dublin, with a combined population of around 190,000. Zone 7 supplies local groundwater and imported surface water to water retailers, and also manages groundwater extractions by the retailers.

Zone 7 uses an extensive land use monitoring program to assess the potential for changes in land use to affect recharge and groundwater quality and to keep its water balance and salinity models up to date (Katen 2012). It reviews land use plans at the city and county level to identify three categories of land uses (urban, agriculture, and gravel mining) and three types of applied water (groundwater, surface water, and reclaimed water), in order to model salt loads and recharge to the basin. It also monitors land use using annual aerial photography, interviews with landowners, and field observations, compiling this information in a monthly land use change report. The report forms the basis for communicating concerns to city land use planning agencies on developments that may impact the regional water supply or groundwater basin (Jones and Stokes 2005). At the time of writing, Zone 7 is also developing a vulnerability map to better assess the potential for groundwater contamination based on land use, and is converting its maps and reports to GIS format to aid analysis. The success of the program is evident in that city agencies routinely refer developments to Zone 7 for comment – a relatively rare example of coordination between urban planning and water agencies. They do so (in part) in recognition of Zone 7's active interest in land use, which it demonstrates through its monitoring program (Katen 2012). Key challenges to the program include maintaining funding and dealing with gaps in historical data, which make it difficult to accurately calibrate the hydrological models that rely on the data (Katen 2012).

This local planning approach promises to deliver valuable tools for managing the impacts of land use change on the infiltration of diffuse surface water to aquifers. Zone 7's innovative land use monitoring scheme helps establish institutional interactions to connect the management of land and groundwater, as well as providing substantive scientific information on these interactions. It augurs well for the benefits of recharge mapping, soon to be implemented more widely through GWMPs.

2.2.4 Outstanding Challenges

Commentators have repeatedly criticised the narrowness of California's legal integration of groundwater and surface water for allowing undue interference with surface water rights and instream flows (Thompson 2011; Hanak et al. 2010; Sax 2002). Arguably the most important challenge facing California's legal framework for linking groundwater and surface water is ensuring that the law better reflects the physical impacts of groundwater pumping, which extend beyond the narrowly-defined concept of a 'subterranean stream'. However, as the Carmel River example shows, doing so can have significant urban water supply – and therefore economic – impacts. Expanding the reach of the law will therefore encounter strong political resistance in a state that relies heavily on groundwater. This is to put to one side the legal difficulties that would be involved in making such a change, which would require integrating the management of resources that currently are governed under very different legal doctrines.

To create the political space for adopting laws to better integrate groundwater and surface water, governments also must adopt policies to manage and reduce the economic impacts of doing so. Some western U.S. states, such as Washington, Idaho, Colorado, and New Mexico, reduce the economic consequences of better integrating groundwater and surface water by permitting – conditional on offsetting the depletion to the stream – a groundwater pumper to withdraw groundwater that would otherwise unlawfully deplete a connected stream. Offset programs can be very complex, aiming not only to replace the volume of streamflow removed by groundwater pumping, but also to replace it at the time that the groundwater pumping has effect and address impacts to stream water quality and temperature. Offset methods include purchasing, or leasing and retiring, or not using a surface water right; conserving water; or 'pumping and dumping' water from an unconnected source into the affected stream (Nelson 2012). Municipal suppliers that prefer to pump high quality groundwater to avoid the cost of treating poorer quality surface water can benefit from such programs: they can maintain access to their preferred water supply while complying with requirements to reduce their impacts on connected surface waters and on the holders of water rights in those waters. California could consider implementing a formal offset scheme to increase the political viability of better connecting groundwater and surface water systems, and offer groundwater pumpers a clear and certain market-based approach to offset the impacts of groundwater pumping.

In contrast to its integration of groundwater and surface water in law, strongly-held principles of localism and voluntarism characterise California's water planning-based approach to linking groundwater and surface water. GWMPs are not mandatory. Their formation and implementation relies on a large number of autonomous local agencies coming together and maintaining a high degree of cooperation with the support of state funds. Though successful instances of this certainly occur, maintaining cooperation between agencies and local stakeholder entities – through sometimes lengthy consultation and implementation processes – is a noted challenge

(Stanford Water in the West 2012). Questions over the continued availability of state bond funds for water projects also threaten GWMP programs.

2.3 *Using Managed Aquifer Recharge for Urban Water Supply*

2.3.1 The Californian Regulatory Approach: Special Purpose Institutions and Water Quality

Some western U.S. states, for example Arizona, New Mexico, Oregon, and Washington, have developed regulatory frameworks for MAR, which involve issuing permits for recharge, storage, and recovery activities and establishing reporting requirements. California is yet to regulate MAR in this broad sense, though California courts have confirmed that public agencies have the basic right to augment water storage underground and withdraw the amount stored, minus losses (Foley-Gannon 2008). California legislation also provides for the establishment of local districts to carry out MAR (Hanak et al. 2011; CWC §§60,000-60,622).

Broader legal frameworks can facilitate MAR in numerous ways: by protecting the rights of water storers to prevent others from pumping the stored water; by protecting third parties and the environment from potential adverse impacts of the MAR activities; and by clarifying control over aquifer storage rights. Established frameworks can also reduce the transaction costs of undertaking complex MAR projects that involve multiple agencies, or aquifers that are connected to one or more other aquifers or surface waters. In such cases, accounting for the stored water is complicated by the need to coordinate between multiple parties, and to consider the timing and magnitude of natural losses (Contor 2009). Californian water agencies and scholars have identified a lack of legal certainty in these and other areas as a barrier to developing MAR projects to their full potential (see, e.g., Foley-Gannon 2008; GEI Consultants 2009; Thomas 2001, 2005; Water Transfer Workgroup 2002). Indeed, the U.S. National Research Council has found that MAR projects "are among the most complex to implement unless a state has addressed these issues in a statutory scheme that was created specifically for the regulation of these projects" (NRC CSUSRW and NRC WSTB 2008). Despite the lack of a broad regulatory or legal framework for MAR, local agencies in California have significant experience in developing MAR projects. The state also funds, and encourages local agencies to construct, recharge facilities under GWMPs and IRWMPs (CWC §§79,560-79,565; California Department of Water Resources DWR 2012).

California does regulate one element of MAR – albeit unevenly between water input methods – the potential water quality impacts of MAR. Both state and federal water standards apply to source water used for injection (42 U.S. Code Service §§300 h to 300 h-8; CWC §§13,260(a)(3), 13,263). In California, the point of water quality monitoring for MAR (using direct injection as an input method) is deemed to be at the injection well prior to movement of the water through the subsurface, which therefore calls for highly treated source water (Drewes et al. 2010).

In contrast to injection, water quality for MAR using infiltration ponds is not regulated at the federal level, and is not specifically regulated by California. In theory, if percolating low-quality water threatened the achievement of regional water quality objectives, California's regional water quality control boards could regulate percolation activities for MAR by imposing waste discharge requirements on operators (CWC §13,263). However, there is no evidence that the regional boards are doing this in a systematic way. Water quality monitoring for surface spreading applications occurs following extraction from the groundwater aquifer (Drewes et al. 2010). On a favorable interpretation, this implicitly recognises water quality improvement in the vadose zone and thus may allow comparatively lower quality source water to be used. A less favorable interpretation is that these rules have been adopted in a piecemeal way, leading to unintended inconsistency.

Some Californian MAR projects have used, for future potable use, highly treated wastewater effluent as a water source. In the U.S., states develop indirect potable reuse (IPR) water quality standards with no federal regulations governing the practice (USEPA 2004). Since protecting public health is the highest priority when MAR employs recycled water for potable reuse, advanced treatment of wastewater typically is essential prior to surface spreading operations or direct injection to the groundwater aquifer. In California, recycled water safety is regulated by the California Department of Public Health (CDPH). CDPH regulations evolve as more information on public health risks becomes available (USEPA 2004). At the time of writing, CDPH Draft Regulations for Groundwater Replenishment with Recycled Water are under public review (CDPH 2011). These regulations will provide detailed standards for groundwater recharge projects in California. The draft regulations group types of projects according to whether they involve surface spreading, full advanced treatment, or subsurface application, the latter requiring full advanced treatment at the point of injection.

2.3.2 Reusing Water for Drinking and Preventing Seawater Intrusion: The Orange County Water District

The Orange County Water District (OCWD) is perhaps California's most celebrated example of IPR. Migrating inland through the sandy, fresh water aquifer, seawater threatened the groundwater resources of coastal cities served by the OCWD. Such was the level of concern with the situation that the California legislature passed special legislation to create the OCWD and endow it with considerable powers to deal with the groundwater situation. The OCWD may construct works to replenish the basin, import water for this purpose, and charge fees to local groundwater pumpers to recover the costs of these activities (CWC App. 40).

From 1976 to 2004, the OCWD's advanced wastewater treatment and water reclamation facility, Water Factory 21, successfully operated a hydraulic barrier system by injecting treated wastewater into the subsurface (NRC 2012). In creating a seawater barrier, Water Factory 21 also implemented the first direct injection of reclaimed water for potable reuse. The injection water traveled both towards the ocean and inland to the main groundwater basin, where it was extracted for domestic and irrigation supply wells (Mills and Watson 1994). The injection water comprised two parts highly treated reclaimed water to one part groundwater obtained from a deep aquifer (NRC 2012).

After closing Water Factory 21 in 2004, the OCWD upgraded its IPR project to a purification process that it calls the Groundwater Replenishment System, increasing potable reuse capacity from 16 million gallons per day (61,000 m^3/day) to 70 million gallons per day (260,000 m^3/day). In its current operation, secondary treated wastewater from the Orange County Sanitation District is further purified by a series of advanced technologies – microfiltration, reverse osmosis, and advanced oxidation – prior to injection in the subsurface. The underground retention time of injected water prior to subsequent extraction is more than 6 months. The OCWD service area, where groundwater withdrawals comprise about 70 % of the water supply, provides water to a population of more than 2.3 million (NRC 2012; Rodriguez et al. 2009).

Facilitated by its special legal status and powers, the long history and large scale of the OCWD's IPR operations represent a significant success in an often contentious area of urban water supply.

2.3.3 Water Banking Using MAR: The Kern Water Bank

Like the OWCD, the Kern Water Bank takes advantage of groundwater–surface water connections, using MAR to store excess surface water underground. The Kern Water Bank is the largest of several water banking projects on the Kern Fan, an alluvial deposit in the southern Central Valley of California, and one of the largest such projects in the world (Kennedy/Jenks Consultants 2011). It is operated by the Kern Water Bank Authority, a local public agency that has six water districts as members. The Kern County Water Agency is one such member (Kern Water Bank Authority 2012). Among the Agency's functions is wholesaling water to retailers serving the city of Bakersfield.

In the absence of a legal framework for MAR, the Kern Water Bank relies on a complex contractual infrastructure between members, non-member adjoining entities, and outside agencies, which can purchase stored water (Thomas 2001). These arrangements cover matters such as reserving basin capacity for local recharge; accounting for losses caused by migration of the stored water; monitoring groundwater conditions; maintaining buffer areas around recovery wells; setting permissible rates of recovery; and compensating overlying landowners for adverse impacts caused by operating the Bank (Thomas 2001). As a member, the Kern County Water Agency has a right to be offered water for sale before it is offered for sale to non-members, and a first right to extract the stored water (Thomas 2001).

Considerable physical infrastructure is required to operate the Kern Water Bank: 72 recovery wells, with an annual recovery capacity of 300,000 acre-feet (around 370 million m^3); and 20,500 acres (8,296 ha) of land used for recharge, canals for conveyance to the percolation ponds, and habitat areas (Kern Water Bank

Authority 2012). The Bank currently stores around 1 million acre-feet of water (1.23 billion m^3) sourced from the Kern River and two major aqueducts (Kern Water Bank Authority 2012).

The Kern Water Bank is hailed by some as a financial success that also creates valuable habitat (e.g., Thomas 2001). However, the project has also been dogged by controversy, including challenges to the transfer of the project from state to local ownership (Center for Biological Diversity 2010), and claims that the project is adversely affecting neighboring groundwater users. In litigation that is ongoing at the time of writing, neighboring water districts that overlie the same groundwater basin allege that a lack of control over the rate at which banked water is recovered has significantly affected groundwater levels, quality, historic hydraulic gradients, and the regional environment (Petition for Writ of Mandate and Complaint 2010).

The complexity of the private legal arrangements among the Kern Water Bank parties and the issues in contention in litigation about the Bank's operations underscore the legal uncertainty that a broad legal framework for MAR could help to resolve.

2.3.4 Outstanding Challenges

The OCWD and Kern Water Bank cases point to key legal and technical challenges facing MAR in California. The relative ease with which the OCWD has been able to operate and fund its MAR project is due in part to its rare legislative framework, which provides it with specific powers to charge a groundwater 'pump tax' to pay for local basin replenishment. Recent California case law, further described in Sect. 3.3.3, interprets groundwater charges as subject to the constitutionally mandated requirement to gain the approval of local electors before imposing or raising fees. This may present a barrier to MAR, since the processes involved in order to comply with this requirement can be costly and burdensome (Soldani 2000). A further obstacle to MAR in California is posed by the legal requirement to seek the state's permission to change the use of surface water intended to be stored using MAR (i.e., replenishment), since the surface water right was previously permitted for a different, direct use, such as irrigation (CWC §1,701). The lack of a broad legal framework for MAR in California also leaves uncertain other key legal issues outlined in Sect. 2.3.1, the clarification of which could remove obstacles to the expansion of the practice.

MAR for IPR presents special issues. As agencies navigate the regulatory framework and technological requirements for IPR, they now recognise that meeting public communication needs is an ongoing challenge, and that dialogue is essential for successful potable reuse. Although no studies to date have detected risks to public health from recycled water use, pathogens as well as chemicals detected in wastewater streams from the production and use of pharmaceuticals and personal care products ('emerging contaminants') may ignite public health concerns – whether perceived or actual – in the use of recycled water (Toze 2006b).

Several IPR projects have been cancelled or indefinitely postponed due to public concerns and perceptions of health risks, including a well-known San Diego case in which an IPR project was halted in 1998 following public opposition and political pressure (Barringer 2012; USEPA 2004). Signifying both continued water supply pressures in the region and potential change in attitudes towards recycled water for potable use, the City of San Diego recently began operating a one million gallon/day (3,785 m^3/day) Water Purification Demonstration project to test a multiple-barrier treatment process that will produce recycled water of potable quality. In this case, advanced-treated recycled water will be used to augment a reservoir where it will be blended with additional potable water sources prior to withdrawal for additional treatment via a drinking water facility (San Diego 2011). Even more significantly, the U.S. National Research Council, which in 1998 described IPR as an option of last resort for drinking water, now concludes that the human health risks of IPR are similar to those from conventional drinking water sources (NRC 2012). This update represents the substantial growth in experience and knowledge gained from IPR projects implemented globally. While public opposition to IPR may be an ongoing challenge, coalescing scientific opinion promises to inform debates in its favor.

Section 2 has reviewed both negative and positive aspects of groundwater–surface water interaction. On the negative side, there is the potential for wells to deplete streams and for paved urban areas to reduce infiltration to groundwater; on the positive side, there is the potential for MAR, including water banking and IPR, to provide valuable storage and supply. Legal challenges attend aspects of both: legal omissions in the case of wells depleting streams and, in the case of MAR, lack of a comprehensive management framework. Nonetheless, innovative techniques like IPR, multi-agency water banking, and recharge mapping demonstrate the value of recognising these interactions in the urban water setting.

3 Interactions Between Urban and Rural Water Systems

3.1 Background

As populations around the globe urbanise, interactions between water systems in urban and agricultural areas become increasingly important. Water transfers from historically high water-use agricultural areas have the potential to satisfy growing urban demands. Such transfers are the most common form of water transfer in the western U.S. states (Grafton et al. 2009). In the reverse direction, water law can facilitate the use of urban wastewater as a highly reliable water source for agriculture.

In California, where cities and agriculture compete for scarce and variable resources, there are a number of factors that make water markets particularly important. Under the 'prior appropriation' system of water rights that dominates Californian surface water, agricultural water rights are often the most 'senior' (and therefore most secure) rights during periods of scarcity, as they were typically put in place before cities and their water rights appeared. Agricultural water rights account for around 80 % of consumptive water use in California (California DWR 2009), are applied to a comparatively low-value end use, and present a tempting target for conservation. By contrast, cities have a high-value end use, but there is a potential difficulty in dealing with the risk of water shortages associated with junior water rights. The difference in value can be extreme: residents of San Diego can pay up to 75 times more for water than agricultural water users in the Imperial Irrigation District, which is 115 miles away (Grafton et al. 2009). Water transfers from agriculture to cities enable cities to manage risk, and agriculture to realise a higher value for water rights (Thompson 1993; Grafton et al. 2009).

However, agriculture–urban transfers invite concerns about their impacts on the long-term viability of the agricultural communities concerned. In particular, transfers of water produced by land fallowing can decrease demand for farm labor and other inputs and increase per capita irrigation costs (Thompson 1993; Hanak 2003). Worries also arise that transfers move water from already poor inland communities to wealthy coastal communities, and the Owens Valley to Los Angeles transfer in the early twentieth century is a notorious historical precedent of community decline in response to such a transfer (Hanak 2003; Libecap 2009). More intangibly, such transfers contravene an ethic in some communities that a district's water is a local community resource that should be used to promote local agriculture. This discourages elected local district leaders from supporting transfers, which would damage their chances of re-election (Thompson 1993). Other opponents of transfers point to the loss of agricultural return flows and the incidental environmental benefits of agricultural land, for example, rice and cornfields that support waterfowl and endangered snake species (California DWR 2009; Hanak et al. 2011; Water Transfer Workgroup 2002). Section 3.2.2 describes how these concerns were addressed in a large-scale agriculture-to-urban transfer from the Imperial Irrigation District to San Diego. Section 3.2.3 profiles a much smaller-scale, and still developing, water 'exchange' project between agriculture and urban communities, which seeks to improve both water quality and supply.

Partly to accommodate demands to transfer water from agriculture to cities, Californian agriculture is attempting to access new, secure water sources, including urban wastewater; here, the outcome is effectively to transfer water from cities to agriculture. Approximately 30 % of recycled water production in California is already used for agricultural irrigation, the greatest single use of recycled water in the state (SWRCB 2011). In coastal regions, the recycling of water frees up supplies that would otherwise be discharged to the ocean, without impacts on downstream users. In addition to providing water supply, treated wastewater contains nutrients such as nitrogen and phosphorus which, when properly managed, may allow farmers to reduce their use of chemical fertilisers without adverse impacts (USDA-CSREES 2008). Section 3.3.2 describes a celebrated example of water reuse for agriculture in Monterey.

To comply with regulatory requirements, reusing wastewater for agricultural irrigation can be expensive, and involve investment in high-cost technology and distribution facilities, factors which often limit the economic feasibility of the practice (Bischel et al. 2013). These costs have meant that cost recovery has become a central issue, and the establishment of rates for water and recycled water systems is a problem when there are both direct and indirect beneficiaries of a project. The case of the Pajaro Valley Water Management Agency, described in Sect. 3.3.3, demonstrates an ongoing struggle to address this issue.

3.2 Transferring Water from Agriculture to Cities

3.2.1 The Californian Regulatory Approach: Managing the Potential for Adverse Impacts

California water policy explicitly encourages, and heavily regulates, some water transfers (California DWR 2009). The SWRCB regulates the transfer of permitted water rights, where that transfer involves changing the point of diversion, place of use, or purpose of use of the water right. Long-term (greater than 1-year) transfers must not cause 'substantial injury' to other water users, nor 'unreasonably affect' fish, wildlife, or other instream beneficial uses (CWC §1,736). Transfers must also comply with California's environmental impact assessment legislation. To prevent transfers contributing to groundwater overdrafting, the state restricts the ability of water suppliers to transfer surface water and then increase pumping of groundwater (CWC §1745.10). Aspiring transferors must also secure permission to convey ('wheel') the water to its new location, which can involve meeting federal and state economic and environmental requirements for using north–south conveyance facilities (CWC §1,810(d)). Finally, counties and individual irrigation districts also regulate water exports using ordinances and bylaws, which often seek to manage indirect economic effects on local communities (Hanak 2003).

3.2.2 Making a Large Agriculture–Urban Transfer in Southern California: The Imperial Irrigation District to San Diego

One of the most significant, complex, and controversial agriculture-to-urban transfers in California occurred in 2003, with the signing of a series of agreements to transfer 200,000 acre-feet (247 million m^3) of water per year from the Imperial Irrigation District (IID) to the San Diego County Water Authority (SDCWA). The initial period of the transfer was 45 years, at an initial cost of US$248 per acre-foot (US$0.20/m^3). The IID, which uses water from the Colorado River for irrigation, has committed to transfer to the SDCWA a portion of water conserved by fallowing land, increasing on-farm efficiency, and lining a major canal (Davids Engineering Inc. et al. 2007).

Responding to local concerns over the socioeconomic effects of the transfer, the IID and SDCWA each contributed to a fund (US$40 million and US$10 million, respectively) designed to mitigate these impacts. This fund has been recognised as a significant innovation that provides a useful model for future large-scale transfers (Hanak et al. 2011). Projects proposed for funding are independently reviewed and rated according to predetermined criteria, and must either directly mitigate negative socioeconomic impacts that result from land fallowing, or indirectly stimulate the regional economy (IID 2010). A construction company, a public education and job placement service, a non-profit socioeconomic justice and community development organisation, and a non-profit food bank are among recent successful applicants (IID 2011).

Environmental concerns also attend the transfer. Reducing irrigation in the IID decreases the volume of crop runoff that flows to the Salton Sea, a vast man-made lake that now harbors endangered birds and fish (Hanak et al. 2011). The transfer has the potential to impact lake levels, water quality, habitat, and air quality, due to dust created as the lakebed becomes exposed and dries, and requires approval under federal and state environmental legislation. Various environmental mitigation measures are required, including providing water directly to the Salton Sea to maintain salinity levels, building bird roosting sites, creating replacement habitat, and relocating stranded fish (IID 2003). The transfer has been subject to intense and ongoing multi-party litigation, particularly in relation to who should pay for these mitigation measures (Quantification Settlement Agreement Cases 2011; Davila 2012).

The IID–San Diego transfer shows the potential complexity of linking agriculture and urban water sources, made even more challenging by environmental and community concerns; to address these concerns, an innovative approach involving economic and environmental mitigation measures has been implemented.

3.2.3 Exchanging Water Sources Between Agricultural and Urban Users in the Central Valley

In addition to large-scale, quantity-focused water trades, unconventional small-scale water 'exchanges' between urban and agricultural areas can be used to achieve both water quality and supply benefits. The Alta Irrigation District (AID) in California's Central Valley is presently developing the Eastside Water Quality and Reliability Project, which aims to do exactly that. The AID overlies the Kings Sub-basin, a state-designated critically overdrafted basin, the groundwater levels of which have decreased by over 40 ft in the past century (AID 2010). The project aims to use water recharge facilities and an agriculture–urban water exchange to provide drinking water to local disadvantaged communities, which presently rely on low-quality local groundwater; it also aims to increase the availability of irrigation water for AID farmers, who currently rely chiefly on surface water (AID 2010; Kapheim 2012).

One component of the project will involve artificially recharging the aquifer underlying the AID using two state-funded recharge ponds that will receive stormwater and floodwater. Around 85 % of this stored water will be permitted to be supplied to AID farmers using AID-owned wells and a new pipe or canal network (AID 2010; Kapheim 2012). Under the second component of the project, the AID will exchange its water right from the Kern River, which runs through the district, for an allocation of water (owned by a third party) from the Friant-Kern Canal, which runs closer to the nearby town centers of Cutler and Orosi. AID will then construct a new treatment plant to treat the imported surface water, and provide it for urban uses to reduce urban reliance on contaminated groundwater (AID 2010; Kapheim 2012).

The project aims to provide both water quality and supply benefits. The first component will increase the availability of irrigation water by harvesting currently uncontrolled water and reducing losses from conveying surface water to farmers who will switch to using stored groundwater. Cutler and Orosi will obtain access to better quality treated surface water. Finally, the project aims to improve regional groundwater levels and groundwater quality through the 15 % of good quality, artificially recharged water that will remain unavailable for extraction.

While the two recharge ponds are close to being fully operational at the time of writing, the project must still undergo a lengthy series of public health, environmental, and finance approvals before it may proceed, following the conclusion of an ongoing (as of 2013) feasibility study (Kapheim 2012).

Projects that link and benefit urban and rural communities simultaneously are uncommon in the Central Valley. This one was driven, at least in part, by links formed between urban and rural water users and a local community water justice organisation during the preparation of the AID's GWMP (Kapheim 2012). This again confirms the value, in linking institutions, of a water planning framework, as it facilitates projects that not only recognise interactions but use them in a way to create multiple benefits for urban and rural communities alike.

3.2.4 Outstanding Challenges

Many challenges continue to limit agriculture-to-urban transfers in California. Since 2007, restrictions on pumping in the Sacramento–San Joaquin Delta to protect endangered species have prevented some water transfers. Previously, such restrictions caused the price of transfer water to increase significantly (California DWR 2009). Concerns over other environmental and third-party impacts continue, the latter particularly in relation to transfers made possible by land fallowing (California DWR 2009; Hanak et al. 2011). The complexity of gaining approval for large water transfers that involve wheeling agreements can also be problematic.

Nonetheless, opportunities for innovation present themselves. Transfers of water can be timed to coincide with species' needs for high flows (California DWR 2009). Transfers can also create dual water quality and supply benefits. Funding mechanisms can be used to mitigate any adverse socioeconomic effects of transfers and create long-term employment opportunities in disadvantaged farming communities. Water transfers can also be used in the context of MAR (as in the case of the Kern Water Bank) to create innovative water banking projects, whereby rural landowners store water in underground aquifers and sell it to urban communities when required.

3.3 *Transferring Urban Wastewater from Cities to Agriculture*

3.3.1 The California Regulatory Approach: Recycled Water Standards

California began using reclaimed water for agriculture in the 1800s, and has since developed stringent regulations to govern its treatment and use. Following widespread use, and with a proven safety record, state-based regulatory frameworks for agricultural reuse are generally well established in the U.S.A., although irrigation of raw-eaten food crops is prohibited in some states (NRC 2012). Recycled water regulations in California, codified in Title 22 of the California Code of Regulations ('Wastewater Reclamation Criteria'), seek to ensure public safety and require high water quality whenever contact with humans may occur; isolated distribution from potable sources must occur via parallel 'purple pipe' infrastructure (USDA-CSREES 2008).

3.3.2 Meeting Water Demands and Alleviating Overdraft: Establishing Irrigation Using Reclaimed Water in Monterey

In an early, widely acclaimed example of transferring urban wastewater to agriculture, the Monterey Regional Water Pollution Control Agency began using recycled water for agricultural irrigation as part of the Castroville Seawater Intrusion Project in 1998. Facing the threat of saltwater intrusion due to extensive withdrawal of groundwater for irrigation, California's Central Coast Regional Water Quality Control Board recommended using reclaimed wastewater for crop irrigation and river enhancement in its 1974 water quality management plan (Asano et al. 2007; Engineering-Science 1987). Using reclaimed water for irrigation near Monterey also provided a potential tool for securing consistent yields and quality of produce from the agricultural region of the lower Salinas Valley in Monterey County. The concern about saltwater intrusion and the opportunity to use recycled water in agriculture constituted a unique set of environmental and economic incentives to implement an agricultural water reuse project. However, in light of the uncertainty about the health effects and the public's opinion about reclaimed wastewater used for raw-eaten crops, the management plan recognised the importance of proving the safety of reclaimed wastewater prior to regional implementation.

An extensive set of full-scale field studies, part of the Monterey Wastewater Reclamation Study for Agriculture (MWRSA), commenced in 1980 and continued with water, soil, and plant tissue analysis over 5 years. This analysis was accompanied by public health, cost, and feasibility assessments. The 1987 MWRSA final report recommended full use of the projected 30 million gallons/day (1.3 m^3/s) flow of recycled water for the duration of each 8-month irrigation season in the Castroville area (Engineering-Science 1987). The MWRSA report declared irrigating raw-eaten food crops with filtered secondary municipal wastewater to be as safe as irrigation with well-water.

As the first large-scale water reuse investigation of its kind, the MWRSA study design has been used widely as a standard for planning the use of reclaimed wastewater for agricultural irrigation (Asano et al. 2007).

3.3.3 Paying for Reclaimed Water in Agricultural Reuse Programs: The Pajaro Valley Case

Like Monterey's program, a major driver for using recycled water in the Pajaro Valley was to prevent seawater intruding into the groundwater basin by providing an alternative to coastal groundwater pumping. The program operates as a partnership between the Pajaro Valley Water Management Agency (PVWMA) and the City of Watsonville, which owns the wastewater treatment plant. Financial woes and legal battles have troubled the program since its inception.

In 1984, the PVMWA was created to manage surface and groundwater resources in the Pajaro Valley (CWC App. 124; McNiesh and Wichelns 2004). A series of lawsuits filed from 2002 onwards has challenged the legality of rate increases approved by the PVMWA's governing board to pay for its water distribution and recycled water scheme. Early disputes arose over the issue of the 'zone of benefit', which defines who receives benefit from improvement projects and therefore should be billed for project costs. Determining this zone is required by California's Constitution (Proposition 218, Articles XIII C and D), under which agencies must prove that they comply with certain substantive and procedural requirements before imposing or increasing certain fees and charges. The PVWMA did not comply with these requirements, leading inland farmers, who were charged increased rates but did not directly receive project water, to question the legality and equity of the chosen rate structure (Wagner 2008).

In 2007, the California Court of Appeals reversed a series of judgments that had upheld the PVWMA fees, invalidating the increased charge and forcing the PVWMA to pay compensation and legal fees (PVWMA v. Amrhein 2007). Subsequently, the PVWMA expended considerable effort to comply with the provisions of Proposition 218 to increase rates in 2010. Further lawsuits followed, challenging fees related to the use of recycled water, but were ultimately decided in PVWMA's favor (PVWMA 2011). In their current form, the rates distribute costs among all Pajaro Valley water users to pay for operating and maintaining the supplemental water scheme (which includes recycled water), with differing fees for those inside and outside water delivery zones.

The PVWMA example illustrates that Californian projects establishing links between urban and agricultural water systems must contend not only with water laws, but also with burdensome administrative requirements. More broadly, it shows how the potentially high cost of water reuse programs, and the need to fairly distribute these costs, can prove a significant obstacle to building this interaction.

3.3.4 Outstanding Challenges

Despite established state frameworks and successful water reuse programs – such as the Monterey project – the absence of regulatory guidelines or limits for emerging contaminants, discussed above in Sect. 2.3.4 in the context of IPR, has led to uncertainty over the health risks of recycled water in other reuse contexts. Management concerns regarding water quality halted a recycled water project for streamflow augmentation (Plumlee et al. 2011) and arose as a public concern in Redwood City, California, for a landscape irrigation project (Ingram et al. 2006).

Nonetheless, following the implementation of a wide range of projects, many communities have become more familiar with the advantages of water reuse, such as its reliability as a water source (NRC 2012). Disseminating research regarding water quality and human health risks from agricultural irrigation may also help combat negative public perception regarding the health risks of using recycled water for food crops (USDA-CSREES 2008). This will help the public to understand the concentrations of chemicals that may be damaging to agricultural yields or to soil conservation, as well as the appropriate treatment technologies applied to minimise human health risks.

In addition to resolving human health concerns, expanding the use of recycled water for agricultural irrigation requires consideration of how applied recycled water interacts with the surrounding environment. Salinity problems in agricultural drainage, for example, could require a reduction in the amount of recycled water applied for irrigation (Toze 2006a). Opportunities should be identified and implemented to reuse drainage water, either for more salt-tolerant crops or for decentralised collection and treatment (Reinhardt et al. 2005; USDA-CSREES 2008).

Even where water quality concerns have been resolved, the high costs of developing schemes to reuse urban wastewater for agriculture may stand in the way of expanding the practice. To reduce overall costs, reusing wastewater for agricultural irrigation commonly requires treatment facilities located in close proximity to the recycled water end-users (USDA-CSREES 2008). In California, opportunities to reuse water for agriculture, the so-called 'low-hanging fruit', have quite often already been identified and established (California DWR 2003). A legal framework that facilitates linking new recycled water sources with geographically proximate heightened water demands – perhaps through IRWMPs – could help to expand water reuse. Corresponding technological innovations that introduce new strategies for distributed recycled water facilities would complement such a framework. For example, distributed water reuse technologies such as membrane bioreactors show promise for tailoring recycled water treatment schemes with irrigation water quality requirements (Meuler et al. 2008).

Section 3 has illustrated how linking rural and urban water systems can create benefits, but also controversy. Californian water laws and institutional arrangements tend to make large-scale agriculture-to-urban transfers very complex, as shown by the IID–San Diego transfer. Strong community concerns can accompany such transfers, as well as urban-to-agriculture wastewater reuse projects, despite successes such as the MWRSA. Additional administrative requirements to hold local elections to raise fees can also stand in the way of projects to create wastewater transfers, as has happened in the Pajaro Valley.

4 Conclusion

Hydrological reality has long challenged water law, which traditionally has failed to reflect the interconnection among the various parts of water systems. This chapter has shown how sophisticated laws and policies can, in the urban setting, reflect the complexity of water systems in two key areas: groundwater–surface water interaction, and agriculture–urban water system interactions.

Laws that control the impacts of pumping groundwater on surface water can restrict the water supplies available to urban utilities, as Cal-Am experienced keenly on the Carmel River. An important issue in this context is how to draw the line, for legal purposes, between connected and unconnected water sources in order to control the impacts of groundwater pumping on streams. Innovative mitigation programs could help reduce the economic costs of controlling these impacts. Water policy can encourage utilities to better manage these connections under voluntary groundwater management plans, such as Zone 7's, which maps recharge areas. On a more positive note, groundwater–surface water interactions can provide agencies with a significant opportunity to increase their storage capacity by percolating or injecting stored water into aquifers. Legal frameworks for undertaking MAR and water banking using MAR can facilitate such projects, like those of the OCWD and the Kern Water Bank, by reducing key sources of legal uncertainty that can otherwise stifle projects. A utility's ability to fund groundwater replenishment using pump taxes has been key to the success of the OCWD, whereas funding obstacles have proved particularly problematic in the case of the PVWMA's water reuse scheme.

State and local laws can facilitate agriculture-to-urban transfers and help to manage their potentially adverse socioeconomic and environmental effects. Environmental and socioeconomic mitigation measures such as those used in the IID to San Diego transfer can help alleviate concerns that attend such transfers. The combination of water rights laws, water wheeling requirements, environmental laws, and socioeconomic concerns can make transfers fiendishly complex, and may stifle the ability of cities to satisfy new urban demands with agricultural water. On the other hand, the AID's Eastside Water Quality and Urban Reliability Project suggests that innovative water exchanges can be designed to achieve benefits for both urban and agriculture communities. Turning the typical form of this interaction in reverse, agricultural irrigation projects using urban wastewater can successfully meet regulatory requirements for water quality – even for raw-eaten food crops, as shown by the MWRSA and PVWMA. However, public acceptance is an ongoing challenge, particularly with respect to emerging contaminants. Ensuring adequate funding for these projects, which can be quite expensive, is another major issue, particularly where local landowners receive indirect benefits but may not wish to shoulder costs, which proved highly contentious for the PVWMA.

A water planning approach is a frequent element of law and policy mechanisms for managing these core interactions. Such an approach helps make connections across different water management institutions, water sources, and parties to water banking arrangements, as well as between land and water and between urban and rural communities.

The interactions discussed in this chapter represent areas of physical and political complexity with which water law, policy, and institutions have, traditionally, most struggled. Indeed, as well as successes and innovations, this tour of California laws and projects has shown many ongoing challenges in dealing with these interactions. Yet refocusing laws and policies on these areas of interaction promises great benefits – safeguarding instream flows from inadvertent depletion, maximising water supplies, storage, and reliability, and creating synergies in water management – benefits that are critical in an urbanising, water-scarce, and ecologically strained world.

Acknowledgments Support for this project was provided by the Dow Sustainability Fund of the United States Studies Centre at the University of Sydney, the S. D. Bechtel, Jr., Foundation, and the NSF Engineering Research Center for Re-inventing the Nation's Urban Water Infrastructure (ReNUWIt).

References

Alta Irrigation District (AID). (2010). *Amended groundwater management plan*. Dinuba:Alta Irrigation District.

Asano, T., Burton, F. L., Leverenz, H. L., Tsuchihashi, R., & Tchobanoglous, G. (2007). *Water reuse: Issues, technologies, and applications*. New York: McGraw-Hill.

Barringer, F. (2012, February 9). As 'yuck factor' subsides, treated wastewater flows from taps. *The New York Times*.

Bekelea, E., Toze, S., Pattersona, B., & Higginsone, S. (2011). Managed aquifer recharge of treated wastewater: Water quality changes resulting from infiltration through the vadose zone. *Water Research, 45*(17), 564–577.

Bischel, H. N., Simon, G. L., Frisby, T. M., & Luthy, R. G. (2012). Management experiences and trends for water reuse implementation in Northern California. *Environmental Science & Technology, 41*(1), 180–188.

Bischel, H. N., Lawrence, J. E., Halaburka, B. J., Plumlee, M. H., Bawazir, A. S., King, J. P., & Luthy, R. G. (2013). Renewing urban streams with recycled water for streamflow augmentation: Hydrologic, water quality, and ecosystem services management. *Environmental Engineering Science, 30*, 455–479.

Bouwer, H. (2002). Artificial recharge of groundwater: Hydrogeology and engineering. *Hydrogeology Journal, 10*(1), 121–142. doi:10.1007/s10040-001-0182-4.

California Department of Water Resources (DWR). (2003). *Water recycling 2030: Recommendations of California's recycled water task force*. Sacramento: California DWR.

California Department of Water Resources (DWR). (2009). California water plan update 2009: Integrated water management – Resource management strategies. *Bulletin 160–09* (Vol. 2). Sacramento: California DWR.

California Department of Water Resources (DWR). (2012). Integrated regional water management grants: IRWM Archives. http://www.water.ca.gov/irwm/integregio_archive.cfm. Retrieved 31 Jan 2012.

Caponera, D. A. (2007). *Principles of water law and administration* (ed.: M. Nanni, 2nd rev. ed.). London: Taylor & Francis.

CDPH. (2011). *Draft regulations for groundwater replenishment with recycled water*. California Department of Public Health (CDPH), Title 22, California Code of Regulations; Division 4, Environmental Health; Chapter 3, Recycling criteria.

Center for Biological Diversity. (2010, July 2). *Suit filed to reverse one of the biggest ripoffs in California history: Groups seek return of Kern water bank to public control*. Press release.

Contor, B. A. (2009). Groundwater banking in aquifers that interact with surface water: Aquifer response functions and double-entry accounting. *Journal of the American Water Resources Association, 45*(6), 1465–1474. doi:10.1111/j.1752-1688.2009.00378.x.

Davids Engineering Inc., Keller Bliesner Engineering, LLC, CONCUR, Inc., Davey-Cairo Engineering, Inc., GEO/Graphics, ITRC, Western Resource Economics, . . . Utah State University. (2007). *Efficiency conservation definite plan. Final report*.

Davies, L. L. (2007). Just a big, "hot fuss"? Assessing the value of connecting suburban sprawl, land use, and water rights through assured supply laws. *Ecology Law Quarterly, 34*(4), 1217–1296.

Davila, A. (2012, January 12). *County of Imperial to continue legal challenge to water transfer*. Imperial Valley Press. California: El Centro.ZZZ. http://www.ivpressonline.com/news/ivp-county-of-imperial-to-continue-legal-challenge-to-water-transfer-20120112,0,287636.story

de Vries, J. J., & Simmers, I. (2002). Groundwater recharge: An overview of processes and challenges. *Hydrogeology Journal, 10*, 5–17.

Drewes, J. E., Anderson, P., Denslow, N., Olivieri, A., Schlenk, D., & Snyder, S. (2010). *Monitoring strategies for chemicals of emerging concern (CECs) in recycled water: Recommendations of a science advisory panel*. Sacramento: California State Water Resources Control Board.

Engineering-Science. (1987). *Final report: Monterey wastewater reclamation study for agriculture*. Berkeley, CA: Prepared for Monterey Regional Water Pollution Control Agency.

Foley-Gannon, E. (2008). Institutional arrangements for conjunctive water management in California and analysis of legal reform alternatives. *Hastings West-Northwest Journal of Environmental Law and Policy, 14*.

GEI Consultants. (2009). *Consolidated Irrigation District groundwater management plan* California: Selma.

Grafton, R. Q., Landry, C., Libecap, G. D., O'Brien, J. R. (2009). Markets – Water markets: Australia's Murray–darling basin and the US Southwest. International centre for economic research. working paper No. 15/2009.

Hanak, E. (2003). *Who should be allowed to sell water in California?: Third-party issues and the water market*. San Francisco: Public Policy Institute of California.

Hanak, E. (2005). *Water for growth: California's new frontier*. San Francisco: Public Policy Institute of California.

Hanak, E., Lund, J., Dinar, A., Gray, B., Howitt, R., Mount, J., & Suddeth, R. (2010). Myths of California water: Implications and reality. *Hastings West-Northwest Journal of Environmental Law and Policy, 16*, 3–73.

Hanak, E., Lund, J., Dinar, A., Gray, B., Howitt, R., Mount, J., Thompson, B. H., et al. (2011). *Managing California's water: From conflict to reconciliation*. San Francisco: Public Policy Institute of California.

Howard, J., & Merrifield, M. (2010). Mapping groundwater dependent ecosystems in California. *PLoS ONE, 5*(6), e11249.

Imperial Irrigation District (IID). (2003). *Imperial irrigation district water conservation and transfer project mitigation, monitoring and reporting program*. California: Imperial.

IID. (2010). *IID local entity mitigation 2010: Competitive grant program request for proposals*. California: Imperial.

IID. (2011). Competitive applicants approved by the local entity board at its meeting April 5, 2011. http://www.iid.com/Modules/ShowDocument.aspx?documentid=4693. Retrieved 28 Feb 2012.

Ingram, P. C., Young, V. J., Millan, M., Chang, C., & Tabucchi, T. (2006). From controversy to consensus: The redwood city recycled water experience. *Desalination, 187*, 179–190. doi:10.1016/j.desal.2005.04.078.

Jones & Stokes (for Zone 7 Water Agency). (2005). *Groundwater management plan for Livermore–Amador Valley groundwater basin*. Sacramento.

Kapheim, C. (2012, March 6). Telephone interview with Chris Kapheim, General Manager, Alta Irrigation District, regarding establishment of the Eastside water quality and reliability project.
Katen, M. (2012, February 17). Telephone interview with Matt Katen, Zone 7 Water Agency, regarding land use monitoring program of the Zone 7 Water Agency.
Kennedy/Jenks Consultants (Prepared for Kern County Water Agency). (2011). *Tulare lake basin portion of Kern county integrated regional water management plan: Final update*. California: Oxnard.
Kenny, J. F., & U.S. Geological Survey. (2009). *Estimated use of water in the United States in 2005* (1st ed.). Reston: U.S. Geological Survey.
Kern Water Bank Authority. (2012). *Notice of preparation of an environmental impact report for the Kern water bank conservation and storage project.* Bakersfield, California: Kern Water Bank Authority.
Libecap, G. D. (2009). Chinatown revisited: Owens Valley and Los Angeles – Bargaining costs and fairness perceptions of the first major water rights exchange. *Journal of Law, Economics, and Organization, 25*(2), 311.
Little Hoover Commission. (2010). *Managing for change: Modernizing California's water governance*. Sacramento: Little Hoover Commission.
McNiesh, C., & Wichelns, D. (2004, February 3 & 4). *A buyer's perspective on the California water market.* Paper presented at the California Plant & Soil Conference, Visalia.
Meuler, S., Paris, S., & Hackner, T. (2008). Membrane bio-reactors for decentralized wastewater treatment and reuse. *Water Science and Technology, 58*(2), 285–294. doi:10.2166/wst.2008.356.
Mills, W. R., & Watson, I. C. (1994). Water-factory 21: The logical sequence. *Desalination, 98*(1–3), 265–272. doi:10.1016/0011-9164(94)00151-0.
National Research Council (U.S.A.), Committee on Sustainable Underground Storage of Recoverable Water (NRC CSUSRW), National Research Council (U.S.A.), Water Science and Technology Board (NRC WSTB), & National Research Council (U.S.A.), Water Science and Technology Board (NRC WSTB). (2008). *Prospects for managed underground storage of recoverable water*. Washington, DC: National Academies Press.
Nelson, R. (2011). *Uncommon innovation: Developments in groundwater management planning in California.* Water in the West Working Papers. Palo Alto: Stanford University.
Nelson, R. (2012). *Instituting integration: Findings of the Comparative Groundwater Law and Policy Program's Workshop 1.* Water in the West Working Papers. Palo Alto: Stanford University.
North Gualala Water Company v. SWRCB, 139 Cal. App. 4th 1577 (1st Dist. 2006).
NRC. (2012). *Water reuse: Potential for expanding the nation's water supply through reuse of municipal wastewater*. Washington, DC: National Research Council, National Academy of Sciences.
Otto, B., Ransel, K., Todd, J., Lovaas, D., Stutzman, H., & Bailey, J. (2002). *Paving our way to water shortages: How sprawl aggravates the effects of drought*. Washington, DC: Smart Growth America.
Pajaro Valley Water Management Agency (PVWMA). (2011). *Tentative ruling OKs Pajaro valley water fees,* from http://www.santacruzsentinel.com/ci_19371947. Accessed 7 Nov 2013.
Petition for Writ of Mandate and Complaint for Injunctive and Declaratory Relief, Rosedale–Rio Bravo Water Storage District and Buena Vista Water Storage District v. California DWR, Kern Water Bank and Others, Case No. 270635-KCT (Superior Court of the State of California, County of Kern 2010).
Plumlee, M. H., Gurr, C. J., Reinhard, M., Ashktorab, H., John, P., & Larabee, J. (2011). *Attenuation of emerging contaminants in streams augmented with recycled water*. Alexandria: WateReuse Foundation.
Postema v. Pollution Control Hearings Board, 142 Wn. 2d 68 (2000).
Purkey, D. R., & Mansfield, E. M. (2002). *Estimating the potential for in lieu conjunctive water management in the Central Valley of California*. Berkeley: Natural Heritage Institute.
PVWMA v. Amrhein, 150 Cal. App. 4th 1364 (2007).
Quantification Settlement Agreement Cases, 201 Cal. App. 4th 758 (Cal. App. 3d Dist. 2011).
Reinhardt, M., Gachter, R., Wehrli, B., & Muller, B. (2005). Phosphorus retention in small constructed wetlands treating agricultural drainage water. *Journal of Environmental Quality, 34*(4), 1251–1259. doi:10.2134/jeq2004.0325.

Rodriguez, C., Van Buynder, P., Lugg, R., Blair, R., Devine, B., Cook, A., & Weinstein, P. (2009). Indirect potable reuse: A sustainable water supply alternative. *International Journal of Environmental Research and Public Health, 6*, 1174–1209.

Sax, J. L. (2002). *Review of the laws establishing the SWRCB's permitting authority over appropriations of groundwater classified as subterranean streams and the SWRCB's implementation of those laws: Final report*. Sacramento: State Water Resources Control Board (SWRCB).

Soldani, D. A. (2000). The impact of proposition 218 on California's water delivery system: Should water be "above the law"? *San Joaquin Agricultural Law Review, 10*, 183.

Stanford Water in the West. (2012). *Learning from best practices*. Palo Alto: Stanford University.

SWRCB. (1995). Order on four complaints filed against the California–American water company: Carmel River, Monterey County (Order No. WR 95–10).

SWRCB. (2009). *Cease and Desist Order WR 2009–0060*. Sacramento: SWRCB.

SWRCB. (2011). *California State Municipal Wastewater Recycling Survey in 2009*. Sacramento: SWRCB.

Thomas, G. A. (2001). *Designing successful groundwater banking programs in the Central Valley: Lessons from experience*. Berkeley: Natural Heritage Institute.

Thomas, G. A. (2005). The future of water law reform in California a quarter century after the governor's commission. *McGeorge Law Review, 36*, 495–533.

Thompson, B. H., Jr. (1993). Institutional perspectives on water policy and markets. *California Law Review, 81*, 673–764.

Thompson, B. H., Jr. (2011). Beyond connections: Pursuing multidimensional conjunctive management. *Idaho Law Review, 47*.

Toze, S. (2006a). Reuse of effluent water: Benefits and risks. *Agricultural Water Management, 80*(1–3), 147–159. doi:10.1016/j.agwat.2005.07.010.

Toze, S. (2006b). Water reuse and health risks: Real vs. perceived. *Desalination*, 187, 41–51.

USDA-CSREES. (2008). *Opportunities and challenges in agricultural water reuse: Final report* (Eds.: J. Dobrowolski, M. O'Neill, L. Duriancik, & J. Throwe). Washington, DC.

USEPA (2004). *Guidelines for water reuse 7* (EPA/625/R-04/108). Washington, DC: U.S. Environmental Protection Agency and U.S. Agency for International Development.

Wagner, P. (2008, April 2). The Watsonville water wars: How the Pajaro Valley Water Management Agency was brought to its knees. *Metro Santa Cruz Weekly*.

Water Transfer Workgroup. (2002). *Water transfer issues in California: Final report to the California State Water Resources Control Board*. Sacramento: SWCRB.

Winter, T. C., Harvey, J. W., Franke, O. L., & Alley, W. M. (1998). *Ground water and surface water: A single resource* (US Geological Survey Circular 1139). Denver: U.S. Geological Survey.

Chapter 23
Integrated Management of Urban Water Supply and Water Quality in Developing Pacific Island Countries

Ian White and Tony Falkland

1 Introduction

The general fragility and unique vulnerability of small island countries in the Pacific (see Fig. 23.1) to climatic, demographic, economic, and development pressures, as well as to natural hazards and extreme events, has been widely acknowledged for a long time (UNDESA 1994). So too has been the diversity of their geography, geology, and sources of available freshwater (Table 23.1). These are is particularly evident when it comes to managing the freshwater supply systems in urban centres on small islands whose typical land area is only 1–10 km^2. Some population centres in the Pacific and Indian oceans are close to – or have already exceeded long-term sustainable water extraction limits (White and Falkland 2011). Frequent severe droughts the El Niño–Southern Oscillation (ENSO), floods, cyclones, tsunamis, earthquakes, and volcanic eruptions coupled with limited resources and the capacity to respond to extreme events, compound freshwater supply problems.

Sparse island communities have demonstrated their remarkable resilience in the face of climatic extremes and natural hazards over the last 1,000–12,000 years. Despite limited financial, technological, and infrastructure resources, the well-developed local institutions, resilient social systems, sensitivity to environmental change, and the high value placed on equity in Pacific islands have provided capacities for adapting to threats and change and have allowed low density subsistence populations to survive (Barnett 2001).

I. White (✉)
Fenner School of Environment and Society, The Australian National University, Frank Fenner Building 141, Canberra 0200, Australia
e-mail: ian.white@anu.edu.au

T. Falkland
Island Hydrology Services, 9 Tivey Place Hughes, 2605 Canberra, ACT, Australia
e-mail: tony.falkland@netspeed.com.au

Q. Grafton et al. (eds.), *Understanding and Managing Urban Water in Transition*, Global Issues in Water Policy 15,
DOI 10.1007/978-94-017-9801-3_23

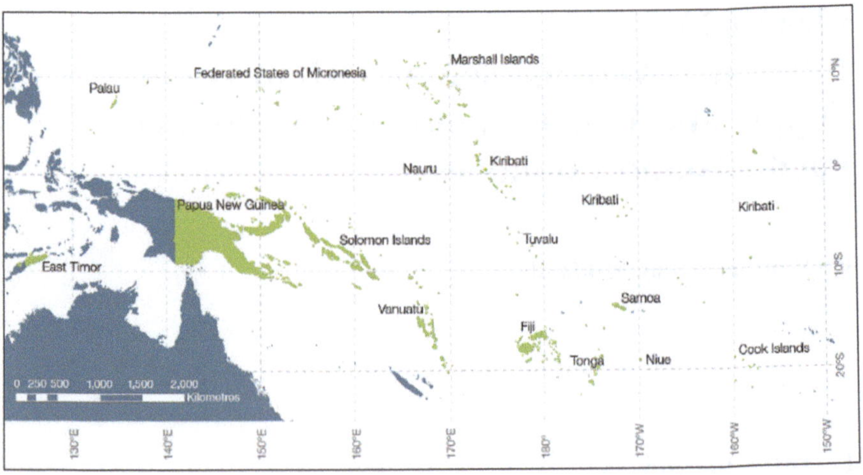

Fig. 23.1 Location of some of the Pacific Island countries in the central and south Pacific (PCCSP 2010)

Traditional coping mechanisms and customary rights and values, however, appear mismatched to the demands, responsibilities, and altered social dynamics of high-density urban centres, many of which are in an interactive phase between subsistence and urban living (Jones 1997; White et al. 1999a, 2008). To compound problems, many smaller Pacific island countries (PICs) have few trained water and sewage management personnel, and limited resources with which to tackle the challenges of urban living. The predicted effects of climate change, particularly sea-level rise (BOM and CSIRO 2011), add to the already complex challenges faced in urban centres in PICs.

Our initial entrée into this sector in the Pacific was as hydrologists and water engineers. Over the years it has become increasingly apparent that many factors beyond the technical contribute to the success or failure of water and sanitation projects in the region, and require attention. In this chapter, we outline the interactions between technical, climatic, social, and cultural factors in urban water supply and water quality in PICs. In working in many PICs over the last 35 years, our experience has been that to improve urban water supplies, policy makers, donors, and practioners need to recognise and address these complex interactions in an integrated way. Single-issue, infrastructure-focused solutions that have not engaged local communities have had a poor success rate in the region.

1.1 Water Security in PICs

A recent report (Falkland 2011) compared risks to water security from both climate change and non-climate factors in East Timor and 14 selected PICs – Cook Islands; Federated States of Micronesia (FSM); Fiji; Kiribati; Nauru; Niue; Palau; Papua

Table 23.1 Geographical and geological characteristics of Pacific Island countries and East Timor, and main sources of freshwater (Adapted from Falkland 2011)

Country	Land area (km^2)	Number of atolls or islands	Highest elevation (m)	Island geology	Main freshwater sources[a]
Cook Islands	237	15	652	Volcanic, limestone, atolls, mixed	SW, GW, RW
Federated States of Micronesia	701	607	791	Volcanic, atolls, mixed	SW, GW, RW
Fiji	18,273	322	1,324	Volcanic, limestone, atolls, mixed	SW, GW, RW, D (resorts)
Kiribati	811	33	87	Atolls & reef islands, limestone island	GW, RW
Nauru	21	1	71	Limestone	RW, GW (limited)
Niue	259	1	68	Limestone	GW, RW
Palau	444	~250	213	Volcanic, with some limestone	SW, GW, RW
Papua New Guinea	462,840	~600	4,509	Volcanic, limestone, atolls, reef islands, mixed	SW, GW, RW
Republic of the Marshall Islands	181	34	10	Atoll and reef islands	SW, GW, RW, D (emergencies)
Samoa	2.785	10	1,857	Volanic	SW, GW, RW
Solomon Islands	30.407	922	2,335	Volcanic, limestone, atolls, reef islands	SW, GW, RW
Tonga	650	176	1,033	Volcanic, limestone, reef islands, mixed	GW, RW, SW (limited)
Tuvalu	26	9	6	Atolls	RW, GW (limited), D (emergency)
Vanuatu	12,281	82	1,877	Volcanic with coastal sands and limestone	SW, GW. RW
East Timor	14,922	3	3,033	Mixed igneous, metamorphic, and sedimentary	SW, GW, RW

[a]*SW* surface water, *GW* groundwater, *RW* rainwater harvesting, *D* seawater desalination

New Guinea (PNG); Republic of Marshall Islands (RMI); Samoa; Solomon Islands; Tonga; Tuvalu; and Vanuatu (see Fig. 23.1). It concluded that, throughout the region out to the year 2030, the non-climate factors: increasing water demand; water pollution due to expanding populations; leakage from pipe systems and unaccounted-for water; poor water governance; and inadequate management; pose much greater risks to water security than does climate change. The most vulnerable areas identified were: densely populated urban and peri-urban settlements; remote communities; and communities in low-lying areas, particularly those in low coral atolls and carbonate islands with no fresh groundwater resources.

1.2 Urbanisation and Water Quality

Across the Pacific region, the percentage of urban dwellers in total populations ranges from about 20–100 %. About 50 % of islanders in the region are able to access improved water supplies although not all urban centres have treated, reticulated water supplies, as evidenced by the health statistics for water-borne illnesses (WHO 2005). Increasing urban population growth and inward migration (Ward 1999) mean that population densities are often high, with some exceeding 10,000 people/km^2. These densities, inadequate sanitation, the added burden of waste from domestic animals (especially pigs), and rudimentary waste disposal systems mean that the quality of shallow groundwater or neighbouring streams, on which communities depend, is often compromised. As a result, death rates and diseases due to preventable water-borne illnesses are tragically high in PICs, particularly among infants and the elderly (Fig. 23.2).

Following a major diarrhoeal outbreak in children in urban South Tarawa, the capital of Kiribati (Fig. 23.1), where treated freshwater remains in chronic short supply, the government of the republic requested assistance in upgrading water supply and sanitation systems. A field appraisal of the situation by the Australian International Assistance Bureau's (AIDAB) Pacific Regional Team (AIDAB 1993) concluded that the problem in South Tarawa was critical and should be addressed in as comprehensive manner as possible if sustainable and effective development was

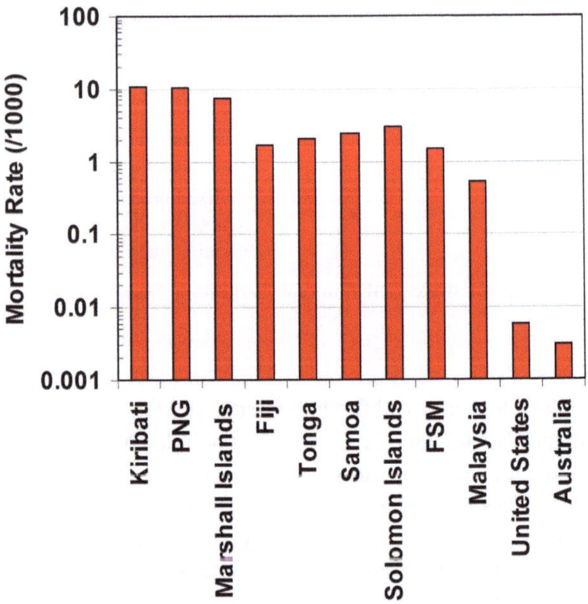

Fig. 23.2 Comparison of infant (<5 years) mortality rates per 1,000 due to diarrhoeal diseases for selected countries (data from WHO 2005)

to be achieved. It recommended a multidisciplinary, coordinated, long-term (at least 10 years) program with multiple components, including community health, education, water supply, sanitation, appropriate technology, and institutional strengthening and management, all conducted in parallel and with annual budgets of several million (1990) Australian dollars (AU$).

Sadly, this early recommendation for strategic integrated water resource management (IWRM, see e.g. Carpenter and Jones 2004), which embraced the cultural, social, governance, economic, geographic, and climatic contexts was shelved. Nonetheless, the Regional Team's appraisal exemplifies the fact that responses based on infrastructure alone are inadequate to meet the complex interacting factors encountered in urban water supply systems in the Pacific.

1.3 Scope of This Chapter

We concentrate in this chapter on freshwater management in urban settlements in PICs, recognising the important climatic, hydrogeological, cultural, and social contexts within which water supply is managed. Water supply systems in urban areas in the Pacific are extremely diverse, ranging from reverse osmosis desalination plants, through fully treated reticulated and treated water supply systems, to household rain tanks and shallow household groundwater wells bailed out with a bucket. Many households use multiple water sources of varying water quality, including seawater to supply their water needs. Despite this diversity, there are some common themes which apply across most urban areas.

Application of simple single-issue solutions, such as water infrastructure development or imposition of water tariffs, to complex urban water supply challenges in PICs have tended to be unsuccessful. Approaches are required which recognise the geographic, climatic, cultural, social, and economic contexts and the strengths and resilience of Pacific islanders and address the key issues broadly. We examine in this chapter issues connected with island development and water, water governance, assessment and monitoring of water resources, management and protection of water sources, management of demand and losses, capacity building, and the role of regional organisations and community empowerment in coping with climate variability and change.

2 Development, Water, and Aid

It has long been recognised that water plays an important role in creating wealth and in advancing national development (see e.g. AusAID 2003). Donor agencies have consequently funded many large water infrastructure projects in small island communities. It has been pointed out, however, that these have resulted in the loss of householder responsibility for water conservation and protection (Crennan 2002).

In other developing regions, widespread criticism of aid infrastructure projects has led to calls for 'pro-poor' development growth assistance with emphasis on improved management efficiency, institutional reforms, and community participation. Poverty in the Asia–Pacific is claimed to stem from institutional weakness and policy failures that prevent countries from taking advantage of the opportunities of globalisation and coping with its risks (Kaosa-ard 2003).

Lessons from other regions in the world that rely on freshwater both for subsistence and development are pertinent to PICs. In the Mekong, six factors have been identified that could cause relative poverty or disadvantage to increase (Kaosa-ard 2003).

1. Continuing deterioration of natural resources, on which the poor depend;
2. The conflict between national laws and customary rights;
3. Spending of public resources on large infrastructure projects rather than on social investment;
4. Inherent inability to transfer the opportunities of globalisation to the poor and its relation to property rights;
5. Institutional failures, conflicting jurisdictions and agendas between government agencies; and
6. Reluctance to empower community participation in natural resource management.

Some of these factors are apposite to PICs. It must be emphasised, however, that most Pacific islanders own or have access to land, so that true poverty as understood by Pacific people in the sense of landlessness is seldom encountered in the region (Barnett 2005; Pacific Forum Secretariat 2007). Also, the potential for small island communities to participate in the benefits of globalisation through improved irrigation are limited by restricted land areas, acute water shortages during dry times, and their isolation from markets. Some niche markets exist, such as the production of squash pumpkins in Tonga (van der Velde et al. 2007), but these are mainly confined to larger, higher, volcanic islands with fertile soils. Tourism could increase wealth generation in atolls, as in the Maldives, but it has high per capita water demands.

Aid and donor programs addressing institutional and governance weakness, policy failure, property and customary rights, community empowerment, and the deterioration of water and associated land resources – together with providing the broad community with appropriate knowledge and information – may provide the greatest benefits in PICs (White et al. 2007a). These, however, require long-term commitment and partnerships with regional, local community, and non-government organisations.

3 Water Governance

Impaired governance is claimed to be the main obstacle to better and more equitable water sharing, and improved water supply and services, in many water-stressed countries (Solanes and Jouravlev 2006; UNWWAP 2006). Importantly, improved governance is one of the four key priorities of the 2005 Pacific Plan (Pacific Islands Forum Secretariat 2007). One of the Plan's key governance strategies is 'improved transparency, accountability, equity and efficiency in the management and use of resources in the Pacific'. This strategy recognises that there are significant challenges in the governance of natural resources in the region which are especially evident in the water sector. This is emphasised in the Pacific Regional Action Plan (RAP) on Sustainable Water Management (SOPAC and ADB 2003) which identifies common themes across the Pacific region and calls for:

1. National water sector assessments;
2. Broadly-based national water vision;
3. National water action agenda and plans;
4. Empowerment of communities;
5. Design of capable institutions;
6. Integrated investment plans;
7. Regional support; and
8. Dialogue with investors and donors.

Problems identified by the RAP included the general absence of national government water policy; implementation plans; water resource legislation; and a lack of capacity and resources to develop and implement them. Public policy is an authoritative response by government to public issues or problems that provides leadership, direction, coordination, and resources (Bridgman and Davis 2004). The absence of national policy, legislation, implementation plans, and whole-of-government and community national steering committees in some PICs means that government priorities in the sector remain unspecified; resources are not directed towards particular needs; the roles and responsibilities of government agencies are not clearly defined; and, in many cases, there is no legal protection for water sources. In addition, there is a general absence of peak national water and sanitation bodies made up of members drawn from relevant ministries, agencies, and community organisations to advise government. A summary of the current state of progress in water governance in 14 PICs as well as East Timor is shown in Table 23.2.

Several of the countries in Table 23.2 have, with external assistance, had draft policies, plans, and legislation prepared, some for several decades, but they have not been submitted to parliament for consideration and endorsement. Table 23.2 reflects a general reluctance to announce national water policies and plans, enact national water legislation, define rights and responsibilities, adopt whole-of-government approaches, and involve communities in planning and managing water resources and related land resources. Part of this reluctance stems from the fact that, until recent urbanisation, water supply was largely an extended family or clan responsibility, while sanitation was an individual concern. Many PICs, however, have recent intermediate- to long-term sustainable development policies, plans, and vision statements in keeping with the Pacific Plan, most of which reflect the general community aspiration for improved and safer water supplies and appropriate sanitation services (see e.g. NDS 2011).

Table 23.2 Summary of water governance progress in 14 selected PICs and East Timor (modified from Falkland 2011). Dark grey – absence of instrument; light grey – draft instrument exists; no shading – instrument exists

Country	Water and sanitation policy	Water legislation	IWRM plans or similar	National water and sanitation committee or similar
Cook Is	(light grey)	(light grey)	(dark grey)	(light grey)
Federated States of Micronesia	(dark grey)	(dark grey)	(dark grey)	(dark grey)
Fiji	(light grey)	(light grey)	(light grey)	
Kiribati		(dark grey)		
Nauru		(dark grey)	(light grey)	
Niue	(light grey)	(light grey)	(dark grey)	(light grey)
Palau	(dark grey)	(dark grey)	(dark grey)	(light grey)
Papua New Guinea	(light grey)	(light grey)	(dark grey)	(light grey)
Republic of the Marshall Islands	(light grey)	(dark grey)	(light grey)	(dark grey)
Samoa		(light grey)	(light grey)	(light grey)
Solomon Is	(light grey)	(light grey)	(dark grey)	(light grey)
Tonga	(dark grey)	(light grey)	(dark grey)	(light grey)
Tuvalu	(dark grey)	(dark grey)	(light grey)	(light grey)
Vanuatu		(light grey)	(light grey)	(light grey)
East Timor	(light grey)	(light grey)	(light grey)	(light grey)

3.1 *Resource Ownership and Customary Rights*

Part of the reluctance to endorse national water policy, plans, and legislation stems from wide-spread customary rights and the tradition that land ownership implies resource ownership, including adjacent surface water and underlying groundwater (White et al. 2007a). In many PICs, virtually all land is owned by traditional owners. This means the creation of water reserves on privately owned land to protect streams, catchments, reservoirs, or groundwater sources creates conflicts between governments and landowners, sometimes resulting in vandalism of water infrastructure (White et al. 1999a), disruption of services, or blockades, leading to the abandonment of suitable public water or hydropower sources (Low 2011).

Land ownership is fundamentally important in many PICs and is complex and very diverse even within single countries (Foukonga 2007). It is the primary source of wealth. Traditionally it confers subsistence rights, including fishing rights, on land owners and also provides a social security system for parents through the prospect of inheritance by their children who are therefore obliged to care for their parents (Jones 1997). The concept of 'public use' – of governments controlling private family lands for water supply to communities and villages remote from the water reserve – is often considered to be to the detriment of local obligations to the extended family and the local community. 'Public use' is a foreign concept in many island communities. From the landowner's view, their long-term relationship with the land and its role in providing daily food for subsistence living for family members far outweighs any need of the government for public good (Crennan 1998).

In many PICs, constitutional law has supplanted customary law. The deep-seated belief, however, that land ownership also confers water ownership means that governments are reluctant to enact water policy or legislation which specifies that freshwater resources belong to the government, or to ban land uses with the potential to pollute water sources for fear of infringing landowner property-use rights. This issue is of extreme political sensitivity. As a consequence, in some PICs there is no legal protection of water from over-extraction or from contamination or misuse. In some cases, more than adequate water resources exists in remote regions of an island, but its transfer to urban areas in times of severe deficit, or its use for hydropower production, is virtually impossible because of land ownership issues (Low 2011).

The impact that traditional land and resource ownership has on national development has been recognised throughout the Pacific. The Prime Minister of Vanuatu, together with the Council of Chiefs, has recently commenced nation-wide negotiations on land reforms (C. Ioan, pers. comm., October 2013).

Many landowners also question the right of government to charge for water abstracted from private land. There is a general view that water is 'a gift from God'. This view makes controlling demand through water tariffs a contentious issue and one which threatens the financial sustainability of urban water systems in several PICs (Low 2011; White 2011c).

While customary rights in relation to water ownership was appropriate in low-density, subsistence conditions it is ill-suited to high-density, urban populations where one household's actions can immediately and directly impact both the quantity and quality of water available for its neighbours. Because customary rights are ingrained, it has been proposed that behavioural change is necessary for conserving and protecting water through longer-term education, awareness, and community engagement (Crennan 2002).

3.2 *Development of Policy, Plans, and Legislation*

An implicit assumption underpinning many attempts to assist PICs to develop national policy and plans is that the plethora of water policy frameworks, and policy and planning "tool kits", available from developed world countries (see e.g. GWP 2003) are directly and rapidly transferable to developing countries. The assumption here is that such approaches are context-independent. But experience has shown that quick, developed-world formulaic solutions, which take no account of island priorities, traditions, strengths and practices that have evolved over millennia, are often politely but firmly ignored.

There are no easy prescriptions for the rapid translocation of relatively recent water governance reforms and water management frameworks from developed countries to developing PICs. There is often an underlying presumption that there are well-developed policy processes in place. In some PICs, policy processes are vague, not clearly defined, or even non-existent. In addition, they assume that there are adequate means to implement policy, plans, and legislation. In developed countries, there are frequently hundreds of people engaged in the implementation of policy, planning, and management of water resources whose ownership by the state is clearly legally defined. In this case, the major priorities are addressing the environmental impacts of water supply extraction and effluent treatment systems, controlling demand through long-established market mechanisms, and accommodating the predicted impacts of climate change.

In many small PICs, there are often only one or two trained water professionals whose tasks may range from replacing washers in domestic taps, replacing pumps, and unblocking clogged sewers to advising the minister and representing the country at international meetings. The major daily challenges in water governance they face are maintaining, even intermittently, supplies of adequate quantities of safe freshwater to growing populations – with very limited resources, no economies of scale, and coping with the complex cultural, social, and institutional changes necessary to move from subsistence to urban living.

Attempts to assist small island states to develop water policy, plans, and legislation must recognise local contexts, capacities, and inherent strengths of island communities (Barnett 2001) and be based on successful public policy principles. Some of the keys to effective, efficient, and widely-accepted public policy are (adapted from Sabatier and Mazmanian 1979):

1. The policy must be supported by the government;
2. Policy and associated implementation mechanisms are based on sound knowledge;
3. The policy, the law that gives effect to the policy, and its implementation contains clear policy directives;
4. Those responsible for implementing the policy have appropriate managerial political skills, information, and resources;
5. Policy and associated implementation mechanisms are actively supported by constituent groups; and
6. The relative priority of the policy's legal objectives is not undermined by other laws, policies, and implementation mechanisms.

Where there are multi-level or multi-agency governance arrangements, an additional key is:

7. Appropriate management structures exist that facilitate negotiations, agreements, and monitoring of implementation.

In assisting PICs to develop water policy and plans, a key strategy is to build on inherent island strengths using the well-developed local institutions, resilient social systems, and sensitivity to environmental change (Barnett 2001, 2005). Fortunately, extensive national consultations (Carpenter et al. 2002) prior to development of the Pacific RAP provided opportunities for island communities to raise their concerns and priorities over water and sanitation, so in most PICs there is already widespread community support for change and improvement. Challenges, however, arise at the organisational level of government, where agencies have a strong tendency to act as 'silos'.

3.3 Peak Water Bodies

Urban water resource management in PICs encompasses health, environmental, economic, social, cultural, infrastructural, economic, and technical issues. Because of this, it is essential in small island states that all government agencies, community organisations, and businesses who have responsibilities and interests in freshwater participate in deliberations on policy and plans and in reviewing outcomes. The establishment of a whole-of-government, agency and community-representative peak bodies (such as a National Water and Sanitation Coordination Committee reporting to cabinet, or the designated lead government water agency) provides a PIC-relevant mechanism to help drive the policy development process, as well as to oversee policy implementation, progress, reporting, and dissemination (White et al. 2009a).

Debates have recently centred on the efficiency of engaging communities in water planning, especially in developed countries (see e.g. Daniell 2012). In island life, negotiation and consensus are fundamentally important in reaching decisions, so although the development of policy can be a lengthy process it is fundamentally necessary to gain widespread support. The process used in the Republic of Nauru (Fig. 23.1) is shown in Fig. 23.3, with the peak government–community water and sanitation body, CPSC, playing a pivotal role in driving an iterative policy development process (White and Falkland 2012).

3.4 Adaptive Process for Policy Development

In assisting Solomon Islands and the republics of Nauru and Kiribati to develop their national water policies and implementation plans, a 5-phase adaptive planning process (adapted from Ackoff 1999) was used with the respective peak bodies. This process is summarised in Table 23.3. The first two phases of the process identify priority issues to be addressed and to corresponding policy goals. The remaining three phases help construct the policy implementation plans.

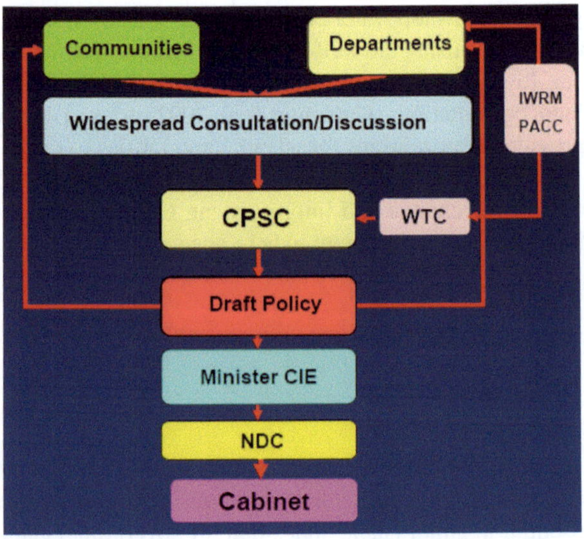

Fig. 23.3 Iterative process used in the Republic of Nauru for developing a draft national water, sanitation, and hygiene policy. CPSC is the Department of Commerce, Industry and Environment's (CIE) whole-of-government and agency–community Project Steering Committee. CPSC is advised by the whole-of-government and agency Water Technical Committee (WTC) convened by the Integrated Water Resource Management (IWRM) and Pacific Adaptation to Climate Change (PACC) project coordinators. Once the draft policy has been approved by CPSC and the responsible minister, it is submitted to the heads of departments and agencies National Development Committee (NDC) for consideration of cabinet

This process is easily understood, even in countries with loosely defined or non-existent policy processes, and assists in identifying pressing problems, priority goals, and the steps necessary to achieve them. Common issues appear in many PICs: a lack of knowledge of the amount, quality, or use of resources being managed; inadequate water supplies; unreliable or compromised water quality; financially unsustainable water supply systems; absence of effective demand management or unacceptable water losses; lack of community participation in the conservation and protection of water sources; poor or inadequate governance; and lack of resources and capacity to manage water resources and supply systems. Many of the pressing current and future problems in PICs appear to be addressed through eight policy objectives (White and Falkland 2011):

1. Improve understanding, assessment, and monitoring of water resources and their safe yields, quality, and sectoral use;
2. Protect and manage sources of freshwater;
3. Increase access to safe and reliable water supplies and appropriate sanitation;
4. Achieve financially, environmentally, and socially sustainable water resource and supply management;
5. Increase community participation in water management and conservation;

Table 23.3 The five phases of interactive planning used in developing national water policy and implementation plans (Adapted from Ackoff 1999)

Phase	Objective	Components	Principal outputs
I. Formulation of the issues	Determine issues, problems, and opportunities	Previous actions and policies; recognised issues; problems, opportunities, and their interactions; constraints to effective management	Issues to be addressed by policy, plans, legislation
II. Ends planning	Determine where you want to be and the gaps between that and now	Extract vision, principles, goals, and objectives to achieve the desired ends	Policy principles, policy goals and objectives
III. Means planning	Choosing mechanisms to achieve goals and objectives	Develop and select actions for achieving goals and objectives and indicators for completion of actions	Implementation plan actions
IV. Resource planning	Determine resources required for planned actions	Define resource needs and identify if resources are available or how they will be generated or acquired	Needs for implementation plan resources
V. Implementation and control	Determine responsibilities and schedules for implementation	Identify who is responsible for actions, when they are to be implemented, and how implementation is to be monitored	Implementation plan schedule and responsibilities for implementation. Inclusion in ministerial operations plans

6. Improve governance in the water and sanitation sector, use the principles of integrated water resource management, and regularly review policy outcomes and planning milestones;
7. Improve management of risk in natural hazards and extreme events; and
8. Provide resources, training opportunities, and mentoring for staff in the sector.

4 Resource Assessment and Monitoring

The first key principle in the list of Sabatier and Mazmanian (1979) for successful public policy (Sect. 3.2) is that policy and associated implementation mechanisms should be based on sound knowledge. In some PICs this remains a challenge. The full extent of their water resources, their quality and fitness for use are often only partly known; the sustainable yields are poorly characterised; the impacts of climate variability, water extraction, and land use on the resources are inadequately characterised; the demand for and use of freshwater by sectors is incomplete; and the impact of management regimes and policy decisions only partly recognised. A range of techniques, from simple approximations through to sophisticated geophysical techniques and modelling, are available for assessing the extent of water resources, their sustainable yield, and the quality of the water extracted (UNESCO 1991; Falkland 2002a, b).

These knowledge gaps are especially evident for vulnerable low, small island and atoll groundwater systems supplying growing urban areas, a situation where thorough resource assessment is required along with a commitment to ongoing monitoring, analysis, and reporting. In these systems, the salinity of the extracted groundwater is a result of the dynamic balance between rainfall recharge, pumping rates, and seawater intrusion. Regular monitoring ensures that sources remain viable (Falkland 2002a, b; White and Falkland 2010). PICs in the central and central western Pacific exhibit extreme variability in rainfall, which is closely correlated with sea surface temperature (Fig. 23.4). Faced with this variability, monitoring of rainfall, groundwater level, and salinity, as well as water extraction rates, is vital.

While monitoring is always contentious even in developed countries, the importance of monitoring is exemplified in one small urbanised island during the widespread 1998–2001 ENSO-related drought. A national state of disaster was declared by the government when urban supplies diminished, despite the fact that the main groundwater source contained more than adequate freshwater to meet reasonable demand. The lead water agency had stopped monitoring groundwater during the

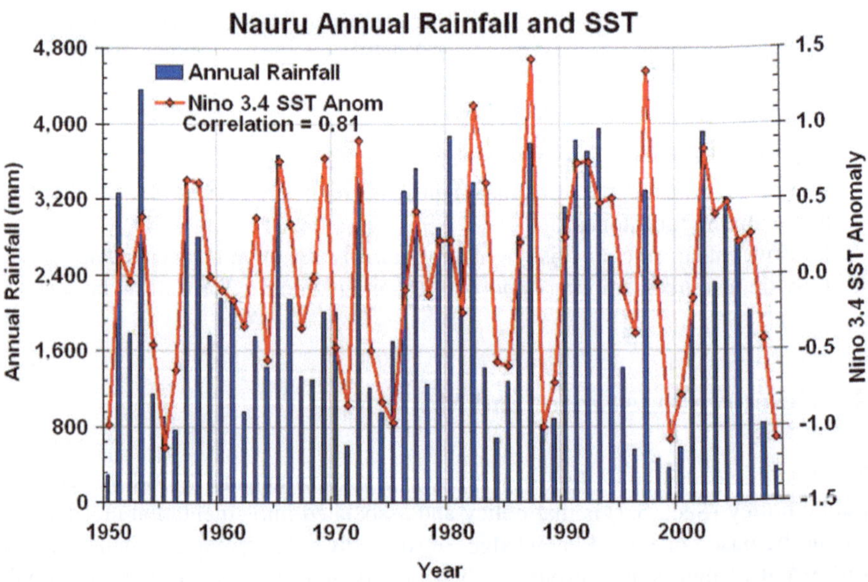

Fig. 23.4 In Nauru there is a strong correlation (0.81) between the large variability of annual rainfall and the variability of annual sea surface temperature anomaly in the Niño 3.4 region of the central-western Pacific (White 2011a)

drought, so the actual availability of water was unknown. The real problem was an inadequate and leaking urban water distribution system. Limited human and financial resources in many PICs mean that assessment, monitoring, reporting, and dissemination of information, even to government, are major challenges (van der Velde et al. 2007).

The establishment of peak government–community water and sanitation committees, with responsibilities for oversight of water resources and for reporting directly to government (Sect. 3.2), can promote regular monitoring, analysis, reporting, and wider dissemination of information; the committees need terms of reference that specify regular inter-agency discussions and reporting on the condition and use of water resources.

5 Protection of Water Sources and Water Quality

Island communities rely on a variety of public and private sources of surface water and groundwater, harvested rainwater, bottled water, brackish water, and even seawater. Strategies for ensuring that the quality of water supplied from surface and groundwater sources for communities is fit for consumption are based on the concept of maintaining multiple barriers to contamination between source and consumption. Protection of water sources from contamination and misuse is a prerequisite first step which requires effective land use planning and zoning, and regulation of land uses in areas around surface water sources and on land overlying groundwater sources. Human settlements, agriculture involving the raising of livestock, the use of chemicals and fertilisers, mining forestry, waste disposal, and especially defective or inappropriate sanitation systems all increase the risk of water contamination. Creation of water reserves surrounding water sources, however, are often problematic owing to land and resource ownership issues (Sect. 3.1) and unclear legal rights and responsibilities.

5.1 Water Reserves

The establishment of water reserves, by current governments and former colonial powers, in order to protect water sources has often failed to appreciate local community needs, culture, land tenure, and land use requirements. Such decisions have led to conflicts between local communities and government agencies, which have sometimes manifested in vandalism of infrastructure, disruption of supply, and demands for compensation by landowners (Low 2011). The impact of water resource development on the health and productivity of neighbouring traditional crops grown on or near water reserves is of social and economic concern to landowners, particularly where land use and access rights have been forbidden or severely limited (White et al. 1999a).

The tensions between 'public use' private family lands by governments for water supply and the landowers' need for unrestricted access to their land for subsistence leads to long-standing conflicts. The resolution of land use conflicts requires appropriate legal, administrative, and financial provisions and the involvement of the local community in managing the reserve, or in some PICs even the local water supply system. Where land is limited, the provision of social amenities such as non-polluting sports fields on reserves appears possible (White et al. 1999a).

Because of the central, social importance of landownership, many governments believe that outright purchase of water reserves is not politically feasible. Instead, some PIC governments pay commercial rentals or annual compensation to land owners. Rental or compensation payments have three distinct disadvantages. First, they impose a heavy, continuing financial burden on water supply systems. Second, they tend to encourage misuse of the reserves since they are viewed as 'government' land with common property uses, including gravel mining and water disposal, and they generate disputes within the communities. Finally, there appears to be little overall long-lasting social benefits from cash payments.

5.2 *Water Reserve Management Committees*

In order to overcome misuse of reserves it has been suggested that, rather than pay water reserve landowners rental as compensation for loss of amenity, they be paid as water reserve managers with a contract to ensure that reserves are well cared for (White et al. 1999a). Involving local landowners in water reserve management committees can be a successful strategy in engaging island communities in the protection and care of water sources and in identifying land uses that minimise impact on water sources. Village-level water committees have proved successful in rural areas in Tonga and Samoa and offer a potential model for other PICs. These, however, need to be underpinned by legal protection for water sources from contamination and misuse, and the agency or agencies responsible for managing water reserves require a clear mandate and legal authority to enforce laws.

5.3 *Appropriate Technology*

Isolation, coupled with the generally corrosive conditions in PICs, dictates that the infrastructure used to extract, store, and supply water must be robust, and simple to operate, maintain, and repair (UNESCO 1991). In addition, the unique hydrogeology of groundwater in low, small islands requires the use of appropriate technology to maximise freshwater yield. The delicate hydrostatic balance between freshwater and the surrounding and underlying seawater in small islands is easily disturbed by inappropriate groundwater development. The most common method of accessing groundwater uses hand-excavated dug wells typically 2–3 m deep and approximately

1 m below the groundwater level. Groundwater is abstracted by buckets, hand pumps, or small electric pumps. Such systems work well at household levels, provided abstraction rates are low.

For public water supply pumping systems, single or multiple dug wells or drilled boreholes have been used on some small islands. These vertical abstraction systems can cause upconing of underlying brackish water or seawater, causing increases in salinity of the abstracted water to levels that are sometimes too high for potable use. Pumping from long, horizontal infiltration galleries (up to 300 m long) or skimming wells has proven to be a far better abstraction method, particularly in islands with thin freshwater lenses (Falkland and Brunel 1993).

Infiltration galleries skim fresh groundwater from the surface of a freshwater lens, and thus distribute the pumping drawdown over a wide area. In so doing, they avoid excessive local drawdown and upconing of saline water associated with pumping from vertical boreholes. Infiltration galleries are used for public water supply in Tarawa and Kiritimati atolls, Kiribati; Majuro and Kwajalein atolls in the Marshall Islands (Peterson 1997); Aitutaki island, Cook Islands; and Lifuka island, Tonga (Fig. 23.1). On Lifuka, replacement of boreholes with infiltration galleries significantly lowered the salinity of the water supply (Falkland 2000).

For raised limestone islands, where depths to the groundwater table are greater than 10 m and up to 50 m or more, such as in Tongatapu in Tonga, abstraction using vertical drilled boreholes is currently the most practical method of developing freshwater lenses. In the future, directional drilling from the surface may be an option for installing horizontal infiltration galleries on these islands.

5.4 *Sewage and Waste Contamination*

The tragically high infant death rates shown in Fig. 23.2 are partly due to contamination of water by faecal material (WHO 2005). Appropriate sanitation in small island urban centres is a major concern. Urban sanitation systems in PICs commonly use rather basic septic tanks and pit latrines. Domestic animal wastes, especially from pigs, add to biological contamination of water sources, particularly household groundwater wells, and pose major health risks in urban areas on many islands and even in designated water reserves (Fig. 23.5).

In parts of heavily populated atolls such as Majuro (Marshall Islands) and Tarawa (Kiribati), piped sewerage systems that use seawater for flushing (to conserve limited freshwater supplies) have been installed to address the problem of sewage contamination of groundwater. Squatter settlements in urban centres, however, do not have access to these systems. Compost toilets protect water sources as well as conserving scarce water resources, and these have been trialled in a number of PICs including Kiribati, Tonga, Tuvalu (Crennan and Berry 2002), and Nauru. While compost toilets have many advantages and have been accepted in some communities, cultural attitudes have so far limited their widespread use in others. Other technical solutions are available, including improved septic tanks and relatively simple

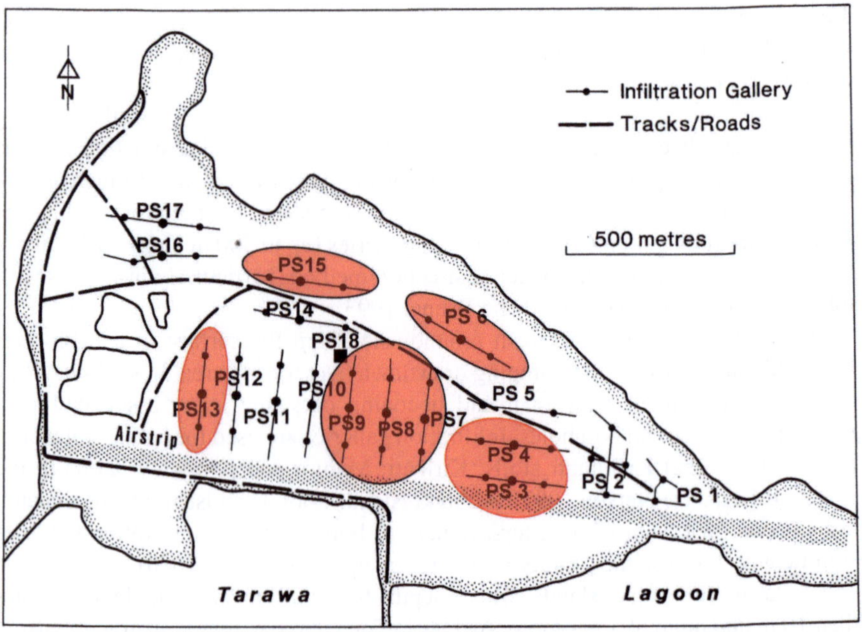

Fig. 23.5 Shaded ellipses show distribution of positive E. coli water samples taken from urban water supply pumping galleries (PS) on Bonriki water reserve, Tarawa Atoll, Kiribati. Positive samples were found to correspond with a graveyard, pig pens, market gardens using animal manure, and squatter huts (White et al. 2007b)

effluent disposal and treatment systems (Bower et al. 2005; WHO 2008). It is emphasised here that selection of appropriate sanitation systems in small island urban areas can be a matter of life and death.

The most appropriate strategy is to ensure that human settlements, including their sanitation and waste disposal systems and domestic animals, are placed well away from freshwater sources used for public water supply (see Sect. 5.1). A study using bromide tracer in Lifuka, Tonga, concluded there was no safe distance between pit latrines or septic tanks and water supply wells in urban areas because of the density of sanitation facilities and the high permeability of the unconsolidated aquifer (Crennan et al. 1998). Instead, the study recommended that alternative strategies such as source control of pollutants using composting toilets and water treatment are required. This is an area requiring major research.

Waste disposal in many developing urban centres in PICs is a continuing challenge for governments. The geographic isolation of PICs means that there are limited opportunities for recycling of wastes. Potentially polluting wastes, such as petroleum products, batteries, and sewage sludge, remain a problem and the separation of waste dumps from water sources is of utmost priority.

5.5 Water Treatment

The level of water treatment required for urban water supplies depends on the quality of the source water (WHO 2004). In PICs, water treatment systems vary from highly sophisticated, reverse osmosis plants; flocculation, sedimentation, filtration, and disinfection systems; through to the boiling of rainwater and household well water. Where water is reticulated to households, treatment is often required at source and close to delivery due to microbial build-up in water pipes and storages. Microbial build-up is a particular problem in systems that deliver water supply intermittently. In some urban centres, water is only supplied every third day.

In some PICS, the resources, supplies, and capacity necessary to manage water treatment facilities is limited. Because of this, a minimum water treatment strategy has been recommended for small island water supply. This strategy emphasises simple designs which minimise mechanical equipment; use of chemicals; and operation and maintenance (UNESCO 1991). One particular problem in PICs is that electric power supplies are sometimes intermittent, so that treatment systems that require power may not be fully functional. Recent advances in simple-to-operate, low-pressure, membrane filtration systems appear to have significant advantages for treating turbid and polluted stream water, groundwater, and stored rainwater, as has been demonstrated in Kiribati, Fiji, and East Timor (Skyjuice 2008, 2011).

The resources and technical skills required to manage wastewater treatment for human re-use is currently not a viable option for most PICs, with some discharging untreated sewage direct to surrounding oceans (Falkland 2011). Waste household grey water is, however, re-used for watering domestic animals, household vegetable plots, and fruit trees.

Many urban households in PICs source water from a range of sources including rainwater tanks, water wells, and the reticulation system, as well as seawater for both potable and non-potable use. Water quality from non-reticulated sources is a significant issue. In some urban areas, groundwater can be so polluted that it is unfit for any human use. In such areas, testing of local groundwater upon which so many households rely should be, but seldom is, routine. Rainwater tanks can also be contaminated and continuing householder education campaigns in the design, construction, maintenance, and care of rainwater harvesting systems are required (UNESCO 1991; SOPAC 2004). Local area water management committees could play a central role in such campaigns.

6 Management of Demand and Losses

Falkland (2011) identified increasing demand and continuing water losses from pipeline systems as major risks to water security in PICs to the year 2030, especially in urban areas (Sect. 1.1). This increasing demand is driven by natural increase in populations, inward migration to urban centres (Ward 1999), and increases in

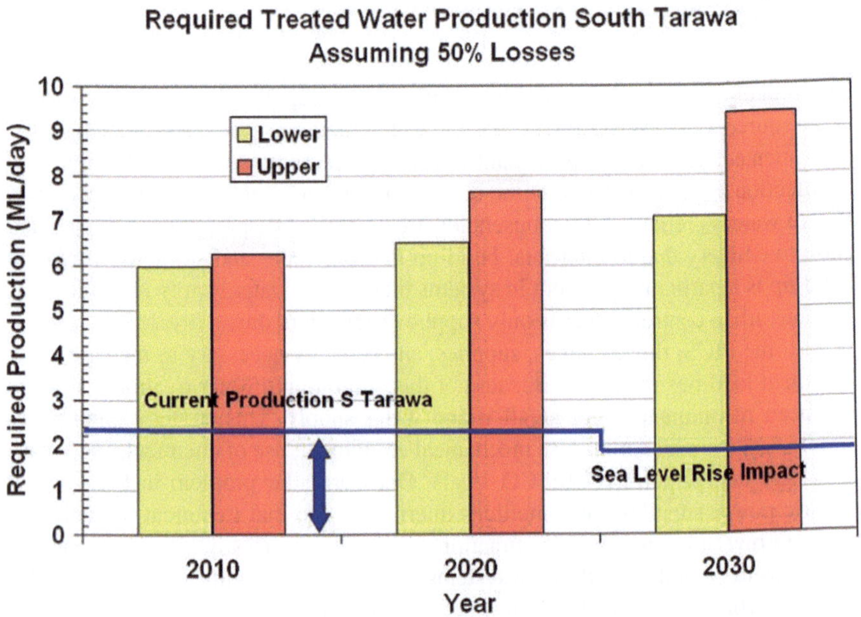

Fig. 23.6 Bars show estimates of the required daily production of treated freshwater required to supply the demand of the growing population of urban South Tarawa, Kiribati, assuming there is a conservatively estimated 50 % loss from the reticulation system. The estimates are compared with the sustainable yield of treated groundwater sources (*horizontal line*), and include a projected worse-case impact of sea level rise (White 2010)

standards of living. In some urban areas in the region, annual population growth rates are as high as 6 %, which places major strain on water supply and sanitation systems as well as urban services. To illustrate this risk, Fig. 23.6 shows the estimated water production required to meet both demand and the estimated current pipeline losses for the urban capital of Kiribati, South Tarawa, compared with the estimated sustainable yield of treated water to the year 2030 (including the projected impact of sea level rise). Households meet the current deficit in production by using water from untreated sources such as rainwater tanks and from unsafe, polluted household wells. Managing water demand and controlling losses are clearly major current and future challenges in PICs.

6.1 Estimating Demand

Designs for urban water supply systems require adequate information on the expected demand for water as well as the losses from the system (unaccounted for, or non-revenue water). Population growth rates and per capita water use are needed for demand projections. Only a few urban centres in PICs have adequate pipeline

Table 23.4 Summary of the estimated per capita freshwater demand for the urban capital, South Tarawa, Kiribati. Toilet flushing is excluded as it is assumed to be sourced from the seawater system (Adapted from White 2010)

Year	Total per capita demand (L/person/day)
1973	18
1975	10–37
1978	14–27
1982	34–35
1986	47
1992	50
1996	40[a]
2000	50[a, b]
2002	40[a]
2003	40[a]
2010	68[c]

[a]Assumes only 80 % of all households supplied with piped water

[b]Includes institutional, commercial, and industrial demand and 30 % assumed losses

[c]Assumes 6 L/person/day ICI demand and an extra 2 L/person/day due to increase in temperatures due to climate change

metering and metered household connections. Estimating demand and losses therefore present challenges. Because of inadequate data, some water supply system designs have been based on estimated minimum per capita island freshwater requirements based on a 'consensus' of needs in developing countries (Table 23.4) or merely on a share of the estimated available water yield. Household surveys suggest that these minimum requirements underestimate total water use, and make no allowance for future increase in per capita water consumption with increasing living standards or with increased temperatures due to climate change. In Table 23.4 the estimate for South Tarawa, Kiribati, in 2010 was based on household surveys from a similar urban environment in a distant atoll, together with a broad estimate of the institutional, commercial, and industrial water use (and a small allowance for increased water consumption due to projected higher temperatures from climate change). Data on institutional, commercial, and industrial water use is often difficult to find. In the absence of metered water supply systems, surveys of water use appear to be the best alternative for estimating demand.

6.2 *Non-Revenue Water*

Supply system design requires estimates of both demand and system losses. Leakage from water supply pipelines and other losses contribute to 'non-revenue water', the difference between the volume of water extracted from a source and the volume of billed, authorised consumption. In PICs these other losses include illegal

Table 23.5 Estimated non-revenue water in selected urban centres in island countries

Urban centre	Country	Water source	Non-revenue water
South Tarawa[a]	Tarawa Atoll, Kiribati	Shallow groundwater	50–70 %[i]
London[b]	Kiritimati Atoll, Kiribati	Shallow groundwater	20–50 %[i]
Nauru[c]	Nauru	Desalination	75–90 %[i]
Nuku'alofa[d]	Tonga	Groundwater	75 %[i]
Honiara[e]	Solomon Islands	Surface and groundwater	50 %
Majuro[f]	Marshall Islands	Rainwater	50 %
Dili[g]	East Timor	Surface and groundwater	85 %
Auki[e]	Solomon Islands	Surface water	49 %
Noro[e]	Solomon Islands	Surface water	54 %
Tulagi[e]	Solomon Islands	Surface water	77 %
Malé[h]	Maldives	Desalination	<10 %

[a]White (2011c), [b]ADB (2007), [c]White (2011b), [d]White et al. (2009b), [e]SINIIP (2013), [f]SOPAC (2007), [g]Falkland (2011), [h]M. Didi, pers. com., March 2013
[i]Estimated value, incomplete or adequate water metering

connections, theft, non-metered connections, and uncontrolled overflows at community or household tanks in urban centres and larger rural villages. Non-revenue water is a major problem for urban water authorities in PICs (see e.g. SOPAC 1999; SIWA 2013). In many urban centres, losses are difficult to assess because of the absence or malfunction of water meters. Losses equal to or exceeding 50 %, and even as high at 90 %, have been measured or estimated in a number of PIC urban water supply systems (see Table 23.5).

There are two contrasting values for non-revenue water from desalination systems in Table 23.5. Nauru's desalination plant, which supplements rainwater harvesting for the island's 10,000 people, consumes one-third of the total national electricity production per year. The magnitude of the non-revenue water there represents a major national economic loss. In Malé, the capital of the Maldives, the Malé Water Authority, a public–private partnership, has a tiered tariff system in place for strictly metered desalinated water supply to the island's 100,000 residents. Malé has very low non-revenue water.

The magnitudes of non-revenue losses in PICs in Table 23.5 are extremely costly and wasteful. They add to water shortages and cause intermittent supply, even in non-drought periods. Reduction of losses and systematic leakage control to reduce shortages, improve services, increase financial sustainability, and delay the need for future investment in new water infrastructure is a regional priority (SOPAC and ADB 2003). Despite this, detection and reduction of losses has been a lower priority for donor and loan agencies than new infrastructure projects.

Regional capacity building programs in leak detection have been run, but many countries have not adopted leakage control as a mainstream activity. This is largely due to inadequate financial resources and insufficient trained personnel. While an adequate water metering network is an essential first step in identifying losses, there is a strong continuing need for training in leak detection and reduction and support for such programs.

6.3 Rainwater Harvesting

As shown in Table 23.6, most of the main urban centres in the regions have annual rainfalls exceeding 1,500 mm. The exceptions are Dili in East Timor and Port Moresby in Papua New Guinea, both in the monsoon belt. Household and public rainwater harvesting has the potential to supplement, and in some cases supplant, limited reticulated supplies and reduce system demand, particularly in peri-urban areas (SOPAC 2004). Rainwater harvesting is vital in urban centres with no surface or groundwater sources, such as Nauru and Funafuti in Tuvalu. Large rainwater storages fed from bigger public buildings such as churches, schools, meeting halls, and government buildings can be used for public water supply, especially during dry periods when household tanks fail, as in Tuvalu.

Rainwater harvesting is attractive in PICs because its water quality may be safer than other sources and the supply is directly under household control. Rainwater harvesting is not, however, the universal answer to urban water supply in PICs. Estimated unit production costs for large rainwater harvesting schemes in Tarawa, Kiribati (with assumed lifetimes of 50 years and a 5 % failure rate) were five times greater than those for shallow reticulated groundwater pumping and over three times those of desalination (with an assumed lifetime of 10 years) (White 2011c). Unit production costs for smaller household systems in PICs are even higher.

Cost is not the only factor in the viability of rainwater harvesting. The variability of rainfall (Fig. 23.4) and the length of regular dry seasons (Table 23.6), coupled

Table 23.6 Mean annual rainfall, coefficient of variability, and distribution of dry and wet season rainfalls in selected urban centres of island countries (Falkland 2011)

Country	Capital/urban centre	Mean annual rainfall (mm)	Coefficient of variation (CV)	Mean percentage in 6 month dry/wet seasons
Cook Islands	Rarotonga	2,000	0.2	35/65
Federated States of Micronesia	Pohnpei	4,700	0.15	45/55
Fiji	Suva	3,000	0.19	37/63
Kiribati	South Tarawa	2,000	0.47	39/61
Nauru	Yeren	2,100	0.54	40/60
Niue	Alofi	2,100	0.24	34/66
Palau	Melekeok	3,700	0.13	41/59
Papua New Guinea	Port Moresby	1,100	0.24	20/80
Republic of the Marshall Islands	Majuro	3,300	0.15	43/57
Samoa	Apia	2,900	0.20	30/70
Solomon Islands	Honiara	2,000	0.20	32/68
Tonga	Nuku'alofa	1,700	0.24	38/62
Tuvalu	Funafuti	3,500	0.20	42/58
Vanuatu	Port Vila	2,100	0.27	33/67
East Timor	Dili	900	0.32	20/80

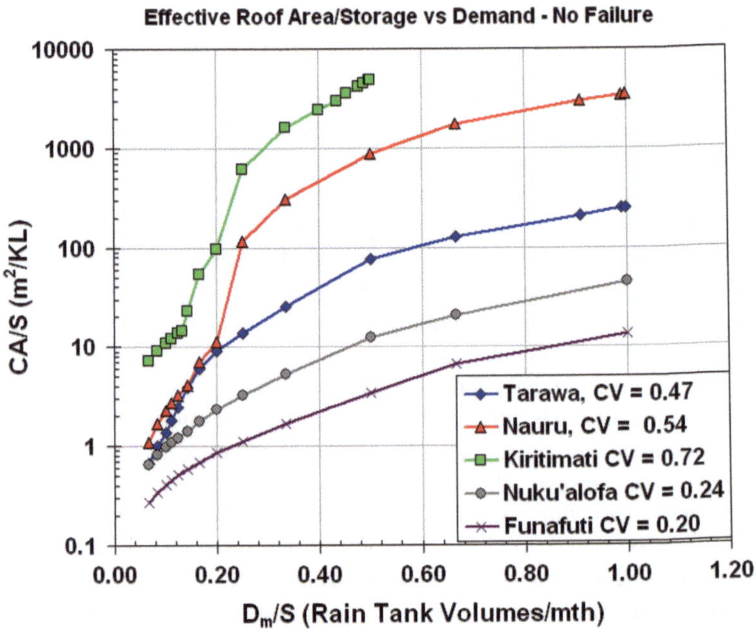

Fig. 23.7 Effective roof area, CA, per rain tank capacity, S, required to meet a monthly demand, Dm, per rain tank capacity for selected PIC urban centres with variable annual rainfalls. The parameter C is the efficiency of capture of rainfall ($0<C<1$)

with available roof catchment area, affordable rain tank capacity, and household demand, need to be considered. The impact of all these factors on the effective roof area per kilolitre of rainwater storage capacity required to supply total household demand relative to storage volume is shown in Fig. 23.7 using historic rainfall records.

Figure 23.7 demonstrates that urban centres in the central Pacific Intertropical Convergence Zone (PICZ) with large annual coefficients of variability (CV) of annual rainfall –South Tarawa, Nauru, and London in Kiritimati atoll – are, because of the variability, generally not able to supply all household demands for freshwater without failure. The size of roof catchment areas and rain tank capacities required to meet reasonable demands without failure, when households can contain up to 31 people, is financially and in some cases physically not possible. For countries with lower annual CV, outside the PICZ, such as Funafuti in Tuvalu, rainfall harvesting is viable even if expensive.

Because of the benefits of supplementing water supplies in urban centres with safe rainwater, some countries have established mandatory building codes to ensure installation of adequate rainwater collection and storage facilities on new buildings. Others, with donor support, have established revolving loan schemes to assist households with purchase of rain tanks and guttering. In order to guarantee safe stored rainwater quality, maintenance and management of rainwater harvesting and storage tanks schemes are essential (SOPAC 2004).

6.4 Demand Management

In some PICs with rapidly increasing populations, demand management is essential but nonetheless difficult. There is an aversion to pay for water, partly because it is viewed as the government's responsibility to provide water, partly because it is considered a common property resource (Sect. 3.1), and partly because of unreliable supplies (White 2011c). Some PICs have no means of controlling overall demand other than by supplying water intermittently for a few hours every 2 or 3 days. This strategy has four disadvantages: households leave taps open to collect the spasmodic water flows in small containers, causing high losses; customers are very reluctant to pay for an intermittent water supply; intermittent supplies are prone to bacterial build-up in supply lines (SOPAC 1999); and households closer to the supply source are able to intercept more water than those further down the pipeline.

At least one government has tried continuous, trickle-feed systems designed to mechanically control demand and give a more equitable supply. These supply a limited volume of water to a 500 L household tank, distributed to most households, over a 24–48 h period (Fig. 23.8). With an average of about 10 people per household this was designed to meet an assumed per capita demand for treated water of 25–50 L/day (see Table 23.3), the 'average' minimum daily requirement. This system was designed to permit continuous pipeline flows. Householders, however, tampered with the trickle-feed valves to increase discharge. As a result, households closer to the source of supply intercepted more water while others at the pipeline terminus received none at all, and the system reverted to intermittent flow. Without legal restrictions on tampering with the water supply, without enforcement of regulations, without water meters, and without general community disapprobation of tampering, such purely mechanical attempts to control demand seem destined to fail.

Fig. 23.8 Trickle feed system for a 500 L household water supply tank designed to limit demand by supplying a fixed daily volume of water to meet the assumed daily household demand. The system failed due to removal of the trickle feed by householders

PICs which have unmetered reticulation or supply systems use either a bulk monthly water charge or no charge at all. In many cases the bulk charge is only a small fraction of the actual cost of water production, distribution, operation, and maintenance, so systems are financially unsustainable and provide no control on excessive or inequitable use or wastage. Even in systems with metered connections, unaccounted for water (water losses) can be as high as 90 % (Table 23.4).

In severe droughts it may be necessary to shut down leaky domestic reticulation systems and supply water either from fixed distribution centres, such as village tanks or by tanker delivery (White 2011c). This reduces the large losses from the domestic pipe lines as well as allowing revenue collection from distribution centres, as is often done for bulk water tanker deliveries. This, of course can disadvantage urban squatters so that a base survival allocation may be necessary.

Efforts to control demand in urban areas by non-tariff means have led to situations where distribution of water is inequitable, where wastage is not controlled, where anti-social and even illegal actions are ignored, and where the water supply is financially unsustainable. Some PICs do meter consumption and have tiered tariffs in urban areas to discourage profligate use and wastage. These strategies tend to occur in larger PICs, where there are stronger governance structures or which have established local institutions such as the *matai* (chief) and *fono* (village council) systems in Samoa. Even in the larger countries with tariff systems, non-revenue water can have a major impact on being able to meet demand (Table 23.5).

One of the problems faced in PICs which have water tariffs is the number of large households with limited capacity to pay. Metering consumption and tiered tariffs appear the only sensible long-term strategies for controlling excessive demand and waste, provided it is well managed and contains measures to accommodate the disadvantaged. The success of pre-paid meters for household electricity supply in some PICs suggests that a similar system for water may be viable. However, water tariffs are only part of the process. A determined, long-term campaign to promote behavioural change, increase awareness, and accompanied by school education programs are required. The safety and adequacy of freshwater as a shared resource, and the importance of water conservation, need to be emphasised. Backing up such a campaign there need to be regulations and community disapprobation for tampering with water supply systems, meters, and pipelines, and stringent enforcement.

6.5 Urban Growth

Controlling per capita demand is only one part of the required strategy. The high urban population growth rates in several PICs are as threatening as climate change. Migration towards urban centres is a world-wide phenomenon, with the attraction of improved social services, increased amenity, and better employment prospects. In addition, the obligation of householders in PICs to their extended family mean that some households in urban areas can contain 30 or more people as distant rural relatives move in. Providing adequate and safe water and sanitation services to such large households and to attendant squatter populations is a universal problem (Kaosa-ard 2003). Strategies such as distributed, multiple growth centres have been proposed but have proved difficult to establish, and population growth remains a major, and largely officially overlooked, threat to the sustainability of urban and peri-urban areas in PICs.

7 Capacity Building

While human and financial resource capacities in the water and sanitation sector in PICs vary widely across the region, the Pacific RAP (SOPAC and ADB 2003) acknowledges the sector's general regional shortcomings in these areas. One of the key factors set out by Sabatier and Mazmanian (1979) for the success of public policy (Sect. 3.2) was that *those responsible for implementing the policy have appropriate managerial political skills, and information, as well as resources.* Capacity and resources are fundamental to improvements in the sector.

As has been discussed throughout this chapter, water agencies are often very under-resourced, partly as a result of financially unsustainable water supply systems, and often because of insufficient professional and technical staff to conduct routine operations. Water improvement projects are generally beyond the financial and human resource capacity of many local agencies. As a result, external development aid is required for planning, design, and implementation. Externally organised and funded projects can place impossibly large additional burdens on local staff, especially when multiple projects are running concurrently. These projects can distract them from important but routine tasks (Falkland 2011). In addition, more lucrative opportunities elsewhere make it difficult to retain trained staff.

It is evident that there is an urgent and long-term need for capacity building and training within water and sanitation agencies and departments. Both in-country and external training and development programs for technical and professional staff, combined with appropriate external courses, are required to build the capacity of these agencies and their staff.

Regional organisations, such as the Applied Geoscience Division (SOPAC) of the Secretariat of the Pacific Community (SPC), the Secretariat of the Pacific Regional Environment Programme (SPREP), the University of the South Pacific (USP), and the Pacific Water and Wastes Association (PWWA) – which pool expertise, share local experience, and provide training opportunities – have a continuing long-term role in supporting PICs through capacity building in urban water and sanitation management and operation. These regional organisations are invaluable in aid, loan, and donor bi-lateral programs in the water and sanitation sector in PICs. They can provide both a long-term memory of successful regional and local strategies and independent advice for in-country staff.

8 Community Empowerment

Another of the key planks of public policy success (Sect. 3.2; Sabatier and Mazmanian 1979) is that *policy and associated implementation mechanisms are actively supported by constituent groups*. In PICs, the active support of the general and local community is at the heart of successful water reforms. Well-developed local institutions, resilient social systems, sensitivity to environmental change, and high degree of equity in Pacific islands (Barnett 2001) provide a basis for change and improvement in the water and sanitation sectors – provided the community is empowered to participate at all levels. Strategies for facilitating community participation have recently been reviewed by Daniell (2012), and there is potential for investigating their use in island communities.

Including community representatives on national peak water and sanitation bodies, emphasised in Sect. 3.3, is a strategy which builds on unique island strengths. In one PIC, this proposal was strongly resisted by senior bureaucrats, who believed that water was 'government business' (White et al. 2009a). When finally community representatives were included in the committee, real progress on development of national policy and implementation plans occurred. These committees, however, require support and training in their role and responsibilities, some of which may require definition in law and at least in policy.

In Sect. 5, the complex interaction of land ownership, customary rights, and public good were raised in relation to protecting water sources. Although underpinning regulations are essential for providing legal protection of water resources, they are largely ineffective unless land owners and local communities are actively involved in the protection and management of reserves (White et al. 1999a). Local water reserve management committees (Sect. 5.2) involving representatives of land owners, local communities, and relevant government agencies appear a useful way of increasing protection and decreasing vandalism. In Tonga and Samoa, the existing *matai* and *fono* village council and village water committee systems responsible for local water supply management are useful Pacific models which build on island community strengths. Such committees require information, resourcing, and training and need to be adapted to the diverse cultural situations throughout PICs. While debate in developed countries continues on the efficacy of community engagement in water planning (see e.g. Daniell 2012), our argument for its value in PICs is based on the fundamental importance of negotiation and consensus in island life, and on examples of successful village or island-owned and managed systems in Samoa, Tonga, and the Maldives.

To control increasing demand (the main risk to water security in the Pacific), as well as to promote water conservation and reduction of wastage, community attitude change and community participation are essential. These values are not part of the old 'extended-family-centric' tradition of subsistence in the Pacific. Behavioural change is necessary for transitioning to urban living (Crennan 2002), and it is a long-term process that requires wide-ranging education programs, focused particularly on school children. Even in urban environments, it is possible that local area water committees, similar to those operating in Tonga and Samoa, could assist in reducing demand and wastage and promoting conservation, again so long as adequate training and resourcing is provided.

9 Coping with Climate Variability and Change

Over the past 1,000–12,000 years, low-density island communities have demonstrated remarkable resilience in coping with major natural hazards and calamities. As they move to high-density urban centres, however, their traditional coping mechanisms are not well-suited and their vulnerability to imposed threats increases. The prime threat from climate change to low-lying islands and atolls is sea level rise. Major threats posed by climate change to PICs have been detailed by the Intergovernmental Panel on Climate Change (Mimura et al. 2007). Some authors, however, conclude that the global focus on climate change has distracted PICs from addressing the actual, local sustainability problems facing island communities (Connell 2003), particularly managing vital freshwater resources (White and Falkland 2010). The conclusion in Sect. 1.1 – that non-climate factors, particularly increasing demand and losses, pose greater risks to water security in PICs out to the year 2030 than predicted climate change impacts (Falkland 2011) – adds weight to Connell's conclusion. The historic variability of rainfalls in the Pacific, and especially in the Intertropical Convergence Zone in the central and central-western areas (see Fig. 23.4), already poses major headaches for island water supply managers, particularly when losses are so high (Table 23.4).

9.1 Climate Variability

Variations of annual or seasonal rainfalls in urban centres in the Pacific are particularly large in the central and central-western Pacific. Nauru and the South Tarawa and Kiritimati atolls of Kiribati (see Fig. 23.4) have the highest variation, followed by the capitals of Vanuatu, Niue, PNG, and Tonga (Table 23.6). Variability of rainfall in these urban centres presents particular difficulties for rainwater harvesting (see Sect. 6.2). Roof areas and rain tank capacities are generally small, and household size and demand are usually large. Even in the capital of Tuvalu, Funafuti atoll, which has a much lower variability of annual rainfall and is heavily dependent on rainwater harvesting, some household rain tanks are exhausted after only a week with no rain.

Key drivers of high rainfall variability in the Pacific Intertropical Convergence Zone are the strong correlations between rainfall, local sea surface temperatures (Fig. 23.4), and major ENSO events. It is that strong coupling between rainfall and sea surface temperature which results in frequent, long, and severe droughts. In Nauru, the median interval between meteorological droughts relevant for rainwater harvesting is only 5 years, and the median drought duration is 19 months (White

2011a). Similar conditions prevail in Kiribati's capital, South Tarawa (White et al. 1999b), and in the urban area of Kiritimati atoll, Kiribati.

In order to replace the current (albeit inadequate) reticulated groundwater supply rate in South Tarawa of 2,000 m^3/day from communal rainwater harvesting, with zero risk of failure, we would require a rainfall collection surface of 1 km^2 and a storage capacity of 600,000 m^3 (White 2011b). In crowded urban South Tarawa, there is no free space available for such large rainfall collection and storage areas. More importantly, the estimated unit production cost of communal rainwater harvesting, using cost recovery over a 20 year lifetime, is over AU$22/$m^3$, far in excess of most other options including desalination, whose estimated unit production cost over a 10 year lifetime is around AU$5/$m^3$ (White 2011b). In these centres, rainfall harvesting remains a valuable supplementary source of freshwater with limited reliability in droughts.

The very strong relation between sea surface temperature, Southern Oscillation Index (SOI), and rainfall in several PICs (see Fig. 23.4) provides a basis for predicting the probability of below average rainfall several months in advance. The SCOPIC program developed by the Australian Bureau of Meteorology for PICs (BOM 2012) is designed to provide seasonal climate forecasts in the Pacific 3 months in advance. These predictions, and the Climate Outlook Programme coupled to drought contingency plans, provide a basis for coping with the frequent central Pacific droughts.

Significant ENSO-related droughts and floods have impacts across the whole region, with major impacts on stream flows (e.g. Terry and Raj 2002), power generation, groundwater availability, water supplies, agriculture, and health. In PNG, the 1998–2000 El Niño drought affected over 70 % of the PNG population, with severe impacts on water, food, agriculture, education, health, and other sectors (World Bank 2009). These climatic extremes are most severely felt in small PICs, who have only one or two water specialists and very limited financial and physical resources to deal with emergencies, especially during cyclones. Cyclones often cause severe wind damage, floods, landslips, erosion, downstream sedimentation, and saline intrusion, and produce major damage to infrastructure (including water supply infrastructure), buildings, and agriculture, as well as loss of life (Terry 2007).

Fresh groundwater lenses on small, low-lying islands can be inundated with seawater during island overtopping by cyclone-generated waves. Many months are required to naturally 'flush' saltwater from groundwater lenses and restore water supply to potable conditions (Terry and Falkland 2010). One major fear concerning climate change is that it will increase the frequency and intensity of climatic extremes in the region, including droughts, floods, cyclones, and island overtopping.

9.2 Climate Change

In searching for current evidence of change in the climate records of PICs there are three major limitations: the limited nature of the data (for example almost no information on evapotranspiration, solar radiation, or short duration rainfall intensity);

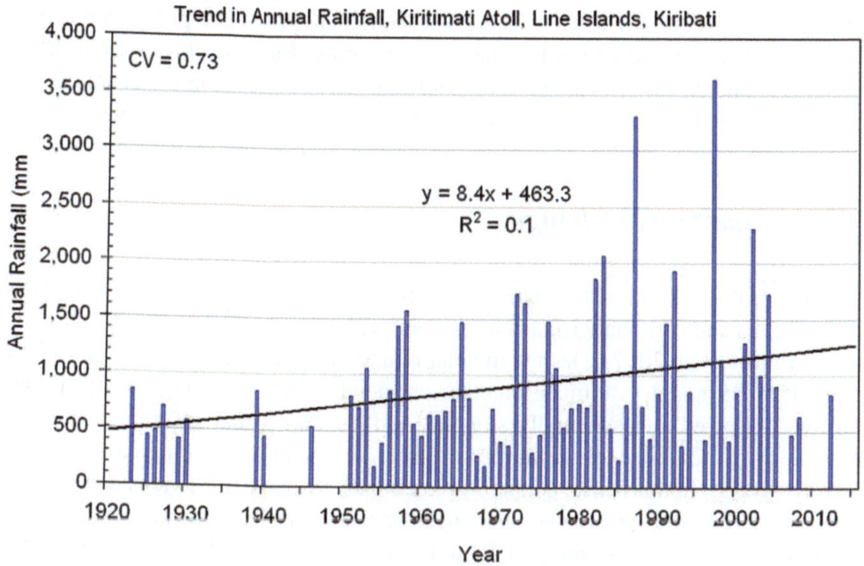

Fig. 23.9 Line shows the increasing trend of 8 mm/year in annual rainfall in the urban centre, London, Kiritimati atoll, the growth centre in the eastern Line Island Group of Kiribati. The bars show the large annual variability which is closely linked to sea surface temperatures and ENSO events

the inherent variability of the records; and the generally short climate records. The last two make identification of trends difficult. As an example, Fig. 23.9 illustrates the weak increasing trend (+8 mm/year) in measured annual rainfall for Kiritimati atoll, the growth centre in the eastern Line Island Group of Kiribati in the central Pacific, embedded in an exceptionally variable annual record (coefficient of variation=0.73) which is strongly correlated with ENSO events. This trend is split evenly (about 4 mm/year) between the wetter (January to June) and drier (July to December) parts of the year. A similar weak increasing annual trend is found in the urban capital of Tonga (White et al. 2009b). Increasing rainfall in small island countries is generally expected as sea surface temperature increases.

The Pacific Climate Change Science Programme (PCCSP 2011) used the outputs from 18 global climate models (GCMs) for two CO_2 emission scenarios to provide projections (relative to the baseline 20 year period centred on 1990) for a number of climatic and ocean parameters for selected PICs out to 2030. For the most likely emission scenario, most countries (in preliminary estimates) were predicted to have only small changes in mean annual rainfall, with only Kiribati, Nauru, and PNG predicted to have larger increases; potential evaporation for the region was also expected to increase by 1–3 % by the year 2030 (Falkland 2011). However, since GCMs do not handle cloudiness well, the predicted impact on actual evapotranspiration is difficult to assess.

There was no consensus between models on the predicted change in frequency of cyclones, and GCMs do not, in general, predict ENSO events. Given that these are key drivers, the projected risks to urban water security in PICs posed by climate change are difficult to assess. The risks, however, posed by population-driven increased demand and continuing large unaccounted-for water losses from urban systems are much easier to assess (Falkland 2011). The conclusion is that addressing basic issues of demand and losses will assist adaptation to climate change.

9.3 Adaptation to Change

The threats posed by climate change to urban areas in PICs are daunting, particularly in low atolls and carbonate islands threatened with partial inundation from sea level rise (Mimura et al. 2007). Despite the magnitude of these risks, some are optimistic that PICs can adapt to climate change, provided, firstly, that the rate of climate change is slowed and eventually stopped, and, secondly, that PICs can achieve a high level of domestic sustainability to promote social and ecological resilience (Barnett 2005). In this sense, adaptation to climate change can be viewed as but one aspect of addressing the broader challenges of sustainability, which are exacerbated in small island states because of their fragility and vulnerability to natural and anthropogenic pressures (UNDESA 1994).

In the face of uncertainties surrounding the magnitude and timing of climate change (Barnett 2001), as well as its impacts and the lack of detail about ecosystem functioning in PICs, it has been concluded (Barnett 2005) that the only rational adaptation strategy is "to develop the general capacity of a society to cope with change by building up its institutional structures and human resources while maintaining and enhancing the integrity of ecosystems". He concluded that the inherent local strengths in PICs are the basis for considerable capacity to adapt to climate change.

Dovers (2009) has argued that challenges faced in adapting to climate change are not new. Humans have had to cope with climate variability for a long time, and he cites examples in developed countries covering water management, local and regional economic vulnerability, biodiversity, health and well-being in remote communities, energy reform, and emergency and disaster management.

A significant problem exists in higher density urban areas in PICs, especially those subject to inward migration with large squatter populations. In these, many of the traditional coping mechanisms, local institutions, and social systems are weak, ineffective, or almost absent. The challenge of maintaining and enhancing the integrity of ecosystems around urban centres in the Pacific is significant, particularly when governments in the region have been reluctant to address the pressing problem of urban population growth over the past 40 years (Hughes 2011).

10 Conclusions

Urban centres across the Pacific region have an extreme diversity of fresh water resources, supply systems, and institutions and markedly different environmental, climatic, geographic, social, cultural, and economic contexts. Even within countries, there is remarkable diversity. In this chapter we have concentrated on identified common issues the region has in terms of urban water systems. These are central concerns about the adequacy and safety of water supplies and the impact of climate on them.

Some of the challenges faced in PICs are common to other developing areas throughout the world. Others are more regional and specific. Urban water resource issues in the Pacific are some of the most challenging and complex in the world. Their isolation, vulnerability to natural and human impacts, hydrogeology, limited resources and capacity, and high urban growth rates,, are intertwined with customary rights and underlying subsistence traditions. Because of the complexity, we have argued that simplistic approaches, focusing on single issues such as infrastructure, water tariffs, and governance tool kits have been largely unsuccessful. Instead we have sought to adopt a broader approach in this chapter, one consistent with the identified regional risks to urban water security.

It has been argued elsewhere that movement towards ecologically sustainable human development is the only rational adaptation strategy to climate change. We believe that this strategy has broader application beyond climate change. Here we have applied it to development in the water and sanitation sector in PICs, where there are difficult interactions between subsistence and urban life styles.

A general lack of government leadership, priorities, laws, plans, and structures has been identified as a regional issue. In three PICs, however, there is now an adaptive policy and plan development process carried out through a whole-of-government and community peak sector committee. This process is broadly acceptable and easily understood, and is one which engages the community strengths. It is particularly useful in countries with limited policy processes.

The first key factor for successful public policy is that policy and implementation mechanisms be broadly supported by government. This remains a challenge in areas where water and sanitation are largely seen as responsibilities of the extended family, the household, or the individual. Inadequate sanitation remains the largest threat to water quality and human health in urban centres in PICs. In many respects it is a 'wicked' problem and affordable solutions remain elusive.

The second key factor is that policy and related mechanisms be based on sound knowledge. Sound knowledge is essential because in PICs there is a delicate balance between water inputs and outputs, and that balance impacts on water availability and quality, and quality in turn has major health implications. Water resource assessment, use of appropriate technology, monitoring, analyses of data, and reporting provide vital information for improved management of freshwater sources. Peak sector government–community committees can play important roles in coordinating information and in sharing and disseminating it to government, relevant agencies, and to the broader community.

Limited land areas in many PICs mean that urban settlements encroach on and contaminate water sources. Protection of water sources, as well as water treatment, are key steps in a 'multiple barrier' approach to water safety. Traditional ownership of land in water reserves, source areas, and catchments, coupled with customary rights, remains a politically very sensitive issue. Engaging landowners and local communities in the care and management of water source areas appears to hold promise of increasing protection and reducing conflict, but it requires widespread consultation.

The two greatest threats to the water security and safety of urban water supplies in PICs are burgeoning demand and large losses from water reticulation and storage systems. Reducing non-revenue water losses has the potential to double the amount of water available in most urban centres across the region. These should be addressed first before any additional investment in new water sources occurs.

Rainfall harvesting has the potential to provide safer water to supplement household water supplies, especially in peri-urban areas. However, limited roof areas, small tank capacities, and large households mean that, with few exceptions, rainfall harvesting cannot meet all demands for freshwater across the region. There will be times when tanks run dry, especially in the central Pacific Intertropical Convergence Zone. Continual maintenance and good management of rainwater harvesting systems is essential for providing safe water.

Controlling demand presents one of the greatest challenges to PIC governments, partly because of customary beliefs and traditional rights, and partly because of unsustainable urban growth rates. There are no easy solutions here. Tiered water tariffs have proven successful in some counties but in others determined, long-term campaigns to promote behavioural change and increase awareness are required. Education in schools to underline the importance of safe and adequate freshwater has long-term value. In addition, increasing community disapprobation of interfering and tampering with water supply systems, meters, and pipelines, and backed up by enforceable regulations, would reduce inequities in community water supply.

A major regional priority is continued capacity building and training within water and sanitation agencies and departments. Regional organisations have played, and need to continue to play, an important role in assisting with capacity building across the region.

Well-developed local institutions, resilient social systems, sensitivity to environmental change, and the high degree of equity in Pacific islands provide basic strengths to build water reforms upon and to adapt to climate variability and change, particularly when island communities are empowered and informed. Successful models of village water supply systems are based on engaged and supported communities. Throughout this chapter we have emphasised the fundamental roles that the community can play, at all levels, in peak sector bodies, in water reserve management committees, and local area water committees; together these efforts can bring about behavioural change in the use, conservation, and management of water. Finally, with an eye to the future, school education campaigns are a key ingredient to the sustainable development and improved health and well-being of island communities.

References

Ackoff, R. L. (1999). *Ackoff's best: His classic writings on management*. New York: Wiley. 356 pp.

ADB. (2007). *Water Supply and Sanitation Project. Proposed Loan and Technical Assistance, Kiritimati Island Growth Center, Republic of Kiribati*, Draft Report and Recommendation of the President to the Board of Directors – for the Loan Fact Finding Mission, Project Number: TA KIR No.4456, December 2007, Asian Development Bank, Manilla, 110 pp.

AIDAB. (1993). *Field appraisal, Kiribati: South Tarawa sanitation and project request.* Pacific Regional Team, Centre for Pacific Development and Training, June 1993. Canberra: Australian International Assistance Bureau.

AusAID. (2003). *Making every drop count: Water and Australia Aid.* Canberra: Australian Agency for International Development.

Barnett, J. (2001). Adapting to climate change in Pacific Island countries: The problem of uncertainty. *World Development, 29*(6), 977–993.

Barnett, J. (2005). Titanic states? Impacts and responses to climate change in the Pacific Islands. *Journal of International Affairs, 59*(1), 203–219.

BOM. (2012). *Pacific islands: Climate prediction project, SCOPIC.* Australian Bureau of Meteorology. Available at http://www.bom.gov.au/cosppac/comp/scopic/. Accessed 1 Dec 2013.

BOM, & CSIRO. (2011). *Climate change in the Pacific: Scientific assessment and new research.* Vol. 1: Regional overview; Vol. 2: Country reports. Melbourne: Australian Bureau of Meteorology and CSIRO.

Bower, R., Crennan, L., & Navatoga, A. (2005). *The sanitation park project Fiji.* SOPAC Technical Report No. 386, SOPAC, Suva.

Bridgman, P., & Davis, G. (2004). *The Australian policy handbook* (3rd ed.). Sydney: Allen and Unwin. 198 pp.

Carpenter, C., & Jones, P. (2004). *An overview of integrated water resources management in Pacific island countries.* A national and regional assessment status report for GWP and Integrated Water Resources Management (IWRM), SOPAC Miscellaneous Report 554, Jan 2004, 37 pp.

Carpenter, C., Stubbs, J., & Overmars, M. (Eds.). (2002). *Papers and proceedings of the Pacific regional consultation on water in small island countries.* Sigatoka: Co-ordinated by SOPAC and ADB, 29 July–3 Aug 2002, Suva: South Pacific Applied Geoscience Commission, (CD-ROM).

Connell, J. (2003). Losing ground? Tuvalu, the greenhouse effect and the garbage can. *Asia Pacific Viewpoint, 44*(2), 89–107.

Crennan, L. (1998). *Water resource conflicts in Tarawa.* Report to UNESCO, SC/API/98/D/0011.

Crennan, L. (2002). Theme 3. Awareness. In C. Carpenter, J. Stubbs, & M. Overmars (Eds.), *Proceedings of the Pacific regional consultation on water in small island countries.* Sigatoka, 29 July–3 Aug 2002. Suva: Asian Development Bank and South Pacific Applied Geoscience Commission, (CD-ROM).

Crennan, L., & Berry, G. (2002). Review of community-based issues and activities in waste management, pollution prevention and improved sanitation in the Pacific Islands Region. In A. Wright, & N. Stacey (Eds.), *Issues for community based sustainable resource management and conservation: Considerations for the Strategic Action Plan for the International Waters of the Pacific Small Island Developing States.* IWP Technical Report 2002/03. Apia: The International Waters Programme, South Pacific Regional Environment Programme.

Crennan, L., Fatai, T., & Fakatava, T. (1998). *Groundwater pollution study, completion of Phase Two, Lifuka, Kingdom of Tonga.* Apia: UNESCO Office of Pacific States.

Daniell, K. A. (2012). *Co-engineering and participatory water management: Organisational challenges for water governance* (International Hydrology Series). Cambridge, UK: CUP/UNESCO-IHP. 346 pp.

Dovers, S. (2009). Normalising adaptation. *Global Environmental Change, 19*(1), 4–6.

Falkland, A. (2000). *An outline of recent water supply improvements for Pangai–Hihifo, Lifuka, Kingdom of Tonga.* Prepared for Tonga Water Board, Aug 2000.

Falkland, A. (2002a). Tropical island hydrology and water resources: Current knowledge and future needs. In J. S. Gladwell (Ed.), *Proceedings, second international colloquium on hydrology and water resources management in the humid tropics.* Panama, 22–26 Mar 1999. UNESCO-IHP-V *Technical documents in hydrology*, No. 52 (pp. 237–298). Paris: UNESCO.

Falkland, A. (2002b). Water resources management. Theme 1 paper. In C. Carpenter, J. Stubbs, & M. Overmars (Eds.), *Proceedings of the Pacific regional consultation on water in small island countries.* Sigatoka: Coordinated by SOPAC and ADB, 29 July–3 Aug 2002, 76 pp.

Falkland, A. (2011). *Draft report on water security and vulnerability to climate change and other impacts in Pacific Island countries and East Timor*. Pacific Adaptation Strategy Assistance Program, Department of Climate Change and Energy Efficiency, GHD Pty Ltd, 133 pp.

Falkland, A., & Brunel, J. P. (1993). Review of hydrology and water resources of humid tropical islands. In M. Bonell, M. M. Hufschmidt, & J. S. Gladwell (Eds.), *Hydrology and water management in the humid tropics* (pp. 135–163). Cambridge/Paris: Cambridge University Press/UNESCO.

Foukonga, J. D. (2007). Legal aspects of customary land administration in Solomon Islands. *Journal of South Pacific Law, 11*, 64–72.

GWP. (2003). *Sharing knowledge for equitable, efficient and sustainable water resources management. Tool box integrated water resources management, version 2*. Stockholm: Global Water Partnership. 155 pp.

Hughes, T. (2011). *Unfinished business*. KAPII Implementation Completion Report, Kiribati Adaptation Programme, Pilot Implementation Phase (KAP II), Government of Kiribati, 34 pp.

Jones, P. (1997). *The impact of the socio-cultural order on urban management in the Pacific: A case study of South Tarawa*. PhD thesis, Republic of Kiribati: Department of Geographical Sciences and Planning, University of Queensland.

Kaosa-ard, M. (2003). Poverty and globalisation. In M. Kaosa-ard & J. Dore (Eds.), *Social challenges for the Mekong River* (pp. 405–440). Bangkok: Chiang Mai University/White Lotus.

Low, M. (2011). *Solomon Islands water, sanitation and climate outlook*. Suva: SPC–SOPAC. 43 pp.

Mimura, N., Nurse, L., McLean, R., Agard, J., Briguglio, L., Lefale, P., Payet, R., & Sem, G. (2007). Small islands. In M. L. Parry, O. F. Canziani, J. P. Palutikof, P. J. van der Linden, & C. E. Hanson, (Eds.), *Climate change 2007: Impacts, adaptation and vulnerability* (pp. 687–716). Contribution of working group II to the fourth assessment report of the intergovernmental panel on climate change. Cambridge: Cambridge University Press.

NDS. (2011). *National development strategy 2011–2020*. Honiara: Ministry of Development Planning and Aid Coordination. 73 pp.

Pacific Islands Forum Secretariat. (2007). *The Pacific plan for strengthening regional cooperation and integration*, 2005. Revised Oct 2007. Suva: Pacific Islands Forum Secretariat, 52 pp.

PCCSP. (2010). *Pacific climate change science program*. Brochure produced for the 41st Pacific Islands Forum, Aug 2010, Port Vila, Vanuatu. AusAID, Bureau of Meteorology, and Department of Climate Change and Energy Efficiency. Available at http://www.pacificclimatechangescience.org/wp-content/uploads/2013/06/PCCSP_-41-PacIsland-forum-2010.pdf. Accessed 1 Dec 2013.

PCCSP. (2011). *Interim climate projections, 25 Mar 2011*. Canberra: CSIRO and Bureau of Meteorology, Pacific Climate Change Science Program, Department of Climate Change and Energy Efficiency, Australian Government.

Peterson, F. L. (1997). Hydrogeology of the Marshall Islands. Chapter 20. In H. L. Vacher & T. M. Quinn (Eds.), *Geology and hydrogeology of carbonate islands* (Developments in Sedimentology 54, pp. 611–636). Amsterdam: Elsevier.

Sabatier, P. A., & Mazmanian, D. A. (1979). The conditions of effective implementation of public policy: A guide to accomplishing policy objectives. *Policy Analysis, 5*(4), 481–504.

SINIIP. (2013). *Solomon Islands national infrastructure investment plan final report*, May 2013. Honiara: Ministry of Development Planning and Aid Coordination, Honiara, 196 pp.

SIWA. (2013). *Solomon water. The Solomon water development plan, 2013–2015*. Honiara: Solomon Islands Water Authority. 82 pp.

Skyjuice. (2008). Report on the provision of low cost water purification units by the Skyjuice Foundation Inc. for Kiribati Ministry of Public Works and Utilities and the Pacific Islands Applied Geoscience Commission (SOPAC). Skyjuice Foundation Inc, 4 pp.

Skyjuice. (2011). *SkyHydrant™ Water Purification Unit*. Skyjuice Foundation Inc. Available at http://www.skyjuice.com.au/skyhydrant.htm. Accessed 1 Dec 2013.

Solanes, M., & Jouravlev, A. (2006). *Water governance for development and sustainability* (Rescursos Naturalles e Infraestructura Serie, Vol. 111). Santiago: UN. 84 pp.

SOPAC. (1999). Water demand management workshop, Nadi, June 1999. In H. Schölzel, & R. Bower (Eds.), *Miscellaneous Report 345*. Suva: South Pacific Applied Geoscience Commission, 41 pp.

SOPAC. (2004). *Harvesting the heavens: Guidelines for rainwater harvesting in Pacific Island countries*. SOPAC Joint Contribution Report 178, prepared by the South Pacific Islands Applied Geoscience Commission and the Secretariat of the Pacific Community (SPC).

SOPAC. (2007). *National integrated water resource management diagnostic report, Republic of the Marshall Islands*. Sustainable Integrated Water Resources and Wastewater Management in Pacific Island Countries, SOPAC Miscellaneous Report 639, 51 pp.

SOPAC, & ADB. (2003). *Pacific regional action plan on sustainable water management*. Prepared following regional consultation at Sigatoka, in preparation for the 3rd World Water Forum, Kyoto. South Pacific Applied Geoscience Commission and Asian Development Bank, Suva.

Terry, J. P. (2007). *Tropical cyclones: Climatology and impacts in the South Pacific*. New York: Springer. 210 pp.

Terry, J. P., & Falkland, A. (2010). Responses of atoll freshwater lenses to storm-surge overwash in the Northern Cook Islands. *Hydrogeology Journal, 18*, 749–759.

Terry, J. P., & Raj, R. (2002). The 1997–98 El Niño and drought in the Fiji Islands. In J. S. Gladwell (Ed.), *Hydrology and water management in the humid tropics*. Proceedings of the second international colloquium, 22–26 Mar 1999, Panama, IHP-V Technical documents in hydrology No. 52 (pp. 80–93). Paris: UNESCO, pp. 80–93

UNDESA. (1994). *Report of the global conference on the sustainable development of Small Island Developing States (SIDS)*, Bridgetown, 25 Apr–6 May 1994, UN Department of Economic and Social Affairs, Division for Sustainable Development. 77 pp. Available at http://www.un.org/documents/ga/conf167/aconf167-9.htm. Accessed 1 Dec 2013.

UNESCO. (1991). Chapter 4, Water resource assessment and monitoring in small islands. In A. Falkland (Ed.), *Hydrology and water resources of small islands: A practical guide*. (Studies and reports in hydrology, Vol. 49). Paris: UNESCO. 435 pp.

UNWWAP. (2006). *Water, a shared responsibility*. The United Nations World Water Development Report 2. Paris/New York: UNESCO/Berhann Books, 584 pp.

van der Velde, M., Green, S. R., Vanclooster, M., & Clothier, B. E. (2007). Sustainable development in small island developing states: Agricultural intensification, economic development, and freshwater resources management on the coral atoll of Tongatapu. *Ecological Economics, 61*, 456–468.

Ward, R. G. (1999). *Widening worlds, shrinking worlds, the reshaping of Oceania*. Pacific Distinguished Lecture 1999. Canberra: Centre for the Contemporary Pacific, Australian National University.

White, I. (2010). *Tarawa water master plan: Te Ran-Maitira ae Kainanoaki, future water demand*. Prepared for the Kiribati Adaptation Programme Phase II Water Component 3.2.1, Canberra: Australian National University, Dec 2010, 58 pp.

White, I. (2011a). *Nauru: Vulnerability of water resources to climate variability*. Canberra: Pacific Integrated Water Resources Management (IWRM) National Planning Programme, ANU, 3 pp.

White, I. (2011b). *Review of water governance and institutional arrangements for the management of water and sanitation*, May 2011, Pacific Integrated Water Resources Management (IWRM) National Planning Programme. Development of a National Water, Sanitation and Hygiene Policy Framework and Implementation Plan for the Republic of Nauru, SPC-SOPAC, 22 pp.

White, I. (2011c). *Tarawa water master plan, 2010–30*. Prepared for the Kiribati Adaptation Programme Phase II Water Component 3.2.1. Canberra: Australian National University, Feb 2011, 49 pp.

White, I., & Falkland, A. (2010). Management of freshwater lenses on small islands in the Pacific. *Hydrogeology Journal, 18*(1), 227–246.

White, I., & Falkland, T. (2011). Reducing groundwater vulnerability in carbonate island countries in the Pacific, chapter 6. In H. Treidel, J. L. Martin-Bordes, & J. J. Gurdak (Eds.), *Climate change effects on groundwater resources: A global synthesis of findings and recommendations*. No. 27 (pp. 75–109). IAH International Contributions to Hydrology. Boca Raton: CRC Press.

White, I., & Falkland, A. (2012). Practical responses to climate change: Developing national water policy and implementation plans for pacific small island countries. In K. A. Daniell (Ed.), *Water and climate: Policy implementation challenges*. Proceedings of the second practical responses to climate change (pp. 439–449), Engineers Australia, Barton, ACT.

White, I., Falkland, A., Crennan, L., Jones, P., Etuati, B., Metai, E., & Metutera, T. (1999a). *Issues, traditions and conflicts in groundwater use and management*. UNESCO-International Hydrological Programme, Humid Tropics Programme. IHP-V Theme 6. Technical Documents in Hydrology No. 25. Paris: UNESCO, 37 pp.

White, I., Falkland, A., & Scott, D. (1999b). *Drought in small coral islands: Case study, South Tarawa, Kiribati*. UNESCO-International Hydrological Programme, Humid Tropics Programme. IHP-V Theme 6. Technical Documents in Hydrology No. 26. Paris: UNESCO, 57 pp.

White, I., Falkland, A., Perez, P., Dray, A., Metutera, T., Metai, E., & Overmars, M. (2007a). Challenges in freshwater management in low coral atolls. *Journal of Cleaner Production, 15*, 1522–1528.

White, I., Falkland, A., Metutera, T., Metai, E., Overmars, M., Perez, P., & Dray, A. (2007b). Climatic and human influences on groundwater in low atolls. *Vadose Zone Journal, 6*, 581–590.

White, I., Falkland, A., Metutera, T., Katatia, M., Abete-Reema, T., Overmars, M., Perez, P., & Dray, A. (2008). Safe water for people in low, small island Pacific nations: The rural–urban dilemma. *Development, 51*, 282–287.

White, I., Falkland, A., Rebgetz, M., Overmars, M., Katatia, M., & Metutera, T. (2009a). Development of national water resources policy and implementation plans for the Kiribati Adaptation Program. In *Proceedings, 32nd hydrology and water resources symposium* (pp. 1603–1618), 30 Nov–3 Dec 2009, Newcastle. Engineers Australia. ISBN 978–08258259461.

White, I., Falkland, A., & Fatai, T. (2009b). *Groundwater evaluation and monitoring assessment, vulnerability of groundwater in Tongatapu, Kingdom of Tonga*. SOPAC/EU EDF8. Reducing the Vulnerability of Pacific APC States. Canberra: Australian National University, June 2009, 351 pp.

WHO. (2004). *Guidelines for drinking-water quality* (3rd ed.), including addenda in 2006 and 2008. Geneva: World Health Organization.

WHO. (2005). *Country health information profiles*. Geneva: World Health Organization.

WHO. (2008). *Sanitation, hygiene and drinking-water in the Pacific island countries: Converting commitment into action*. Geneva: World Health Organization/SOPAC.

World Bank. (2009). *Disaster risk management programs for priority countries*. Global Facility for Disaster Reduction and Recovery, World Bank and International Strategy for Disaster Reduction, 306 pp.

Chapter 24
The "Renaturation" of Urban Rivers: The Case of the St Charles River in Quebec

Alexandre Brun

1 Introduction

In industrialized and developing countries during the nineteenth and twentieth centuries, most urban rivers were either channelled into canals, buried, or otherwise confined. The approach was designed both to improve urban hygiene and protect cities from flooding. In the 1950s, the growing use of cars in cities led to river banks being transformed into high-speed traffic lanes (e.g. the Manzanares in Madrid, the Seine in Paris). Due to pollution and the fact that river banks had become increasingly difficult to access, traditional uses (boating, bathing, fishing) disappeared. Cities gradually turned their backs on the rivers that they once relied on for their prosperity. Only major water shortages and floods reminded local authorities and residents of the presence (or relative absence) of water in the city.[1]

The management of river banks and beds and, on a larger scale, the urbanization of drainage basins, has had a wide range of consequences. In quantitative terms, these processes have altered the natural dynamic whereby groundwater is replenished and river levels are regulated. In qualitative terms, the capacity of water courses to clean themselves – although sometimes overestimated – has been reduced. In the 1960s and 1970s, in industrialized countries, substantial investments were made in storm water and sanitation schemes. But progress in this field has

[1] This was the case in Prague (Czech Republic) in August 2002 when the Vltava broke its banks, causing catastrophic floods, and in Spain where water shortages threaten the economies of entire regions.

A. Brun (✉)
Paul-Valéry University Montpellier 3, Université Montpellier 3,
Site Saint-Charles, 34090 Montpellier, France
e-mail: alexandre.brun@univ-montp3.fr

Q. Grafton et al. (eds.), *Understanding and Managing Urban Water in Transition*, Global Issues in Water Policy 15,
DOI 10.1007/978-94-017-9801-3_24

failed to offset damage to aquatic ecosystems caused by human interference.[2] In sum, it is not enough just to "wash the water". Public policies which focus exclusively on the fight against water pollution have revealed their limitations (Brun and Lasserre 2012).

Legislative bodies in most countries now consider water as both a 'resource' and part of the 'environment'. The destruction of fish habitat due to a range of operations (weirs, dikes, cleaning operations, the excavation of river beds, channelization, etc.) is recognized as a major cause of the degradation of natural aquatic environments. Development has also led to river beds becoming deeper, thus perturbing the spatio-temporal evolution of the rivers themselves, generating a large number of short-, medium-, and long-term consequences (e.g. impacts on hydraulic evolution, development of flora, use of drinking water, etc.). That is why, beginning in the 1970s and 1980s, contractors were obliged to carry out environmental impact studies, and, in the 1990s, environmental agencies focused on the central importance of guaranteeing the physical processes of river hydrosystems.

This change of paradigm can be at least partially explained by a growing interest in the 1980s and 1990s in the conditions and resources required to restore physical milieus, a research approach largely based on work initially carried out in the United States in the 1950s (Petts and Calow 1996; Malavoi and Bravard 2010). This research generated a great many promising projects, even if the notion of the 'renaturation' of rivers means very different things in different drainage basins.

There are opportunities for rehabilitation of urban rivers as cities have changed their strategies over water. Since the 1990s, the emergence of sustainable development has contributed to greener, more water-efficient buildings. Engineers have developed new storm water management techniques. Cities have rediscovered the value of the rivers around which they were originally organized and developed: the river was a strong element structuring urban development, particularly in the industrial era. This is why the ports and docks were rehabilitated during urban renewal in France (Nantes, Rouen, Paris…) and elsewhere in Europe (London, Bilbao, Lisbon, Genoa, Copenhagen, Oslo, Hamburg, Frankfurt Main…). Outside Europe, cities have transformed their port wastelands into vibrant urban spaces.

This article focuses on the reasons behind decisions to 'renature' urban rivers and on the conditions in which such projects are carried out. In effect, many municipalities launch renaturation projects not so much with a view to improving the quality of aquatic biotic milieus, but as a part of urban projects closely associated with the rivers running through their cities.

---

[2] The aquatic ecosystem or hydrosystem is "the specific eco-system of the aquatic milieus generally defined in reference to the living beings which are a part of it, the nature of the bed and the banks, the characteristics of the drainage basin, and the hydraulic regimes and the physico-chemical qualities of the water" (source: http://www.glossaire.eaufrance.fr/concept/ecosyst%C3%A8me aquatique). The term describes a different reality to the one covered by the river hydrosystem. "The authors of this concept, elaborated and tested on the Rhône, explain that it is particularly suitable for major water courses, or at least those disposing of a flood plain large enough to accommodate a mosaic of geomorphological units and eco-systems, notably the sub-systems of the high water bed" (Malavoi and Bravard 2010, p. 10).

After defining the concept of renaturation and presenting the issues that it encompasses, the case of the St Charles River in Quebec will be described with a view to illustrating the advantages and limitations of urban development projects linked to water. In its study of the renaturation of the lower reaches of the St Charles River in Quebec, the Franco-Canadian IDEAUX research programme (2008–2012) demonstrated that if the interests of the actors of urban planning (mayors, promoters, etc.) converge with those of water users, spectacular projects can be implemented.[3] The fact nevertheless remains that, in the absence of a global approach and taking into account the costs and duration of these kinds of urban projects, it would perhaps be more opportune to concentrate on the renaturation of rural rivers rather than urban ones.

2 The Concept of Renaturation: Definition, Evolution, and Issues

In the 1980s and 1990s, a number of restoration projects – first in rural milieus, then in urban settings – were successfully carried out, notably in Germany (Boutet 2006; Semrau and Hurck 2008). Since then, public and private contractors have viewed urban rivers as a factor in sustainable urban development. However, audacious programs – such as returning a great river to its original bed in the centre of an urban agglomeration, or bringing a buried stream back to the surface – are rare.

2.1 *Renaturation: Definition and Principal Characteristics*

Renaturation is defined as a management process whose objective is to restore to an aquatic milieu (a stream, a river, etc.) its main ecological and dynamic functions. Renaturation thus does not necessarily designate a return to a state considered as 'natural' for the water course – a process that presupposes a state of reference – but an approach designed to re-establish its natural functions (Adam et al. 2006; Bunusevac 2007; Brun 2011).

[3] The objective of IDEAUX – *Pour une Intégration des politiques de Développement, de l'Eau, d'Aménagement et d'Urbanisme en faveur des milieux aquatiques* (For the Integration of Policies on Development, Water, Planning and Urbanism in Favour of Aquatic Milieus) – is to study the mutual influence of so-called 'territorial' policies and water management policies. The project was initiated in 2008 by the Sogreah-Artelia consulting firm with the technical and scientific support of the Art-Dev Mixed Research Unit in response to the Eaux et Territoires call for projects launched conjointly by the French Ministry of Ecology, the CNRS, the Cemagref, and the Ministry of Natural Resources and Fauna of Quebec. IDEAUX is one of the 12 interdisciplinary projects selected for the Eaux et Territoires program (2008). The project is subsidized by the Ministry of International Relations of Quebec and the French General Consulate in Quebec. It also receives a subsidy from the Jacques Cartier Centre (2009).

Renaturation is an operative concept which has not yet been fully defined, an empirical approach at the crossroads between water management and the ecology of aquatic milieus. This explains why authors working in the field suggest a variety of typologies relative to the physical restoration of aquatic milieus. Nevertheless, four major levels of intervention can be schematically distinguished:

1. **Complete restoration** consists of redirecting the river back to its historical course in its original hydrodynamic conditions.[4] Insofar as major water courses are concerned, the objective is to provide the water course with a greater degree of freedom[5] by demolishing lateral (dams, weirs) and longitudinal (concrete anti-flood walls, levees) obstacles.
2. **Re-creation** is an approach applied to situations in which the original bed cannot be located, to ensure that the river follows its original course as closely as possible.
3. **Compromise** is an approach used when roads, buildings, etc., make scenarios such as those described above impossible. The objectives of this approach are to diversify habitats and achieve a dynamic equilibrium. Some structures (dikes, dams, etc.) can be demolished if no surrounding buildings are impacted.
4. **Diversification** applies to situations characterized by strongly constraining factors in which, for example, there is no freedom to build laterally (the areas on either side of the river have already been urbanized). Conditions can only be improved by introducing structured elements into the channel – by adding slightly sloping banks, placing obstacles in the channel, introducing flow deflectors, etc.

An analysis of projects in France and around the world reveals a number of constants. Renaturation is a voluntary, planned approach. In most circumstances, projects are organized and funded by municipalities, government agencies, or local residents associations. The cost of renaturation projects varies significantly depending on the characteristics of the site in question. For example, in Western Europe, the cost of studies and works per linear metre varies between a few dozen euros in rural areas to almost 1,000 euros in an urban context. The renaturation of an urban river is expensive, in part because it requires thoroughgoing studies encompassing a multitude of issues (real estate, hydraulics, ecology, etc.). Such projects also involve the acquisition of parcels of land next to the water course and, in many cases, compensation of local residents and companies whose properties and premises have been expropriated.

The renaturation of urban rivers is generally accompanied by other development projects (optimizing the management of rain water, modernizing waste water net-

[4]The approach consists in defining a point of reference based on core drilling and the diachronic analysis of old maps and plans. The present-day river bed is blocked off and the river redirected to its original bed.

[5]The high water bed within which the river channels provide lateral outlets for the movement of sediments, as well as hosting aquatic and terrestrial eco-systems. Source: http://www.glossaire.eaufrance.fr/concept/espace-de-libert%C3%A9-d'un-cours-d'eau

works, renovating residential buildings, public spaces, and public facilities, etc.). That is why such operations take a number of years to complete and are particularly complex in terms of organization. Furthermore, some projects are preceded by public information campaigns or the decontamination of land previously polluted by industry.

The concept of renaturation covers a wide variety of projects, ranging from redirecting rivers once channelled into canals back into their original courses, to infinitely more modest programs. In certain cases, projects described as exemplary – on the grounds that they focus on ecology rather than civil engineering – can be counter-productive. For example, when the banks of the Erveratte, in Switzerland, were reinforced using vegetal techniques (vegetal embankments, fascining), energy once dissipated laterally was concentrated into the channel of the water course. In fact, the river has dug its own bed so deeply that the banks have been undermined.

In France, it might have been expected that, due to the adoption of the European Framework Directive on Water (2000), attitudes to the concept of renaturation would have evolved towards an approach based on the efficient functioning of ecosystems. But an initial analysis reveals that, on the contrary, traditional concepts based on good water flow and flood protection continue to hold sway.[6] In places, traditional hydraulic projects are still funded by the French water agencies within the framework of the new renaturation policy.

Meanwhile, Idarraga (2010) lists over 160 completed and ongoing renaturation projects worldwide, three-quarters of them in Europe (79) and North America (43).[7] Nonetheless, the world's other regions are also active.[8] For example, in 2002, in Seoul, the then mayor, Lee Myung-bak, took the decision to launch a project to rehabilitate the Cheonggyecheon, a tributary of the Han River. Work on the project was commenced in 2003 after the demolition of the main road built over the tributary in 1968. A 5.8 km stretch of the river serving the city's commercial districts[9] has once again become a popular site for promenades and leisure activities.

Renaturation is a concept developed by ecologists, biologists, and water managers. However, the ecological and social aspects are inseparable, especially in urban areas where human densities are high and usage conflicts are many. "Actions to conserve or restore the river must maintain or restore the dynamics of the river to the greatest extent possible, identify and protect healthy parts of the ecosystem, link social patterns with ecological patterns, and develop management alternatives that are robust in the face of future uncertainty" (Gregory 2012).

[6] Working within the framework of the IWRM-Net Forecaster project in 2009, B. Morandi listed 597 sites or projects relevant to the field of river restoration. According to Carré et al. (2009), the differences between such projects are mirrored by the geographical disparities of the sites on which they are carried out.

[7] An analysis of the data collected by Idarraga (2010) for Canada and France suggests that the author has underestimated the number of projects in the field. Nevertheless, his work does provide an insight into the number of projects undertaken in individual countries and the disparities between various continents.

[8] Idarraga (2010) lists 21 projects in Asia, 12 in Latin America, 9 in Oceania, and 1 in Africa

[9] Source: http://shadowrun.over-blog.com/article-18979076.html

Fig. 24.1 The banks of the Rhône in Lyon were built for cyclists and pedestrians, Brun 2010

2.2 The French Experience: Renaturation Limited to the Landscaping of River Banks in Urban Milieus

Case studies carried out in France show that, unlike in rural territories, projects carried out in urban contexts tend to be limited to 'landscaping' the river banks. For example, in Lyon, 5 km of Rhône quays were recently redeveloped for €44.1 million by the Greater Lyon municipality in partner with the agency, In Situ, creating accessible pedestrian walkways with a large number of trees and providing shelter from the roadways built over them. The redevelopment of the quays, which took their current form in the nineteenth century, was rated a success by the public because of the linking of certain of the city's public facilities (see Fig. 24.1).[10] Indeed, the city has initiated a study for a project entitled Rives de Saône, involving artists and landscape designers working on the other major river that flows through Lyon. According to projections, 15 km of the 22 km project will have been completed by 2013.[11]

Between 2000 and 2009, in Bordeaux in south-west France, after a preparatory period of 20 years during which numerous old riverside buildings were demolished, the quays on the left bank of the Garonne were redeveloped at a cost of tens of millions of euros. Overseen by the landscape designer, Michel Corajoud, the 45 ha development made a major contribution to the urban renewal of the Bordeaux

[10] Interview with the geographer, J.-P. Bravard, Le Rhône: plus qu'un atout de marketing urbain, Diagonal, No. 177, 2008, pp. 34–36.

[11] http://www.grandlyon.com/Rives-de-Saone.3531.0.html. Accessed 6 Mar 2012.

agglomeration.[12] There are similar examples in the Paris region where most major urban projects involving the Seine include the redevelopment of the river banks. This is true of the Ardoines (280 ha) in Vitry-sur-Seine developed by the Orly-Rungis-Seine Amont Public Development Establishment (Brun and Adisson 2011).[13] And it is also true of the river eco-neighbourhood of Ile-Saint-Denis (22 ha) developed by the Plaine Commune, and of the eco-neighbourhood of the Docks Park in Saint-Ouen (12 ha) developed by Séquano Aménagement, an agency dependent on the Paris municipality.

In most cases, these riverbank redevelopment projects are reminiscent of approaches generally applied by urbanists to public spaces: the use of suitable public lighting, the development of safer and more peaceful access-ways, the planting of riverside vegetation, the creation of new urban neighbourhoods, etc. The list is not exhaustive. Such projects in fact generate results that combine aesthetic, functional, and sewerage objectives with the intention of returning cities to their rivers, often focusing on formerly industrial riverside areas (Docklands in London) or on outstanding heritage sites (the Rideau Canal in Ottawa). However, these projects do not strictly involve a process of complete renaturation, since the density of urban developments often makes it impossible to redirect rivers to their original courses. This problem rarely arises in rural situations.

2.3 An Operational Concept in Rural Milieus, But the Number of Major Projects Remains Low

The Drugeon is a 35 km long stream in the east of France. Its renaturation was one of the most successful projects of recent years. In 1951, under pressure from the farming community, local municipalities, in conjunction with the French Ministry of Agriculture, commenced drainage work on the marshland around the stream. Although the Drugeon lost 30 % of its length, the project ended in failure: only around 200 ha of land, or 10 % of the initial objective, were reclaimed. More than 40 years after it had been channelled into a canal, local municipalities, working within the framework of an inter-municipal structure, decided to direct the Drugeon back to its original course. The renaturation of the Drugeon began in 1993 and was completed in the early 2000s. Initial assessments were positive in that experts were able to observe that certain species were gradually recolonizing the sections of the stream that had been redirected to its original course. The ecological re-conquest was, moreover, preceded by a reappropriation of the river's banks on the part of local inhabitants and fishermen.[14]

---

[12] Ville d'Angers, La reconquête des quais à Bordeaux. http://www.angers.fr/fileadmin/plugin/tx_dcddownloads/La_reconquete_des_quais_a_Bordeaux.pdf May 29, 2010. Accessed 26 Jan 2013.

[13] The overall cost of the Ardoines project, which will take 20–30 years to complete, has been estimated at over a billion euros.

[14] The advantages and disadvantages of partially renaturing the Drugeon have been discussed in a large number of scientific articles. Representatives of local missions have also contributed valuable comments regarding the phasing of the project and its financial and administrative organization

The renaturation of the Drugeon was exceptional in terms of the scope of the project. However, in Europe, almost all rural rivers have been channelled into canals, either for agricultural purposes or as an anti-flooding measure. After a long period of land acquisition in the 1990s, a few kilometres of the Ouche, a small river in Burgundy, were reintroduced to the river's original course. In the case of the Veyle, a river flowing 67 km across a plain to the north of Lyon, local municipalities created a new river bed that was as natural as possible for a water course that had been destroyed due to the extraction of gravel in the minor bed, a process that had commenced in the 1960s.[15]

Thus, a number of ambitious projects have been successfully completed in rural milieus. However, in regard to the number of water courses in France and the rest of Europe, relatively few rivers have been the object of hydromorphological restoration programs and, with rare exceptions such as the Isar (Germany), the rivers concerned were small. In European cities, renaturation, even when combined with urban projects, as in Lyon and Bordeaux, is often confused with the landscaping of river banks. In effect, the urbanization of riverside lots limits the legal, technical, and financial margins of manoeuvre of contractors.

The situation is similar in the United States, where dozens of projects have been carried out, particularly in cities on the shores of oceans and major lakes. The same applies in Canada, where, for example, 6 km of the banks of the River Outaouais in Gatineau have been redeveloped.[16] This project, the intention of which was to bolster trade and tourism, differs from the renaturation of the upriver section of the St Charles River in Quebec (1996–2009), an approach which marks a turning point in water policies at the municipal level (Brun 2011).

In the 1990s, the City of Quebec came to the conclusion that projects carried out in the past (the construction of embankments, the concreting of the banks, etc.) had perturbed the functioning of the aquatic ecosystem and failed to have the desired effect on potential users, who continued to avoid the riverbank. The municipality therefore decided to develop a restoration policy for the St Charles River, while simultaneously pursuing the urban renewal of the riverside neighbourhoods in the old town neighbourhood of the city. The project was a success. The ecological and landscape improvements made to the river resulted in a jump in the value of river-

(see http://www.liferuisseaux.org/rencontre_colloques/Colloque_2009/10-06-09/restauration_drugeon_Resch.pdf).

[15]The objective of the project was to reduce the impact of a former gravel pit located at Saint-Denis-les-Bourgs on the river's ecosystems and hydrosystems. Costing an estimated €860,000, the project consisted in redirecting the course of the river into a minor bed running in parallel for approximately 2,000 m. Source http://www.veyle-vivante.com/. Accessed 26 Jan 2013.

[16]The project, entitled Destination Gatineau, is located not far from Ottawa. Supported by the mayor of Gatineau, Marc Bureau, it is set to be completed in time for the 150th anniversary of the Canadian confederation in 2017. The project will require an initial investment of CA$135 million and will attract an estimated three million visitors a year to the Quebecois side of the Outaouais River. According to its promoters, the project will generate CA$170 million per annum. Source: http://www.radio-canada.ca/regions/ottawa/2012/03/01/002-destination-gatineau-devoilement.shtml. Accessed 26 Jan 2013.

side property. In addition, the experience is remarkable because it precedes the birth of several years of water policy in Quebec.[17]

3 The Case of the Renaturation of the St Charles River in Quebec

The St Charles River occupies a drainage basin of 550 km². The area is among the first to have been colonized by European settlers and is one of the most densely populated areas of its kind in Quebec (350,000 inhabitants). Its 35 km length, from its source in Lake St Charles to its mouth in the St Lawrence, traverses various geological formations from the Canadian Shield to the St Lawrence Lowlands (Figs. 24.2 and 24.3). The course of the river can be divided into three segments: the

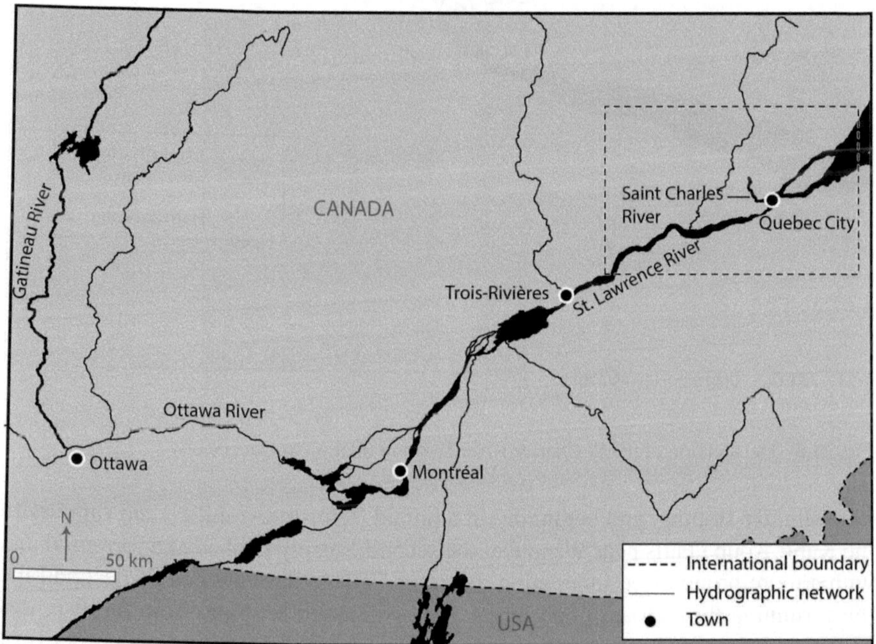

© S.C. - ART-Dev - UMR 5281 - CNRS / 2012
Source : adapted from "Département de Géographie, Université Laval".

Fig. 24.2 The location of the St Charles River in Canada

[17] Among other things, the Politique Nationale de l'Eau du Québec (Quebec National Water Policy) introduced water master plans (portraits of drainage basins and recommendations) for 33 sub-basins of the St Lawrence, suggesting that local drainage basin bodies should encourage the development of basin contracts (action programs) in those sub-basins. Since then, the Quebec government has set up Integrated Management Zones (2009) to cover the whole of southern Quebec, while simultaneously recognizing the growing role of drainage basin bodies. On 11 June 2009, the government also passed a law "affirming the collective nature of water resources aimed at reinforcing their protection".

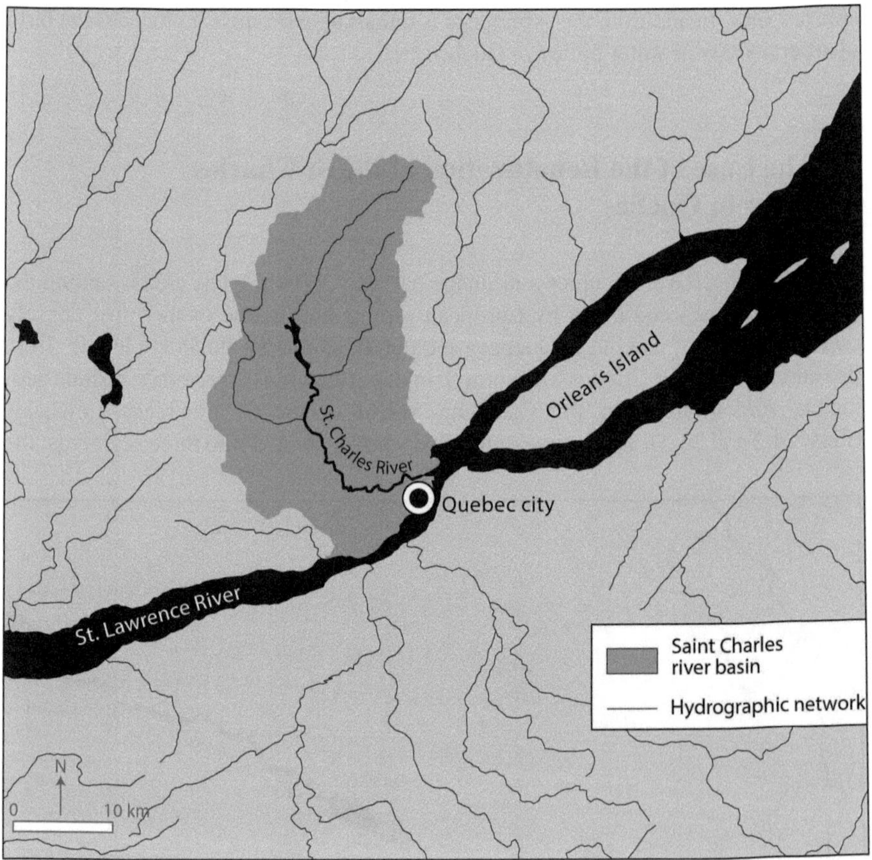

Fig. 24.3 The location of the St Charles River Basin in relation to Quebec city

first, slightly sloping, and beginning in a humid, semi-forest milieu and running to the Kabir Kouba Falls near Wendake; the second, steeply banked and torrential and with a more pronounced inclination, from the Falls to the Parc des Saules; and the third, running through an urban milieu and with a gentle slope, from the Parc des Saules to the river mouth. It is this third segment which, in undergoing a process of renaturation, has constituted a major axis of urban development.

3.1 The Lower St Charles: The Major Axis of Urban Development

Samuel de Champlain founded Quebec City in 1608, almost a century after Jacques Cartier first set foot in the area. The settlement initially developed around the Abitation (Place Royale), and later along the banks of the St Charles River (Pouliot

2005). The upstream section of the river was the object of successive projects associated with the development of the lower part of Quebec City. In the second half of the seventeenth century, the river bed was excavated in order to provide adequate depth for shipping, and the banks were partially reinforced following the construction of the first dockyards. After American independence and following Napoleon's continental blockade, the British Empire chose to secure timber supplies in eastern Canada and industrialize shipbuilding in Quebec.

Between 1820 and 1899, 1,760 ships were built in the dockyards of the St Charles estuary. During this period, the population of Saint-Roch (a town near the St Charles River) increased from 829 inhabitants in 1795 to 10,760 in 1842. Floating timber and industrial and domestic waste negatively impacted the quality of the river's water and significantly reduced biological diversity. From 1870, the dockyards were gradually replaced by factories, and in 1878, with the opening of the Canadian Pacific railway station, the neighbourhood of Saint-Roch became more attractive. Stench and disease were attributed to the poor quality of the river water. Drinking water was the reason the municipality built, in 1872, the first section of an aqueduct from Lake St Charles.

In an alarming report published in 1936, the Surveyor Commission described the river as one of the most polluted in the province (Quebec Chamber of Commerce 1962). The main outlet for the agglomeration's waste water and a backyard for declining industries, the St Charles River became a thorny issue for the municipality beginning in the inter-war period.[18] In 1966, in the wake of a tri-partite agreement between the municipal, provincial, and federal governments, a decision was taken to concrete over the banks (Dumont 1998). After a major embankment construction project (1957) designed to increase the amount of land available for construction purposes, the municipality of Quebec City concreted over the banks of the St Charles River between 1969 and 1974 in an attempt to make them more attractive.[19] The municipality also erected an anti-tidal dam in the early 1970s (Samson Dam) in order to stabilize the water table in the urban area.

The objective of these projects was to improve water management in terms of hygiene, while simultaneously modernizing the historic neighbourhoods that the river runs through. At the time – the time of the Gréber-Fiset Plan – functionalism was the order of the day. Between Scott Bridge in the west of the city and the confluence with the St Lawrence in the east, roads supplanted the natural corridor represented by the St Charles River. Urban development strategies carried out in Quebec City echoed European 'models' such as the quayside roadways in Paris. "We have stopped dreaming in colour; now we're dreaming of concrete", Gilles Lamontagne, mayor of Quebec City, said at the time.[20]

---

[18] The tributaries of the St Charles River were also of concern to the public authorities during this period (see: Fuites souterraines inattendues. Enfouie dans les années 1960, la rivière Lairet laisse ses traces à l'Hôpital Saint-François d'Assise, *Le Soleil*, Quebec City, 24 May 2012).

[19] The municipality was also responsible for the construction of a purification station and for improving the management of urban water in order to address the manifest shortcomings in the field of sanitization in the Saint-Roch and Limoilou neighbourhoods of the Old Town (Quebec City Town Planning Department 1973).

[20] Lamontagne: Québec deviendra l'une des plus belles capitales du monde, *Le Soleil*, Quebec City, 21 November 1966. Cited by Dumont (1998).

Thanks to the projects completed in the *Trente Glorieuses* – the period of economic growth between 1945 and 1975 – the most urban section of the St Charles River was no longer the "gutter transporting waste and debris, exposing a stinking, muddy bed" described by Gréber and Fiset in (1956, p. 49). However, the water quality was still very bad due to poor management of storm water. Furthermore, far from making the river more attractive, development projects had succeeded only in making it less so. The municipality of Quebec City therefore resolved, 30 years on, to make another attempt by pursuing the same objectives while adopting a radically different approach.

3.2 From the "Kabir-Kouba Plan" to the "Renaturation" of the St Charles River

A development plan by the name of Kabir-Kouba (a Huron-Wendat name meaning "river of a 1,000 meanders") was conceived in 1974 with the aim of once more placing the river at the centre of the city's urbanization project. In this regard, a number of ecological objectives were added, and the plan was updated in 1995 (Paulin 2001). From the technical point of view, civil engineering had to some degree been replaced by vegetal engineering: the 8 km stretch of river bank concreted over for CA$16 million a few decades earlier was demolished and 'renatured' (City of Quebec 1996; Paulin 2001).

The project consisted in developing a variety of fauna habitats and encouraging the redevelopment of riverside vegetation, while at the same time reinforcing the river's role as a centre for recreation accessible to the local population (Paulin 2001; Pronovost 2009). A 32 km network of pedestrian walkways, frequently accompanied by cycle paths, now links the restored sectors to Lake St Charles. Partaking of a more traditional approach, the 12 rainwater collection basins (underground concrete structures) near the outlets in the wooded areas, now capture 90–95 % of the volume of water overflowing the gutter system, thus helping to reduce the number of times water is released into the river in cases of heavy rain from 55 to 4 times a year. The overall cost of the project, shared between the municipality of Quebec and the provincial and federal governments, was approximately CA$115 million (Figs. 24.4 and 24.5).

Infrastructure designed to collect and transport rain water to the main storage facilities accounted for over 80 % of public expenditure. Rain water and waste water networks were diverted and modernized; according to the technical services of the City of Quebec, municipal management of water in the riverside neighbourhoods was substantially improved. On the other hand, according to a number of environmental associations, in regard to the tributaries of the St Charles River, waste water treatment and the management of rain water still leaves a good deal to be desired.[21]

[21] To our knowledge, no precise assessment of the project has been carried out in terms of urban sanitation concerning the renatured section, the other upstream sections, and the tributaries, some of which are still used as outlets for domestic and industrial waste.

Fig. 24.4 Downstream of the St. Charles River in Quebec City, Brun 2010

The urban scheme developed by the City of Quebec between 1996 and 2009 after a series of public consultations included a number of spectacular projects (Pronovost 2009). If the 'social' benefits are hard to quantify, the presence of an increasing number of promoters on the plots of industrial wasteland sites next to the St Charles Linear Park bears witness to an urban revitalization (Fig. 24.6); the Old Town is being transformed into a mixed-use, pedestrianized area; to the social habitat of the working class neighbourhoods will be added more opulent plots with 'a river view' (Boutet 2006). The scheme encompasses the usual imperatives of security and accessibility for the handicapped which pertain in major cities and agglomerations. For example, particular attention is paid to public lighting and to the height of vegetation, so that people taking a stroll always have a clear and unimpeded view. The urban furniture is adapted to inclement weather, flooding, and vandalism. The inclination of the paths along the river banks are sufficiently slight that they can be easily used by people with reduced mobility (Pronovost 2009).

The urban project benefitted from a 'window of opportunity' which opened in the 1990s. Leading up to the 400th anniversary of the founding of Quebec City (2008), the Provincial government had, in effect, selected three priority projects in the spring of 2000, foremost among which was the depollution of the St Charles River. The capital of the Francophone province was to become the shop window of the nation in the upcoming international festivities, and large amounts of funding were made available with the aim of removing the river from the blacklist of the most polluted water courses in North America. Around the world, rivers were regu-

Fig. 24.5 Property developers return on the banks of the St. Charles River to build housing, Brun 2009

larly featured in major events which gave a boost to improvement work (Fig. 24.7). For example, the transformation of the quays of the Thames underwent two sudden accelerations, the first due to Year 2000 celebrations, the second to the 50th anniversary of the reign of Queen Elizabeth II (Lemonier 2003).[22]

[22] In the west of Canada, the small town of Whitehorse, the capital of the Yukon Territory, partially renovated its waterfront to mark Elizabeth II's Golden Jubilee (Jubilee Park was opened on 25 May 2002).

Fig. 24.6 A trail connects the upstream and downstream of the St. Charles River, Brun 2010

The St Charles project also benefitted from experience acquired by the municipality a few years earlier from meetings on the urban renewal of the Saint-Roch neighbourhood. The symbol of a decline in the city's commercial industrial centre which lasted until the early 1990s, Saint-Roch now provides an example of negotiated urbanism. Following the election of Jean-Paul L'Allier's Rassemblement Populaire party in 1989, the municipality succeeded (after having vainly attempted for a number of years to modernize the neighbourhood by means of 'grands projets' based on a highly controversial top-down model of urban planning) in completing a number of projects in conjunction with local actors. The new government emphasized a policy based on popular consultation and focused on developing the

Fig. 24.7 Whitehorse waterfront, Brun and Lasserre 2012

neighbourhood based on its existing architectural heritage and on its history, rather than by parachuting in major infrastructural projects. During consultations on the renaturation project, the municipality followed this precept, even going so far as to appropriate some of the ideas of critical associations at the expense of its own technical departments (as the report of the association, Rivière Vivante, suggests; 1999, p. 6).

Generally speaking, the Quebec City municipal government was able, thanks to the favourable political context that pertained in the 1990s, to introduce a new approach. According to Scherrer (2004), writing on the subject of the increase in the value of riverside plots in urban areas, the new approach consisted in limiting anthropic pressure on a reconstituted natural milieu and considering the overall quality of the urban river as an inherent factor in urban amenities. Although the Samson Dam in Quebec City prevents fish from migrating, the return of certain species bears witness to the ecological interest of the renaturation of the downstream section of the St Charles River. The physico-chemical quality of the surface water will probably get better over the course of the next few years thanks to gradual improvements in rain water collection systems. This is also true for the rest of Quebec (Duchesne 2009).

Furthermore, the creation of islands, beaches, and rocky breakwaters are now among the most audacious projects in an urban water course that is very 'corseted'.[23]

[23] In comparison, the drainage and reforestation project carried out in the valley of the Saint-François River over the past 30 years, described by Castonguay and Samson (2010), seems to be lacking in ambition.

However, while such projects mark turning points in terms of urban development in Quebec, the issue of urban sprawl characterizing North American cities remains problematic. Badly managed urbanization, particularly at the head of a drainage basin, can have negative consequences on water quality and, consequently, on the attractiveness of the upstream sector of the drainage basin in which the public authorities have invested a great deal of time and money.

3.3 The Fight Against Urban Sprawl: A New Environmental Priority for the Municipality of Quebec City?

Between 1950 and 2000, the surface area of the *continuous urban habitat zone* (measured according to the 1:50,000 topographical maps published by the Canadian Ministry of Natural Resources) of the Quebec City region increased by 630 % (from 36.9 to 269.3 km^2), while its population grew by only 35 % in the same period (Mercier 2006). In certain areas at the head of the drainage basin, periurban development has expanded dramatically. This is true of the sub-basins of the Hibou River and of the stream in the St Charles River basin (Conseil du bassin-versant de la rivière Saint-Charles 2007). Many well-off households have been attracted by the nearby ski slopes and the 'natural' character of the Quebec countryside, which is relatively well connected to the city. "[However], the approach employed by developers, who build houses in the middle of plots, takes up a lot of space", says a local politician, who adds that "demand must be met".[24]

According to environmental associations, "there are many environmental consequences of urban sprawl in the drainage basin of the St Charles River" (e.g. domestic waste at the origin of the degradation of the quality of the water course, acute riparian erosion after the clearing of plots of land following the destruction of woodland, destruction of fauna habitats, etc.). Most lakes in the drainage basin of the St Charles River have also been impacted due to residential developments. At certain periods of the year, the bacteriological quality of the water in Lake Durand makes bathing impossible. The lake's water is not used for drinking purposes, and local residents use their own systems to cater to their needs. Not all local residences are linked to a municipal system for taking away waste water and therefore need to employ individual systems (Gérardin and Lachance 1997; Bolduc 2002). Indeed, it is impossible to verify the conformity of all these installations.

Upstream, forested areas in the drainage basin of the St Charles River are becoming increasingly scarce. While residential development and recreational tourism have caused a number of ecological problems, the same can be said for newly built roads. For example, in 1973, the reinforcement of 500 m of the banks of the River Berger north of the Boulevard Père-Lelièvre as part of the construction of the Le Vallon motorway and, later, the canalization of another part of the river following

[24] Source: interview with a local politician in charge of planning in January 2010 in the city of Quebec.

the construction of the La Capitale motorway, modified the natural direction of the water course, destroyed the fish habitat, and rendered riverside land more vulnerable to flooding (Conseil du bassin-versant de la rivière Saint-Charles; 2007, p. 22).[25] According to the association, Vivre en Ville, "Over the course of the last 10 years, 75 km of new streets have been built in Quebec City".[26]

The municipality is responsible for sanctioning or prohibiting specific approaches to land use and building projects, while taking into account their environmental and public safety impacts on nearby rivers and lakes, and flood plains. In taking the issue of public safety into account, legislation on the regulation of flood plains also subjects land use to severe constraints. The municipality's urban land use plan includes provisions for dividing its territory, establishing use categories, deciding which construction projects to allow or reject, and decreeing prohibitions or rules – and all of these vary according to locations within the territory, established categories, or any combination of those selection criteria. Nevertheless, to paraphrase a regional planning expert, everyone agrees that "today, it's the law on the protection of agricultural land that provides the most effective rampart against the urbanization of rural areas".[27]

The absence of an overall territorial strategy covering the metropolitan area as a whole is, from the point of view of associations like Vivre en Ville, patently obvious, particularly in terms of urban public transport which is limited to bus services. Furthermore, according to a politician from the Quebec City suburbs, this explains why, "since we haven't really got to grips with the urbanization of the territory, we have opted for more ecological buildings in the Quebec region."[28] The reasons for the difficulty of implementing territorial planning based on solidarity between municipalities should be sought in the predominance of the city centre, which wants "everything its own way".[29] Furthermore, "there's no shortage of space; that's why it's not easy to change the way developers act and how municipalities think… densification is a dirty word in North America!".[30]

In this context, in spite of the implementation of a theoretically 'integrating' local governance system for water, development actors find it hard to accept the principles of water management based on drainage basins dictated by the Politique Nationale de l'Eau du Québec (Quebec National Water Policy) presented in 2002 by

---

[25] See too Brodeur, C., Cuff, D., Dionne, N., Laberge, V., Labrecque, R., Trepanier, J., & Turmel, P. (2012). Portrait Watershed Capital. Body Watershed Capital. Published in March 2012, ongoing review. http://www.obvcapitale.org/plans-directeurs-de-leau-2/2e-generation/diagnostic

[26] Source: interview with the director of the association "Vivre en Ville" in Quebec City in January 2010.

[27] Source: interview with a planner of the Ministry of Planning and Municipal Affairs Quebec in January 2010 in Quebec City.

[28] Source: interview with a local politician in charge of planning in January 2010 in the city of Quebec.

[29] Source: interview with the director of the association "Vivre en Ville" in Quebec City in January 2010.

[30] Source: interview with the director of the association "Vivre en Ville" in Quebec City in January 2010.

the government led by Bernard Landry (Parti Québécois). To simplify greatly, for the municipal government the local governance of water does not constitute a lever for development in any way. Firstly, the drainage basin is still relatively unfamiliar to local politicians and developers, whence the difficulty of combining urban planning with water planning.[31] Secondly, local water governance is based on associative organizations whose financial, human, and technical resources pale in comparison with those of professional development and urban planning bodies.[32] Lastly, those associative organizations have no political legitimacy since they come under the aegis of the Ministry of the Environment and their members are not elected. That explains the limited role of the St Charles River Drainage Basin Council.

4 Discussion

In the case of the drainage basin of the St Charles River, an Association for the Protection of the Environment (APEL) was set up in 1980 by volunteers living near Lake St Charles, 20 years before the creation of the St Charles River Drainage Basin Council. In 1997 the APEL began a gradual process of 'professionalization' and now has several full-time employees. The Association's budget varies between CA$250,000 and 500,000 per annum. It carries out maintenance work and manages the Northern Swamps of Lake St Charles, which it co-owns with the City of Quebec. Founded in 2003, the St Charles River Drainage Basin Council has, up until now, been more of an observer than an actor, since the government has restricted it to that role. According to its statutes, the principal mission of the Council is to provide

[31] The Law on Planning and Urban Development (LAU) introduced in 1979 does not include a section specifically focusing on water management based on drainage basins. On the other hand, it does include the elaboration of the land use development plans published by all the Regional County Municipalities (MRCs), which often facilitate the articulation between urban planning and planning in the field of water management.

[32] The Government of Quebec, the main provider of funds for the drainage basin bodies, underestimated the needs of basin organizations, some of which, in spite of the recent increase in government subsidies, are no longer operating due to a lack of money. As a result, the bodies are not only experiencing operational difficulties (limited professional travel for employees, insufficient access to conferences and socio-professional meetings, etc.), but also difficulties in terms of recruitment (only people with relatively little experience are now hired due to the kind of salaries that can be offered). Furthermore, the financial fragility of the drainage basin bodies has had the effect of increasing competition between them. Initially announced in 2002, and then again in 2004 by the former Liberal minister, Thomas Mulcair, then again in the 2006–07 and 2010 budgets, the water charge was finally introduced on 1 January 2011 after a draft regulation was passed by the Council of Ministers on 2 December 2010. Quebec was thus late in introducing charges for water use, doubtless because measures of this kind continue to be unpopular. It is hard, especially in a country in which water is abundant and almost free, to make people pay for it in one way or another. However, the fact remains that water management based on drainage basins, which has been a central platform of public water policy in Quebec for some years now, demands that a scale of charges be implemented.

information to users and managers. Thus, the Council wields less influence with local actors than the APEL.[33] Like any other association, it has the right to point out potential incoherencies in municipal policies, but even this capacity is limited by the fact that some of its members (politicians, developers) represent municipalities.

Furthermore, the City of Quebec is unwilling to delegate competences to the drainage basin body since, internally, the municipal unions are strongly opposed to it: water management is the responsibility of the city government. This means that the Council can neither act as contractor, nor as delegate contractor, even if it disposed of the necessary human, material, and financial resources. Neither does the Council make any substantial contribution to the development of urban planning tools. Consultations with local people and users (in terms of information meetings) have been organized by the municipality rather than the Drainage Basin Council. Furthermore, a number of questions need to be answered: What kind of participation should be encouraged? Who should organize it? How can local people, concerned about depreciation in the value of their properties or an attack on their property rights, be reassured?

In 2008, the Federation of Canadian Municipalities (FCM) awarded a prize to Quebec City (sewage). Quebec should nevertheless develop indicators relating to the rehabilitation program of the St Charles River. There is a clear gap in terms of evaluation. What expertise does it develop? Should it be limited to ecological indicators or should it experiment with economic and social indicators? In coming years, the City of Quebec must also 'finish the job' by recommending that wastewater must not be discharged into the tributaries of the St Charles River (the de Lorette, for example). Should untreated effluents be discharged, indicators of water quality in the St Charles will always be bad.

In the City of Quebec, the inability of poorly funded water associations to play the role of contractor in the development and management of water courses means that they wield little influence on local politicians and urban developers, some of whom do not even know that they exist. In other words, the local governance of water as gradually implemented since the turn of the millennium has had little influence on better established and politically more legitimate municipal executives.

Beyond Quebec too, municipalities rather than the actors of the water sector are responsible for urban river renaturation projects. How should they be encouraged to develop urban projects which are more integrated with their environment (short-term), as well as a form of territorial planning which makes it possible to fight more

[33] The APEL notably contributed to the adoption, on 8 November 2010, of *Interim Control Measure to limit human intervention in the drainage basins of the water supply points of Quebec City installed in the Saint-Charles River and the Montmorency River*. This shows that the theme of the preservation of water as a resource is now a matter of concern to the city government. Although the objective of the Interim Control Measure is not to put an end to new housing developments, it does impose precise technical and architectural parameters in terms of sanitation with a view to affording a higher level of protection to water supply points in the Quebec City region. Does this suggest that the municipal policy is now favouring a more integrated approach to water management? Will it provide a genuine articulation between planning instruments used in the water sector and town planning policy?

effectively against urban sprawl while respecting the principle of ecological continuity (long-term)? The question is not a marginal one, since the urbanization of strategic sectors (flood risk zones, drinking water catchment areas, etc.) may have the effect of reducing the current emphasis on renaturation programs.

Lastly, since it is difficult to do more than simply diversify urban rivers (the *compromise* scenario presented in Sect. 2), perhaps more emphasis should be placed on rural milieus in which more ambitious projects are possible. But, then again, perhaps not. As interviews with architects, urbanists, municipalities, developers, etc. working within the framework of the IDEAUX programme suggest, the success of renaturation projects focusing on urban rivers may encourage municipalities to take a systematic attitude towards such approaches and optimize urban water management by developing projects combining water and territory to a greater degree.

Acknowledgements I thank the following for their contribution: politicians and water managers in Quebec City, members of the St Charles River Drainage Basin Council, APEL, fishing associations, and environmental protection organisations; also the planners, architects, and engineers who agreed to work with me through the IDEAUX program. Thanks also to Stephane Coursière, Michael Lavin, and Jamie Linton.

References

Adam, P., Malavoy, J.-R., & Debiais, N. (2006). *Retour d'expérience d'opérations de restauration de cours d'eau et de leurs annexes, menées sur le bassin RMC, rapport et annexes, Agence de l'Eau Rhône et Corse*, Document No. 05.079-ETU-101, Lyon, 133p.

Bolduc, F. (2002). *Diagnose des lacs Durand et Trois-Lacs, Cantons-Unis de Stoneham et Tewkesbury*. Pro Faune report, 56 pp + 3 appendices.

Boutet, G. (2006). *Le changement de forme des berges de la rivière Saint-Charles à Québec : l'explication de l'approche culturelle de la géographie*. Master's thesis, Laval University, Quebec, 162 pp.

Brun, A. (2009). L'approche par bassin versant: le cas du Québec. *Options Politiques*, Juillet, 36–42. http://policyoptions.irpp.org/fr/issues/canadas-water-challenges/lapproche-par-bassin-versant le-cas-du-quebec/

Brun, A. (2010). Reconquête écologique et aménagement urbain: faut-il réaménager le Lez à Montpellier. *Revue d'Économie Méridionale, 58*(229–230), 85–107.

Brun, A. (2011). Politique de l'eau et aménagement urbain. La renaturation de la rivière Saint-Charles à Québec. *Revue Norois, 219*, 89–107.

Brun, A., & Adisson, F. (2011). *Renouvellement urbain et risque inondation: le plan-guide Seine-Ardoines*. Cybergeo http://cybergeo.revues.org/24751

Brun, A., & Lasserre, F. (Ed.). (2012). *Gestion de l'eau*. Approche territoriale et institutionnelle, Presses de l'Université du Québec, Quebec.

Bunusevac, M. (2007). *Renaturation des cours d'eau, restauration des habitats humides*. Gouvernement du Grand Duché du Luxembourg, 97 pp.

Carré, C., Deroubaix. J. -F., Deutsch. J. -C., Hague, J.-P., de Gouvello, B., Belaidi, N., & Charrier, A. (2009). *Elaboration et mise en œuvre des projets de renaturation: la quête de l'improbable consensus*, Programme PIREN-Seine, working paper, 19 pp.

Castonguay, S., & Samson, H. (2010). Les productions de l'environnement riverain au Québec: industrialisation, décontamination et embellissement de la rivière Saint-François au 20ème siècle. *Géocarrefour, 85*, 241–251.

Chambre de Commerce de Quebec. (1962). *Mémoire sur la rivière Saint-Charles*, La Chambre, Quebec, 33 pp.

Conseil du bassin-versant de la rivière Saint-Charles. (2007). *Portrait de bassin de la rivière Saint-Charles*, 216 p.

Duchesne, S. (2009). *La gestion des eaux usées et pluviales en milieu urbain au Québec: problématiques et défis*. La gouvernance de l'eau dans les Amériques, Observatoire des recherches internationales sur l'eau, Laval University, Quebec, 15–16 Oct.

Dumont, J. -P. (1998). *Une rivière dans la ville; l'usage urbain de la rivière Saint-Charles: origines et perspectives*. Mémoire de maîtrise, Laval University, Quebec, 88 pp.

European Framework Directive on Water. (2000). Directive 2000/60/EC of the European parliament and of the council establishing a framework for the community action in the field of water policy. http://ec.europa.eu/environment/water/water-framework/index_en.html

Gerardin, V., & Lachance, Y. (1997). *Vers une gestion intégrée des bassins versants. Atlas du cadre écologique de référence du bassin versant de la rivière Saint-Charles, Québec, Canada*. Ministère de l'Environnement et de la Faune – Ministère de l'Environnement, Canada, 58 pp.

Greber, J., & Fiset, E. (1956). *Projet d'aménagement de Québec et sa région*. Commission d'urbanisme et de conservation de Québec, Québec, 72 pp.

Gregory, S. (2012). *Trajectories of change and the challenges of future conservation and restoration of major rivers*. In Proceedings of 1st international conference on integrative sciences and sustainable development of rivers, Lyon.

Idarraga, F. L. F. (2010). *Urban river restoration in Colombia*. Milano: Building and Environment Science and Technology Department, Politecnico di Milano. 313 pp.

Lemonier, M. (2003). Les berges, territoires de projets. *Diagonales*, No. 163, 39–42.

Malavoi, J.-R., & Bravard, J.-P. (2010). *Élements d'hydromorphologie fluviale*. Paris: Onema. 224 pp.

Mercier, G. (2006). La norme pavillonnaire: mythologie contemporaine, idéal urbain, pacte social, ordre industriel, moralité capitaliste et idéalisme démocratique. *Cahiers de géographie du Québec, 50*(140), 207–239.

Paulin, M. (2001). *Mise en valeur et renaturalisation de la Saint-Charles*. Techniques Sciences Méthodes, No. 3, 29–37.

Petts, G., & Calow, P. (Eds.). (1996). *River restoration*. Oxford: Blackwell Science. 231 pp.

Pouliot, Y. (2005). Le corridor naturel de la Saint-Charles, un milieu exceptionnel. *Naturaliste canadien, 129*(2), 78–81.

Pronovost, R. (2009). *La renaturalisation des berges de la rivière Saint-Charles. Rendez-vous sur la gestion intégrée des ressources en eau*. Quebec: University of Sherbrooke.

Rivière Vivante. (1999). *Le cas de la rivière Saint-Charles à Québec: le rôle d'un organisme de rivière, succès et embuches*. Report delivered by the association to the Bureau d'audiences publiques sur l'environnement during the Public Consultation on Water Management in Quebec, 46 pp.

Scherrer, F. (2004). *L'eau urbaine ou le pouvoir de renaturer*. Cybergéo, http://cybergeo.revues.org/pdf/1496

Semrau, M., & Hurck, R. (2008). *The reconstruction of Emscher-system – An integrated planning process according to the WFD*. In Proceedings of IVth International Conference on River Restoration, Venice.

Service de l'Urbanisme de la Ville de Québec. (1973). *Sauvegarde des abords des rivières Saint-Charles et Duberger*. Ville de Québec: Division de l'aménagement. 86 pp.

Ville de Québec: Rapport des Commissaires. (1996). *Commission pour la mise en valeur du projet de dépollution et de renaturalisation de la rivière Saint-Charles*, Ville de Québec, 83 pp.

Chapter 25
Adapting to Climate Change in Urban Water Management: Flood Management in the Rotterdam–Rijnmond Area

M. van Vliet and J.C.J.H. Aerts

1 Introduction

Many large coastal cities are located in deltas. Proximity to rivers and sea, low elevations, and subsidence makes them vulnerable to floods (de Moel et al. 2011; Nicholls et al. 2008). Increases in flood damage are a common trend, and this is largely due to increases in people and industry on the floodplain, combined with a rise in property values (Dawson et al. 2011; Nicholls et al. 2008; Maaskant et al. 2009). This trend is expected to continue: in 2007 half the world's population lived in cities and the number is increasing (UN-Habitat 2008). In addition, the magnitude and frequency of flooding in many river basins and deltas is expected to increase due to climate change and associated sea-level rise (IPCC 2007). For example, higher peak discharges are expected for the river Rhine in north-western Europe (Hooijer et al. 2004; Te Linde et al. 2010).

The use of structural measures alone, such as levees, does not provide sufficient protection as there is always the chance they will fail, while the costs from a major flood are rising because of population and economic growth (van Herk et al. 2011; de Moel et al. 2011). Therefore, recent research has advocated that additional measures in managing flood risk be implemented, combining both structural and non-structural measures (Aerts and Botzen 2011; Bubeck et al. 2012a; Dawson et al. 2011; Evans et al. 2004; Schelfaut et al. 2011; Stephenson 2002; Wheater and Evans

M. van Vliet (✉)
Public Administration and Policy Group, Wageningen University,
P.O. Box 8130, 6700EW Wageningen, The Netherlands

Movares Nederland B.V., P.O. Box 2855, 3500 GW Utrecht, The Netherlands
e-mail: mathijs.vliet@gmail.com

J.C.J.H. Aerts
Institute for Environmental Studies, VU University – Faculty Earth and Life Sciences,
C.J.H. Aerts De Boelenlaan 1087, 1081 HV Amsterdam, The Netherlands

Q. Grafton et al. (eds.), *Understanding and Managing Urban Water in Transition*, Global Issues in Water Policy 15,
DOI 10.1007/978-94-017-9801-3_25

2009; Begum et al. 2007; de Moel et al. 2011). To reduce flood damage, several studies have shown the effectiveness of adaptive, non-structural measures and of increasing adaptive capacity (Botzen et al. 2013; Dawson et al. 2011; Kreibich et al. 2011; Bubeck et al. 2012b; de Moel et al. 2013). Some authors have even advocated that adaptive measures are more sustainable than structural ones, as they are more flexible, involve less commitment of resources, and do not exclude options for future generations (Kundzewicz 2002).

As the number of vulnerable assets in a floodplain determines the potential damage from floods, spatial planning can be a very powerful tool to reduce flood impact (Schelfaut et al. 2011; van Herk et al. 2011). However, for spatial planners water safety is often not a priority. Many issues play a role in urban development, some of which tend to oppose water safety, e.g. developing attractive waterfront houses for higher incomes in flood-prone areas. Houses bordering water often carry higher prices (Luttik 2000), illustrating the stronger demand for houses overlooking water. A large embankment in front of such houses can lead to lower real estate values. Adaptive measures, such as dry-proofing, flexible and removable barriers, and multi-functional embankments can be used to maintain the attractiveness of water-side properties (Botzen et al. 2009). The different interests between water management, housing, business, harbour activity, and recreation have to be balanced in the spatial planning process. Many authors therefore stress the need for participatory approaches to spatial planning and water management (e.g. Healy 1997; Pahl-Wostl 2007; van Herk et al. 2011). At the same time, municipalities should be able to enforce flood safety measures when necessary.

1.1 Problem Description

In the City of Rotterdam – located on the North Sea coast of the Netherlands along the river Rhine (*Nieuwe Maas*) – flood risks are steadily increasing due to restructuring and redevelopment of unembanked areas. Until now, new developments in unembanked areas were allowed in Rotterdam if they were elevated to a safe level. This policy causes problems, however, when former harbour areas are redeveloped in phases, as it results in neighbouring buildings having large differences in ground level. It is also very expensive when all infrastructures (including underground infrastructure) have to be elevated as well. Several older buildings have to be saved as cultural heritage, which causes further problems with elevation differences.

Due to past sea level rises, the ground level of several existing urban areas is now too low. When rebuilding occurs in these areas, similar complications emerge. Furthermore, not all buildings will be rebuilt; many will be renovated or given new facades (see Fig. 25.1). Instead of elevating ground floor levels or installing embankments, other options need to be studied, including spatial planning, flood zoning, and building codes.

Fig. 25.1 Redevelopment plans in Feijenoord, Rotterdam consisting of new buildings (nieuwbouw), renovation (renovatie), selling former rental houses (verkoop bestaande woningen) and work on the facades (gevelaanpak bestaande woningen)

Some research has been done to show the risk-reducing effect of certain spatial planning measures in the Netherlands (e.g. Aerts et al. 2008; Runhaar et al. 2012), but adaptive (non-structural) flood measures are still little used. One reason could be the lack of supporting policies or legal barriers. So far, there has been very little attention to the impact of policies, laws, and regulations on the actual implementation of adaptive flood measures.

Therefore, the main research question of this chapter is: what is the effect of policies, laws, and regulations on the potential to reduce flood damage through spatial planning measures, flood zoning, and building codes? How can municipalities stimulate the use of adaptive measures and, if necessary, enforce them? Data was gathered via a literature review and a series of interviews and two workshops with policy makers, scientists, consultants, and lawyers.

The chapter starts with a description of the case study and adaptive measures. Subsequently, current policies, laws, and regulations with respect to building codes and zoning in the Netherlands are reviewed, leading to an analysis of how they affect the implementation of adaptive measures. We will describe the barriers we encountered against implementation, and how these are affected by the legal framework and historical background of the Netherlands.

2 Case Study Description

Rotterdam contains relatively large unembanked areas within the port zone, most of which were elevated when first developed. The newest parts of the harbour of Rotterdam, developed in 2008–2013, are elevated to 5 m above sea level. Some of the oldest harbour areas are located in the city centre and were only elevated 2.5–3 m above sea level, which results in regular flooding on the quays. Currently, 4.1 m above sea level is advised for new developments (van Veelen 2012). Flood frequencies in the case study area range from 1 in 10 years on the quays to 1 in 10,000 years in the higher parts (Fig. 25.3). The last time that large parts of the unembanked areas were flooded was in 1953. Smaller parts, mainly located on the quays, do flood more often, most recently in December 2011 (Fig. 25.2).

A model study on damage in residential areas showed expected damage in the unembanked area of Rotterdam ranging from nearly two million euro (for a 1 in 100 years) to 14 million euro (1 in 10,000 years) (Veerbeek et al. 2010). Currently, floods lead to serious damage from around the 1 in 50 year return period, with a sharp increase in damage for 1 in 3,000 year floods (Veerbeek et al. 2010). Due to Rotterdam's proximity to the sea, sea level rise will increase flood frequencies and

Fig. 25.2 Flooding of the low lying quays of the Noordereiland in Rotterdam (picture courtesy to Pieter de Greef)

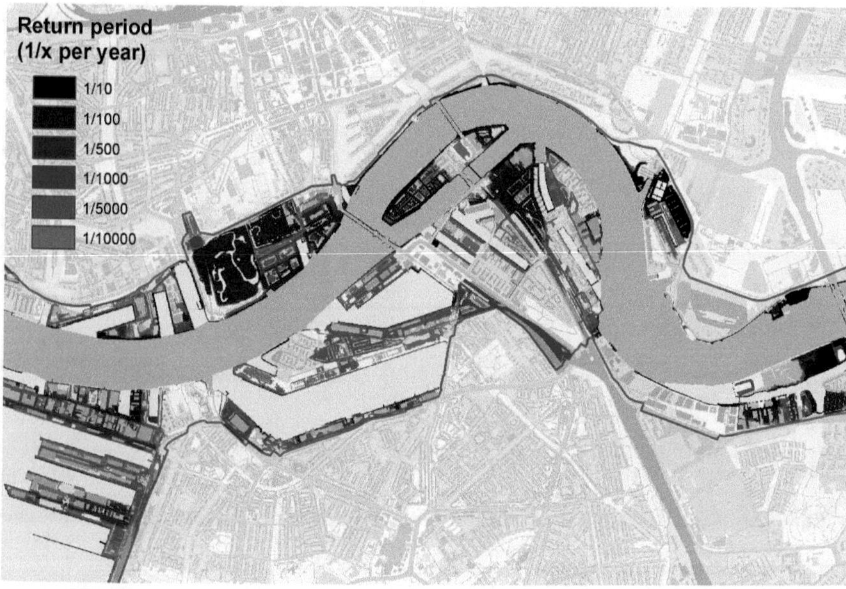

Fig. 25.3 Flood extent in the main case study area of the HSRR3.1 project, associated with the predicted return periods for the current conditions in part of the Rotterdam unembanked area (based on Huizinga 2010) (Figure courtesy to Hans de Moel, IVM)

damage: a 60 cm increase in sea level changes a 1 in 1,000 year flood into a 1 in 100 year flood, and a 1 in 100 year flood into a 1 in 10 years flood (Slootjes et al. 2011).

As harbour activities are moving out of the city centre to the newer harbour areas, the old harbour areas in the city centre are being redeveloped into residential and commercial areas (Fig. 25.4). Due to this redevelopment the number of people living in unembanked regions of the larger Rotterdam–Rijnmond area is expected to strongly increase, from 65,000 now to 80,000–100,000 in 2050 (Wardekker et al. 2010). The new developments, in conjunction with sea level rise, mean that adaptive measures become necessary in the unembanked areas.

2.1 *Adaptive Measures*

The project started with developing a list of possible adaptive measures that could be used to decrease potential flood damage, both in existing residential areas as well as in new developments. This list of measures was used to create a 'toolbox', essentially a matrix of nine categories of measures – embanking, elevating, dry-proofing,[1]

[1] Dry-proofing: preventing flood water from entering a building by reinforcing walls, windows, and doors and making them impermeable.

Fig. 25.4 New developments in the former harbour area Kop van Zuid in Rotterdam

wet-proofing,[2] temporal adaptation, floating, evacuation, regulations, communication – and four different scales on which they could be implemented – building, building block, neighbourhood, city district (Table 25.1). Measures could be implemented in different ways and at different scales. For example, a single building can be wet-proofed, but so too can a neighbourhood's entire public space. Evacuation can be done within a building (towards higher floors), within an unembanked area (to higher places or a flood-proofed shelter), or to an embanked area; which option is most viable depends on local characteristics.

Single buildings or a whole city district can be elevated. When a single building or building block is elevated the architect will have to deal with height differences at the entrance to the building. When larger areas are elevated the surrounding infrastructure is often elevated as well. This makes the entrance to buildings easier, but care has to be taken that transitions to other parts of the city are smooth.

Instead of keeping water out, one can also allow water to flow into an area. A side-effect is an increase in the inhabitants' flood awareness. Often, water is only allowed to enter public spaces. Such public spaces will need to be wet-proofed to minimise damage, for instance by using a ground cover that cannot be easily washed away. Utilities like power houses will need to be elevated or dry-proofed, and sewerage openings must be closable. To minimise disturbance it is important that water can drain away quickly after a flood. Another option is wet-proofing of buildings. In this case electrical outlets have to be placed high on walls, tile floors are needed instead of wooden ones, and wall panels should be placed horizontally instead of vertically.

[2] Wet-proofing: flood water is allowed to flow into the building, but the building is built of materials that are not damaged by flood water and are easy to clean.

Table 25.1 Toolbox of adaptive measures studied in the Feijenoord, Rotterdam case study (based on van Veelen 2012)

	Building	Building block	Neighbourhood	City district
Embankments		Embankment around building block	Embankment around neighbourhood	(Secondary) embankment around unembanked area
Elevating	Elevating building	Elevated building block	Elevating neighbourhood, including infrastructure	
Dry-proofing	Dry-proofed building	Dry-proofed outside walls to protect building block	Dry-proofed outside walls as part of embankment around neighbourhood	
Wet-proofing	Wet-proofed garden and building	Wet-proofed quays	Wet-proofed public space	Flood proof public space
Adapting	Temporary adaptations to buildings	Temporary closing open spaces in building block	Temporary protection of neighbourhood	
Floating	Floating building	Floating building block	Floating neighbourhood	
Evacuation	Evacuation within building	Evacuation to higher grounds in unembanked area		Evacuation to embanked areas
Regulations	Building codes and regulations for use		Zoning within unembanked areas	Zoning within unembanked areas
Communication	Personal communication on flood risk and evacuation		Large scale communication and risk maps	

Temporary adaptations can be taken when a flood hits. This way there is little disturbance in times of normal water levels and adequate protection in times of flood. With temporary embankments, however, there is always the chance they will not be ready in time or other human errors may occur. On a building they can take the form of temporary baffles in front of doors or windows (Fig. 25.5), or on a larger scale they can be fitted between gaps in embankments or make an existing embankment higher (Fig. 25.6). Other temporary adaptations are, for instance, seasonal permits for riverside pubs or newspaper stands, which need to be moved in the flood season.

Fig. 25.5 Placeholders for demountable barriers in front of basement windows in the city of Kampen

Fig. 25.6 Placeholder for demountable barriers (in the pillar, fence will be replaced by wooden boards) at the riverside quay in Kampen. In the pavement a hidden piston embankment can be seen (highlighted for better visibility)

Fig. 25.7 Floating pavilion in the Rijnhaven in Rotterdam

In several places in the Netherlands experiments have been made with floating buildings. Rotterdam has a floating pavilion (Fig. 25.7) and there are plans for a floating neighbourhood in the same harbour. Amsterdam already has a small neighbourhood of floating houses. In several places people live on 'houseboats', which often do not resemble a boat at all (Fig. 25.8).

To effectively evacuate unembanked areas, there should be evacuation plans, evacuation routes, and shelters. People need to be able to get quickly to safe areas, which can be in shelters, higher ground within unembanked areas, or embanked areas. At times of normal water levels, shelters can be used as schools, gyms, or concert halls. People can also move to the higher floors of multi-storey buildings. Shelters on the top floor can perform other functions when water levels are normal, such as a restaurant or gym.

Regulations can be made for single buildings, for instance in the form of building codes, or for whole neighbourhoods in the form of land use zoning and evacuation plans. Building codes can specify dry- or wet-proofing, and set out appropriate styles for floating houses, buildings on poles, and escape routes. Another option is to use ground floors for less vulnerable functions like shops or car parks. On a neighbourhood level, sensitive functions can be placed in those parts that are not flood-prone and evacuation routes can be secured. Flood-zone maps can give guidelines about which type of regulation is needed in which location.

Fig. 25.8 A houseboat in Nijmegen

Communication is important as it gives inhabitants of unembanked areas better knowledge of risks. Flood-risk maps can be a good tool for doing this (de Moel et al. 2009). Municipalities can inform residents of the measures they have taken (e.g. evacuation routes and location of shelters) and inform citizens of the options there are for minimising property damage. Currently, roughly half the population in the unembanked areas of the larger Rotterdam–Rijnmond area do not know if they live in an unembanked area (de Boer et al. 2012). This shows the need for improved communication, although it has to be done with extreme care in order not to provoke unnecessary anxiety.

For the Rotterdam project, the toolbox was used in workshops for discussing which measures could best be used in which part of the case study area. During the design phase of the project, the focus was mainly on spatial quality. Sketches were made of embankments that are able to retain the relation of waterfront houses to water, such as embankments in the form of low benches or elevated boulevards (van Veelen 2012). In later phases, aspects such as support among the local population and business owners, and financial feasibility, were discussed.

At the end of the workshops, three different possible flood management options were evaluated: keeping water out, letting water in, and an intermediate option. In the 'water out' option, different forms of embankments are used. These embankments were developed in such a way that contact with the water could be maintained. Embankments should therefore not be higher than approximately 1 m, so that owners of houses overlooking the water can still see it when standing in the living room (van Veelen 2012). The embankment was shaped as a low bench, or a higher promenade with a long slope on the inside. This height is, however, not enough to maintain current safety levels in the long run (>50 years; van Veelen 2012). In the intermediate option, the flood safety standard would be 1 in 100 years (rather low for Dutch standards), and more attention was given to evacuation and communication strategies. In the 'water in' option, flooding will occur regularly, especially if sea levels continue to rise. Therefore a lot of buildings need to be dry-proofed, and much attention needs to be given to evacuation. Wherever possible, highly flood-sensitive functions (such as energy plants, drinking water plants, and hospitals) are moved to higher parts of the unembanked areas.

3 Building Code and Zoning Policies in the Netherlands

Below, we review the policies and legislation in the Netherlands which affect adaptive measures in terms of building codes and zoning policies. Table 25.2 gives an overview of these policies and legislation, a short description of their content, and their implications for adaptive flood measures.

3.1 European Legislation

As the Netherlands is a member state of the EU, European regulations also affect Dutch regulations. After a number of severe floods in Europe, the European Union developed the European Flood Directive (Directive 2007/60/EC). It offers a framework for assessing flood risks and gives guidelines for decreasing the negative consequences of floods. It requires member states to develop flood hazard and flood risk maps, as well as flood risk management plans (Flood Directive website 2011). The Netherlands is currently in the process of developing these maps. The Dutch Water Act of 2009 incorporates the EU Flood Directive.

3.2 The Netherlands

3.2.1 Policy

The Major Rivers policy states that the primary function of the major rivers is to discharge water to the sea. New activities in the unembanked areas may not impede the rivers' discharge, as that would increase flood risks to the embanked areas (Ministry of Transport Public Works and Water Management 2006). The policy employs a five-step approach to determine if new developments are allowed in the major rivers' unembanked areas. In the first four, it distinguishes between the type of activity and the place where it is executed. A division is made between areas having a storage regime (which mainly store water during high flows) and a discharge

Table 25.2 Overview of policies and legislation on different government levels in the Netherlands, with a short description and implications for adaptive measures

	Policies and guidelines	Description	Implications for adaptive building	Legislation	Description	Implications for adaptive measures
European Union	Flood directive[a]	Risk approach,	No direct impacts,			
		Assess protection and damage reduction measures	Flood risk maps can create awareness			
		Develop flood damage and flood risk maps	Incorporated in Dutch Water Act			
National	Guidelines major rivers[b]	Development in floodplain[c] restricted; only if discharge is not changed	Little new development in floodplain	Rules for spatial planning[d]	Enforcement of guideline major rivers, implemented by municipalities via land use zoning plans	Little new development in protected floodplains
		Damages for own risk	No refund of damages gives some incentive for adaptive building			No refund of damages gives some incentive for adaptive building
			Rotterdam's unembanked areas are excluded	National building code[e]	Does not address flood proofing or elevation	Municipalities cannot demand stricter norms making enforcement of dry-proofing troublesome

				Spatial planning Act[f]	Municipalities must develop land use zoning plan each 10 years and must have a spatial policy	Contains rules for new developments
						Possibility to ask for minimal ground floor level for new developments on low lying floodplains
				Water Act[g]	Maintaining national flood safety levels	Does not allow for building on or near embankments
Province of South-Hollandh	Risk guideline[i]	Guidelines for casualties and social disruption for new developments	Municipalities should test land use plans for minimal casualties and social disruption	Land use planning rules (*Verordening Ruimte*[j])	Secures provincial interests	none
			Can integrate adaptive measures		No special rules for unembanked areas	
Municipality	Spatial policy	Document showing the desired future land uses	Spatial policies can show the need for adaptive measures and create awareness	Land use zoning plan	Shows which land uses are allowed on which location	Zoning possibilities (mainly effective in new to develop areas)
					Can include regulations that need to be fulfilled before land used can be implemented (linked to spatial planning Act)	Gives possibility to demand minimal ground floor elevation in low lying areas, but little used

(continued)

Table 25.2 (continued)

	Policies and guidelines	Description	Implications for adaptive building	Legislation	Description	Implications for adaptive measures
	City vision Rotterdam[k]	Long term vision for Rotterdam	States the need for taking future sea level rise into account, without detailing how			
		Includes redevelopment of harbours near city centre into housing and commercial area				
	Rotterdam water plan[l]	Plans for water infrastructure, developed together with water boards	Includes plans for floating buildings and small neighbourhoods, but not detailed			
Waterboard	Several policy documents	Focused on embanked areas	No implications for building in unembanked areas	'*Keur*'[m]	Regulations on embankments and water courses	No building on or near embankments

[a]website Flood directive (2011)
[b]Ministry of Infrastructure and the Environment (2006)
[c]Here: the unembanked areas of the floodplain.
[d]Ministry of Infrastructure and the Environment (2011)
[e]Ministry of Internal affairs (2011)
[f]Ministry of Infrastructure and the Environment (2008)
[g]Ministry of Transport Public Works and Water Management (2009)
[h]Policies differ per province; Rotterdam is part of the province of South-Holland.
[i]Huizinga et al. (2011)
[j]Provincie Zuid-Holland (2011)
[k]Gemeente Rotterdam (2007)
[l]Gemeente Rotterdam et al. (2007)
[m]e.g. Morreau (2009)

regime (areas through which the river discharges). The fifth step lists a number of conditions that should be met.

Step 1 allows for small, temporary activities and river management activities.
Step 2 allows for all types of activities in the storage regime areas.
Step 3 states that in areas with a discharge regime only river-related activities are allowed.
Step 4 allows for (*a*) activities that need to be realised in the unembanked areas, (*b*) activities for ground-based agriculture that are needed on compelling operational interests, (*c*) changes in functional use of existing buildings, and (*d*) activities that increase the volume that can be occupied by the river.
Step 5 states that all developments mentioned in the previous steps are only possible when they comply with the following conditions:

- The activities are situated in such a way that the safe functioning of the river and its flood defence systems is maintained.
- There is no obstacle for increasing the rivers' discharge capacity in the future.
- The increase in water level or decrease in storage capacity should be minimised.
- For activities taken under steps 2–4, the increase in water level and decrease in storage capacity should be compensated by measures to decrease water level and increase storage capacity.

3.2.2 Legislation

The Major Rivers policy is enforced via the Dutch Spatial Planning Act and Regulations (Ministry of Infrastructure and the Environment 2011) and the Water Act (Ministry of Transport Public Works and Water Management 2009). The Spatial Planning Act states that municipalities should develop land use zoning plans ('Bestemmingsplan') every 10 years. In these plans they have to incorporate legal rules from higher tiers of government. By law, the land use zoning plans should always be financially feasible. Municipalities can demand funding from construction companies for measures taken in the public space. They can, however, only demand funds for a limited set of measures, which is decided upon by the state. The list only includes more traditional measures. Municipalities can negotiate on a contract for funding for other types of measures.

In the national Spatial Planning Regulations, extra rules and regulations are given that municipalities should follow when developing land use plans. It also protects national interests, like maintaining the discharge capacity of the major rivers.

The Water Act protects water courses and embankments. The state is, by law, responsible for water safety in the embanked areas. When a flood is declared a disaster, the state must compensate people for flood damage in the embanked areas. The state is not responsible for compensation in the unembanked areas, where the building owners take their own risk. After the main storm event of 1953, flood insurance was not available as the state had initiated works on the delta and hence insurers concluded that the potential losses from a large flood were uninsurable. This also means that no flood insurance is available in the unembanked areas (Bouwer et al. 2007; Botzen and Van den Bergh 2008).

The national building codes contain norms for buildings, escape routes, strength of walls, etc. The building codes have a preemptive effect: other government layers cannot enforce norms that are stricter than the building codes' norms.

3.3 *Provinces*

3.3.1 Policy

Each of the 12 provinces is allowed to develop their own spatial and water policies within the framework of national policies. This has resulted in different approaches to flood management in unembanked areas. The province of Flevoland has, for instance, designated embankments in unembanked areas; the water board will have to maintain the current safety level. The province of Overijssel focuses on flood safety in the embanked areas, and ensures that flood risks are taken into account in land use zoning plans (Neuvel et al. 2011). In Utrecht, new developments have to meet the safety level of 1 in 4,000 year floods. The province of South-Holland, which includes our Rotterdam case study, has recently developed a new instrument to guide development of the unembanked areas. This so-called 'risk method' (Huizinga et al. 2011) takes a flood risk instead of a flood probability[3] approach. With a GIS-based model, municipalities can check the effect of spatial plans on the expected number of casualties and social disruption. The target levels can be met via various methods such as elevating, evacuation plans, or flood-proofing. The risk method will be implemented in 2013. Although municipalities are urged to use it, it is a voluntarily tool. It seems it should increase the awareness of flood risks among municipalities.

3.3.2 Legislation

The provinces not only develop policy documents, but also enforce the most important aspects via spatial planning rules. The province of South-Holland has no specific rules for unembanked areas. The risk method it is not enforced via the province's spatial planning rules because several assumptions behind it are not yet scientifically proven.

[3] Flood probability only looks at the chance of water levels becoming higher than the embankments, thus causing a flood. Flood risk also takes the *consequences* of a flood into account (that is, flood risk = probability × consequences).

3.4 *Municipality of Rotterdam*

3.4.1 Policies

In their spatial planning vision, the municipality of Rotterdam has planned several redevelopments of unembanked harbour areas into residential and commercial areas (Gemeente Rotterdam 2007). Most of the unembanked areas of Rotterdam are excluded from the 'Main Rivers' policy guidelines, making these new developments much easier. Rotterdam has put a lot of emphasis on becoming a 'green' city. It has for instance started the Rotterdam Climate Initiative (http://www.rotterdamclimateinitiative.nl/en) which promotes green roofs, floating buildings, and energy efficiency. There are plans to build a small neighbourhood of floating houses in one of the old harbours (Gemeente Rotterdam et al. 2007).

3.4.2 Legislation

For new developments a new land use zoning plan needs to be made. These land use zoning plans are legally binding. They show which land use is allowed in which place, but it cannot be used to enforce changes. It can only contain spatially relevant measures which, by law, need to demonstrate 'good spatial planning'.

Since land use zoning plans allow desired changes, if a municipality wants to make commercial activities possible, they can re-zone an area and designate it as commercial. When the current function is different, e.g. residential, the owner can still maintain that function and nothing will change. It is however not possible to start an industry or any function other than commercial if the current owner of the residential building leaves. Enlargement of the current function becomes difficult.

In order to demonstrate good spatial planning, it is sometimes necessary to demand that extra measures be taken. Jurisprudence shows at least one case in which a plan was expunged because flood risk in an unembanked area was not taken into account (for instance, by elevating ground level). It is, however, not possible for municipalities to enforce building codes that have stricter norms than the national building codes.

Requests for new developments, or large redevelopment of current buildings, will always be evaluated against the current land use zoning plan. If the plan allows for the new development, the municipality has no right to refuse the building permit. Of course, the building plans should also fit other legislation such as the national building codes and the water legislation ('keur') administered by the water board (see below).

Normally, the municipality of Rotterdam asks the national water services ('Rijkswaterstaat') for a minimum ground level for new developments in unembanked areas. This works well on new land that can easily be elevated using sand, but problems arise in existing built up areas. For these areas it is difficult for municipalities to enforce the use of adaptive building techniques, such as flood proofing, as they are not allowed to enforce stricter building norms than those in the national building code.

3.5 *Water Boards*

3.5.1 Policies

Water boards are a separate government layer in charge of water management: they maintain most embankments and the watercourses within the primary embankments. The management of the major rivers and part of the primary embankments are managed by the state water services (Rijkswaterstaat). To increase the attention on water management in the spatial planning process, water boards must be consulted during the development of land use zoning plans. As the water boards are responsible for embanked areas, they are often less concerned with water safety issues in unembanked areas.

3.5.2 Legislation

Water boards develop rules and regulations which can be found in the 'keur'. The 'keur' describes what activities are allowed in and near watercourses and on or near embankments. In principle, no buildings are allowed on or near embankments.

4 Implications of Policies and Laws on Adaptive Flood Management

From a study of the literature it became clear that policies, laws, and regulations do not forbid the use of adaptive measures. At the same time, regulations sometimes hinder government *enforcement* of some of these measures. Also, there are few or no policies or regulations to promote adaptive measures in unembanked areas, except for the more or less traditional ground level elevation.

The national 'Major Rivers' policy discourages new developments in most unembanked areas, but Rotterdam's unembanked areas are excluded. The province of South Holland has issued a 'risk method' which also takes adaptive approaches into account. The effect of this policy is not yet apparent, as it is still being tested. In principle, it can bring to light the implications of other adaptive measures on casualties and social disruption, which could increase the use of these adaptive measures. Rotterdam has a stated desire to use more innovative adaptive flood measures, but, except for a floating pavilion, this has not yet been implemented in practice. There is thus some enthusiasm to look beyond current practices, but this has not led to real action.

A problem for the implementation of new building codes and flood zoning is that land use plans are mainly effective for flood zoning in new areas. Up until now, there has been relatively little attention given to flooding in land use zoning plans. The main flood-proofing strategy was to elevate ground levels. Enforcing a minimum ground elevation via the land use plan can only be done for new developments, and only if it is clearly shown that the new development will not demonstrate good spatial planning. Otherwise it is possible that an agreement could be made with the construction company. Once again, it is difficult to enforce adaptive building techniques like dry-proofing, as municipalities are not allowed to demand stricter norms than those in the national building codes.

During a series of workshops and interviews following the literature study, a number of additional problems became clear. The discussions showed that although several actors had a good knowledge of adaptive measures, they hardly ever used them. In Rotterdam, only elevation of ground level is widely used, mainly because the municipality asks developers to do so. A number of reasons can be found for why adaptive measures are little used.

4.1 *Little Experience*

Because most adaptive measures are little used in the Netherlands, there is little experience with them and how they play out in the Dutch environment. This forms a barrier for construction companies to apply them, especially in uncertain economic circumstances. A representative of a construction company explained in a workshop that they compared dry-proofing techniques with elevation of the ground floor level for a new building that they were about to build. Both methods cost more or less the same, but because there was more uncertainty about dry-proofing they chose elevating the ground floor level.

4.2 *Little Reason for Adaptive Building*

In the Netherlands, there is relatively little need for adaptive measures, as the embanked areas have a very high safety level and most inhabited unembanked areas have been elevated in the past, often to a quite high safety level (>1 in 100 years).

Water and flood management issues only play a moderate role in spatial planning and other interests often seem to have priority. As municipalities have little right to withhold a permit for developments that fit the land use zoning plan, there is often no need for a developer to use adaptive building techniques. Moreover, some adaptive measures have negative consequences for the safety of the neighbourhood; for example, if ground floors are not used for housing, there is less social control on the streets and the crime rate might rise. The same goes for buildings built on poles, which have the added disadvantage that homeless people prefer to sleep below these buildings as it protects them from the rain. This was an issue that was raised by several participants in workshops and questionnaires.

Inhabitants and property owners have little or no experience with flooding, which could otherwise function as a pressure for more adaptive measures. It has been shown (Bubeck et al. 2012a, b) that owners only start to take action after one or two flood events. So perhaps with sea level rise, inhabitants will experience more flood events and start to take more measures, as is already the case in some of the very low unembanked areas that flood regularly.

4.3 *Laws and Regulations Are Unclear*

As there is relatively little experience with adaptive measures, the laws and regulations are often not clear on their use and impact. Land use zoning plans, for instance, should show good spatial planning by law, but it is not clear if that includes flood zoning. Most likely, flood zoning in new areas will not prove problematic, but it is much harder in existing areas where sensitive functions like day care, homes for the elderly, and industries are situated in flood-prone areas. After all, a land use zoning plan can only steer new developments, not change current functions. Removing current functions is only possible if the current owners are willing to move. Only for very urgent and important causes –those needed to serve the general public interest – can the government force them to move. Most likely a chance of flood damage will not qualify as such.

4.4 *Most Adaptive Measures Cannot Be Enforced*

It is difficult for municipalities to enforce adaptive measures, as permits can only be denied when they do not meet land use zoning plans or national building codes. Land use zoning plans can only contain spatially relevant rules, and there is no option to ask for higher safety norms than those in the national building codes. There is, therefore, no way of enforcing dry- or wet-proofing of buildings. Any contract that forces developers to dry- or wet-proof is illegal. The only possibility is to get a voluntarily agreement with the developer, or give subsidies to stimulate the use of adaptive building measures.

Spatial relevance is important for enforcing additional measures for new developments via land use zoning plans. The 'risk method' of the province of South Holland can demonstrate this spatial relevance by showing that adaptive measures are important for reducing risk, which makes it more likely that they could be enforced via land use zoning plans. Dry- and wet-proofing, however, will only become possible when the national building codes allow for it. Wet-proofing is likely to remain too difficult to enforce, as it would require officials to enter the buildings to see if wet-proofing measures are in place. This seems to be too much of a breach of privacy.

As it is difficult to enforce adaptive measures on buildings, a municipality could take measures in public spaces, like temporary embankments. They can, however, only implement such measures if their budget allows it. In case of new developments they can use part of the developer's profit margin, if the measures ensure that the developers will make a profit. For example, the developers of the large 'Zuid as' business district in Amsterdam pay for water retention areas elsewhere in Amsterdam because there is no room for water retention in the business district, although it is required by law. However, the money can only be used for a limited number of measures, which does not include adaptive flood management.

Furthermore, some measures might also meet opposition among society. This makes it harder to enforce them via, for instance, spatial plans, which are subject to a participatory process.

4.5 *No Clear Safety Norms and Responsibilities for Compensation*

Another issue is the lack of clear norms or guidelines that can be used for testing adaptive measures. As there is still little knowledge available on the effectiveness of adaptive measures, it is difficult to satisfy the requirements of a building permit by means of adaptive measures.

Finally, it is not clear who is accountable for damage compensation in unembanked areas. The state has clearly said that they will not pay for flood damage, but for municipalities and provinces it is not clear if they can or cannot be held responsible. If they can be held responsible, it is an incentive for them to finance adaptive measures for current buildings and to become much more cautious about allowing new developments. For new developments, they would also have more reason to take extra measures. As the financial plan should be sound, they could demand an extra sum from the developer's profits. Municipalities can, however, only demand funds for a limited set of measures, which is decided upon by the state. The list only includes more traditional measures. Municipalities can negotiate on a contract for funding for other types of measures.

5 Recommendations and Conclusions

In this chapter we have presented several adaptive flood management measures that can be used to lower flood damage in the unembanked areas of Rotterdam. They are already used incidentally in some places in the Netherlands, but their use is hampered by current policies, laws, and regulations, as well as a lack of experience. Policies are slowly starting to change from a 'flood probability' approach to a more 'flood risk' based approach. The laws and regulations, however, are still very much tailored to structural flood defences in the form of levees and dykes. This makes it difficult to implement adaptive measures, which in turn decreases the support for these measures among major stakeholders. The government could opt for enforcing changes, but this is not always possible due to the current legal framework.

Municipalities can try to enforce adaptive measures via land use zoning plans. Currently it is not clear how much could be enforced this way as it has not been done before. By making such efforts, new jurisprudence will become available that will provide more clarity. The state could help by incorporating dry-proofing techniques into specific 'unembanked areas'-section of the national building codes.

One might expect that the national policy of not compensating flood damage in unembanked areas would have led to widespread use of adaptive measures. As we have seen, this is not the case. One reason is that most parts of the unembanked areas do not flood very often. Most of the time only quays are flooded and houses are unaffected. Another important reason for neglecting flood risk might be that there has been little or no communication to residents in unembanked areas about flood risks. In fact, in most areas little communication is the norm: research shows that more than half the populations in unembanked areas do not know if they are living inside or outside the dykes (de Boer et al. 2012). Municipalities could stimulate adaptive flood measures by communicating risks and providing subsidies for adaptive measures, but some municipalities are hesitant to do so as they are afraid that companies and more affluent people might leave the area. A good example of how flood risks can be communicated is given by the city of Dordrecht, where demountable barriers are tested yearly, keeping inhabitants aware of the flood risk and reminding them of the function of barriers and slots (Zevenbergen et al. 2008). A well designed participatory process could assist in communicating risks while also allowing for discussion of solutions. Both in spatial planning (Healy 1997; Van Assche et al. 2010; Mannigel 2008), as well as in water management (Kok and van Vliet 2011; Mouratiadou and Moran 2007; Raadgever and Mostert 2005), a lot of attention has been put on participatory processes. Participatory processes can lead to social learning and can aid in paradigm change (Pahl-Wostl 2007; van Herk et al. 2011). Moreover, participatory processes are widely acknowledged as important components of adaptive management (Folke et al. 2005; Pahl-Wostl et al. 2007; Vervoort et al. 2010). Hopefully, they can lead to a paradigm change among all important stakeholders, which would make the implementation of measures a lot easier.

Although there still are several problems and barriers to be solved, there are also positive trends:

- In the Netherlands increasing attention is being given to flood measures other than embankments. Recently, more attention is being given to the area behind the dykes. A 'three layer' approach is studied in which the first layer is formed by prevention (structural methods). The second layer is formed by smart spatial planning and building codes in the embanked areas, reducing the impact when a flood occurs. The third layer is formed by evacuation and preparedness strategies, which are mainly focused on minimising casualties. The layered approach is, however, not yet implemented in regulations.
- A consortium of construction companies has developed a set of norms for floating buildings, based on the national building codes. This could be a first step for including floating buildings in national building codes.
- In 2012, new national building codes came into force. They open up the possibility of demanding reinforced walls in buildings to be built near railroads used for

the transport of hazardous goods. This provides hope that a similar option is possible for dry-proofing norms in unembanked areas.

- There is growing experience with floating buildings (e.g. IJburg in Amsterdam and plans in Rotterdam).
- There is an increasing level of experience with adaptive building (e.g. Hamburg, Stadswerven in Dordrecht). Several research projects have worked on adaptive flood management for the Netherlands (e.g. climate change, spatial planning, and urban flood management).

This chapter has looked at a wide range of adaptive flood measures and how they are affected by Dutch policies, laws, and regulations. It is clear that several adaptive measures, like dry-proofing and flood zoning, are difficult to enforce, especially in existing urban areas. Due to the way spatial planning is organised in the Netherlands, it is easier to enforce these measures for new developments. Because of the integrated nature of adaptive flood management, the options are also affected by factors other than flood risk, policies, and laws. Aspects like spatial quality and social safety also play a role in spatial planning. To achieve a more adaptive flood management, a paradigm change is needed among spatial planners, water managers, and policy makers at the different government levels involved, as well as among major stakeholders. Better communication of flood risks, and subsidies for adaptive measures, might also help implementation.

Acknowledgments This research was part of three research projects within the Knowledge for Climate project HSRR3.1: "Adaptive strategies for the unembanked area of Rotterdam", "Theme 1: Water safety" and "Theme 7: Governance of adaptation to climate change". We want to thank Knowledge for Climate for funding these projects. We want to thank our partners in these projects for fruitful cooperation, and Peter van Veelen in particular for organising the workshops. We also want to thank the anonymous reviewer for excellent comments.

References

Aerts, J. C. J. H., & Botzen, W. J. W. (2011). Flood-resilient waterfront development in New York city: Bridging flood insurance, building codes, and flood zoning. *Annals of the New York Academy of Sciences, 1227*(1), 1–82.

Aerts, J. C. J. H., Botzen, W., van der Veen, A., Krywkow, J., & Werners, S. (2008). Dealing with uncertainty in flood management through diversification. *Ecology and Society, 13*(1), 41.

Begum, S., Stive, M. J. F., & Hall, J. W. (2007). *Flood risk management in Europe: Innovation in policy and practice*. Dordrecht: Springer.

Botzen, W. J. W., & Van den Bergh, J. C. J. M. (2008). Insurance against climate change and flooding in the Netherlands: Present, future, and comparison with other countries. *Risk Analysis, 28*, 413–426.

Botzen, W. J. W., Aerts, J. C. J. H., & van den Bergh, J. C. J. M. (2009). Willingness of homeowners to mitigate climate risk through insurance. *Ecological Economics, 68*(8–9), 2265–2277.

Botzen, W. J. W., Aerts, J. C. J. H., & van den Bergh, J. C. J. M. (2013). Individual preferences for reducing flood risk to near zero through elevation. *Mitigation and Adaptation Strategies for Global Environmental Change, 18*(2), 229–244. doi:10.1007/s11027-012-9359-5.

Bouwer, L. M., Huitema, D., & Aerts, J. C. J. H. (2007). *Adaptive flood management: The role of insurance and compensation in Europe*. Amsterdam: Institute for Environmental Studies.

Bubeck, P., Botzen, W. J. W., & Aerts, J. C. J. H. (2012a). A review of risk perceptions and other factors that influence flood mitigation behavior. *Risk Analysis, 32*(9), 1481–1495. doi:10.1111/j.1539-6924.2011.01783.x.

Bubeck, P., Botzen, W. J. W., Kreibich, H., & Aerts, J. C. J. H. (2012b). Long-term development and effectiveness of private flood mitigation measures: An analysis for the German part of the river Rhine. *Natural Hazards and Earth System Sciences, 12*(11), 3507–3518. doi:10.5194/nhess-12-3507-2012.

Dawson, R. J., Ball, T., Werritty, J., Werritty, A., Hall, J. W., & Roche, N. (2011). Assessing the effectiveness of non-structural flood management measures in the Thames Estuary under conditions of socio-economic and environmental change. *Global Environmental Change –Human and Policy Dimensions, 21*(2), 628–646. doi:10.1016/j.gloenvcha.2011.01.013.

de Boer, J., Botzen, W., & Terpstra, T. (2012). *Percepties van burgers over binnen – en buitendijks wonen*. Rotterdam: Gemeente Rotterdam (Rotterdam Climate Initiative).

de Moel, H., van Alphen, J., & Aerts, J. C. J. H. (2009). Flood maps in Europe: Methods, availability and use. *Natural Hazards and Earth System Sciences, 9*(2), 289–301.

de Moel, H., Aerts, J. C. J. H., & Koomen, E. (2011). Development of flood exposure in the Netherlands during the 20th and 21st century. *Global Environmental Change, 21*(2), 620–627.

de Moel, H., Van Vliet, M., & Aerts J. C. J. H. (2013). Evaluating the effect of flood damage-reducing measures: A case study of the unembanked area of Rotterdam, The Netherlands. *Regional Environmental Change*. doi:10.1007/s10113-013-0420-z.

Evans, E., Ashley, R., Hall, J., Penning-Rowsell, E., Sayers, P., Thorne, C., & Watkinson, A. (2004). *Foresight: Future flooding. Scientific summary* (Managing future risks, Vol. 2). London: Office of Science and Technology.

Flood Directive. (2011). *A new EU floods directive*. http://ec.europa.eu/environment/water/flood_risk/index.htm. Accessed 20 Oct 2011.

Folke, C., Hahn, T., Olsson, P., & Norberg, J. (2005). Adaptive governance of social-ecological systems. *Annual Review of Environment and Resources, 30*(1), 441–473. doi:10.1146/annurev.energy.30.050504.144511.

Gemeente Rotterdam. (2007). Stadsvisie Rotterdam, ruimtelijke ontwikkelingsstrategie 2030.

Gemeente Rotterdam, Waterschap Hollandse Delta, Hoogheemraadschap Delfland, & Hoogheemraadschap Schieland en de Krimpenerwaard. (2007). Waterplan 2, werken aan water voor een aantrekkelijke stad.

Healy, P. (1997). *Collaborative planning: Shaping places in fragmented societies*. Houndsmill/London: MacMillan.

Hooijer, A., Klijn, F., Bas, G., Pedroli, M., Os, V., & Ad, G. (2004). Towards sustainable flood risk management in the Rhine and Meuse river basins: Synopsis of the findings of IRMA-SPONGE. *River Research and Applications, 20*(3), 343–357.

Huizinga, J. (2010). *Flood risk in unembanked areas. Part A flooding characteristics: Flood depth and extent*. Report number KfC 022A/2010, ISBN 978-94-90070-22-9.

Huizinga, J., Nederpel, A., de Groot, K., & Batterink, M. (2011). Risicomethode buitendijks: Methodiek ter bepaling van risico's als gevolg van hoogwater: provincie Zuid-Holland.

IPCC. (2007). *Climate change 2007: Impacts, adaptation and vulnerability*. Contribution of working group II to the fourth assessment report of the intergovernmental panel on climate change. Cambridge: Cambridge University Press.

Kok, K., & van Vliet, M. (2011). Using a participatory scenario development toolbox: Added values and impact on quality of scenarios. *Journal of Water and Climate Change, 2*(2–3), 87–105. doi:10.2166/Wcc.2011.032.

Kreibich, H., Christenberger, S., & Schwarze, R. (2011). Economic motivation of households to undertake private precautionary measures against floods. *Natural Hazards and Earth System Sciences, 11*(2), 309–321. doi:10.5194/nhess-11-309-2011.

Kundzewicz, Z. W. (2002). Non-structural flood protection and sustainability. *Water International, 27*(1), 3–13.

Luttik, J. (2000). The value of trees, water and open space as reflected by house prices in the Netherlands. *Landscape and Urban Planning, 48*(3–4), 161–167.

Maaskant, B., Jonkman, S. N., & Bouwer, L. M. (2009). Future risk of flooding: An analysis of changes in potential loss of life in South Holland (The Netherlands). *Environmental Science & Policy, 12*(2), 157–169. doi:10.1016/j.envsci.2008.11.004.

Mannigel, E. (2008). Integrating parks and people: How does participation work in protected area management? *Society & Natural Resources, 21*(6), 498–511.

Ministry of Infrastructure and the Environment. (2006). *Spatial planning key decision: Room for the river*. Explanatory memorandum.

Ministry of Infrastructure and the Environment. (2008). Spatial Planning Act. In *Stb. 2008-159.*

Ministry of Infrastructure and the Environment. (2011). Besluit algemene regels ruimtelijke ordening.

Ministry of Internal Affairs. (2011). Bouwbesluit 2012. In *Stb. 2011, 416.*

Ministry of Transport Public Works and Water Management. (2006). Beleidslijn grote rivieren (*Policy guideline on Major Rivers*).

Ministry of Transport Public Works and Water Management. (2009). Waterwet. In *Stb. 2009, 107.*

Morreau, I. (2009). Beleidsnota waterkeringen: Beleidsregels inzake de vergunningverlening op grond van de Keur voor waterschap Hollandse Delta, ed. waterschap Hollandse Delta.

Mouratiadou, I., & Moran, D. (2007). Mapping public participation in the Water Framework Directive: A case study of the Pinios River Basin, Greece. *Ecological Economics, 62*(1), 66–76.

Neuvel, J. M. M., Boxman, A. M. C., & Stoffer, H. J. (2011). Handreiking overstromingsrisicoparagraaf: Een handreiking bij het verantwoorden en meenemen van overstromingsrisico's in de ruimtelijke planvorming: provincie Overijssel.

Nicholls, R. J., Hanson, S., Herweijer, C., Patmore, N., Hallegatte, S., Corfee-Morlot, J., Chateau, J., & Muir Wood, R. (2008). *Ranking port cities with high exposure and vulnerability to climate extremes: Exposure estimates*. OECD environment working papers, no. 1. OECD Publishing.

Pahl-Wostl, C. (2007). Transitions towards adaptive management of water facing climate and global change. *Water Resources Management, 21*(1), 49–62.

Pahl-Wostl, C., Sendzimir, J., Jeffrey, P., Aerts, J., Berkamp, G., & Cross, K. (2007). Managing change toward adaptive water management through social learning. *Ecology and Society, 12*(2), 30.

Provincie Zuid-Holland. (2011). *Verordening Ruimte*. Available at http://decentrale.regelgeving.overheid.nl/cvdr/XHTMLoutput/Actueel/Zuid-Holland/72386.html

Raadgever, G. T., & Mostert, E. (2005). Public participation in information management: Overview of participatory tools and their contribution to adaptive river basin management. In *Deliverable 1.2.2 of the NeWater project*. Delft: RBA Centre, TU Delft.

Runhaar, H., Mees, H., Wardekker, A., van der Sluijs, J., & Driessen, P. P. J. (2012). Adaptation to climate change-related risks in Dutch urban areas: Stimuli and barriers. *Regional Environmental Change, 12*(4), 777–790. doi:10.1007/s10113-012-0292-7.

Schelfaut, K., Pannemans, B., van der Craats, I., Krywkow, J., Mysiak, J., & Cools, J. (2011). Bringing flood resilience into practice: The FREEMAN project. *Environmental Science & Policy, 14*(7), 825–833.

Slootjes, N., Botterhuis, T., Jeuken, A., & Gao, Q. (2011). *Resultaten MHW berekeningen t.b.v.probleemanalyse en verkenning hoekpunten. Voor het Deltaprogramma Rijnmond-Drechtsteden*. Delft: HKV/Deltares.

Stephenson, D. (2002). Integrated flood plain management strategy for the Vaal. *Urban Water, 4*(4), 423–428.

Te Linde, A. H., Aerts, J. C. J. H., Bakker, A. M., R. & Kwadijk, J. C. J. (2010). Simulating low-probability peak discharges for the Rhine basin using resampled climate modeling data. *Water Resources Research, 46(3)*. doi:10.1029/2009WR007707.

UN-Habitat. (2008). *State of the world's cities 2008/2009. Harmonious cities*. Technical report. Nairobi: United Nations Human Settlements Programme.

Van Assche, K., Verschraegen, G., & Salukvadze, J. (2010). Changing frames: Citizen and expert participation in Georgian planning. *Planning Practice and Research, 25*(3), 377–395.

van Herk, S., Zevenbergen, C., Ashley, R., & Rijke, J. (2011). Learning and action alliances for the integration of flood risk management into urban planning: A new framework from empirical evidence from The Netherlands. *Environmental Science & Policy, 14*(5), 543–554.

van Veelen, P. (2012). *Adaptieve ontwikkelingsstrategieën in buitendijks gebied in de Regio Rotterdam*. Report WP2, HSRR3.1.

Veerbeek, W., Zevenbergen, C., & Gersonius, B. (2010). *Vulnerability assessment based on direct flood damages*. Delft: UNESCO-IHE.

Vervoort, J., Kok, K., van Lammeren, R., & Veldkamp, T. (2010). Stepping into futures: Exploring the potential of interactive media for participatory scenarios on social-ecological systems. *Futures, 42*(6), 604–616. doi:10.1016/j.futures.2010.04.031.

Wardekker, J. A., de Jong, A., Knoop, J. M., & van der Sluijs, J. P. (2010). Operationalising a resilience approach to adapting an urban delta to uncertain climate changes. *Technological Forecasting and Social Change, 77*(6), 987–998.

Wheater, H., & Evans, E. (2009). Land use, water management and future flood risk. *Land Use Policy, 26*(Suppl. 1), S251–S264.

Zevenbergen, C., Veerbeek, W., Gersonius, B., & van Herk, S. (2008). Challenges in urban flood management: Travelling across spatial and temporal scales. *Journal of Flood Risk Management, 1*(2), 81–88.

Chapter 26
Transboundary Urban Water: The Case of Singapore and Malaysia

Yue Choong Kog

1 Introduction

The Southeast Asian environmental crisis has long been a matter of concern within the region. Decades of rapid industrialisation and urbanisation, together with a lack of effective environmental management programs, have led to widespread environmental degradation. This is only partially reflected by the deforestation, water quality, and flood problems that have beset many Southeast Asian countries. Water quality standards have deteriorated at a speed which matches the high economic growth rates that have, at least until recently, been recorded. Most countries have, in the absence of adequate treatment and disposal infrastructure, been grappling with how to clean up major rivers that have been despoiled by industries and households using them as waste dumps. In particular, according to DOE (2007, 2009a), many Malaysian rivers are badly polluted because of inadequate enforcement of environmental laws and standards. River systems have also been increasingly subjected to uncontrolled logging and land development. At the same time, water demand in Malaysia has increased substantially because of population growth and improved lifestyles associated with sustained economic growth over the past few decades.

Environmental studies show that cities have ecological footprints that are many times the sizes of their territories. For example, it requires resources from an area many times the size of Singapore to produce food, water, energy, and other resources needed to sustain the country's people and economy. Whereas the ownership and management of natural resources continues to remain a strictly national concern, over time there will be an increased need for trade and sharing of resources among the nation-states of Southeast Asia. With diminishing supplies of resources and

Y.C. Kog (✉)
East West Engineering Consultants, 98 Duchess Road,
Singapore 269022, Singapore
e-mail: eastwestconsult@yahoo.co.uk

Q. Grafton et al. (eds.), *Understanding and Managing Urban Water in Transition*, Global Issues in Water Policy 15,
DOI 10.1007/978-94-017-9801-3_26

contest over them for even domestic needs, tensions are likely to grow, not only within countries but also at the regional level. The banning of rice exports by Vietnam and Cambodia in 2008 to ensure local food security and to boost the value of the grain is a reflection of such tensions (Damazo 2008).

The need for water is ubiquitous. It is needed for industry, for agriculture, and by every living being. It is possible to work on alternative and renewable energy resources when there is an energy crisis, but when it comes to a water crisis, there is no alternative. There are already signs of impending water crises in a number of countries. If water is only needed for public health and hygiene, scarcity is rarely a problem, but it is unusual for demand to be confined to such a base level. Apart from human use, water is also needed to sustain the natural ecosystems found in wetlands, rivers, and the coastal waters into which they flow. In addition, water is needed for agriculture, industry, and economic development. Consequently, water can easily become a scare resource, especially in droughts and especially when gross water consumption continues to steadily increase as a result of population growth, urbanisation, and economic development.

Spatial and temporal variations in precipitation conspire to make water physically scarce in particular areas and at certain periods of time. Scarcity can be treated as an economic issue, involving an imbalance between available supplies and demands. Economics involves the analysis of the production, distribution, and consumption of goods and services, and so the tools of the discipline can be applied to the problem of managing scare resources between alternative uses. When investment in the supply of water lags behind demand, water scarcity is a certainty. Water scarcity is an interplay between the natural supply and demand for water. And of course demand can be affected by the pricing policy of supply agencies.

Many believe that water will be to the twenty-first century what oil was to the twentieth: a precious commodity that determines the wealth of nations. Some also believe that how a country handles its water problem could spell the difference between greatness and decline. The nations that keep their waterworks in top working order and operate them at the lowest cost will have a competitive edge in an increasingly competitive globalised world (Tully 2000).

This paper considers the case of transboundary sharing of water supplies between Singapore and Malaysia where both countries share water resources even though they do not share the same water basin. Singapore has to buy water to meet its domestic and economic needs. These supplies have to be considered in the context of increasingly polluted sources in Malaysia. Given Singapore's rising demand for water and the difficulties encountered by Malaysian authorities in meeting the country's own needs, supplies to the city-state are, judging from past incidents, likely to meet increasing local opposition (Kog 2002). This has previously occurred when bilateral relations experienced problems or when there were dry spells in Malaysia and domestic water shortages and rationing occurred. In the 1990s, attempts by Singapore to negotiate a new water agreement for water supply beyond 2061 were rejected by the Malaysian government led by Dr Mahathir, the then Prime Minister. This led Singapore to aggressively pursue other sources of supply to achieve water self-sufficiency. To meet its water demands Singapore has embarked on the production of 'NEWater', and on the desalination of seawater. 'NEWater' is the brand name given to reclaimed water from treated sewerage effluent produced by the Public Utilities Board (PUB).

2 Malaysian Water Resources

In order to better understand the issues involved, it is necessary to appreciate the problems faced by Malaysia with respect to its water sources. Of particular interest are the rivers in the state of Johor that Singapore relies upon for its imported water.

Malaysia is blessed with abundant water resources. The east coast of Peninsular Malaysia generally has a much higher rainfall than the west coast. However, water demand in the west is higher and has increased more rapidly as a result of faster economic development (Camerlengo et al. 1996). There is therefore a spatial mismatch of resources and demand, even though water resources as a whole are sufficient to meet all foreseeable requirements. The heavy rainfall that occurs almost all year round has produced a dense network of rivers and streams. The rivers are generally short and extremely swift in the upper reaches and have a higher load of silt in the lower reaches. The rainfall pattern ensures perennial flow, but the rain usually arrives in cloud bursts over a short period. This means that much of the water resources cannot be captured for useful purposes unless dams are constructed to regulate the flow.

It is generally recognised that the river basin is the appropriate economic and jurisdictional area for water resource development because within such a geographical and hydrological unit it is easier to gauge economies of scale and to assess the impact of water resource projects and programs in terms of the quantity, quality, and timing of flows. In Malaysia, the state governments have almost absolute rights over all surface water. Malaysia finds itself with a plethora of sector-based water laws, both at federal and state levels, and lacks a comprehensive water law. At present, water legislation is contained within laws that are enforced by various water-related government agencies. Many of these laws could be interpreted as outdated, ambiguous, or in need of revision (Le and Facon 2001). This diverse water legislation is difficult to effectively enforce because it deals only with aspects of water resources and water supply directly related to the responsibilities of particular government agencies.

Because of the divided responsibilities between state and federal authorities for water resources and supply, cooperation between agencies in order to deal with water issues, obtain supporting facts and figures, or develop policies can be slow and time consuming. There is no single agency in the country entrusted with the overall responsibility for holistic planning and management of water. Conflicts in water resources management, such as allocation of water rights, flood management, pollution control, and environmental protection, are resolved through inter-agency coordination and consultation. This implies the need for cooperation between water-related agencies (such as state water authorities) and for coordination at all levels of government, since river basins seldom conform to political and administrative boundaries. Resource use is likely to be suboptimal when existing water laws obstruct the transfer of water between areas and uses, or they hinder the relocation of supplies in order to integrate water quantity and quality control, or to coordinate the management and conjunctive use of surface water.

At the federal level, the National Water Council (in 1992) and the National Water Resources Council (in 1999) were established to pursue more effective water management, including the implementation of interstate water transfers. As a result, many water supply systems in Malaysia are now inter-connected. For example, Negeri Sembilan has been supplying water to Melaka and Kedah to Pulau Pinang (see Fig. 26.1). Another issue that needs serious consideration is more attention to control of demand and the promotion of water use efficiency. According to the President of the Malaysian Water Association, Ahmad Zahdi Jamil, in 2009 the average non-revenue water across the states stood at 36.6 %, with Pahang being the highest at 59.9 % and Selangor the lowest at 19 % (The Star 2011).

The river systems in Malaysia have increasingly been subjected to instability due largely to siltation brought about by uncontrolled logging and land development (DOE 2007, 2009a). Some of the rivers have become tourist attractions and this has prompted the construction of hotels and resorts. As a result, many of the forests

Fig. 26.1 States of Peninsular Malaysia

surrounding the river areas have been cleared. The surrounding soil erodes when rains come and the rivers become turbid, affecting aquatic life. Siltation of these rivers has also adversely affected water treatment operations (DOE 2007, 2009a).

Because Singapore relies on rivers in Johor, pollution in these rivers affects the water supply to the country. For example, the Skudai water treatment works had to be closed down in 1991 while pollution problems in the catchment were being resolved (Kog 2002). River water quality appraisal is based on a water quality index (WQI) consisting of six parameters: biochemical oxygen demand, chemical oxygen demand, ammoniacal nitrogen, dissolved oxygen, suspended solids, and pH. Water quality data collected from monitored rivers are compared with the WQI and the Interim National Water Quality Standards for Malaysia to determine their status: clean, slightly polluted, or polluted, and to classify them accordingly as Class I, II, III, IV, or V, which is an annual rating system.

According to the Department of Environment's (DOE) Malaysian River Basins Water Quality Trend (1990–2003) (DOE 2007), there were nine river basins categorised as polluted in 2003 compared to 14 in 2002. The polluted river basins are Juru and Pinang in Penang, Buloh and Klang in Selangor, and Air Baloi, Segget, Kempas, Danga, and Tukang Batu in Johor. The river basins in Johor constituted the bulk of polluted river basins. The five polluted river basins in Johor in 2003 are also listed as polluted in the DOE's list of 2002. Stations located upstream were generally clean, while those downstream were either slightly polluted or polluted. A major factor for river pollution is sewage. According to the environmental regulations, all waste water from toilets and kitchens should go into a septic tank or directly to a central sewage treatment system. To ensure that a septic tank is working, the owner is required to desludge the septic tank at least once every 2 years, but this does not appear to be happening. The evaluation of Sungai[1] (Sg) Segget in Johor showed that in terms of pollution load, sewage contributed 68.5 % of the pollution load in the river. The other 31.5 % came from restaurants and food courts. The state capital, Johor Bahru, is relatively old and not all of its wastewater goes into the sewage system. Johor Bahru is also a growth area and there is a correlation between the level of river pollution and development (DOE 2007).

In 2008, the majority of rivers (14 out of 21) in the Iskandar Malaysia Region were within Class III, indicating moderate levels of pollution, while five rivers (Pandan, Plentong, Sebulung, Sengkuang, and Tampoi) in the Tebrau catchment were in the more polluted Class IV. Iskandar Malaysia is located just north of Singapore. One river in the Pasir Gudang catchment remained in Class V, indicating severe river water pollution most likely caused by industrial and development activities. According to the Environmental Quality Report 2008 (DOE 2009a), 4 out of the 7 most polluted river basins in Peninsular Malaysia were located in the Johor Bahru area. The four polluted river basins were Sg Danga, Sg Segget, Sg Tebrau, and Sg Pasir Gudang. In 2009, most of the rivers (12 out of 21) scored a slightly polluted level of Class II or III. The most contaminated river basins were Segget, Pasir Gudang, Tebrau, and Air Baloi. Endau, Sedili Besar, Johor, and Mersing river

[1] Sungai means 'river' in Malay.

basins were classified as Class II and were deemed to be clean. River water quality in Mersing and Kota Tinggi had the highest WQI value and was the cleanest in Malaysia. Muar with a value of 0.91 was third, followed by Segamat and Batu Pahat with 0.86 and 0.77 respectively. The Johor Bahru area had the lowest WQI value of 0. This was because rivers in Johor Bahru area, especially Sg Segget and Sg Skudai were contaminated (DOE 2009b).

It is apparent from the above that there was no significant improvement to the general conditions of polluted river basins from 2003 to 2009. Water pollution adds enormously to existing problems of local and regional water scarcity by removing large volumes of water from the available supply. According to the Water Sustainability Index, the availability and usage of water in Malaysia showed a drop from 64 % in 1992 to 33 % in 2002, reflecting the fact that Malaysia's water resources were being rapidly depleted and requiring improved management (The Star 2011).

3 Water Dispute Between Singapore and Malaysia

Singapore has depended on Johor to meet the country's water needs since 1929; currently, about 40 % of its water needs are met by water drawn from rivers in Johor. Supply of water from Johor to Singapore became a security issue when Singapore was separated from the Federation of Malaysia on 9 August 1965. Since then relations between the two countries have been marked by alternating periods of tension and normalcy. Acrimonious disputes over communal questions – such as Malay hegemony in Malaysia[2] in 1963–1965 – have left both sides with a lingering bitter aftertaste (Kog and Ooi 2007).

The Malaysian government guaranteed Singapore's water supply in a Separation Agreement deposited with the United Nations. Archival records show that, in 1965, the then Malaysian Prime Minister, Tunku Abdul Rahman, told Anthony Head, the British High Commissioner in Kuala Lumpur, that "if Singapore's foreign policy was prejudicial to Malaysia's interest, they could always bring pressure to bear on them by threatening to turn off the water in Johor".[3] Such an idea reappeared in 1986 and 1998 when, in attempts to further their policy objectives or protest against perceived slights, some Malaysians called for a review of water links with Singapore. The 1986 episode occurred when Israeli President, Chaim Herzog, was visiting Singapore. The visit sparked off angry protests from anti-Zionist Malaysian activists in Johor who criticised Singapore's perceived irreverent attitude toward the religious and political sensitivities of its predominantly Muslim neighbour. Verbal

[2]During 1963–1965, when Singapore was part of Malaysia, there was political competition between rival politicians from the People's Action Party, United Malays National Organisation, and the Malayan Chinese Association.

[3]A copy of Anthony Head's confidential telegram (no. 1344) from Kuala Lumpur to the Commonwealth Relations Office on 9 August 1965 is in the National Archives of Australia.

attacks launched by the youth wing of the ruling United Malays National Organisation filled the air with corybantic calls for the Malaysian government to terminate the water supply to Singapore. The issue threatened to boil over when crowds apparently moved to forcibly occupy the Singapore-run waterworks in Johor and disrupt the water supply (Huxley 2000).

The second incident occurred in 1998 at a political rally in Johor Bahru when the then Malaysian Prime Minister, Mahathir Mohamad, criticised Singapore's decision to relocate Malaysia's railway station and its Customs, Immigration and Quarantine (CIQ) outpost in the south of Singapore to Woodlands in the north of Singapore. This roused the rally attendees to call for Malaysia to terminate the water links. The water supply from Johor to Singapore was being used as leverage to force the Singapore government to modify its policies (Leifer 2000).

Lee Kuan Yew told Mahathir that Singapore feared that at some time or other – whenever there were differences between Singapore and Malaysia – there could be "a random act of madness" like cutting off Johor's water supplies, an action which some in Malaysia had publicly threatened. In that case, the Singapore Armed Forces would have to move into Johor, forcibly if need be, to restore the water supply if water shortages became urgent (Lee 2000). Some observers believed it was possible that disputes over water supply, if inappropriately handled, could boil over and even ignite armed hostilities between Singapore and Malaysia.

In August 2011, the 1961 Johor River Water Agreement expired, and the 1962 Agreement will expire in 2061. Singapore officials have met their Malaysian counterparts to discuss a new 100 years water agreement after 2061. Singapore officials have requested 350 million gallons a day (1.59 million m^3/d) of raw water and 400 mgd (1.82 million m^3/d) of treated water to be supplied by Johor and Pahang, another state above Johor, to meet its projected demand of 950 mgd (4.32 million m^3/d) for a population of seven million. This request for water beyond 2061 was, however, contingent on Malaysia satisfying its own needs first. Malaysian officials were prepared to supply Singapore the same volume of 350 mgd (1.59 million m^3/d) and asked Singapore to source water elsewhere, such as in Indonesia, or build desalination plants to meet additional water demands. Malaysia could not commit itself to such a quantum in view of the uncertainty of its own situation in the future and Pahang's resources were planned for inter-state water transfer following the 1998 water crisis in the Klang Valley (Singh and Zulfakarin 1999).

Since the early 1990s Singapore has been seriously exploring the possibility of sourcing alternative water supplies from the Indonesian Riau islands. In late November 2000, after an ASEAN Heads of Government meeting, the Indonesian President, Abdurrahman Wahid, revealed at a closed-door meeting at the Indonesian embassy in Singapore that he had suggested to Mahathir that Indonesia and Malaysia jointly withhold water supply to Singapore to teach it a lesson (The Straits Times 2000). This may well have left an indelible mark on the Singaporean psyche. The fact that the lack of water was a key factor that hastened the fall of Singapore to the Japanese serves as a painful reminder of how crucial water supply is to the security of Singapore (Simson 1970).

What makes water a particularly emotive issue among the Malaysian states is that some states suffer from chronic water shortages while others enjoy water surpluses. In March 1990, water rationing was imposed in the northern regions of Johor while water flowed from reservoirs in Johor managed by the PUB to Singapore (The Straits Times 1990). Malacca faced a water crisis in 1991 with water rationing being imposed throughout the state. Malaysians living in Malacca were incensed that, while they had to endure the inconvenience of water rationing, water continued to flow across to Singapore. Eventually, the National Water Council was set up to allow for sharing of water resources between various states (The Straits Times 1992). Future water talks between Singapore and Malaysia will only be held closer to the expiry period of the two agreements (Pereira 1998).

Singapore continues to partly rely on water resources in Malaysia to meet the country's water needs. International law does not allow Singapore to directly intervene in the management of catchments in Malaysia. Water quality is a matter controlled by the environment policies and enforcement practices of the territory in which the water catchments are located. Therefore, pollution control in these catchments is a regional concern and requires skilful diplomatic handling of pollution problems while recognising sovereignty over the territory.

Many countries throughout the world share water resources with their neighbours because rivers and ground water basins transcend national boundaries. Neighbouring countries of the water resource have a natural 'right' to the shared resource. Such rights are recognised in international law and form the basis for countries to negotiate should disputes on the use of a shared water resource arise. The Singapore and Malaysia situation is unique in that the water resources in Malaysia are being shared with Singapore even though Singapore does not share river or ground water basins with Malaysia (Kog 2002).

The most recent widely publicised political negotiations between the two countries on the water supply issue took place between 1998 and 2003. A framework agreement was signed in September 2001, after Singapore agreed to make some major concessions. Among the agreed terms were a higher price for the water Malaysia supplies to Singapore, a compromised location for CIQ, and permission for Singapore's air force to resume its use of Johor's airspace. The agreement also stipulated a new bridge and tunnel to be built, replacing the existing causeway. In January 2002, Mahathir announced the intention for Malaysia to build the bridge itself because there was no agreement. Sharp rhetorical exchanges took place throughout that year. When talks resumed in September 2002, Mahathir sought to renegotiate several issues separately, despite the framework agreement signed a year earlier. Singapore apparently had no choice but contended that the price of water, among other issues, should be discussed within the framework's larger context.

Moves by the Malaysian government have re-focused public attention on the water issue. There was media activity by the government and the publication of a booklet in 2003 entitled "Water: The Singapore–Malaysia Dispute – The Facts". It resulted in January 2003 in the Singapore government taking the unprecedented step of releasing into the public domain correspondence between the two countries on the negotiations (MICA 2003).

Not only is the water issue embroiled in Malaysia's domestic politics but it has also been taken to the regional arena, so it is going to take some time and major efforts from both parties to work out a settlement. Fortunately, the second water agreement will only expire in 2061 and technologies to produce water are available to help Singapore meet the ever increasing demand for more water. The costs of such technologies will also become cheaper over time.

4 Singapore Water Management Policies and Practices

Singapore is a small city-state of about 712 km^2, with limited natural water resources of its own. It is densely populated with 5.08 million people according to the 2010 census. Singapore is classified by the United Nations as a water-scarce country (UNESCO 2006). Even though climate change may bring more rainfall in Singapore, it will remain water-scarce. The constraint faced by Singapore is its limited land area to catch and store rain, and the absence of natural ground water aquifers.

Singapore's first reservoir was the MacRitchie reservoir, which was built by the British in 1867. Subsequently, as Singapore developed into a modern city, more sources of water, including imported water from Johor, were needed to sustain the city's growth. Singapore and Malaysia have signed four agreements for the supply of water from Johor to Singapore. The first, signed in 1927, is no longer in force. The 1961 and 1962 agreements were signed between Malaysia and Singapore when Singapore was a self-governing state within the British Empire and Malaya was already an independent nation. The two water agreements signed between the Johor state government and Singapore's City Council in 1961 and 1962 are collectively known as the Johor Water Agreements. The City Council is the predecessor of Singapore's Public Utilities Board (PUB). Singapore paid for all the costs of reservoirs, dams, pipelines, plant, and equipment in Johor and Singapore continues to pay all the costs of operating and maintaining this infrastructure. The 1961 agreement expired in August 2011 and, in accordance with the agreement, the complete water works has been handed over to the Johor state government. The 1962 and 1990 agreements will expire in 2061. These agreements provide for Singapore to import raw water from Malaysia and, more specifically, for Singapore to draw up to 350 mgd (1.59 million m^3/d) as follows (MICA 2003):

(a) **Sg Tebrau and Sg Scudai Water Agreement** The agreement was officially signed on 2 October 1961 but took effect on 1 September 1961. The 1927 agreement was declared void in this document. The agreement gave Singapore the full and exclusive right to draw off all the water within the designated land at Gunong Pulai, Sg Tebrau, and Sg Scudai for a period of 50 years up until 2011. Singapore was to pay an annual rent of $5 per acre for the land and a charge of 3 cents for every 1,000 gal (4.55 m^3) of water. Singapore also agreed to provide

Johor with a daily supply of treated water up to 12 % of the raw water it drew, subject to a minimum of four million gallons, and at a price of 50 cents per 1,000 gal. If the 12 % provided by Singapore was insufficient, Johor could request for more treated water to be supplied.

(b) **Sg Johor Water Agreement** This gave Singapore the full and exclusive right to draw water from Sg Johor up to a maximum of 250 mgd (1.14 million m^3/d). In return, Johor was entitled to a daily supply of treated water from Singapore up to 2 % of the raw water it supplied. Singapore had to pay rent for the land it used "at the standard rate applicable to building lots on town land". The water prices remained the same as in the previous agreement. After Singapore and Malaysia stopped using a common currency, prices became denominated in Malaysian ringgit (RM).

(c) **Sg Linggui Water Agreement** This was signed on 24 November 1990 between the PUB of Singapore and the Johor state government. It was supplementary to the 1962 pact and also expires in 2061. A separate document was signed on the same day by the governments of Malaysia and Singapore to guarantee adherence to the agreement. Under this agreement, Singapore was allowed to construct a dam across Sg Linggui to facilitate the extraction of water from Johor River, with Johor setting aside about 21,600 ha of land for the project. Singapore agreed to pay RM320 million as compensation for the permanent loss of use of the land and its associated revenue, in addition to a premium of RM18,000 per ha and an annual rent of RM30 for every 1,000 ft^2 of land. The cost of building and maintaining the dam would be borne by Singapore. In return, Singapore could buy from Johor treated water generated by the new dam. This would be over and above the 250 mgd of raw water that it was allowed to draw from Sg Johor under the 1962 agreement. The price of this additional supply would be calculated based on a fixed formula: the weighted average of Johor's water tariffs plus 50 % of the surplus from the sale of this water by PUB to its consumers (after deducting Johor's price and PUB's cost of distribution), or 115 % of the weighted average of Johor's water tariffs, whichever was higher. This agreement was a follow-up to the memorandum of understanding (MOU) signed on 28 June 1988 between the two countries' prime ministers at the time, Lee Kuan Yew for Singapore and Mahathir Mohamad for Malaysia. The signing of the MOU was hailed as a breakthrough in Singapore–Malaysia water relations, the culmination of 6 years of difficult negotiations (Chew 2009).

The 1961 and 1962 agreements were guaranteed by the Malaysian government in the Separation Agreement that established Singapore as a sovereign state in 1965, and were enacted into the Malaysian Constitution by an Act of Parliament (MICA 2003). Under the 1961 and 1962 Water Agreements, Singapore pays the Johor state government 3 cents (RM0.03) for every 1,000 gal (4.55 m^3) of raw water. It costs Singapore RM2.40 to treat every 1,000 gal of water. By selling treated water to Johor at only 50 cents (RM0.50), Singapore absorbs a subsidy of RM1.90 per 1,000 gal of water (MICA 2003). The Singapore government claims that Singapore is subsidising Johor to the tune of RM29 million a year by selling it treated water at a reduced rate. It also noted that Johor is voluntarily buying an average of 37 million gallons of treated water per day from Singapore, more than double the amount of 15 million gallons of water that it has a right to (Sunday Times 1998).

Both Water Agreements contain a provision that allows for a review of water prices in 25 years' time, and arbitration in the event of a disagreement. Prices can be revised in line with the purchasing power of money, labour costs, and cost of power and materials used to supply water. In 1986 and 1987, Malaysia did not revise the water rates because of financial considerations. The Johor government knew that if it were to raise the price of raw water, the price of treated water that Johor buys from Singapore would also go up (MICA 2003).

After Singapore became independent in 1965, the PUB took over the control of basic supplies including water. Its main source of water supply was water imported from Johor, supplemented by water from reservoirs collected from local catchments. To ensure a stable and sustainable water supply for the country's growing economy and population, there were many developments to increase Singapore's water supply such as the Kranji–Pandan Scheme, the Upper Pierce Project, and the Western Catchment Water Scheme.

In the 1970s, a Water Planning Unit was formed in Singapore to develop a Water Master Plan with the objective of integrated planning encompassing land use, environmental concerns, and water matters across various government agencies (Tan 2009). Its aim was to help the various government ministries work effectively together as a public sector in a coordinated way and to prevent water pollution at an early stage. Early prevention of water pollution is a critical factor in successful catchment management. Effective cross-sector coordination among the relevant government agencies in water management in Singapore helped integration and reduced inter-sectoral conflicts of interest (World Bank 2006).

Singapore has two separate systems to collect rainwater and used water. Rainwater is collected through a comprehensive network of drains, canals, rivers, stormwater collection ponds, and reservoirs before it is treated for drinking water. This makes Singapore one of the few countries in the world to harvest urban stormwater on a large scale for its water supply. Two-thirds of the country's surface area is classified as partly protected catchment areas; there are certain restrictions on land use so that rainwater can be collected and used as drinking water. As of 2010, the water supply system included 19 raw water reservoirs, 9 treatment works, and 17 service reservoirs for treated water. The largest reservoir, the Marina Bay reservoir inaugurated in 2008, is in the estuary of a river that has been closed off by a dam, known as the Marina Barrage, across the Marina Channel to keep seawater out. Besides creating the fresh water reservoir, the barrage acts as a tidal barrier that prevents high tides from flooding inland low-lying areas. This unique project provides three main benefits: water supply, flood control, and a new lifestyle attraction. Two similar barrages were commissioned in 2011 in the northeast of Singapore with the damming of Sg Punggol and Sg Serangoon, forming the Punggol Reservoir and the Serangoon Reservoir. Through careful planning and appropriate measures, there have been no reported adverse environmental impacts of closing off these estuaries. Combined with Marina Reservoir (which is the most urbanised catchment of

10,000 ha), Punggol and Serangoon Reservoirs will increase Singapore's water catchment area from half to two-thirds of its land area.

In 2001 the PUB, which had previously been in charge of water supply only, was given the responsibility for sanitation as well, a role that had previously been provided by the Ministry of Environment. The change facilitated the implementation of an integrated water management approach. In 2002, and after a monitoring period of 2 years to ensure safe water quality, Singapore inaugurated its first reclaimed water plant. In 2005 its first desalination plant was opened.

In the 1960s, an intensive sewage development program was established to cater to the demands of Singapore's rapid housing and industrialisation growth. Waste water is collected through a network of underground sewers that leads to sewage treatment plants. As a result, 100 % of Singapore's population is today served by modern sanitation. To further meet the long-term waste water needs of Singapore and to ensure the long term sustainability of the waste water system, PUB has put in place a deep tunnel sewage system (DTSS) and the Changi water reclamation plant (WRP).

The DTSS was conceived as a long term solution to meet the needs for waste water collection, treatment, and disposal and serve the development of Singapore. The project consists of two large, deep tunnels criss-crossing the island, two large centralised WRPs, deep sea outfalls, and a link sewer network. Completed in 2008, Phase 1 of the DTSS comprises a 48 km long deep tunnel sewer running from Kranji to Changi WRP, two deep sea outfalls, and 60 km of link sewers (see Fig. 26.2). The treated effluent from Changi WRP is used in the production of NEWater at Changi NEWater factory. With Phase 1 of the DTSS in place, the existing sewage treatment plants at Kim Chuan, Bedok, and Seletar, along with their pumping stations, have been progressively phased out, freeing land for other developments.

NEWater is ultra-clean water produced by purifying wastewater with advanced membrane technologies. The primary use of NEWater is for non-potable industrial

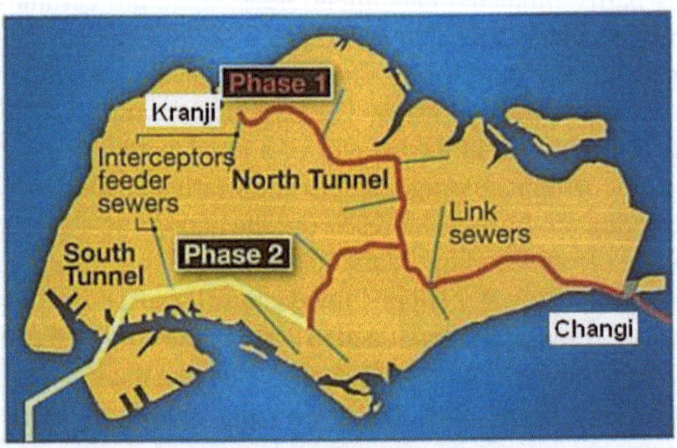

Fig. 26.2 Layout of Singapore's deep tunnel sewage system

purposes. The evolution of NEWater was a journey that spanned three decades or longer. In 1966, the Jurong industrial water treatment plant was commissioned and began to supply industrial water to Jurong. This water was reclaimed from treated sewerage effluent and while of sufficiently high quality for industrial use was not potable. Singapore's first Water Master Plan was drawn up in 1972, and in 1974 PUB built a pilot plant to turn waste water into potable water. This was the precursor of today's NEWater factories. The pilot plant was ahead of its time, but the costs were astronomical and the membranes were unreliable, so the idea was shelved to await further technological advancement. In 1998, the technology had matured and production costs were down. In May 2000, the first NEWater pilot plant was completed. PUB accepted an expert panel's proposal to use NEWater for indirect potable use. This means mixing and blending NEWater with raw water in the reservoirs before the mixture undergoes conventional treatment at the waterworks for supply to the public for potable use. Initially, PUB has started with 3 mgd (13,700 m^3/d) of NEWater, which is about 1 % of total daily water consumption, into raw water reservoirs. The amount increased progressively to about 2.5 % of total daily water consumption in 2011. The first NEWater plants were opened in Bedok and Kranji in 2003, and currently there are four NEWater plants in Singapore. The latest and the largest NEWater plant at Changi, with a capacity of 50 mgd (227,000 m^3/d), was opened in May 2010. With this addition, coupled with the expansion of existing plants, NEWater now meets 30 % of Singapore's total water demand. By 2060, NEWater is projected to meet 50 % of Singapore's future water demand (PUB 2011). Currently, 94 % of 'NEWater' is used at wafer fabrication plant and other non-potable applications.

The reuse of treated used water is not new. In countries with long riverine systems, unplanned indirect potable use occurs when upstream communities use the water and discharge it, after treatment, back into the river, lake, or aquifer that serves as a source of water supply for downstream users. Successive downstream communities then reuse the water several times before the river finally flows into the sea. Examples are cities along the Rhine and Thames in Europe, the Mississippi in the United States, the Yangtze in China, and the Mekong in Indo-China. Until recently, the biggest constraint to greater recycling was cost. Now, new technology has produced superior filters and membranes and significantly reduced the cost of microfiltration and reverse osmosis. It has become economically attractive to recycle used water on a large scale.

Another technology-based water source is desalinated water. Singapore has one of Asia's largest seawater reverse osmosis plants, which produces 30 mgd (137,000 m^3/day) to meet about 10 % of Singapore's water needs. By 2060, PUB intends to ramp up desalination capacity so that, in the long term, desalinated water will meet at least 30 % of Singapore water demand. Singapore's first desalination plant was commissioned in 2005 under a design, build, own, and operate (DBOO) agreement with SingSpring Pte Ltd, a wholly owned subsidiary of Hyflux Ltd. The SingSpring desalination plant is PUB's pioneering public–private partnership project under the DBOO arrangement. Under this partnership, SingSpring will supply 30 mgd (137,000 m^3/d) to PUB for 20 years. In April 2011, PUB and Hyflux signed a 25-year water purchase agreement for Singapore's second and largest desalination plant. The desalination plant, located at Tuas, constructed under the DBOO model commenced operations in 2013, adding another 70 mgd (318,000 m^3/d) to Singapore's water supply. The proper disposal of the waste brines of the desalination plant has been incorporated in the plant design so that there will not be adverse environmental impacts.

Today, Singapore has four different sources of water supply, known as the four national taps. They are:

- Rainfall, collected in reservoirs (about 20 % of supply in 2010)
- Imported water from Malaysia (about 40 % of supply)
- Reclaimed water (NEWater) (30 % of supply), and
- Seawater desalination (10 % of supply).

Water reclamation and waste management will become a huge industry, particularly in view of the predicted impacts of climate change. Almost every society, notably China and India, will have to cope with this problem. Singapore's current technical capacity to be self-sufficient if the need arises is the result of building on existing knowledge. Singapore brought the technologies together and improved on them. It is a collaborative effort, and Singapore is looking to share the technology with others who need it so they too can improve on it.

5 Conclusion

The present study shows that technologies in water reclamation and waste management, used with effective water management policies and practices, are effective and cost competitive in helping humankind face an impending water crisis.

The present study of the transboundary sharing of water resources between Singapore and Malaysia shows that such an arrangement can be a win–win arrangement for both countries. The major benefits for Malaysia can be summarised as follows:

1. In hindsight, the 1961 Agreement is similar to a BOT (build–operate–transfer) arrangement. All the costs of reservoirs in Johor, dams, pipelines, plant, equipment, etc., and all the costs of operating and maintaining this infrastructure from September 1961 to August 2011, were borne by PUB. In August 2011, PUB handed over to the Johor state government, free of charge, a complete developed catchment covering an area of 3,301 ha, two water treatment plants at Skudai and Gunung Pulai, two pump houses at Tebrau and Pontian, administrative buildings, and workers' living quarters. Since January 2011, PUB has trained 60 workers to take over operations. This means a saving amounting to hundreds of millions of RM for the construction of this infrastructure.
2. An annual income to the Johor state government, payable by PUB, of an annual rent of $5 per acre for the land and a charge of 3 cents for every 1,000 gal (4,500 l) of water.

3. For the 21,600 ha of land set aside by Johor under the Sg Linggui Water Agreement, Singapore has agreed to pay RM320 million as compensation for the permanent loss of use of the land and its associated revenue, in addition to a premium of RM18,000 per ha and an annual rent of RM30 for every 1,000 ft^2 of land. This works out to be a premium of RM389 million and an annual rent of RM70 million.
4. Johor is able to meet part of its water needs with good quality treated water produced by the water works operated by PUB in Johor.
5. From September 1961, Singapore has been providing Johor with an average of 37 million gallons of treated water per day (168,000 m^3/d) at a price of 50 cents per 1,000 gal (4.55 m^3) even though the treatment cost is much higher. The Johor state government is selling this water at RM0.98/m^3 (RM4.31 per 1,000 gal). This in turn has generated a profit of RM52 million a year, which includes a subsidy of RM29 million a year by PUB.

The benefits for Singapore can be summarised as follows:

1. From 1927 onwards Singapore has been able to meet its water needs economically with high quality treated water, helping Singapore to compete globally.
2. Because of difficulties, when Mahathir was Prime Minister of Malaysia, of negotiating another water agreement beyond 2061, Singapore was forced to invest heavily in utilising the latest technology in water reclamation and waste management to ensure its water security. The Singapore water story is one of how a small nation's efforts to overcome its constraints in natural resources has moved it from a water-scarce city to a leader in water supply solutions. PUB won the 2007 Stockholm Industry Water Award and was named Water Agency of the Year in 2006 at the Global Water Awards. Since 2008 Singapore has hosted the Singapore International Water Week, an event that brings policymakers, industry leaders, experts, and practitioners together to address challenges, showcase technologies, discover opportunities, and celebrate achievements.
3. With the success of NEWater and desalination, Singapore has become a world leader in water technology. Water reclamation and waste management have become major industries because many countries, especially dry and populous ones, will have to cope with water shortages. With experience gained in Singapore, over 50 local and international companies located there can now offer solutions, making this sector a substantial export industry. At the Singapore International Water Week 2010, the total value of announcements for projects awarded, tenders, investments into Singapore, and MOUs for research and development exceeded S$2.8 billion, up 27 % from last year's S$2.2 billion.

By the time the water supply agreement with Malaysia expires in 2061, and if there is no new water agreement in place, Singapore can still be self-sufficient in water, with reclaimed water meeting 50 % and desalination 30 % of its needs in addition to the 20 % supplied by internal catchments (Tan 2009). At the same time, the target of PUB is to reduce daily per capita domestic water consumption to 147 l by 2020, and 140 l by 2030 (PUB 2011). Optimistically, there is the possibility of

Malaysia and Singapore signing a new water agreement for water supply beyond 2061, especially in view of the more cordial relationships that have grown between Singapore and Malaysia since Najib Razak became Prime Minister of Malaysia. This has been well demonstrated by the recent resolution of the customs–immigration–quarantine issue that had been unresolved for at least 13 years. This optimism is based on some indications from Malaysian officials, extending as far back as 1999, that supply of extra water to Singapore is possible. Maintaining cordial relations between the two countries will help solve remaining problems. Ultimately, it is the total costs of the secured water supply that will determine whether a future water agreement is reached.

References

Camerlengo, A. L., Hisham, M. R., & Nasir, S. M. (1996). On the passage on the Intertropical Convergence Zone in Peninsular Malaysia. *Malaysian Journal of Physics, 17*(4), 155–162.

Chew, V. (2009). *Singapore–Malaysia water agreements*. Singapore: National Library Board. http://infopedia.nl.sg/articles/SIP_1533_2009-06-23.html. Accessed 10 Nov 2011.

Damazo, J. (2008). *Rice shortage: crisis or hype?* Hong Kong: Asia Sentinel. http://www.asiasentinel.com/index.php?option=com_content&task=view&id=1227&Itemid=35. Accessed 15 Nov 2011.

DOE. (2007). *Malaysia environmental quality report 2006*. Kuala Lumpur: Department of Environment, Ministry of Natural Resources and Environment.

DOE. (2009a). *Malaysia environmental quality report 2008*. Kuala Lumpur: Department of Environment, Ministry of Natural Resources and Environment.

DOE. (2009b). *Water quality index for river basins in Iskandar Malaysia, 2005–2008*. Kuala Lumpur: Ministry of Natural Resources and Environment.

Huxley, T. (2000). *Defending the lion city: The armed forces of Singapore*. St Leonards: Allen & Unwin.

Kog, Y. C. (2002). Natural resources management and environment security in Southeast Asia: A case study of clean water supplies to Singapore. In C. G. Kwa (Ed.), *Beyond vulnerability?: Water in Singapore–Malaysia relations* (IDSS monograph, Vol. 3, pp. 14–44). Singapore: Institute of Defence and Strategic Studies.

Kog, Y. C., & Ooi, G. L. (2007). Environmental security and water resources in Southeast Asia. In V. R. Raghavan (Ed.), *India and ASEAN: Non-traditional security threats* (pp. 75–95). Chennai: EastWest Books (Madras).

Le, H. T., & Facon, T. (2001) *From vision to action-a synthesis of experiences in Southeast Asia*. The FAO-ESCAP Pilot Project on National Water Visions, Bangkok, Oct 2001, p. 32.

Lee, K. Y. (2000). *From third world to first: The Singapore story 1965–2000* (p. 276). Singapore: Times Media Pte Ltd.

Leifer, M. (2000). *Singapore's foreign policy: Coping with vulnerability*. London: Routledge.

MICA. (2003). *Water talks? If only it could*. Singapore: Ministry of Information, Communications and the Arts.

Pereira, B. (1998). Water talks: Singapore asking too much. *The Straits Times*, 31 Aug 1998, p. 15.

PUB. (2011). *From one to four plants*. Singapore: Public Utilities Board. http://www.pub.gov.sg/water/newater/Pages/default.aspx. Accessed 11 Nov 2011.

Simson, I. (1970). *Singapore: Too little, too late: The failure of Malaya's defence in 1942*. Singapore: Asia Pacific Press.

Singh, H., & Zulfakarin, M. (1999). *Look elsewhere for extra water, Singapore told*. Kuala Lumpur: The Star.

Tan, Y. S. (2009). *Clean, green and blue: Singapore's journey towards environmental and water sustainability*. Singapore: Institute of Southeast Asian Studies.
The Star. (2011). *Time to review water tariff: MWA*. 23 Apr 2011. http://thestar.com.my/news/story.asp?file=/2011/4/23/nation/20110423180155&sec=nation. Accessed 12 Nov 2011.
The Straits Times. (1990). *Johor MB to critic: Water project not a disadvantage to Malaysia*, 10 Apr 1990.
The Straits Times. (1992). *States agree to set up national water council*, 12 Jun 1992.
The Straits Times. (2000). *Wahid launches tirades against Singapore*, 27 Nov 2000.
The Sunday Times. (1998). *S'pore sells subsidised water to Johor*, 6 Sep 1998.
Tully, S. (2000). Water, water everywhere. *Fortune, 141*(10), 69–78.
UNESCO. (2006). *Second United Nations World Water Development Report*. New York: United Nations Educational Scientific and Cultural Organization.
World Bank. (2006). *Dealing with water scarcity in Singapore: Institutions, strategies, and enforcement, addressing water scarcity*. Background paper no. 4 (China: The World Bank Analytical and Advisor Assistance (AAA) Program). Washington, DC: World Bank, East Asia and Pacific Region, Environment and Social Development.

Chapter 27
The Opportunities and Challenges of Implementing 'Water Sensitive Urban Design': Lessons from Stormwater Management in Victoria, Australia

Karen Hussey and Esther Kay

1 Introduction

Responding to pressure on the natural and built environments from population growth, the need for safe and secure water to support resilient and liveable communities, and the need to adapt to increased climate risk and variability, many cities around the world have sought new ways to develop, manage and sustain their urban environments. Indeed, in recent years, three important and related concepts have emerged in relation to cities and urban development: first, the need to design cities so as to minimise the impact of climatic events on populations, infrastructure and the environment; second, the need to optimise human consumption of scarce natural resources, particularly land, water, energy and nutrients; and third, the need to protect and where possible conserve the natural environment in and between cities. Over time, these concepts have evolved into a variety of different concepts, approaches and 'visions' to guide future land use development. One such concept is 'water sensitive urban design' (WSUD), a term whose definition varies but which is nevertheless now readily accepted internationally as a core aspiration of urban development (Wong and Brown 2011).

In its broadest context, WSUD is the integrated design of the urban water cycle, incorporating water supply, wastewater, stormwater and groundwater management, urban design and environmental protection (NWC 2009). It represents a fundamental

K. Hussey (✉)
Fenner School of Environment and Society, The Australian National University,
Canberra 0200, ACT, Australia
e-mail: karen.hussey@anu.edu.au

E. Kay
Environment & Land Management Pty Ltd, 28 Royle Court,
Woodend, VIC 3442, Australia
e-mail: kay.elm@bigpond.com

Q. Grafton et al. (eds.), *Understanding and Managing Urban Water in Transition*, Global Issues in Water Policy 15,
DOI 10.1007/978-94-017-9801-3_27

shift in the way water and related environmental resources and water infrastructure are considered in the planning and design of cities and towns, at all scales and densities (see also Barraque, Chap. 27, this volume). However, the transformational nature of WSUD poses enormous challenges to the existing paradigm, which is characterised by a complex and entrenched web of institutional and regulatory arrangements that very distinctly see the administrative and planning aspects of land use, flood mitigation, water quality management, urban development, and environmental protection managed separately (Brown 2008), with numerous and different actors involved but kept separate by very different regulatory regimes. Current institutional arrangements also reflect earlier thinking that critical infrastructure for water should be designed and managed centrally, with a strong preference for 'hard' engineered supply solutions such as dams, city-level waste water treatment plants and reticulation. In contrast, features of WSUD include decentralised, on-site treatment of stormwater through bio-retention pits and rainwater tanks, the establishment of wetlands within city boundaries, and the development of dual reticulation systems to exploit both potable and 'grey' water in all new developments.

The focus of this chapter is on the challenges of implementing WSUD, specifically with respect to the stormwater management aspects of WSUD. The chapter is based on a significant research project undertaken by the authors on behalf of the Government of Victoria, Australia, to examine the compliance and implementation challenges of that state's regulatory arrangements for WSUD and stormwater management. The research was designed to identify the barriers to, and 'enablers' of, compliance with the relevant law, so as to provide insight into how stormwater management through WSUD can be achieved. While this chapter focuses on the experiences in the State of Victoria, the findings and conclusions are highly relevant to other cities grappling with the implementation of WSUD.

In the next section, we briefly explore the rationale for, and objectives of, WSUD. In Sect. 3 we introduce the background to, and methodology for, the Victorian case study, with particular reference to the regulatory and legislative context for WSUD and stormwater management in Victoria's planning regime. In Sect. 4 we present the research findings, focusing particularly on the 'enablers' and 'barriers' to successful implementation in the planning process. Based on the findings from the Victorian case study, in Sect. 5 we offer a number of insights into the conditions necessary to effectively implement WSUD, in Australia and elsewhere.

2 Water Sensitive Urban Design

'Water sensitive urban design' (WSUD) is based on the premise that the process of urban development and (importantly) redevelopment needs to address adequately the sustainability of the water environment. It aims to see all streams of water being managed as a resource, as they have quantitative and qualitative impacts on land, water and biodiversity, and the community's aesthetic and recreational enjoyment of waterways (NWC 2011). Australia's joint standing committee on water sensitive cities defined the overall objectives of WSUD to include (JSCWSC 2009)[1]:

- Reducing potable water demand through demand and supply side water management, incorporating the use of water efficient appliances and fittings as well as a fit-for-purpose approach to the use of potential alternative sources of water;
- Minimising wastewater generation and treatment of wastewater to a standard suitable for effluent reuse and/or release to receiving waters;
- Treating stormwater to meet water quality objectives for reuse and/or discharge;
- Restoring or preserving the natural hydrological regime of catchments;
- Improving waterway health by the management of the previous two objectives;
- Improving aesthetics and the connection with water for the residents of developments where it is applied; and
- Promoting a significant degree of water related self-sufficiency within a development by optimizing the use of water sources from within the development to minimise potable water inflows and water outflows from a development, both stormwater and wastewater.

To achieve these objectives, a planning and design approach is adopted in WSUD that integrates the following opportunities into the built form of cities and towns (see JSCWSC 2009; ARQ 2004; PMSEIC 2007):

- A 'diverse portfolio' of water sources, including rainwater, stormwater, wastewater, groundwater and grey water reuse as alternative sources of water to conserve potable supplies;
- Localised wastewater treatment and reuse systems at the household or suburb-level to reduce potable water consumption and minimise environmentally harmful wastewater discharges;
- Detention, rather than rapid conveyance, of stormwater, through rainwater tanks, urban wetlands, swales etc.;
- Reuse, storage and infiltration of stormwater, instead of drainage system augmentation;
- Use of vegetation for stormwater filtering purposes, for example through swales or bio-retention pits;
- Water-efficient landscaping to reduce potable water consumption;
- Protection of water-related environmental, recreational and cultural values by minimising the ecological footprint of a project associated with providing supply, wastewater and stormwater services;
- Provision of stormwater or other recycled urban waters (in all cases subject to appropriate controls) to provide environmental water requirements for modified watercourses;

[1] For a comprehensive definition of WSUD and its relationship with integrated water cycle management, see Wong and Brown 2011.

Clearly, a very prominent feature of WSUD is the design of localised water storage, treatment and reuse technologies. Such approaches, which are often referred to as distributed systems, can involve the application of these alternative technologies at lot, neighbourhood or district residential scales or for commercial/industrial/high rise developments. The focus on distributed systems is also thought to increase the built environment's resilience to climate change impacts. Brown et al. (2007) also stress, importantly, that a key feature of WSUD should be flexible institutional arrangements designed to (i) cope with increased uncertainty and variability in climate, and (ii) conduct ongoing monitoring, evaluation and review of policies and regulations for WSUD. The research that informs this chapter was undertaken with that objective in mind.

The departure from traditional, dominant and entrenched philosophies on urban development towards the principles enshrined in WSUD, represents a fundamental and transformative shift in how we *conceive, plan for, and manage* our cities, their functions, and the environments in which they exist. To achieve such a transformational shift, a high degree of societal adaptation – i.e. behavioural change – is required. We know from experience that societal adaptation occurs at multiple social scales, including (i) individual behaviour, (ii) formal laws, incentives and governance arrangements, and (iii) evolving norms, attitudes and understanding (Hussey et al. 2013). Each of these scales is reinforcing i.e. formal laws can bring about shifts in norms and attitudes, which can in turn bring about changes in individuals' behaviour. The reverse is also true: shifts in norms, attitudes and understandings (for example within water agencies and government departments), can usher in new laws and governance arrangements which in turn invoke changes in individual behaviour.

3 The Case Study: Stormwater Management in Victoria, Australia

Using the State of Victoria as a case study, this research focused on level (ii), and the extent to which the introduction of one new governance arrangement to implement a feature of WSUD – Clause 56.07-4 relating to stormwater management – is achieving the intended changes in behaviour, and thus achieving its stated objectives. In the next section we briefly outline the background to the study and our research methodology.

3.1 Policy and Legislative Context

In Australia, the *2004 National Water Initiative* formalised the ambition to achieve 'water sensitive Australian cities', requiring all Australia's states and territories to identify and implement innovative options for WSUD in new developments and remove regulatory and institutional impediments to achieving it (Section 92, NWI 2004). In the last decade, all states and territories have introduced provisions for WSUD in their planning regimes, though to varying degrees (NWC 2011). The State of Victoria has been praised as particularly progressive and far-sighted in its attempts to integrate WSUD in its regulatory and planning regimes, with the most significant reforms in 2006 when amendments to the Victoria Planning Provisions were made to encourage better planning outcomes for residential land development. Known as the *Sustainable Neighbourhood* provisions, the new Clause 56 modified the approach to the design and development of land for subdivision. Clause 56.07 of the *Sustainable Neighbourhood* provisions introduced the concept of integrated water management into the planning system, with Clause 56.07-4 introducing provisions for stormwater management through WSUD. Clause 56.07-4 has three objectives:

1. to minimise damage to properties and inconvenience to residents from urban runoff
2. to ensure that the street operates adequately during major storm events and provides for public safety
3. to minimise increases in stormwater run-off and protect the environmental values and physical characteristics of receiving waters from degradation by urban run-off.

WSUD is particularly pertinent with respect to objective three, such that urban stormwater management systems must be designed:

- to meet the current best practice performance objectives for stormwater quality, and
- to ensure that flows downstream of the subdivision site are restricted to pre-development levels unless increased flows are approved by the relevant drainage authority and there are no detrimental downstream impacts.

To achieve these objectives, a variety of different water 'assets' or infrastructure needs to be incorporated into the development site, ranging from rainwater tanks, to swales, bio-retention pits, and community-level urban wetlands – though the precise type of infrastructure required, and under what circumstances, remains unclear, a point developed in the analysis section below.

Clause 56.07-4 currently applies to each new residential subdivision that creates three or more vacant lots. The subdivision of land zoned Residential 1, Residential 2, Residential 3, Mixed-Use, Township, Comprehensive Development and Priority Development must meet the objectives of clause 56.07-4 unless the lots being created contain an existing dwelling or a car parking space. Importantly, under Clause 56 councils are responsible for application and enforcement of the measures in the planning provisions based on the design and water management requirements of the relevant drainage and water authority. However, there have been varied levels of success in implementing Clause 56 more broadly, including Clause 56.07-4. This research project was borne from the understanding that while implementation of Clause 56.07-4 is successful in some parts of Victoria, implementation is at best 'patchy' and improvement is needed across Victoria. However, as is the case with regulatory compliance generally, there is little understanding as to where the barriers exist or what improvements might be made.

3.2 A Framework for Analysis

Ascertaining the extent to which any law – in this case Clause 56.07-4 applying WSUD to residential sub-development – is effective in achieving its objectives involves consideration of a range of issues. In particular, there is a clear distinction to be drawn between the existence of legal arrangements designed to enable or require an action, such as implementing WSUD; the capacity of these legal arrangements to do so in practice; the differentiated capacity of these legal arrangements to be enforced; and ultimately the extent to which the implementation of these legal arrangements actually achieve the outcomes for which they have been designed.

In short, the effectiveness of Clause 56.07-4 depends on more than simply *compliance*; its effectiveness in achieving stormwater quality objectives and, ultimately, improved river health, depends to a very great extent on the context in which it operates. This research therefore considered the policy and planning issues which may inhibit effective implementation, such as:

- information and knowledge gaps,
- overlap, ambiguity, or contradictions in legislative requirements or processes,
- inappropriate scale or scope of implementation and regulatory arrangements,
- incentive gaps and conflicts for private and public sector actors that risk impeding implementation,
- conflicting strategic policy goals, which create unintended outcomes,
- the availability or lack thereof of human, financial and other (i.e. technical) resources,
- inappropriate working environments which impede application of the Clause (i.e. professionals not physically located in the same building).

To facilitate the identification of barriers to, and enablers of, compliance with Clause 56.07-4, we structured the research around the three distinct phases in the approval and implementation of Clause 56.07-4 as shown in Fig. 27.1.

The **pre-planning application phase** includes the first steps in organisations and individuals being aware of the negative environmental implications of traditional, piped drainage systems and a commitment to changing processes and behaviours to shift focus to a sustainable stormwater pathway. In the context of this research, this has generally been expressed by councils as senior management support for use of WSUD, both as a municipal objective and through on-ground physical implementation, and appropriate staffing to deal with assessments of subdivision design submitted for approval against the requirements of Clause 56.07-4. Subsequent work in this phase includes clarity around responsibilities for water resource decisions in the planning system, an area which still seems to cause confusion, and moving toward integrated water management planning for regional, sub-regional and local areas. Industry buy-in and technical competencies complement the success of this phase.

The **planning application and approval phase** is where Clause 56.07-4 is used to benchmark planning assessments in terms of stormwater management and

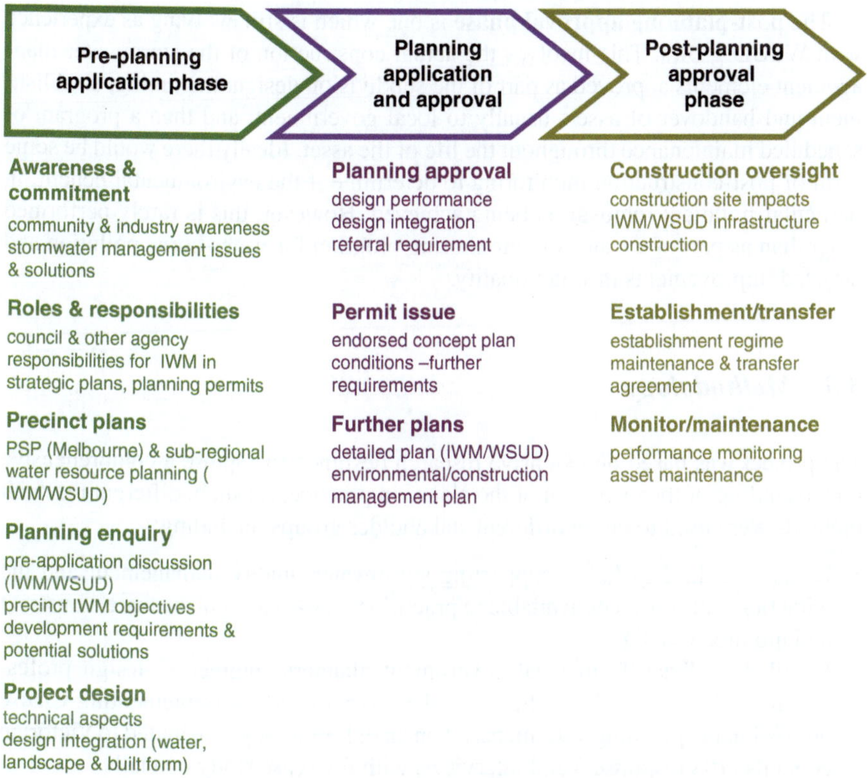

Fig. 27.1 Integrated water management: approval and implementation phases of Clause 56.07-4

compliance with the best practice environmental management objectives.[2] There is management software to determine numerical compliance. Either projects comply, in some cases through use of offsets, or they do not. Planning permit conditions require more detailed plans to be submitted and approved prior to the actual start of land development works. It is at this stage that there can be a need for more advanced technical expertise, to ensure that the constructed technologies are designed correctly and will perform as has been assumed in the numerical compliance assessment.

[2] The best practice environmental management objectives for stormwater are referenced as a compliance measure in Clause 56.07-4 and these objectives are supported by industry compliance software. The objectives are published in the Victoria Government's Urban Stormwater Best Practice Management Guidelines (Victorian Stormwater Committee, CSIRO Publishing, 1999) providing a consistent numerical compliance standard for all Victorian WSUD projects. The post-construction standard requires pollution reduction when compared to the typical urban load of 80% of Suspended solids, 45% of Total nitrogen, 45% of Total phosphorus, 70% of litter and discharges for the 1.5 ARI maintained at the pre-development level.

The **post-planning approval phase** is one which is still evolving as experience with WSUD grows. This involves the actual construction of the stormwater management elements approved as part of the subdivision design; the timing, establishment and handover of assets usually to local government; and then a program of scheduled maintenance throughout the life of the asset. Ideally there would be some form of post-construction monitoring to determine if the environmental benefit, in particular pollutant removal, is being achieved. However, this is rarely performed other than as part of a waterway monitoring program for water-borne pollution and targeted improvements in water quality.

3.3 Methodology

The project was based on extensive research designed to capture a comprehensive understanding of the operation of the clause in practice. As such, different research methods were used to target different stakeholder groups, including:

- Review of the legislation supporting stormwater quality management and the wide range of resources available to practitioners who design, evaluate, construct and monitor WSUD.
- Detailed feedback from local government planners, engineers, design professionals and sustainability officers on their experiences in implementing Clause 56.07-4 in the planning system, based on an online survey distributed to Victorian councils (158 responses) and interviews with four case study councils.
- Detailed feedback from land developers and their consultants on their experiences with the clause when preparing subdivision designs, based on an online survey (48 responses) with Victorian members of the Urban Development Institute of Australia and the Association of Land Development Engineers.
- Interviews with Victorian representatives from peak bodies such as the Institute of Public Works Engineering Australia, Urban Development Institute of Australia and Association of Land Development Engineers, and the Municipal Association of Victoria (11 in total).
- Interviews with government stakeholders such as the Office of Living Victoria, Melbourne Water and the Department of Planning and Community Development (7 in total).
- Discussions with stakeholders in the water industry such as staff in Melbourne's water retailers and a regional water authority.

4 Research Findings

In this section, we present the factors which have enabled successful compliance with Clause 56.07-4 in some jurisdictions, as well as the factors which – as articulated above – pose barriers to successful implementation in the various phases of the planning process. We begin with an overview of the overall perceptions of the rationale for, and effectiveness of, Clause 56.07-4, including the extent to which stakeholders believe the clause is being implemented successfully.

4.1 *Overall Perceptions of Clause 56.07-4*

The most important finding from the research was that there is overwhelming support from all stakeholders in the case study for integrated water management more broadly, and stormwater quality treatment specifically. None of the responses indicated concerns with the best practice environmental management objectives for stormwater quality including their current configuration or their use as a standard in the planning system. Moreover, almost all respondents stated there is significant awareness of Clause 56.07-4 within their organisations and almost 80 % of council respondents felt that Clause 56.07-4 can make a difference to the environment. Furthermore, councils see it as reasonable and/or necessary that private land should contribute to meeting stormwater quality objectives.

In contrast, over 80 % of industry respondents felt WSUD should be undertaken on public land such as parks and local reserves, rather than on private development on private land, which they see as an unnecessary and unfair burden on private developers (see Sect. 4.3). Nevertheless, more than two-thirds of industry respondents were confident that the Victorian development industry can deliver and sustain innovation and individual solutions for WSUD.

There was strong consensus and support from all stakeholders, including industry, for the Victorian government's plans to extend the clause to industrial, commercial and infill developments, though with revisions necessary (see below). Some councils are already applying stormwater quality objectives to non-residential developments with cooperation from developers of these sites. There was also support for broadening the objectives of Clause 56.07-4 to include other public benefits, such as stormwater harvesting and water reuse, which is a significant finding in support of WSUD.

When asked whether Clause 56.07-4 is effective in meeting best practice objectives for storm water runoff at the subdivision level, the responses were only very slightly different between council staff and industry stakeholders (see Fig. 27.2).

However, there was considerable doubt expressed about *how* the effectiveness of Clause 56.07-4 can actually be measured and understood, i.e. in the absence of regular monitoring of, and/or communication about, the quality of receiving waters, how are council staff, relevant government agencies, developers, and the broader public to know whether the clause is meeting its objectives? It is worth noting that scientific analysis of the effectiveness of WSUD (such as constructed wetlands, vegetated swales and biofiltration through vegetated media such as sand or soil), indicates that while it has been largely successful in reducing loads, few studies have reported the effectiveness of load-reduction in actually protecting or restoring the ecological health of streams (Burns et al. 2012). Attributing water quality to any

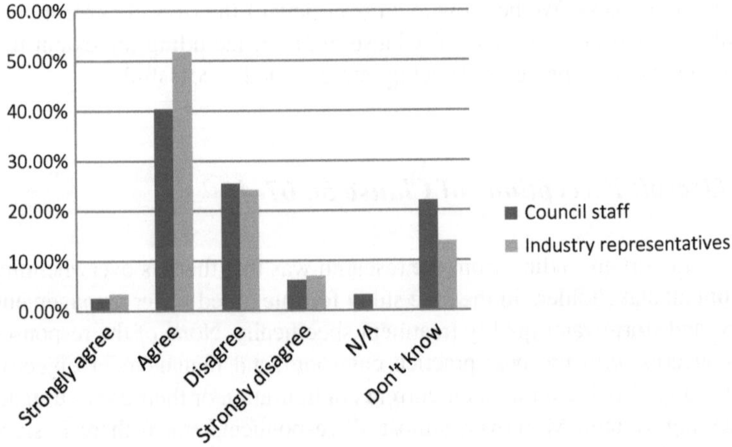

Fig. 27.2 Council staff and industry representative responses to the question 'From your experience, Clause 56.07-4 is effective in meeting the best practice objectives for stormwater runoff in residential subdivisions'

one factor is also, clearly, a considerable constraint in accurately assessing the effectiveness of Clause 56.07-4. Therefore, on-going, longitudinal analysis of the health of Victoria's waterways is an essential element in establishing whether WSUD is meeting its overall objectives, which would directly respond to some doubts expressed by respondents as well as provide the Victorian Government with 'hard' evidence to inform future policy.

While perceptions of the success of Clause 56.07-4 varied between stakeholders, the research also highlighted that significant diversity exists across Victoria in how the clause is implemented. Some councils are pursuing very progressive and impressive strategies for WSUD (e.g. with dedicated teams formed and funding allocated to the development and implementation of WSUD approaches across their jurisdiction, together with initiatives aimed at raising awareness of WSUD in their community), while other councils are struggling to understand when the clause applies, and how to implement it so as to achieve its intended objectives over the long run. This is clearly expressed in the contrasting quotes below:

> Our Council is very committed to stormwater quality improvements/management. We have a strategy that is focussed on SWQ and WSUD implementation as a priority consideration for all drainage initiatives. Even maintenance teams have come around and are actively involved in cross discipline project meetings and contribute to ideas/innovations to ensure ease of maintenance (but we are still learning). We have a strong "across dept/technical discipline alliance" in the office areas, some work still needed with the construction (operational teams).

> At an executive level there is little understanding & support. It is being delivered in an ad hoc manner without any overall coordination.

Interestingly, variability in the implementation of the clause also exists *within* councils, with some council staff pointing to inconsistencies in application:

Whilst Council generally has the capacity to achieve all of the above [best practice environmental management objectives], the real practice is not always consistent, being influenced by the scale of development and willingness or unwillingness of applicants to work with Council.

Some respondents suggested that expectations for developments to meet water quality objectives differ according to the site proposed for development:

Our council also invests more effort in developments that occur within a "high value catchment" in accordance with our organisation's strategic policy documents - expectations to meet water quality objectives are much higher in a high value catchment as opposed to a degraded catchment (which is seen as more good will or opportunistic to get better outcomes).

[My council is...] Largely disinterested, with occasional interest in particular cases, as a result of where the site is located within the landscape.

These findings support earlier research which found that 'Councils across Victoria have had varying levels of success and failure with WSUD and IWM' (Clearwater 2012: 5).

4.2 *Factors for Successful Compliance*

A number of councils in Victoria are leading the way in implementing WSUD, with some councils already considerably advanced in the development of integrated water management plans and 'whole of council' strategies to capitalise on *all* the benefits of WSUD. From this research, several factors which enable successful compliance with Clause 56.07-4 were identified in each phase of the approval and implementation process. We explore each in turn.

4.2.1 Pre-planning Application Phase

Interviewees indicated that strong support from the upper echelons of council is critical to ensure WSUD is readily accepted and strongly enforced. For example, where such support exists, the incorporation of WSUD and IWM in council strategic plans is strongly encouraged, and those plans are in turn utilised in pre-application planning discussions and relied on to provide the 'mandate' for high expectations (from council staff *and* developers) in relation to WSUD and compliance with Clause 56.07-4.

Similarly, interviewees indicated that high-level strategic planning must be sufficiently detailed as to include stormwater management at the subdivision stage, ensuring that stormwater considerations are integrated across the water, land and built forms, which in turn enables the planners, engineers and others involved in pre-application discussions to maximise win-win solutions across the catchment. Depending on the council, this high-level, integrated planning can be in the form of catchment management plans, 'master' or 'strategic' plans for the whole jurisdiction, or in some cases 'integrated water cycle management' plans. In some councils this process is just beginning:

> Council's internal processes ensure timely review of applications and resolution of issues, but this does not yet include best practice storm water management. Council is about to undertake & develop an Integrated Water Cycle Management plan. This plan should enable Council to adopt a water sensitive city planning & development model. Council respondent.

As is the case with other policy domains, interviewees also pointed to the important role played by 'champions' in the pre-application phase. 'Champions' can exist at all levels of seniority, but they typically play an important role in convincing colleagues, developers and the broader community about the benefits of WSUD: where there is a high level of industry and community acceptance of WSUD in the planning system, the approval and implementation process is greatly facilitated. As one interviewee put it:

> What you need is willing developers and willing councils.

However, councils need more than champions to successfully implement the Clause 56.07-4 requirements, and often champions do not have sufficient engineering background and skills to play a hands-on role in the detailed design of WSUD and assessment of planning applications. In addition to high-level support for WSUD within councils, access to 'fit for purpose' technical resources is also vital to ensure pre-application advice is informed and appropriate, especially with respect to which design solutions are most feasible in any given context. Interviewees consistently pointed to the important role played by Melbourne Water in providing expertise – through guidelines as well as human resources – in supporting councils within its jurisdiction to comply with the clause. While not explicitly mentioned by research participants, Melbourne Water's periodic 'needs assessment' of capacity within councils together with the *Living Rivers* initiative are no doubt supporting elements in capacity building efforts in metropolitan Melbourne.

In regional Victoria, interviewees indicated that they share information, manuals and guides to draw from each other's experience. Other regional councils are also participating in this self-initiated program with a common infrastructure design manual for developers published online. The resources, workshops and advice provided through the Victorian Government's Clearwater program was also identified as a very important enabling factor, particularly for those councils outside the Melbourne Water jurisdiction.

4.2.2 Planning Application and Approval Phase

A number of enabling factors are evident in the 'planning application and approval phase', with interviewees from leading councils stressing the need for integrated and inclusive approval processes to ensure development proposals are considered by not only the planners, but also council engineers, sustainability officers, urban designers, landscape architects and asset managers. A number of the case studies pointed to well-developed handover procedures within councils as proving particularly valuable in identifying potential solutions, and mitigating potential problems.

Amongst the leading councils, there was also evidence of closer relationships between water authorities and councils to deliver integrated water management – this was especially true for metropolitan Melbourne councils where Melbourne Water plays a critical role in providing advice and encouragement.

The importance of using council engineers adequately was stressed by several interviewees, which in turn relies on a council commitment to provide a reasonable level of engineering staff resources to provide technical support to planners. Not surprisingly, the extent to which councils are prepared to commit resources to WSUD and approval processes relies to a very great extent on whether there is high-level support for WSUD in the first place.

Interviewees also identified technical experience of WSUD amongst council engineering (or architecture) functions as vitally important; such experience can quickly identify appropriate solutions for compliance with Clause 56.07-4, together with knowledge on the long term costs for maintenance and management of the WSUD assets.

Survey responses from council staff reinforce the argument that planning processes and in-house expertise is critically important for successful compliance. This reflects the need for good internal council processes where technical support is available to the statutory planner responsible for the application who is dependent on this support for the technical compliance assessment and often the negotiations that take place with the developer. Forty-nine respondents from that survey agreed or strongly agreed that, in their experience, the stormwater quality objectives of Clause 56.07-4 are being met, indicating that those staff have favourable experiences with the clause.

4.2.3 Post-planning Approval Phase

To a very great extent, success in the 'post-planning approval phase' is determined by the adequacy of the processes and expertise in-place in the previous two phases. In other words, with clearly articulated strategic plans in place, and well informed council staff who have a high level of expertise and experience with Clause 56.07-4, the post-planning application phase will have been considered prior to approval.

A number of the industry respondents pointed to body corporate arrangements as being particularly effective in maintaining and managing the WSUD assets over the long term. Clearly, such arrangements are most feasible in greenfield developments, brownfield developments and larger building complexes.

4.3 Barriers to Successful Compliance

Notwithstanding the impressive progress made by some councils in implementing Clause 56.07-4, the overwhelming response from this research was that significant barriers exist which undermine the stormwater quality objectives of the clause, but which also present significant impediments to the Victorian government's broader objectives in the *Living Melbourne, Living Victoria Plan for Water.*

4.3.1 Barriers in the Pre-planning Application Phase

Limited Scope of Clause 56.07-4 to Residential Subdivisions

The most significant barrier identified was the inconsistency and limited scope of Clause 56.07-4 as it only applies to residential subdivision. This was considered by respondents as a barrier for several reasons. In the first instance, interviewees indicated that the limited scope of Clause 56.07-4 creates ambiguity amongst council staff and developers around when the clause applies. A council staff member stated,

> Currently there is obviously complexity in the implementation of Clause 56 and it is not utilised by inner urban councils. There are a number of major gaps in this space with inner urban councils currently only enforcing WSUD implementation through planning permits without actually even having a mandate to do so. There is real opportunity to extend Clause 56 and taylor [sic] it to account for infill and industrial type development. In our council area we have been advocating strongly for extra powers to mandate WSUD in all development broadening Clause 56 would be but one important component of this.

While an industry respondent put it thus:

> I believe that Developers are confused about where and when they are required to provide for stormwater quality devices. For example, on some projects they are not required to do anything (i.e. the drainage scheme provides all the treatment, so nothing is required of the developer), then on their next project, they need to allocate a large chunk of land for water quality treatment.

There was also considerable uncertainty around how the objectives of Clause 56.07-4 relate to other water-related legislation in the planning scheme, with 88 % of industry respondents identifying that ambiguity as a significant barrier to successful implementation.

The limited scope of Clause 56.07-4 also means that some councils use other 'triggers' to incorporate stormwater quality objectives in developments:

> We don't generally apply Clause 56 as we are an inner urban Council and as I understand it this means that generally Clauses 54 and 55 are usually applied and this then means it is complex to use 56 as well. Most of the planning influence we wield is through the sustainable design in the planning process (SDAP) this is triggered by planning permit applications not subdivision

There was also a distinct feeling amongst all stakeholders that the failure to include other land developments in Clause 56.07-4 undermines the environmental credibility of the clause. Many respondents expressed exasperation that such a small component of development in Victoria (i.e. residential subdivision) triggers Clause 56.07-4:

> As an established municipality we effectively have no subdivision applications that require a Clause 56 assessment. It is Council policy that new dwellings (i.e. dual occupancies) must have a planning permit and construction be at frame stage before the Council will support the subdivision – hence an assessment against Clause 56 is not required.

> Clause 56 is the only part of the planning provisions that mandates integrated water objectives, which means all the developments under Clauses 54 and 55 and exempt from WSUD unless the council finds some other way to insist on it. It is farcical to have so much of the burden for stormwater management objectives fall on residential subdivisions - what about all the other development going on?
>
> It's not covering half of our growth which is in-fill!

And from industry:

> ...can't see why industry should not have to comply. Potentially, runoff is greater and pollution more concentrated than residential. There should be no discrimination if the objectives need to be met.

The limited scope of the clause was also seen by some as placing an unfair burden on residential subdivision developers who 'carry the cost' of broader stormwater quality objectives, while other developers (and, arguably, government agencies) are exempt from the obligation to build WSUD assets:

> It is an unfair burden on contributors to a scheme with a narrow basis of collection to subsidise greater community needs BSP#12.

That sentiment was echoed in the industry survey, which found that 80 % of respondents thought WSUD should be undertaken on public land, rather than private land.

Lack of 'Fit for Purpose' Technical Expertise for Pre-application Advice

There was significant awareness amongst participants of the many resources on Clause 56 available to council staff and industry, and those councils that perceive the clause as meeting its objectives utilise those resources very readily. Council staff from within the Melbourne Water region also relied on expertise within the Melbourne Water organisation to assess development proposals vis à vis Clause 56.07-4. However, the research indicated for those stakeholders working outside metropolitan Melbourne, the existence of resources was either not known, or there was considered to be so much information that it was 'hard to see the wood for the trees'. Almost without exception, the industry respondents indicated that a well-organised, single, centralised, state-based site would be beneficial in helping them to comply with Clause 56.07-4.

There was considerable variability in how stakeholders perceive the availability of pre-application expertise. Almost two-thirds of council staff respondents indicated they thought pre-application advice on Clause 56.07-4 is adequate. In contrast, 80 % of industry respondents said pre-application advice was inadequate. However, when asked about their utilisation of various supporting documentation, the majority of industry respondents said the resources available were sufficient, appropriate and in the main useful. Nevertheless, a number of respondents from the Industry survey and the case studies pointed to the need for information on WSUD options that include:

- Long term maintenance and management issues with each option.
- Worked examples for different solutions in different contexts.

- Regular updates of the STORM tool because it is currently out of date – in terms of the WSUD solutions it supports, particularly with reference to plumbing of rainwater tanks to internal uses.
- A single 'point of call' for all consultants, developers and council staff to refer to across the State.
- Standardised solutions – to ease compliance, with solutions that are relatively 'simple' to construct and maintain, have a reasonable life cycle cost and provide multiple sustainable stormwater management benefits including on-site detention where this is required.

On the latter point, two illustrative examples from industry:

> Have a set of 'off the shelf' design solutions and educate Councils when and how these can be applied. For example we now have many cases for infill development where we seek to utilise rainwater tanks to fulfil a storm water quality and retardation function - some Councils promote this and others accept this concept (and the science behind it) whereas some Councils don't accept this at all. The science is the same in every case so this is just an attitudinal issue. Same goes with other good ideas that get stonewalled. Industry respondent.

> While a standardised WSUD guideline (which includes the current WSUD guideline information plus recommended MUSIC inputs/checklists, soil/media parameters and design recommendations, maintenance schedules and checklists, details on hydraulic calculations etc) would make innovation less attractive in the short-term, I believe it would improve the acceptance among LGAs (especially in regional Victoria) of WSUD measures. The acceptance and encouragement of WSUD will ultimately lead to innovations which at this stage are not feasible. Industry respondent.

Uncertainty About, Long Term Management and Maintenance Costs of the Built Infrastructure and the Absence of a Market for WSUD

As previously indicated, there is considerable variation in the perceptions and implementation of Clause 56.07-4 across councils; some councils have increasingly innovative approaches to WSUD, while others are lagging. This research highlighted one particular concern – the costs of maintaining WSUD assets – that contributes significantly to a reluctance amongst councils and developers to comply with Clause 56.07-4. The survey of councils found that three-quarters (74 %) of council staff consider the cost of maintaining WSUD assets as a barrier to the successful implementation of Clause 56.07-4. Nearly three-quarters (73 %) of industry respondents agreed or strongly agreed that there was little appetite amongst councils to pursue WSUD, with a number indicated that concerns around maintenance costs was a contributing factor. Interestingly, almost the same number (76 %) of industry respondents consider WSUD to be expensive to design, construct and maintain.

On a related point, the overwhelming majority of industry respondents indicated that, in their experience, councils did not have effective asset management and maintenance programs in place for constructed WSUD.

Critically, almost all (93 %) of industry respondents 'agreed' or 'strongly agreed' with the notion that 'there are limited incentives for developers to pursue best-practice

WSUD'. In interviews with industry stakeholders, there was keen interest in establishing an annual prize for innovative WSUD, in the hope that such a prize and publicity surrounding it would stimulate market demand.

Clearly, state-wide implementation of Clause 56.07-4 will not be achieved until the cost concerns – from a supply and demand perspective – of WSUD assets are addressed.

4.3.2 Barriers in the Planning Application and Approval Phase

Lack of Adequate Processes Within Councils to Successfully Implement Clause 56.07-4

The research highlighted a number of barriers around the processes and capacity within councils to successfully implement Clause 56.07-4, including:

- A lack of sufficient resources and technical expertise within councils to review submitted plans and for construction oversight.
- Unclear coordination between the building approvals system for planning permits with on-site WSUD.
- Internal planning approval processes which fail to incorporate all the significant functions within councils i.e. asset managers, landscape architects, parks and maintenance staff, etc.
- Variable attitudes within and between the various functions in council around the rationale, and appropriate solutions for, WSUD.

Overall, industry respondents were sceptical about the adequacy of council processes to support implementation of Clause 56.07-4.

Lack of Enforcement

Related to the deficiencies of planning processes within councils, is the more general but no less significant failing of enforcement powers in Clause 56.07-4. At present, there are no punitive measures in place if a development does not comply with Clause 56.07-4. As one council respondent put it:

> "Should" requirements do not hold up in VCAT[3]

That same respondent went on to say,

> Please do something with the clause. Planners across the state will continue to fight an uphill battle until Government takes an assertive stance to do something.

Interestingly, 25 % of industry respondents said they had had a subdivision application refused because it didn't fully comply, while 43 % of respondents said

[3] VCAT is the Victorian Civil and Administrative Tribunal.

they had not had a subdivision application refused despite the fact that it did not comply. The remaining proportion indicated they had never been in that position.

4.3.3 Post-planning Approval Phase

Lack of Oversight and Auditing of Construction Phase

Of all the issues raised throughout the course of this research, the challenge of overseeing construction, management and maintenance of WSUD assets was the most prevalent from councils and industry. Industry responses – through interviews and the on-line survey – revealed significant distress about the willingness and capacity of councils to manage and maintain the built infrastructure once it has been handed over to councils. Several interviewees referred to the assets as being 'orphaned' or 'lost' assets, owing to the inability of councils to look after them. One respondent stated,

> It is quite distressing to design and construct a feature, which complies with the law, only to drive past that asset a year later to see that what was once a bioretention pit is now being used as a car park.

Another broader policy stakeholder related the following anecdote,

> I had one phone call from a resident asking whether her children could swim in the retention pit in her development. I say 'no!', but clearly there is a complete lack of awareness or understanding in the public about what these assets are for…

Concerns around ongoing maintenance and management of WSUD infrastructure also raised the issue of council staff (often parks staff) receiving training in asset maintenance. This comment illustrated the issue:

> Further education of local government in maintenance challenges, managing handover expectations and creating a more practical understanding of the transition from construction of subdivisions to construction of homes on lots. Control of silt / run off from individual building sites by local government is an ongoing issue which Councils tend to ignore and push the cost of maintenance back onto the developer rather than policing their streets properly. Industry respondent.

5 Discussion

Based on the experience in Victoria, Australia, the research findings presented in the previous section provide a comprehensive overview of the challenges of implementing a key feature of WSUD. Importantly, the research provided insight across all three stages of the planning phase, thus highlighting critical issues that need to be considered before, during and after the planning assessment phase. While the research was focused specifically on the experience in Victoria, the findings resonate with, and have implications for, the design and implementation of WSUD in other jurisdictions. In this section we discuss four broader 'lessons' from the Australian case study that could prove useful for other cities attempting to introduce WSUD.

Lesson 1: The Need for Clarity and Consistency in Roles, Responsibilities, Expectations and Policy Intent
There is a very great need to provide clarity and consistency in key policies, strategies, plans, and regulations, as well as around the roles and responsibilities of agencies and departments with respect to WSUD. A key strength of the Victorian system is the use of a single set of numerical performance measures for designing and assessing the (theoretical) success of WSUD, and the software to measure compliance. The wide acceptance of these measures has eliminated any ambiguity about the minimum performance requirements of projects, although the measures now need to be revised or expanded to take account of the fuller possibilities of integrated water management.

Despite the success of the state-wide performance measures, the regulatory drafting has been ad hoc and this has resulted in a patchwork system of compliance, where the performance measures are mandated for residential subdivision only and a number of local councils have introduced policy guidance for other land use types. Industry participants as well as councils have indicated a preference for consistent requirements for all development applications. The skills required to develop appropriate WSUD solutions are transferable across developments, as the same principles are applied in each case. Indeed, an expanded repertoire of designed, constructed and tested solutions would accelerate the expertise available within the industry.

Most importantly, the policy and institutional settings must support the overall desired strategic policy outcomes. Implementation can be assisted when there is a single lead agency with responsibility for setting policy and arbitrating on policy questions. There can be a positive opportunity for knowledge transfer where different agencies operate in the same policy or geographic space, however, at the same time, if confusion arises about roles and responsibilities this sets the scene for conflicting advice and 'buck-passing' between organisations. Assessing the 'best fit' roles and responsibilities may require overturning traditional organisational (and industry and public) assumptions of legal status and power to successfully underpin WSUD policy with new institutional governance arrangements.

Lesson 2: The Need for 'Fit-for-Purpose' Information and Resources
There are a number of activities that can strongly support the implementation of WSUD. The first port of call for those assessing approvals should be a comprehensive resource (such as a dedicated web site) to ensure the statutory provisions and compliance assessment requirements are well understood. While recognising a range of professions will be involved, the statutory assessor bears the responsibility for ensuring integration across disciplines as well as integration of WSUD with other regulatory requirements, e.g. siting, lot coverage, vegetation protection or landscaping provisions. Clarity for the assessor is essential, and only when the statutory issues are spelled out are the guidance provided by case studies and technical manuals of benefit.

Where it is possible to provide standardised solutions ('deem to satisfy'), this can assist both smaller developers and councils with limited access to experienced designers and reviewers of WSUD. Clearly standardised solutions will need to take account of a number of factors such as technical design to ensure quality in terms of compliance and functionality, construction ease and cost, maintenance regimes and cost, and what role each device can play in delivering WSUD objectives for particular sites, including stormwater storage and use. Well targeted training courses are also essential with a focus on the professional skills needed to select appropriate solutions for each case, design, construct and maintain WSUD devices, and negotiate 'fit for purpose' outcomes that further the strategic policy objectives. Certification by third parties should also be explored.

Conflicting objectives will need resolution, i.e. use of recycled waste water v stormwater, lot based solutions v precinct solutions. The preferred approach will depend on local conditions and the availability of alternative water supplies. Ideally this should be resolved for nominated geographic locations ahead of development to provide certainty for councils, applicants and water authorities.

Lesson 3: The Need to Understand, and Account for, Costs and Market Drivers

The cost of ongoing maintenance for WSUD has become a significant stumbling block for some council jurisdictions where the maintenance costs are starting to be keenly felt as construction of WSUD, and particularly decentralised WSUD, is growing. There is no cost recovery mechanism for councils to fund the added infrastructure responsibility. Maintenance problems will persist until the industry gains more experience with WSUD lifecycle costs and councils are experienced enough to reject poorly designed and constructed devices which require greater investment to maintain (hence one reason for the interest in standardised designs). The effectiveness of the design and construction process, and potentially identification of additional maintenance resources, needs greater attention if policy makers are to overcome lingering resistance to WSUD and reduce the longer term cost burden to councils.

Land developers identified the lack of a well-developed consumer housing market that values environmental sustainability, and is willing to pay for it, as an impediment to doing more than is required for project approval. This is of note given the desire to expand WSUD requirements more broadly from a regulatory perspective and integrate stormwater more closely with other water sources. The benefits of using new technologies to deliver environmental improvements in stormwater runoff and the creation of a more diverse water supply from local resources is rarely mentioned in public discussion, by governments, news agencies and public figures. Raising public awareness about the benefits of WSUD would assist developers to at minimum recover costs for the environmental improvements they incorporate in their projects.

Lesson 4: The Need for Integrated Strategic and Statutory Planning

While this case study explored one particular aspect of WSUD, the full scope and potential of WSUD requires that both strategic and statutory planning be fully integrated. Beginning with the strategic design of catchments, townships, suburbs and neighbourhoods, through to the approval of infill developments in built up areas, a sustained focus on WSUD can ensure that its delivery is integrated with a full range of resource allocation objectives, and that WSUD assists the delivery of the broader planning objectives that are to be achieved for the area. At its most effective, WSUD encourages involvement by the natural resource and built environment professionals. At the council level, a number of council officers will need to be involved including the planner, WSUD engineer, sustainability officer, landscape architect and architect.

6 Conclusion

The core objectives and features of WSUD simultaneously address population and environmental pressures and climate variability – challenges that afflict all cities and built environments around the world. Indeed, if it were to be implemented to its fullest extent, WSUD would fundamentally transform the way cities are designed, perceived and experienced. Yet, as with all new ideas that advocate a radical departure from the status quo, the barriers to implementation can range from being fierce, complex and daunting on the one hand, to frustrating and mundane on the other.

To recall, our purpose in this chapter was to identify and understand the challenges that exist in implementing one element of WSUD – stormwater management. The case study revealed overwhelming awareness of, and support for, this particular feature of WSUD, amongst industry, government and stakeholder groups. Given the relative 'newness' of the idea, this finding is extremely positive and bodes well for the concept as a whole. However, the case study also exposed a range of impediments to the successful implementation of distributed stormwater management, and the four lessons described in the previous section provide a succinct summary of our findings. The impediments we identified are not unique to this one element of WSUD, nor are they likely to be unique to this particular case study. Indeed, many of the barriers discussed can usefully be described as 'capacity' issues, including financial, human and technical issues, to varying degrees. Similarly, we identified underlying deficiencies in the system that result in conflicting policies and ad hoc application of laws, and ambiguity around the roles and responsibilities of key agencies. Arguably, such capacity issues are easily dealt with if the will is there, but as with all things involving systems and people, the road is a long one, most likely generational.

The findings of this case study demonstrate that as with all significant change, there must be a strong, almost idealistic, commitment when introducing a new concept such as WSUD, a willingness to embrace change, to capitalise on success and learn from failures, and to find unifying measures for bringing the stakeholders together to work toward a common goal.

References

ARQ (Australian Rainfall Quality). (2004). *Australian runoff quality: A guide to WSUD.* Engineers Australia, National Committee for Water Engineering.

Brown, R. R. (2008). Social and institutional considerations. In T. D. Fletcher & A. Deletic (Eds.), *Data requirements for integrated urban water management* (pp. 159–169). London: Taylor & Francis.

Brown, R.R., Farrelly, M., & Keath, N. (2007). *Summary report: Perceptions of institutional drivers and barriers to sustainable urban water management in Australia* (Report No. 07/06). Melbourne: National Urban Water Governance Program, Monash University.

Burns, M. J., Fletcher, T. D., Walsh, C. J., Ladson, A. R., & Hatt, B. E. (2012). Hydrologic shortcomings of conventional urban stormwater management and opportunities for reform. *Landscape and Urban Planning, 105*, 230–240.

Clearwater. (2012). *Building capacity with regional and rural councils – Water sensitive urban design.* Presentation by Anna Jennings to Stormwater Victoria annual conference, Melbourne.

Hussey, K., Price, R., Pittock, J., Livingstone, J., Dovers, S., Fisher, D., & Hatfield Dodds, S. (2013). *Statutory frameworks, institutions and policy processes for climate adaptation: Do Australia's existing statutory frameworks, associated institutions and policy processes support or impede national adaptation planning and practice?* (p. 193). Gold Coast: National Climate Change Adaptation Research Facility. Available at http://www.nccarf.edu.au/publications/statutory-frameworks-policy-processes.

JSCWSC. (2009). *Evaluating options for WSUD – A national guide: Prepared by the Joint Steering Committee for Water Sensitive Cities: In delivering Clause 92(ii) of the National Water Initiative, Joint Steering Committee for Water Sensitive Cities, Canberra.* Available at http://www.environment.gov.au/system/files/resources/1873905a-f5b7-4e3c-8f45-0259a32a94b1/files/wsud-guidelines.pdf. Accessed 21 Aug 2013.

National Water Initiative. (2004). *Intergovernmental agreement on a national water initiative between the commonwealth of Australia and the governments of New South Wales, Victoria, Queensland, South Australia, the Australian Capital Territory and the Northern Territory*, Council of Australian Governments.

NWC (National Water Commission). (2009). *Evaluating options for water sensitive urban design – a national guide*. Prepared by the joint steering committee for water sensitive cities in delivering clause 92(ii) of the National Water Initiative 2004. NWC, Canberra. Available here http://www.environment.gov.au/resource/evaluating-options-water-sensitive-urban-design-%E2%80%93-national-guide

NWC (National Water Commission). (2011). *The National Water Initiative – Securing Australia's water future: 2011 assessment*. Canberra: NWC.

PMSEIC (Prime Ministers Science, Engineering and Innovation Council). (2007). *Water for our cities: Building resilience in a climate of uncertainty*. A report of the PMSEIC working group, June. Available at http://www.innovation.gov.au/science/PMSEIC/Documents/WaterforOurCities.pdf

Wong, T. H. F., & Brown, R. R. (2011). WSUD. In Q. Grafton & K. Hussey (Eds.), *Water resources planning and management* (pp. 483–504). New York: Cambridge University Press.

Chapter 28
How Does Energy Efficiency Affect Urban Water Systems?

Steven J. Kenway and Paul A. Lant

1 Introduction

Between 2005 and 2007, five major Australian cities faced crisis-level shortages of water. Headlines warned of "Armageddon"; cities had to "desalinate or die" (Stolz 2007; ABC News 2007). Climate change was widely held responsible (Howe et al. 2005; Bernstein et al. 2007; WSAA 2005).

The 'millennium drought' crisis spawned a significant surge in urban water infrastructure investment. In 2005–06, a then record AU$2.4 billion was invested in major cities alone for water (and wastewater) infrastructure. Most funding was aimed at increasing the proportion of climate-independent sources such as desalination or wastewater reuse (WSAA 2008). This investment grew to almost $4 billion in 2007–08, and over $7 billion in 2008–09. Expenditure for 2009–10 was forecast as greater than $14 billion (WSAA 2009), but stabilised around $7 billion, excluding the Melbourne desalination plant (National Water Commission 2011).[1]

Between 2008 and 2013, rainfall alleviated water shortages in eastern and southern Australia. Full water storages have reduced the pressure for investment or management action, but a new challenge has emerged. The energy demand for water in Australian cities is anticipated to grow to 200–250 % of 2007 levels by 2030 (Fig. 28.1) (Kenway and Lant 2011; Kenway et al. 2008). The primary drivers of this increase are the shift to energy-intensive sources of water, such as desalination, coupled with expected population growth. Typically, desalination requires around 4.5 kWh of electricity to produce, at sea level, one cubic meter of water. In 2006–07,

[1] Estimated at around AU$4 billion.

S.J. Kenway (✉) • P.A. Lant
School of Chemical Engineering, University of Queensland,
Level 3, Chemical Engineering Building (74), St Lucia 4072, Australia
e-mail: s.kenway@awmc.uq.edu.au; paul.lant@uq.edu.au

Q. Grafton et al. (eds.), *Understanding and Managing Urban Water in Transition*, Global Issues in Water Policy 15,
DOI 10.1007/978-94-017-9801-3_28

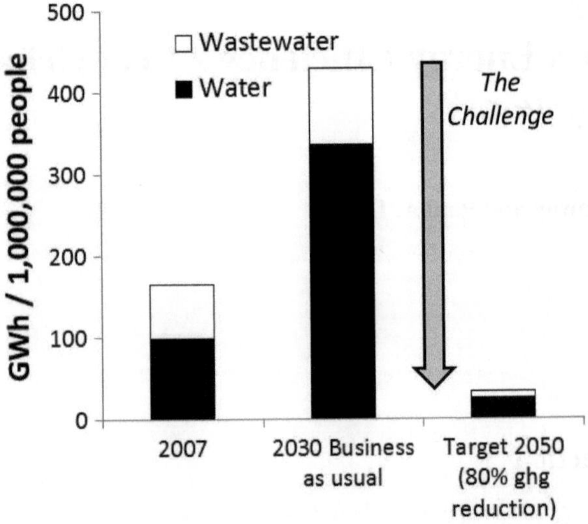

Fig. 28.1 The challenge: reduce energy use for water and wastewater services by more than 90 % of the 2030 forecast (Source: Kenway (2012)). Here, the 2050 energy target assumes no change in the greenhouse gas emission intensity of fuel used; the target arbitrarily shows water and wastewater in proportion to the 2030 forecast

established capital city water supply options in Australia used under 1.0 kWh/m^3 of supplied treated water (Kenway et al. 2011b).

An increase in energy use for urban water supplies is one problem. But when this increase in energy is added to an expected rise in electricity costs ($/MWh), the energy bill for urban water is anticipated to increase 600–800 % for most Australian states (Cook et al. 2012). This now represents a significant business risk to the Australian water sector (Victorian Water Industry Association 2011). Ultimately, these cost increases have to be incorporated into water prices and passed on to consumers.

To compound the challenge, in December 2007 Australia ratified the Kyoto Protocol. This established a national long-term goal to reduce greenhouse gas (GHG) emissions 80 % below 2000 levels by 2050 (Australian Government Department of Climate Change and Energy Efficiency 2012). This means that if the water sector was to contribute proportionately, then energy use for urban water must be reduced by more than 90 % from the projected 2030 levels. Alternatively, an equivalent cut to the GHG emissions of the energy sources traditionally relied upon needs to be achieved.

How are these challenges to be met? And what will be the implications for our urban water systems? It is timely to look at water–energy connections. This is because understanding links between water and energy may well help find practical solutions to reduce water-related energy use and cost. This chapter draws on recent research to outline the broad anticipated directions for management of 'water-related energy' in cities.

2 The Opportunity of Cities

Cities offer considerable scope for solving many problems of resource efficiency (O'Meara 1999). In part, this is because large flows of water, energy, and materials pass through cities; it is also because they are centres of investment and change.

Cities are continually being redesigned and recreated. Australia anticipates a future in which an increasingly large proportion of the population resides in cities. In 2006, 68 % of the Australian population of 21 million people lived within major cities (ABS 2006), and around 90 % live in urban settings (PMSEIC 2010). This population is expected to grow to 25–33 million by 2050 (ABS 2006), the majority of which will live in major cities. This growth will intensify Australia's urbanism and have major implications for infrastructure and related investment decisions. The substantial new infrastructure and redevelopment anticipated in our cities offers the chance of solving emerging water and energy problems. Well planned and governed cities could play a central role in decoupling high quality of life from high energy and water consumption – and high rates of GHG emissions.

There is currently poor understanding of how 'urban systems' perform. Even basic statistics, such as population, can vary by millions of people depending on definitions adopted Sattherthwaite et al. (2008). For more complex attributes such as energy or GHG emissions, knowledge is even patchier (Kennedy et al. 2007). This has flow-on effects for the design and management of our cities. Sattherthwaite (2008), also notes the central role that well planned and governed cities could play in delinking high quality of life from high energy and materials consumption, and GHG emissions.

Despite the importance of water-related energy in cities, there has been a lack of conceptual and analytical models addressing links between the two resources (Conrad et al. 2011). Energy effects have to-date typically been ignored when water decisions are made, particularly during water 'crises'. Improved understanding of water-related energy in cities is necessary for management to improve.

3 Quantification of Water-Related Energy in Cities

To understand water-related energy in cities, it is helpful to consider both the *direct* and *indirect* energy. In this work, the perspective of *direct* energy is adopted from that of water utilities. *Direct* energy impacts occur, for water utilities, when water or wastewater is pumped or treated. *Indirect* energy is considered, for this paper, as energy outside the domain of water service providers that is associated with water. For example, this includes energy use associated with water consumption. Other authors may use the terms differently (deMonsabert and Liner 1998).

Indirect energy impacts are diverse. They can include energy associated with (i) water supply other than from utilities; (ii) water use; and (iii) resource loss. Energy impacts associated with other water sources, for example, include the use of rainwater tank pumps, high rise pumping, or creating and transporting bottled water.

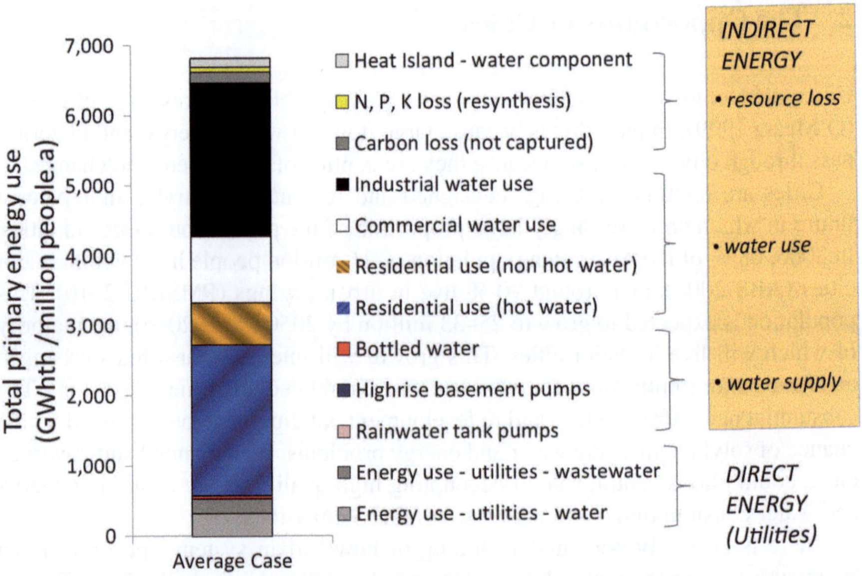

Fig. 28.2 Water-related primary energy use in the average Australian city, based on average data for major capital cities (Sydney, Melbourne, Brisbane, Adelaide, and Perth). GWh-th is thermal (as opposed to electrical) energy use. Three units of thermal energy are approximately equal to one unit of electrical energy (Source: Kenway (2012))

Energy impacts associated with the use of water include heating, cooling, and pressurising water in homes, industry, commerce, and government. More complex indirect connections exist via flows of carbon, nitrogen, and phosphorus resources in wastewater. Energy is needed to return these elements, typically discharged as wastes, to productive purposes. There are also energy consequences of water relating to the urban heat island effect (warming in the core of cities), in the production of food in cities, and elsewhere.

The *indirect* influences of water on energy are very substantial. In 2006–07, water-related energy associated with urban water provision, use, and related resource flows accounted for 6,811 GWh/annum of primary (thermal) energy use per million people in the 'average' Australian city (Fig. 28.2) (Kenway et al. 2011b). This is 6,811 kWh of thermal energy per person: equal to the electricity consumption of running 23 compact fluoro light bulbs of 11 watts each (that is, 253 W), 24 hours a day, 365 days a year. Collectively, this water-related energy adds up to 13 % of Australia's natural gas consumption plus 18 % of its electricity consumption, and accounts for 9 % of the equivalent primary energy use and 8 % of the GHG emissions in Australia in 2006–07 (Table 28.1).

In the average Australian city (Fig. 28.2), energy associated with residential water use comprises 45 % of water-related energy, while water use in the commercial and industrial sectors accounts for 41 %. Consequently, water end-use appears to be the dominant process affecting water-related energy in cities. Utility energy

Table 28.1 Water-related energy in cities and average Australian energy use and GHG emissions per million people (2006–07)

Comparative area	Units	Water-related in cities	Average Australian use	Proportion of Australian average (%)
Electricity use	GWh-e	1,379	10,476	13
Natural gas use	GWh-th	2,675	15,154	18
Primary energy use	GWh-th	6,811	76,384	9
GHG emissions	CO_2-equiv. (1,000 t)	1,914	24,239	8

Note: GWh-e is GWh of electrical energy use; GWh-th is GWh of thermal energy use. About 3 GWh-th is required to produce 1 GWh-e (Source: Kenway et al. (2011b))

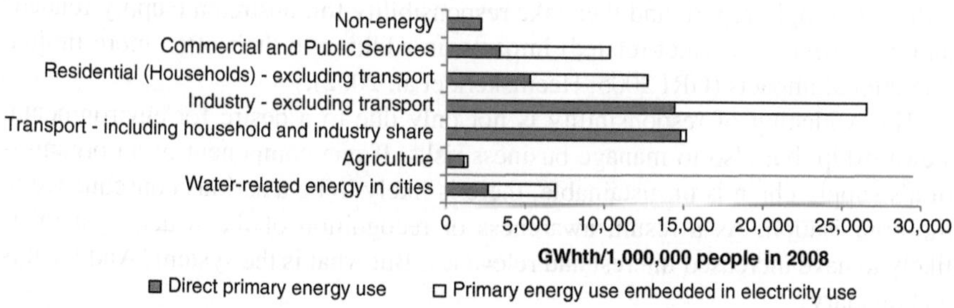

Fig. 28.3 Water-related energy use in Australian cities in 2006–07 per million people (horizontal scale) apportioned by sectoral energy use for the Australian economy in 2008 (bars). Sectoral source is from IEA (2011), with sectors adding up to Australia's primary energy use; data source for water-related energy is Kenway et al. (2011b)

use for water and wastewater service provision accounts for 10 % of the total 'water-related energy'. The water component of the urban heat island effect, and energy related to resource loss (such as carbon, nitrogen, phosphorus, and potassium), account for the balance of energy use and emissions (around 4 %).

To compare energy use when different fuels (such as electricity and natural gas) are used requires conversion of energy to primary energy equivalents (Gleick and Cooley 2009). In a typical thermal power plant, three units of thermal energy (e.g. coal) are required to produce one unit of electrical energy. Two units of energy are lost at the power plant as heat up the stack. Comparing water-related energy in cities with the energy use of other Australian sectors (Fig. 28.3) shows that water-related energy in cities is around five times the direct energy use of Australian agriculture (excluding transport).

Given the significance of water-related energy in cities, it is unusual that it is not more widely considered, monitored, or managed. Part of the reason is undoubtedly due to the fragmented management responsibilities (i.e. it is a cross-sectoral issue). As a consequence, many relevant datasets and information are fragmented. However, part of the reason is also likely to be the lack of conceptual and associated analytical frameworks for understanding or quantifying water-related energy. Frameworks are also needed for dealing with 'organisational' or 'urban' performance.

4 Frameworks for Understanding Organisational and Urban Performance

4.1 Organisational Performance and Management of Water-Related Energy

Many sustainability assessment frameworks, such as the Triple Bottom Line or The Natural Step, help organisations identify and reduce their individual impacts, such as their energy use. However, such frameworks typically say relatively little about the broader impacts of the system within which those organisations play a role. A number of authors think this is about to change, and that governments and industry will increasingly report, and then take responsibility for, upstream (supply-related) and downstream (product-related) impacts, in addition to their often more limited operational impacts (GRI 2005; Heemskerk et al. 2002).

This widening of responsibility is not only due to a desire for environmental stewardship, but also to manage business risks. If one component of an organisation's supply chain is unsustainable, there is likely to be a serious consequence to the organisation. As a result, awareness or recognition of the wider 'system' is likely to have increased interest and relevance. But what is the system? And what is its boundary?

Many water organisations, who currently actively manage their own direct energy impacts, are (A) unaware of the wider energy use, yet (B) influenced indirectly by their own operations or decision making (Fig. 28.4). Considering possible future states, one pathway may involve an organisation reducing its own energy use, yet the pathway may lead to increased energy use of the system that it supports (e.g. Pathway 1 in Fig. 28.4). Conversely, futures are possible where the organisational energy use grows, yet the energy use of the total 'system' reduces (Pathway 2). Futures also exist where both utility energy use and wider water-related energy use both decrease (Pathway 3).

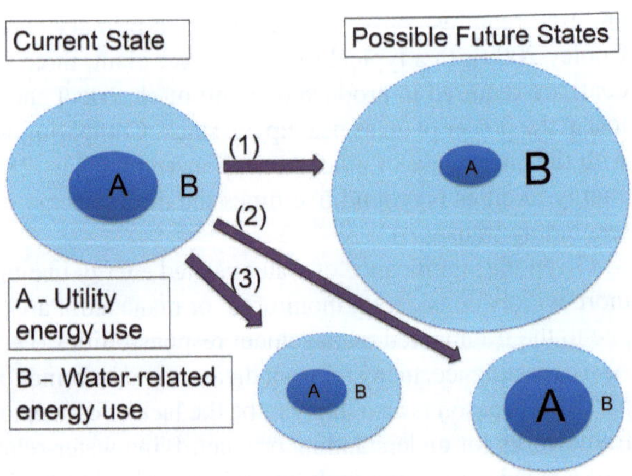

Fig. 28.4 The effect of 'boundary' placement on sustainability performance, from Kenway (2012)

According to French and Geldermann (2005, 2007), placement of the boundary unequivocally affects what is the 'best' option. Several authors note the importance to research conclusions of where the 'system boundary' is placed (Satterthwaite 2008; Decker et al. 2000; Flower et al. 2007; French and Geldermann 2005). System boundary definition is a critical first step in modelling analysis (Sterman 1991). Without knowing the boundary, it is impossible to know which factors should be included in, or excluded from, analysis. Despite its importance, the boundary definition for cities appears absent from many discussions.

4.2 *Connecting Water-Energy and Energy to Urban Performance*

Water and energy are intertwined in the supply of water to cities, as well as in the use and disposal of wastewater (Fig. 28.5). Currently, the management of the water, energy, and material flows through cities is undertaken in isolation. There is little or no recognition that in solving one problem, another may be created. For example, we may improve the water system to perform from level B to level A by using a lot of energy. However, another problem may be created, for example, shifting the energy performance from level C to level D. Here, A, B, and C are indicative of levels of performance, with A being good and B, C, and D progressively poorer levels of performance.

Compartmentalising the performance of cities into many smaller indicators (such as water supplied, energy used, wastes recycled) currently means that decision-making in each sector is also narrowly focused. No consideration is currently given to the overall efficiency of the city. Rather, performance assessments are more commonly found at the levels of organisation, state, or nation.

Fig. 28.5 The current paradigm: urban performance is managed in a fragmented way. Performance indicators are not aligned or co-ordinated, even though water and energy are interconnected outside and also within cities (Source: Kenway (2012))

Newman et al. (1996) articulate two critical, and equally important, principles of urban design and management. The first is to reduce flow-through of resources. The second is to maintain or improve human liveability and ecosystem health. Several authors (e.g., Sahely et al. 2003; Tambo 2002) present similar arguments. Clearly such goals create great challenges for the design and management of our cities. In order for us to meet these challenges it is likely that we will need to develop new models to conceptualise and analyse existing problems and evaluate potential solutions. Urban metabolism is one such model.

5 How Can the Concept of Urban Metabolism Support Better Understanding and Management of Water-Related Energy?

Urban metabolism is a conceptual model used to describe and analyse flows of materials, including water and energy, within cities (cf. Newman 1999; Wolman 1965; Decker et al. 2000). At its simplest, urban metabolism considers the mass balances of all materials flowing into and out of urban systems (Sahely et al. 2003). At its most complex, the concept draws on the rich metaphor that cities can be considered organisms (Wolman 1965; Decker et al. 2000).

Using the wider metabolism metaphor, we can start to consider the water and wastewater systems as the arteries and veins of the city. Water and wastewater treatment plants are akin to the kidneys and liver. Some benchmarking is already happening, comparing the 'organs' of the city with natural systems. For example, the performance efficiency of wastewater treatment plants has been compared with the kidneys (G. Leslie, 2009, personal communication).

The idea of urban metabolism was first put forward by Abel Wolman (1965) with the core intent of addressing contemporary urban resource issues. He suggested metabolism was a means of simultaneously addressing the problem of "shortages of water and the pollution of water and air" as well as the call for "public economic decisions". Wolman defined urban metabolism as all the materials and commodities needed to sustain a city's inhabitants at home, work, and play. This included the construction materials needed to build and rebuild the city.

Decker et al. (2000) used metabolism as a framework to evaluate the flows of energy, water, and materials through megacities. They suggested that consideration of water in the metabolism of cities should be a priority, because of its dominance in the material flows of a city. Fischer-Kowalski et al. (1998) argue that the concept of urban metabolism has become one of the most powerful paradigms and interdisciplinary concepts for the empirical analysis of the society–nature interaction.

5.1 *Application of Metabolism Theory: A Water Mass Balance*

In order to apply a small part of urban metabolism theory, a metabolic water mass balance was developed and applied to four Australian cities (Kenway et al. 2011a). The cities selected were experiencing significant drought at the time. The system boundary was defined to include typical intensive land use classes: residential, manufacturing, industrial, transport, waste treatment and disposal, and intensive horticultural activities. Collectively, this area represented the 'demand area' where water is consumed. It also generally represents the area of highest proportion of impervious surfaces where most stormwater is generated in the city. The boundary focused from 1 m above the roofline and tree canopy to 1 m below ground. For this boundary, all natural (e.g. rainfall, evapotranspiration) and anthropogenic (e.g. piped water) flows were quantified.

The analysis identified large flows of water, which passed, typically unaccounted, through the cities (Fig. 28.6). This included appreciable flows of rainfall, stormwater, and wastewater. The water mass balance demonstrated that our cities are not taking advantage of significant water flows. For example, all cities demonstrated that rain falling on them was at least 250 % of the demand for centralised water. The potential for wastewater flows to meet total water demand was 26–86 % of the demand for centralised water supply. The potential for stormwater recycling to meet

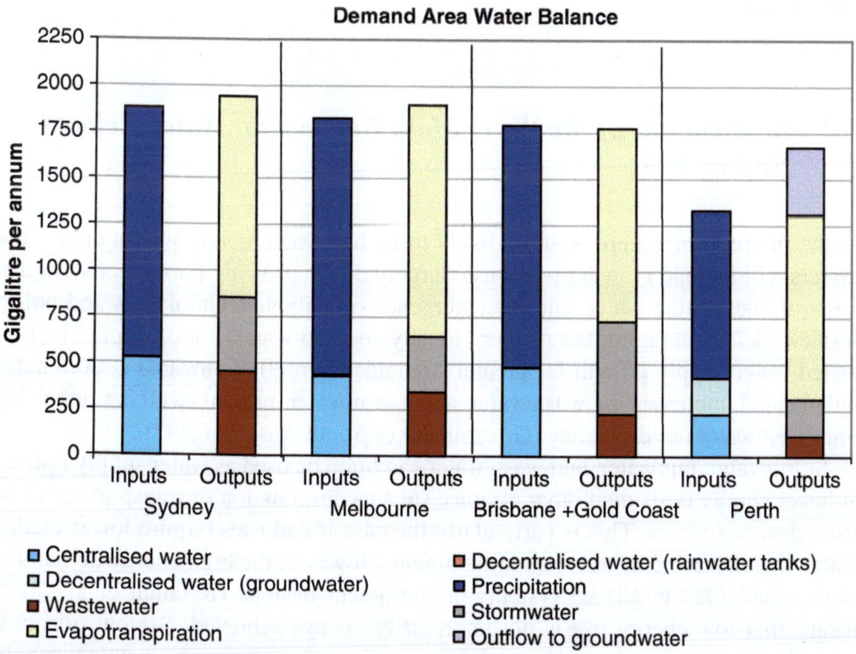

Fig. 28.6 Water mass balances of four Australian cities/metropolitan areas 2004–05 (Data from Kenway et al. (2011a))

Table 28.2 Summary water balance performance in four Australian cities (2004–05)

	Rainfall harvesting (%)	Rainfall potential to meet supply (%)	Wastewater proportion of water use (%)	Stormwater proportion of water use (%)	Reuse proportion of anthropogenic water input (%)
Sydney	0.1	257	86	76	1
Melbourne	0.5	322	79	68	4
Brisbane+Gold Coast[a]	0.1	273	48	104	2
Perth	22	397	26	47	1

[a]The urban footprint of these two areas was considered representative of South East Queensland (SEQ) because it accommodates the majority of the population in the area (Data from Kenway et al. (2011a))

demand was 47–104 %. The approach also demonstrated large inter-city variation (Table 28.2). For example, in 2004–05, Perth used 44-fold more rainfall harvesting than other major Australian cities, largely due to its use of local groundwater.

In the four urban water systems evaluated, water reuse was 1–4 % as a proportion of total water inputs from centralised and decentralised supplies (Table 28.2). In contrast, the human body reuses water some 20 times (2,000 %), at various qualities, in order to conserve both water and energy and thermo-regulate 'the system' (Tambo 2002).

5.2 *Implications of the Water Mass Balance for Water-Related Energy Use*

In the future, much more water is likely to be harvested locally within cities. The drivers will include growing water use (largely due to growing populations), higher costs of distributing water, and the emergence of suitable technologies and policy frameworks encouraging use of more locally sourced water. For example, the projected water supply growth for greater Adelaide from 2014 onwards is dominated by planned increases in wastewater and stormwater use, as well as achieving improved water use efficiency (Government of South Australia 2009).

Stormwater, rainwater, and wastewater can often be used as water supply options at lower energy costs than other options, such as desalination or transporting water from distant sources. This is particularly the case if end uses require lower quality water, thereby avoiding substantial treatment. However, the high energy demand of some small (decentralised) systems or inefficient designs (Retamal et al. 2009) means that low energy use is not necessarily always achieved. System design is critical for reducing energy use. For example, rainwater tank systems can be designed to use less than 0.2 kWh/m^3 if they utilise header tanks, low pressure pumps, and reconfigure toilet valves to operate under low pressure (Cunio and Sproul 2009). The energy required for wastewater reuse can be reduced if land use planning allows industries able to use recycled water to be located close to its point of production. Similar scenarios are likely to exist for stormwater harvesting.

Knowledge of full urban water balances helps quantify other energy-relevant aspects of urban design such as: (i) improved quantification of the urban heat island effect, (ii) potential for energy generation (e.g., micro-hydro), and (iii) potential for thermal storage and recovery.

Finally, understanding the full urban water balance is vital for comparing water-related energy demands from city to city. To take one example, the mass balance revealed major inter-city differences in decentralised water use. Consequently, in cities using a large proportion of decentralised water supply, it would be essential to consider the energy-demands associated with those flows. For example, 46 % of Perth's water supply is sourced from decentralised ground water sources. These supplies have considerably different energy demands to water supplied from other sources such as desalination.

Future urban water strategies will improve energy efficiency if they (i) quantify and consider all urban water flows, and (ii) identify how the overall balance (including end-use) affects the system's energy use.

6 Transitioning Urban Water Systems to Reduce Water-Related Energy

A number of measures can be implemented to reduce the energy demand of existing urban water systems. Far more measures can be implemented to improve our knowledge of water-energy linkages so that management in the future can achieve greater levels of efficiency.

6.1 Reducing Water-Related Energy in Current Systems

Many steps can be taken now to reduce water-related energy use. Many Australian utilities are working on a range of strategies such as replacement of inefficient assets (often pumps) and optimisation of treatment and system management processes. Most water conservation strategies reduce energy use, both at the point of end-use, but also both in the upstream (water supply) and downstream (wastewater) systems. Conservation of hot water (e.g. for warm water within homes or steam within industry) typically has greater energy conservation effects.

Increased recycling of wastewater, for conservation of both water and heat, is also now possible at household, district, and city scales. For example, 'recirculating showers' which treat and reuse water are now available and claim to reduce both water and energy use by up to 70 % of baseline. Energy recovery from sewers is a relatively new technology which is having increased uptake. District heating systems in many European cities now take advantage of waste heat from thermal power plants to provide cheap, clean heating to homes and industry.

In addition to direct 'physical' (technological) strategies, water-related energy use could be reduced with a range of technical and behavioural changes via a mixture of programs or government instruments. For example, targeted information might communicate water use and related energy use. Financial incentives and penalties for particular behaviours or patterns could be introduced. Water pricing could be structured to influence water use, and incentives provided to encourage the uptake of more efficient technologies.

Further assessment is necessary to identify what are the most effective or accessible areas to target, or which are the more effective policy interventions on the scales of households through to city. Consideration needs to be given to scales that include household, district, utility, and city. Least-cost analysis, and the production of cost-abatement curves comparing options within and beyond water utility boundaries, would be helpful. Gleick et al. (2009) point out that options which conserve water will likely out-perform those which encourage continued patterns of high usage. This is particularly the case in a future where water and energy are expected to be more expensive.

6.2 *Understanding and Managing Water-Related Energy in the Future*

Measures are needed to understand, track, and communicate (i) water-related energy, and (ii) overall city water and energy performance. Broadly, these measures include: new frameworks, performance indicators, methods, datasets, and benchmarking and targets. Using the improved information, optimisation studies are needed to inform blueprints for future development and related governance arrangements and business models. Provided below are views, informed by an international workshop, on these changes (Kenway et al. 2013).

New *frameworks* are necessary to help conceptualise and analyse cities. Frameworks should be informed by total urban balances of water and energy to a consistent three-dimensional boundary (Fig. 28.7). Without a consistent metabolic boundary it is likely to be harder to fully understand the energy implications of water management or other urban strategies. Adoption of different boundaries could lead to inclusion or exclusion of very different sets of linkages, and consequently reach very different conclusions about the best solution. For example, if water-related energy in the urban water cycle is considered with, or without, the energy implications of end-use, approximately 90 % of the implications would be missed.

Frameworks should also consider performance of the overall system itself (the city) along with how individual components of the system, such as water and energy utilities, perform in a way which contributes to overall systems efficiency. An example of such a requirement has been established by the new Minister for Water and Energy in Queensland, Australia, where the driver of regional outcomes

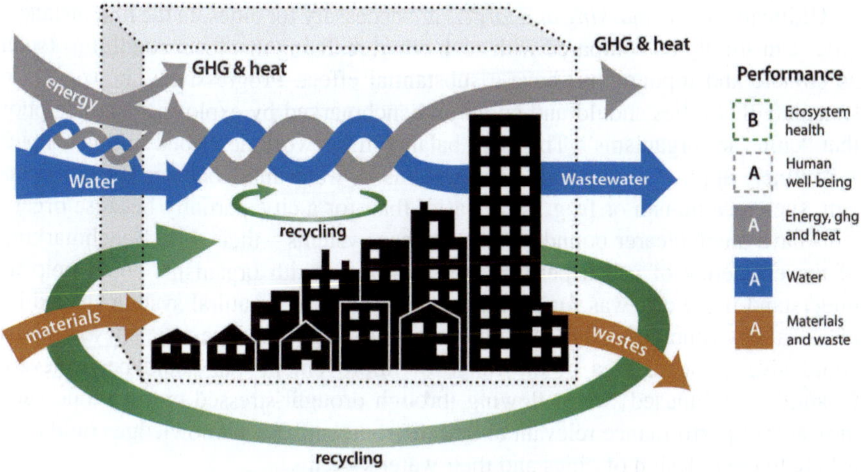

Fig. 28.7 The future paradigm: a framework for co-ordinated and more efficient urban performance management (for comparison with Fig. 28.2) (Source: Kenway (2012))

is embedded in the Statement of Obligations (Queensland Government 2013). Clause 3.4.3 requires the bulk water authority (Seqwater) to "collaborate with all regional bodies and authorities to achieve whole of region benefits … at least cost to the community".

Urban metabolism appears to be a highly suitable starting point. This is because of (i) its foundation in mass balance of 'an entity', (ii) its apparent rich capacity for benchmarking urban systems with natural systems, and (iii) its highly multi-disciplinary nature, giving it relevance across both physical and social sciences.

Conceptual frameworks guide our selection and tracking of *performance indicators*. For example, how do we know how well a city is performing with regard to water, energy, or other material flows? Similarly, how well is the city design and management achieving a high level of human well-being and ecosystem health? It is possible for us to achieve a high level of urban performance (i.e. 'A's) for all indicators; achieving this, however, would require a truly multi-disciplinary, multi-sectoral effort spanning government, industry, research, and community.

Guided by the conceptual framework, improved *methods* are needed to quantify (repeatably) not only water-related energy in cities but also all performance indicators. For example, how can cities accurately know how much water-related energy is consumed by households, industry, business, and government? Such knowledge is essential so that management can recognise if its actions have been solutions that have addressed multiple problems simultaneously (such as water and energy), or have simply moved a problem from one domain to another.

Coordinated and inter-locking *datasets* are likely to be needed in order to enable such performance indicators to be tracked for water, energy, and other related material flows. How can the myriad data sources which underpin an urban water balance (or urban greenhouse gas emissions) be rapidly assembled?

Ultimately *benchmarking and targets* are necessary for cities. In the first instance, cities can simply be compared with each other, realising that local conditions (such as climate and topography) have a substantial effect. Progressively, however, we speculate that cities should and could be benchmarked by exploring the metaphor that "cities are organisms". The mass balance framework described in this chapter is routinely applied to the study of organisms. A water mass balance for an organism, such as a human or frog, is far easier than for a city, partially because organisms have much clearer boundaries than urban systems – their skin. Benchmarking of the efficiency of urban performance compared with organisms could help us understand if the city was functioning as efficiently as a natural system shaped by the local environment and evolutionary pressures. The relative efficiency of cities could give us some idea of the range of improvement that could be achieved. Quantities of 'unused' water flowing through drought-stressed cities is one such indicator of performance relevant to dry environments. Such knowledge could contribute to the redesign of cities and their water systems.

Optimisation studies for water and energy management in cities is currently absent from the literature (Kenway et al. 2011c). Cities and their component districts, buildings, and systems will need to be progressively designed to achieve the highest levels of water and energy efficiency.

Optimisation analysis could guide *blueprints* for urban futures such as regional and city plans and related building codes and standards. Optimisation studies, for example, could determine how to accommodate one million additional people at a certain quality of life, with minimum water and energy inputs.

Ultimately, *governance* arrangements need to be aligned to ensure systemic efficiency through planning and regulatory mechanisms (e.g., regional planning and building approval processes), rather than simply efficiency in individual system operations.

A wide and integrated suite of programs and measures, including complementary government policy instruments, is likely to be needed to progress change. Consideration of the overall institutional frameworks, governance arrangements, and business models for water and energy management in cities may also be necessary. Mechanisms for increasing the connection between service provision and revenue need to be strengthened. Likewise, decoupling dependence of revenue on volumetric sales of water and energy (Lesh 2009) would provide alternatives to governments for their continued dependency on high throughputs of water and energy.

7 Conclusions

This work demonstrates how water-related energy in cities is (i) large, and (ii) able to be influenced by management. Water end-use dominates water-related energy. Energy used by water utilities, while significant, is only a small, albeit growing, component of energy use in Australian cities. To enable an efficient overall system, managing the *indirect* (or hidden) energy component appears important. It is also expected to be important in finding least-cost urban water management solutions.

The concept of urban metabolism appears to provide a rich conceptual basis and context against which the water performance of cities could be investigated. It could lead to an accurate quantification of 'urban performance'. A water mass balance (metabolic) analysis has identified large unaccounted water flows through four Australian cities that were all experiencing a water 'crisis'. The mass balance also demonstrated major differences in the hydrological performance of the cities. The mass balance provides a foundation upon which other water-related balances, including water-related energy, could be based.

To improve management of water-related energy in cities, much change is necessary – and possible. Many existing and new technologies can be deployed. Nevertheless, there is a need for a new and more fundamental framework for understanding water and energy in cities. The framework needs to include clear articulation of the 'system boundary'. This will help identify new indicators, performance reporting techniques, targets, and performance benchmarks. Ultimately, to enable the highest levels of efficiency, new forms of management and governance of water and energy in cities may be necessary.

Wider use, understanding, and reporting of urban metabolism and the water–energy nexus would be highly beneficial to future urban management. It could change how we think about our cities and their water systems. We could view our cities as sources of water, and we could view our urban water systems as part of a solution to constraining energy and greenhouse gas emissions. Such thinking could lead to considerable changes in physical and institutional structures. As pointed out by Abel Wolman in 1965, there is no shortage of water or energy, but, we may add, finding new solutions requires long-term thinking.

Acknowledgements The authors gratefully acknowledge the support given by the Urban Water Security Research Alliance, including the Queensland Government, CSIRO, the University of Queensland (including the School of Chemical Engineering and Advanced Water Management Centre), Griffith University, and the Swiss Federal Institute for Environmental Science and Technology (Eawag). The support of the Smart Water Fund and Australian Research Council is appreciated. We are also grateful for the data provided by the Bureau of Meteorology, Seqwater, Queensland Urban Utilities, AGL, and Energex.

References

ABC News. (2007). *Beattie scraps water poll amid 'Armageddon' situation*. 28 January. http://www.abc.net.au/news/2007-01-28/beattie-scraps-water-poll-amid-armageddon-situation/2180992. Accessed 16 Dec 2013.

ABS. (2006). *Australian social trends 2006*. Canberra: Australian Bureau of Statistics.

Australian Government Department of Climate Change and Energy Efficiency. (2012). National targets. In *Think change*. Canberra: Australian Government. http://www.climatechange.gov.au/sites/climatechange/files/files/climate-change/national-targets.pdf. Accessed Dec 2013.

Bernstein, L., Bosch, P., Canziani, O., Chen, Z., Christ, R., & Davidson, O. (2007). *Climate change 2007: Synthesis report. Summary for policymakers* (Serie Indicadores Ambientales, Vol. 10). Lima: CONAM. 71pp.

Conrad, S., Gelsenhoff, J., Brueck, T., Volna, M., Brink, P., Hall, M., Cook, S., & Kenway, S. (2011). *Decision support system for sustainable energy management*. Denver: Water Research Foundation.

Cook, S., Hall, M., & Gregory, A. (2012). *Energy use in the provision and consumption of urban water in Australia: An update.* A report prepared for the Water Services Association of Australia. Canberra: CSIRO. https://www.wsaa.asn.au/WSAAPublications/Documents/CSIRO%20WSAA%20Energy%20use%20in%20the%20consumption%20and%20provision%20of%20urban%20water%20in%20Australia.pdf. Accessed 16 Dec 2013.

Cunio, L. N. & Sproul, A. B. (2009). Low energy pumping systems for rainwater tanks. In *Solar09,* 47th ANZES Annual Conference, Townsville, 29 Sep–2 Oct 2009.

Decker, E. H., Elliott, S., Smith, F. A., Blake, D. R., & Rowland, F. S. (2000). Energy and material flow through the urban ecosystem. *Annual Review of Energy and the Environment, 25,* 685–740.

de Monsabert, S., & Liner, B. (1998). Integrated water and energy conservation modelling. *Journal of Energy Engineering, 124,* 1–19.

Fischer-Kowalski, M., & Huttler, W. (1998). Society's metabolism: The intellectual history of materials flow analysis, part II, 1970–1998. *Journal of Industrial Ecology, 2*(4), 30.

Flower, D. J. M., Mitchell, V. G. & Codner, G. P. (2007). *The potential of water demand management strategies to reduce the greenhouse gas emissions associated with urban water systems.* In Proceedings, 1st conference on sustainable urban water management and 9th conference on computing and control in the water industry, Leicester, UK.

French, S., & Geldermann, J. (2005). The varied contexts of environmental decision problems and their implications for decision support. *Environmental Science & Policy, 8,* 14.

Gleick, P. H., & Cooley, H. S. (2009). Energy implications of bottled water. *Environmental Research Letters, 4,* 6.

Government of South Australia. (2009). *Water for good: A plan to ensure our water future to 2050.* Adelaide: Department for Water.

GRI. (2005). *Sector supplement for public agencies: Pilot version 1.0.* Boston: Global Reporting Initiative.

Heemskerk, B., Pistorio, P., & Scicluna, M. (2002). *Sustainable development reporting: Striking the balance*. Geneva: World Business Council for Sustainable Development.

Howe, C., Jones, R., Maheepala, S., & Rhodes, B. (2005). *Melbourne water climate change study: Implications of potential climate change for Melbourne's water resources*. Canberra: CSIRO.

IEA. (2011). Statistics and balances for OECD countries. In *Online statistics and balances*. Paris: International Energy Agency.

Kennedy, C., Cuddihy, J., & Engel-Yan, J. (2007). The changing metabolism of cities. *Journal of Industrial Ecology, 11*(2), 43–59.

Kenway, S. J. (2012). *The water–energy nexus and urban metabolism: Identification, interpretation and quantification of the connections in cities*. PhD thesis, School of Chemical Engineering, The University of Queensland, Brisbane.

Kenway, S. J., & Lant, P. (2011). The influence of water on urban energy use. In C. Howe & C. Mitchell (Eds.), *Water sensitive cities*. Amsterdam: International Water Association/UNESCO-IHE.

Kenway, S. J., Priestley, A., Cook, S., Seo, S., Inman, M., & Gregory, A. (2008). *Energy use in the provision and consumption of urban water in Australia and New Zealand.* CSIRO and Water Services Association of Australia. ISBN 9780643096165.

Kenway, S. J., Gregory, A., & McMahon, J. (2011a). Urban water mass balance analysis. *Journal of Industrial Ecology, 15*(5), 693–706.

Kenway, S. J., Lant, P., & Priestley, A. (2011b). Quantifying the links between water and energy in cities. *Journal of Water and Climate Change, 2*(4), 247–259.

Kenway, S. J., Lant, P., Priestley, A., & Daniels, P. (2011c). The connection between water and energy in cities: A review. *Water Science and Technology, 63*(9), 1983–1990.

Kenway, S., McMahon, J., Elmer, V., Conrad, S., & Rosenblum, J. (2013). Managing water-related energy in future cities: A research and policy roadmap. *Journal of Water and Climate Change, 4*(3), 161–175.

Lesh, P. (2009). *Rate impacts and key design elements of gas and electric utility decoupling: A comprehensive review*. Lake Oswego: Graceful Systems.

National Water Commission. (2011). *National performance report 2009–10, Part A: Comparative analysis*. Canberra: National Water Commission.

Newman, P. W. G. (1999). Sustainability and cities: Extending the metabolism model. *Landscape and Urban Planning, 44*(4), 219–226.

Newman, P., Birrell, R., Holmes, D., Mathers, C., Newton, P., Oakley, G., O'Connor, A., Walker, B., Spessa, A., & Trait, D. (1996). Human settlements. In *State of the environment Australia 1996*. Melbourne: State of the Environment Advisory Council, CSIRO Publishing.

O'Meara, M. (1999). *Reinventing cities for people and the planet*. Worldwatch paper 147. Washington, DC: Worldwatch Institute.

PMSEIC. (2010). *Challenges at energy–water–carbon intersections*. Canberra: Prime Minister's Science, Engineering and Innovation Council.

Queensland Government (2013). *Statement of obligations for Queensland Bulk Water Supply Authority*. Issued by the Treasurer and Minister for Trade, and Minister for Energy and Water Supply.

Retamal, M., Abeysuriya, K., Turner, T., Glassmire, J., & White, S. (2009). *The water energy nexus: Investigation into the energy implications of household rainwater systems*. Sydney: Institute for Sutstainable Futures, University of Technology.

Sahely, H. R., Dudding, S., & Kennedy, C. A. (2003). Estimating the urban metabolism of Canadian cities: Greater Toronto area case study. *Canadian Journal of Civil Engineering, 30*(2), 468–483.

Satterthwaite, C. (2008). Cities' contributions to global warming: Notes on the allocation of greenhouse gas emissions. *Environment and Urbanization, 20*(2), 539–549.

Sterman, J. (1991). A skeptic's guide to computer models. Reprinted in G. O. Barney et al. (Eds.), *Managing a nation: The microcomputer software catalog* (pp. 209–229). Boulder: Westview Press.

Stolz, G. (2007). The water crisis: Its desalinate or die. *Courier Mail* (Brisbane), 1 February.

Tambo, N. (2002). A new water metabolic system. *Water, 21*, 67–68.

Victorian Water Industry Association. (2011). *Electricity issues in the Victorian Water Sector*. Melbourne: Victorian Water Industry Association.

Wolman, A. (1965). The metabolism of cities. *Scientific American, 213*, 179–190.

WSAA. (2005). *Testing the water: Urban water in our growing cities. The risks, challenges, innovation and planning*. Melbourne: Water Services Association of Australia.

WSAA. (2008). *WSAA report card 2007–2008: Performance of the Australian urban water industry and projections for the future*. Melbourne: Water Services Association of Australia.

WSAA. (2009). *WSAA report card 2008–2009: Performance of the Australian urban water industry and projections for the future*. Melbourne: Water Services Association of Australia.

Index

Q. Grafton et al. (eds.), *Understanding and Managing Urban Water in Transition*, Global Issues in Water Policy 15,
DOI 10.1007/978-94-017-9801-3

MIX
Papier aus verantwortungsvollen Quellen
Paper from responsible sources
FSC® C105338

If you have any concerns about our products,
you can contact us on
ProductSafety@springernature.com

In case Publisher is established outside the EU,
the EU authorized representative is:
Springer Nature Customer Service Center GmbH
Europaplatz 3, 69115 Heidelberg, Germany

Printed by Libri Plureos GmbH
in Hamburg, Germany